Astrology and Cosmology in Early China

The ancient Chinese were profoundly influenced by the Sun, Moon, and stars, making persistent efforts to mirror astral phenomena in shaping their civilization. In this pioneering text, David W. Pankenier introduces readers to a seriously understudied field, illustrating how astronomy shaped the culture of China from the very beginning and how it influenced areas as disparate as art, architecture, calendrical science, myth, technology, and political and military decision-making. As elsewhere in the ancient world, there was no positive distinction between astronomy and astrology in ancient China, and so astrology, or more precisely, astral omenology, is a principal focus of the book. Drawing on a broad range of sources, including archaeological discoveries, classical texts, inscriptions and paleography, this thought-provoking book documents the role of astronomical phenomena in the development of the "Celestial Empire" from the late Neolithic through the late imperial period.

DAVID W. PANKENIER is Professor of Chinese at Lehigh University, Pennsylvania. His current research interests range from the history of ideas in early China, to archaeoastronomy and cultural astronomy. He is particularly interested in the connection between rare astronomical phenomena and epoch-making political and military events in ancient China.

Astrology and Cosmology in Early China

Conforming Earth to Heaven

David W. Pankenier

Lehigh University

CAMBRIDGE
UNIVERSITY PRESS

University Printing House, Cambridge CB2 8BS, United Kingdom

Cambridge University Press is part of the University of Cambridge.

It furthers the University's mission by disseminating knowledge in the pursuit of education, learning and research at the highest international levels of excellence.

www.cambridge.org
Information on this title: www.cambridge.org/9781107539013

First published 2013
First paperback edition 2015

A catalogue record for this publication is available from the British Library

Library of Congress Cataloguing in Publication data
Pankenier, David W. (David William)
Astrology and cosmology in early China : conforming earth to heaven /
David W. Pankenier.
pages cm
Includes bibliographical references and index.
ISBN 978-1-107-00672-0 (hardback)
1. Astrology, Chinese. 2. Cosmology, Chinese. I. Title.
BF1714.C5P38 2013
133.59231 – dc23   2013005731

ISBN 978-1-107-00672-0 Hardback
ISBN 978-1-107-53901-3 Paperback

落其實者思其樹, 飲其流者懷其源

Plucking the fruit, one thinks of the tree;
drinking from the stream, one is mindful of the source.

Yu Xin 庾信 (fl. *c*.544)

Contents

Figures

Maps

Tables

Foreword

In *Astrology and Cosmology in Early China* David Pankenier has given us a particularly potent way to understand the genius of the ancient Chinese religio-political vision of the universe. I refer to the acute Chinese concern for the interaction of the human and celestial worlds as seen in rare astral omens (what Pankenier felicitously refers to as "astal omenology"). Known to all students of China as the "Mandate of Heaven" (*tian ming*), this theory of portentous heavenly and earthly correlations echoes through all of Chinese history down to the present day, and in the ancient period had interesting parallels with Hebrew theories of a Sky-God's covenantal relationship with his chosen people. These correlations are central to the ancient Chinese worldview or mythic cosmology and are detailed in some of the earliest Chinese texts – as well as being encoded in early architectural structures and other symbolic forms.

What was not fully appreciated until now with Pankenier's work was the astronomical specificity and broad cultural impact of these celestial correlations. Beginning with his breakthrough analysis of unusual planetary massings related to the political foundations of ancient China, Pankenier shows us how the early cosmology was truly formative for almost all significant aspects of Chinese civilization. What he accomplishes here has been hinted at by other scholars, but no one has put it all together in such a technically sophisticated, interpretively imaginative, and brilliantly convincing way. Indeed, this is a work that has broad significance for understanding ancient Chinese tradition. Moreover, as he teasingly suggests in the epilogue, such celestial and cosmological anomalies continue to provoke, mark, and haunt significant contemporary political events in China. Pankenier's findings in this work also have ample implications for many other ancient and contemporary civilizational traditions seen, for example, in the many world-cultural variations on the macro-/microcosmic theme of "as above, so below" (as encoded in the ancient Western hermetic text of the *Emerald Tablet*).

I write not as a sinologist but as a comparative religionist or scholar of the world history of religions. However, I have had much familiarity with aspects of Chinese tradition (especially early Daoism), as well as with the whole history of Western scholarship concerning China (e.g. the pivotal work of the great nineteenth-century scholar of the Chinese classics, James Legge). I know enough, in other words, to recognize real sinological expertise – something that is clearly and abundantly in evidence in this work. My self-appointed task in these brief comments is not, therefore, to rehearse Pankenier's proficiency as a Chinese textual scholar. Rather I want to emphasize his ability to creatively and productively stretch the boundaries of the often philologically and culturally circumscribed modes of traditional Chinese scholarship.

The real grace and power of this book, then, is not just Pankenier's competence as a scholar of early China. Equally remarkable is his careful and critical application of comparative and interdisciplinary methods of interpretive analysis. Most noteworthy in this regard is his use of techniques and insights coming from the highly specialized field of archaeoastronomy, which combines aspects of archaeology, astronomy, philology, history, paleography, and cross-cultural hermeneutics. These disciplinary methods especially draw upon perspectives coming from the comparative history of religions as related to general cultural development. Pankenier in this sense notes that the "ancient Chinese preoccupation with the heavens was hardly unique." This in turn leads him to pay attention to non-sinological scholars (e.g. Raffaele Pettazoni, Georgio de Santillana, Hertha von Dechand, and Alexander Marshak) who have recognized the symbolic language of the sky and astronomical phenomena written into the myths, rituals, and cultural creations of many different early civilizations.

The comparative scholar Mircea Eliade noted in his study of the mythritual "patterns" or "structures" of religious belief that the human symbolic awareness of and imaginative reaction to the natural world – most primordially, profoundly, and transcendently the radiant sky – is always embedded within, and shapes, a culture's fundamental worldview and vision of life. The basic human encounter with, or experience of, the sky and its related phenomena – an experience witnessed in all ancient cultures in relation to various ideas of divinity (such as the ancient Chinese Supernal Lord or Shangdi) – immediately implies feelings of height, flight, transcendence, power, and universal order. This is because the vault of the day and night sky, filled with luminous and constantly changing celestial bodies, is "just there" as something "above" and "beyond" ordinary human existence on the earthly plane. All of these patterns witnessed on high are truly and generally inspirational and potentially symbolic of patterns in earthly existence. This awareness hinges on the archaic and fundamental human ability to see the sky as a sign with a message of existential meaning that calls for a cultural response. The general perception of the astronomical "above" only becomes culturally and humanely *productive* and *significant* therefore in relation to how those experiences are imaginatively (i.e. artistically and technically) embraced, envisioned, communicated, and made real "below" in the stories/myths, actions/rituals, architectural structures/visual forms, and social institutions/political practices that allow men and women to live their lives with meaning.

David Pankenier persuasively shows us the all-pervasive religio-political relationship of the Above and Below in ancient China, but his work is even more broadly and importantly suggestive. As seen by his interdisciplinary methods and sensitivity to comparative cross-cultural perspectives, Pankenier helps us imagine and understand how our response to astronomical phenomena is at the very core of our cultural development as human beings. In many ways for the Chinese as well as for other ancient traditions it was the awesome vision of the sky that inspired our ancestors to create the human world we still inhabit. Contemplation of the sky was originally, as Eliade reminds us, a revelation of the human participation in celestial patterns that define the entire cosmos. Reading Pankenier is likewise a revelation in that we come to see that knowing early China is simultaneously to know the wellsprings of human nature. This is a work that embraces the starry sky and by so doing inspires us to know the whole world more fully. As above, so below.

Norman Girardot
University Distinguished Professor
of Comparative Religions, Lehigh
University

Preface

This book has been long aborning. In the early 1980s I discovered in ancient Chinese historical sources observational records of astronomical phenomena, which if scientifically verified had the potential to establish reliable benchmarks essential in dating China's earliest dynasties. The challenge of reconstructing China's chronology prior to the earliest secure date of 841 BCE has motivated historians since at least the fourth century BCE. At first the historical records I was studying seemed scarcely credible, appearing as they did in sources like the *Bamboo Annals* (*Zhushu jinian*) and *Lost Books of Zhou* (*Yi Zhou shu*) long exposed to the vagaries of textual transmission. Subsequent research demonstrated, however, that similar accounts could be found in too many other reliable early sources to be the result of interpolation, and that at the time of their earliest appearance the Chinese did not possess the ability to retrospectively compute such ancient and complex astronomical phenomena. I concluded that the reports must have survived for a millennium and more and been transmitted in ways that are still poorly understood.

At the same time, it became clear that several of these rare events, dense clusters of the Five Planets in particular, were associated in ancient tradition with dynastic transitions resulting from overthrow or "change of the Mandate" (*ge ming*), as it came to be called, thus solidifying their status as astral omens. It was clear to me that they had the potential to open a new window on the world of thought in Bronze Age China by shedding light on the ancient doctrine of Heaven's Mandate (*tian ming*) and the unique relationship with Heaven (or the Supernal Lord, Shangdi) that the late Bronze Age Chinese believed themselves to enjoy. My ongoing research has focused on the recovery of ancient astronomical concepts and practices through the study of archaeological discoveries, inscriptional sources, language, history, astrology, and cosmology. The upshot is that a new perspective on the role of astronomy–astrology and cosmology in the formation of Chinese civilization has gradually taken shape. This new view, informed by the comparative methodology of archaeoastronomy and cultural astronomy, demonstrates that preoccupation with things celestial manifested itself in many aspects of ancient Chinese civilization in heretofore unappreciated ways.

The initial breakthrough came when I was able to verify that a misdated record of the spectacular planetary massing of 1059 BCE in the transmitted text of the *Bamboo Annals* must be a genuine eyewitness account. Encouraged to probe further by this discovery, I then found that early Bronze Age people had witnessed and preserved in mythic language the even more spectacular planetary massing of 1953 BCE, the densest such cluster of planets in more than 5,000 years. The realization that the ancient Chinese were impressed enough to incorporate astronomical phenomena into accounts of the founding of the dynastic system opened up an entirely new perspective on the genesis of the concept of Heaven's Mandate – the idea that political legitimacy is directly conferred by Heaven on a worthy ruler. My reading of the classical canon had impressed me early on with the centrality of Heaven (lit. "sky," *tian*) in both its cosmological and politico-religious roles as the source of all-pervasive cosmic and spiritual influence. Only after I delved into the ramifications of that early Chinese preoccupation with the sky did the depth and extent to which cosmology exerted a profound formative influence on the civilization become apparent.

Not being a formally trained astronomer, in order to better appreciate the cultural significance of my discoveries I had to immerse myself in an emerging new discipline. The study of the astronomical practices, celestial lore, astral religion, mythologies, and cosmologies of ancient cultures is called archaeoastronomy. It is, in essence, the historical anthropology of astronomy, as distinct from the history of astronomy. In 1983, I presented my early discoveries at the First International Conference on Ethnoastronomy at the Smithsonian Institution, and what I learned from other presentations there convinced me that the ancient Chinese preoccupation with the heavens was hardly unique; indeed, it is a human universal. I continue to be inspired by the burgeoning literature in cultural astronomy and archaeoastronomy which has yielded innumerable insights into how our forebears, at all times and places, have shown intense interest in what transpired in the sky, in the familiar, predictable cycles and in unpredictable, transient phenomena alike.

* * *

Many scholars and friends have offered invaluable suggestions and advice over the years and I have endeavored to acknowledge their work in the book. To them I am deeply grateful for sharing their knowledge in a spirit of collegiality and common endeavor. Inevitably there will be lapses and I apologize in advance for any oversights.

Christopher Cullen took time from his demanding schedule to read the entire draft of my translation of the Appendix, the "Treatise on the Celestial Offices," and offered numerous insightful comments and suggestions. Juan Antonio Belmonte, David P. Branner, Wolfgang Behr, Nick Campion, Li Feng, Norman

J. Girardot, Paul R. Goldin, Lionel Jensen, David N. Keightley, Martin Kern, Liu Ciyuan, John S. Major, Göran Malmqvist, Deborah L. Porter, Michael Puett, Ken-ichi Takashima, Xu Zhentao, Ray White, members of the Columbia University Early China Seminar, and others too numerous to mention have been extraordinarily supportive and helpful, professionally and personally. Nick Campion, Marc Kalinowski, David N. Keightley, and Charles E. Pankenier Jr. read some or all of the manuscript and offered suggestions for improvement. I am particularly grateful to Björn Wittrock, head of the Swedish Collegium for Advanced Study, and deeply honored to have been awarded a Bernhard Karlgren Fellowship, underwritten by the Bank of Sweden's Riksbankens Jubileumsfond in memory of the great Swedish sinologist. The generous support of both institutions enabled me to spend an enormously satisfying and rewarding year at the Collegium in Uppsala during 2010–11 while I finished writing this book.

I have been extraordinarily fortunate in having had the opportunity to study with great minds at an impressionable age: Norman O. Brown, my *qimeng laoshi*, who awakened me to the life of the mind; Hayden V. White, from whom I learned to read historical writing as literature; Göran Malmqvist and Ning-tsu Malmqvist, whose inspired pedagogy and profound love for the language and culture of China set me on the path; and Aisin Gioro Yuyun (Yu Lao), who taught me to read and appreciate the Chinese Classics in the traditional way, and who instilled in me a profound admiration for the depth of his learning and that of those who preceded him in transmitting the teaching. To all the above this book is dedicated with sincere thanks, much affection, and deep respect. I also owe a debt of gratitude to those closest to me, without whose patience and forbearance through long years of study I could not have persevered: Eva Pankenier-Minoura, Sara Pankenier-Weld, Emma Pankenier-Leggat, Sophia Pankenier, Simone Pankenier, Birgitta Wannberg, and my unfailingly supportive wife and native informant, Zhai Zhengyan. Finally, I acknowledge with gratitude the loving kindness of my parents Elsa Wunsch and Charles E. Pankenier Sr., who did their best to indulge my intellectual curiosity.

San Pedro, Ambergris Caye Belize

Acknowledgments

Portions of Chapter 1 are reprinted here with permission from "The Xiangfen, Taosi Site: A Chinese Neolithic 'Observatory'?" (with Liu Ciyuan and Salvo De Meis), in Jonas Vaiškūnas (ed.), *Astronomy and Cosmology in Folk Traditions and Cultural Heritage* (Klaipeda: University of Klaipeda, 2008, *Archaeologia Baltica* 10), 141–8.

Portions of Chapter 3 are reprinted here with permission from "A Short History of Beiji," *Culture and Cosmos* 8.1–2 (2004), 287–308. "A Brief History of Beiji (Northern Culmen): With an Excursus on the Origin of the Character *Di* 帝," *Journal of the American Oriental Society* 124.2 (April–June 2004), 1–26. Ban Dawei 班大为 (David W. Pankenier), "Beiji de faxian yu yingyong 北极的发现与应用 (Locating and Using the Pole in Ancient China)," *Ziran kexueshi yanjiu* 自然科学史研究 27.3 (2008), 281–300. "Locating True North in Ancient China," *Cosmology across Cultures*, Astronomical Society of the Pacific Conference Series, 409 (2009), 128–37. Ban Dawei 班大为 (David W. Pankenier), "Zai tan beiji jianshi yu di zi de qiyuan" 再谈北极简史与 「帝」字的起源," in Patricia Ebrey 伊沛霞 and Yao Ping 姚平 (eds.), *Xifang Zhongguo shi yanjiu luncong* 西方中国史研究论丛, Vol. 1, *Gudai yanjiu* 古代史研究 (ed. Chen Zhi 陈致) (Shanghai: Shanghai guji, 2011), 199–238.

Portions of Chapters 4 and 5 are reprinted here with permission from "Getting 'Right' with Heaven and the Origins of Writing in China," in Feng Li and David Prager Branner (eds.), *Writing and Literacy in Early China* (Seattle: University of Washington Press, 2011), 13–48.

Portions of Chapters 6, 7, and 8 are reprinted here with permission from "The Cosmo-political Background of Heaven's Mandate," *Early China* 20 (1995), 121–76.

Portions of Chapter 9 are reprinted here with permission from "Applied Field Allocation Astrology in Zhou China: Duke Wen of Jin and the Battle of Chengpu (632 BCE)," *Journal of the American Oriental Society* 119.2 (1998), 261–79. "Characteristics of Field Allocation (fenye 分野) Astrology in Early China," in J.W. Fountain and R.M. Sinclair (eds.), *Current Studies in Archaeoastronomy: Conversations across Time and Space* (Durham: Carolina Academic, 2005), 499–513.

Portions of Chapter 10 are reprinted here with permission from "Popular Astrology and Border Affairs in Early Imperial China: An Archaeological Confirmation," *Sino-Platonic Papers* 104 (July 2000), 1–19.

Portions of Chapter 11 are reprinted here with permission from "The Cosmic Center in Early China and Its Archaic Resonances," in Clive L.N. Ruggles (ed.), *Archaeoastronomy and Ethnoastronomy: Building Bridges between Cultures*, Proceedings of the International Astronomical Union (IAU Symposium 278) (Cambridge: Cambridge University Press, 2011), 298–307. "Cosmic Capitals and Numinous Precincts in Early China," *Journal of Cosmology* 9 (July 2010), available at http://journalofcosmology.com/AncientAstronomy100.html.

A portion of Chapter 12 is reprinted here from "Temporality and the Fabric of Space–Time in Early Chinese Thought," in Ralph M. Rosen (ed.), *Time and Temporality in the Ancient World* (Philadelphia: University of Pennsylvania Museum, 2003), 129–46.

Portions of Chapter 14 are reprinted here from "The Planetary Portent of 1524 in Europe and China," *Journal of World History* 20.3 (September 2009), 339–75. Portions of the Introduction are reprinted here with permission from Horowitz, Maryanne (ed.). *New Dictionary of the History of Ideas* (6 Volume Set), 1E. © 2005 Gale, a part of Cengage Learning, Inc. Reproduced by permission. www.cengage.com/permissions.

Chronology of early China

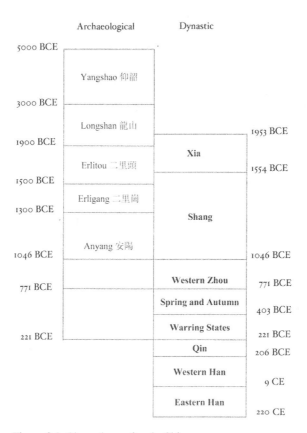

Figure 0.1 Chronology of early China.

Introduction

The awe-inspiring presence of the sky has left a profound imprint on human culture at all times and in all places, to an extent that is seldom appreciated today. It is only very recently that we have managed to construct an environmental cocoon around ourselves. Barely a generation after the advent of street lighting in the United States, Ralph Waldo Emerson wrote,

> One might think the atmosphere was made transparent with this design, to give man, in the heavenly bodies, the perpetual presence of the sublime. Seen in the streets of cities, how great they are! If the stars should appear one night in a thousand years, how would men believe and adore; and preserve for many generations the remembrance of the city of God which had been shown! But every night come out these envoys of beauty, and light the universe with their admonishing smile.[1]

How could Emerson have imagined the current state of affairs in which most urbanites no longer have any idea how stupendous a sight is a truly dark sky on a clear night. Fewer still have ever actually seen the luminous band of the Milky Way, twisting and undulating with the hours and the seasons in its glowing course across the sky. While insulating us from the elements and enabling our modern lifestyle, our artificial surroundings now isolate us from the sky and our primordial heritage to an unprecedented degree. The unfortunate result is an impoverished understanding of the links between earth and sky, and a lack of appreciation for how profoundly astronomical phenomena have historically influenced domains as disparate as art, myth, cosmology, literature, music, philosophy, and the built environment. How, indeed, shall we preserve the remembrance of Emerson's "city of God?" Not for nothing was Emerson a leading light among "Transcendentalists."[2]

---

[1] Ralph Waldo Emerson (1979, Chapter 1).

[2] Surprisingly, in his magisterial survey *Religion in Human Evolution: From the Paleolithic to the Axial Age* (2011), Robert N. Bellah makes no mention of the profound impression the heavens surely must have made on the archaic mind. Although he notes in passing examples like Scipio's dream and Kepler's mystic vision of the cosmos as the fountain of all harmony – "there is something marvelous in the fact that the man who confirmed the Copernican heliocentric theory of the solar system actually 'heard' the music of the spheres" (p. 41) – Bellah overlooks the

Consider the broad rock shelf of the Sibudu Cave in South Africa with its eight-meter-deep record of continuous human occupation extending back 77,000 years.[3] Or Les Eyzies in the Dordogne, in France, whose Font-de-Gaume cave is a "showpiece of Magdalenian engravings and paintings" from around 14,000 BCE, and whose cultural accumulations and abundance of flint tools give evidence of continuous occupation since the Mousterian (c.300,000 BP to 30,000 BP). Even if there are doubts about the cognitive abilities of their distant cousins, the Neanderthals, it is certain that the Cro-Magnons (with brains a third larger than ours) who created that wondrous Magdalenian cave art in the Upper Paleolithic were modern humans no less intelligent than ourselves. They deployed their ample intelligence in meeting the particular challenges of their own time within the constraints of their conceptual framework. What did the people of Les Eyzies think about the spectacular nightly display of luminous patterns wheeling across the sky? When they began to appreciate the seasonal regularity of distinctive stellar configurations, what stories might they have told their children to help them fix in memory the patterns that forewarned of the arrival of the spawning salmon or the reindeer migration on which their lives depended? Might they not have populated the sky with those creatures in their seasonal associations, together with items of daily use, much as the ancient Mesopotamians, Egyptians, and Greeks later populated theirs? Remarkable traces of these imaginings may have survived at Lascaux and elsewhere. As Ludwig Wittgenstein observed in commenting on Frazer's *The Golden Bough*,

That the shadow of a man, which has the appearance of a human being or his mirror image, that rain, that a thunderstorm, the phases of the Moon, the changing seasons, the similarities and differences between animals and people, the phenomena of death, of birth, and of sexual life, in short, everything that a person perceives around him year after year, connected with one another in the most diverse ways, that these will appear (play a role) in his thinking (his philosophy) and his customs is obvious . . . [4]

Raffaele Pettazoni puts the matter in more spiritual terms:

The sky, in its unbounded immensity, in its perennial presence, in its wondrous luminosity, is particularly well suited to suggest to the mind of man the idea of sublimity, of incomparable majesty, of a sovereign and mysterious power. The sky elicits in man the

spectacle of the sky as a source of inspiration for the ancient belief in a "transcendent sphere of existence," as Li Bo (701–62) said, "another heaven and another earth, not of this world."

[3] Balter (2011).

[4] Quoted in Tambiah (1990, 60). The Big Dipper has been variously perceived to form the shape of a ladle, a plow, a bear, a cart, the thigh of a bull, etc. No less an authority that Owen Gingerich mused (1984, 220), "In the widespread mythological connection of the dipper stars with a Great Bear (Ursa Major) we have a hint that a few of the constellations may date back as far as the Ice Ages." See also Joseph (2011).

feeling of a theophany. This is the feeling of a manifestation of the divine, which finds adequate expression in the notion of a Supreme Being.[5]

* * *

In 1894, the prominent Victorian astronomer and father of archaeoastronomy, J. Norman Lockyer, in his the *Dawn of Astronomy: A Study of the Temple Worship and Mythology of the Ancient Egyptians*, was the first to describe the importance of astronomy to the ancient Egyptians and to show that the pyramids were astronomically oriented with a surprising degree of accuracy. His discovery was largely overlooked until the middle of the twentieth century when Gerald S. Hawkins (*Stonehenge Decoded*, 1963), Alexander Thom (*Megalithic Sites in Britain*, 1967; and *Megalithic Lunar Observatories*, 1970), and others began to publish studies of the astronomical alignments they perceived at Stonehenge and other megalithic monuments. Georgio de Santillana and Hertha von Dechend, in their controversial but inspired *Hamlet's Mill: An Essay on Myth and the Frame of Time* (1969), conjectured that astronomical knowledge was anciently encoded and transmitted in myth.[6] In 1972, Alexander Marshak published his groundbreaking study *The Roots of Civilization: Cognitive Beginnings of Man's First Art, Symbol, and Notation*, suggesting the possibility that ancient inhabitants of Europe during the Ice Ages may have been recording lunar phases in rudimentary fashion as early as the Upper Paleolithic.

Archaeoastronomy has matured as a discipline, especially due to the exemplary efforts of leading scholars like Anthony Aveni, Juan Antonio Belmonte, John B. Carlson, Von del Chamberlain, Michael Hoskin, Stanislaw Iwaniszewski, Steven McCluskey, Kim Malville, Michael Rappenglück, Clive Ruggles, Feng Shi, Lionel Sims, Rolf Sinclair, Ray White, Ray Williamson, Tom Zuidema, and many, many others. The more speculative early ideas have gradually been winnowed out, and increasing stress is being placed on the ethnographic evidence and cultural context, on methodological and theoretical rigor, and on the anthropological interpretation of the findings. This is in line with efforts to better define the multidisciplinary approach that archaeoastronomy demands, combining some or all of the methods of anthropology, astronomy, ethnology, history, statistics, and landscape archaeology.[7] With the rapid improvement in digital tools, including accurate astronomical software, computer animation, GPS, satellite imaging, etc., it is now much easier to reliably

---

[5] Pettazoni (1959, 59). Of course, in his essay Pettazoni goes on to qualify this, "on the other hand, the notion of a Supreme Being is not exhausted in the image of the Celestial Being," but his point about the experience of a theophany is by now well established.

[6] E.g. Barber and Barber (2004); for the emergence of spiritual consciousness and mythic thinking, see e.g. Joseph (2011); Donald (1991).

[7] Ruggles (2011).

simulate ancient observing conditions and to better understand the relationships among celestial bodies, constellations, zenith and horizon phenomena, and terrain throughout the year and over the centuries.

Along with the increased focus on history and ethnography, a broader discipline of cultural (or ethno-)astronomy has emerged, concerned with the modern and premodern periods for which considerable ethnographic evidence exists; as, for example, in the case of Polynesian celestial navigation or the accounts of the Spanish chroniclers of Inka and Maya astral lore and calendrics. Scholars worldwide have published on hundreds of impressive architectural sites and cultural phenomena whose designs are manifestly astronomical, so that the geographical coverage by now extends from the Indian subcontinent to China, from Europe to Mesopotamia, from the American Southwest to Central and South America, from Australia to the islands of the Pacific.[8]

In China, the study of astronomy in archaeological contexts is still the province of historians of astronomy, who are seldom equipped to explore adequately the cultural and historical ramifications. Recent landmark studies are Feng Shi's *Zhongguo tianwen kaoguxue* (*Chinese Archaeoastronomy*, 2007), a comprehensive survey of recent archaeological discoveries, practices, and artifacts, and Jiang Xiaoyuan's *Tianxue zhenyuan* (*True Origin of the Study of Heaven*, 2004). As overviews of the vast richness of China's astronomical legacy, however, they lack contextualization from a cultural perspective. A recent exception to this trend is Sun Xiaochun and Jacob Kistemaker's *The Chinese Sky during the Han: Constellating Stars and Society* (1997), which begins to explore the ideological significance of astronomy–astrology under the early empire. In contrast with megalithic Britain and Europe, Egypt, or the Maya and Inka worlds, with their abundance of monumental stone structures and ancient urban agglomerations, the paucity of surviving architectural remains from ancient China, Korea, and Japan means that archaeoastronomical studies in the East Asian cultural sphere will be distinctly different.

Although monumental architectural remains from China's pre-imperial past and the early empire are limited to ancient rammed-earth building foundations, city walls, and tombs, the study of Chinese archaeoastronomy benefits from the time depth and richness of China's written record. For this reason Chinese archaeoastronomical research calls for a combination of sinological expertise and a good grasp of astronomy, Western as well as Chinese. As I hope this book will demonstrate, however, one need not be a formally trained astronomer to pursue this line of research. Given the inherently multidisciplinary character of archaeoastronomy, active engagement in the ongoing conversation among scholars is crucial in order to attune oneself to the many ways astronomy is

---

[8] Ruggles and Cotte (2010); Kelley and Milone (2011); Magli (2009).

reflected in culture.[9] For countless millennia our ancestors in all corners of the world observed the majestic phenomena nightly on display in the sky, most especially after dark when it was not safe to move about and the starry sky was all there was to see beyond a small circle of flickering firelight. Invaluable insights can be gleaned from the great variety of human cultural responses to the spectacle of the heavens as well as from the striking commonalities. One has to become sensitized to the range of possibilities essayed by the ancients.

It is no doubt obvious by now that this book is not a history of Chinese astronomy in the ordinary sense. Nor is it a textbook on the methodology of archaeoastronomy; both have been written by more competent authorities.[10] The present work is instead the product of three decades of basic research in "archaeoastronomy with Chinese characteristics," to coin a phrase. It is intended as an introduction to diverse aspects of an understudied field rather than a work of synthesis. Otto Neugebauer called astronomy the "first of the exact sciences." Here I provide an account of the many profound ways in which astronomy and its earliest application to architecture, astrology, the calendar, cosmology, divination, political ideology, mythology, and religion have shaped Chinese civilization from the very first. As elsewhere in the ancient world, in China there was no positive distinction between astronomy and astrology – even in the West the two did not definitively part company until the late eighteenth century.[11] So astrology (or, more precisely, astral omenology) as it was practiced in early China, and cosmology, will be the focus. Thirty years on, my initial discovery of the historical connection between the impressive planetary cluster of 1059 BCE and the founding of the Zhou Dynasty (1046–256 BCE) is accepted by many, if not by most, as established fact. The historical linkage between astronomical phenomena and the ancient politico-religious concept of the Mandate of Heaven is clear, as is the practice of astral divination from the early Bronze Age on.

In China, co-ordination of human activity with the Sun, Moon, and stars, including the cardinal orientation of structures in the landscape, can be traced back to the Neolithic cultures of the fifth millennium BCE. According to

---

[9] The following are good sources of information: the Center for Archaeoastronomy and the International Society for Archaeoastronomy and Astronomy in Culture (ISAAC); the European Society for Astronomy in Culture (SEAC); the International Conferences on the Inspiration of Astronomical Phenomena (INSAP); Ruggles and Cotte (2010); and the UNESCO world heritage study, available at http://issuu.com/starlightinitiative/docs/astronomy-and-world-heritage_thematic-study.

[10] E.g. Ruggles (1999); Aveni (2008b); Magli (2009); Kelley and Milone (2011); Campion (2012). A popular but informative online sketch of archaeoastronomical methodology is www.greatarchaeology.com/archaeostronomy.htm#2. For a critical survey of the progress of the discipline in recent decades, see Ruggles (2011).

[11] As Nathan Sivin (1990, 181) explained, "The difference between astronomy and astrology was a contrast of emphasis on the quantitative as opposed to the qualitative and on objective motions as opposed to the correlation between celestial and political events."

China's first great historian, Sima Qian (fl. 100 BCE), "ever since the people have existed, when have successive rulers not systematically calendared the movements of Sun, Moon, stars, and asterisms?" By the early Bronze Age, toward the very end of the third millennium BCE, attention had already begun to focus on the circumpolar region as the abode of the sky god Di, and from this time forward the north celestial Pole increasingly became a locus of spiritual significance. The polar-equatorial emphasis of ancient Chinese astronomy took shape, which meant that the ancient Chinese remained largely indifferent to the ecliptic (path of the Sun, Moon, and planets across the sky) and much less interested in heliacal phenomena (such as the first predawn rising of Sirius) than were the Egyptians or Babylonians. A prominent feature of this polar focus was the use of the handle of Ursa Major as a celestial clock hand and the identification of cardinal asterisms with the seasons and their unique characteristics – the Blue-Green Dragon with spring, the Vermilion Bird with summer, the White Tiger with autumn, the Dark Warrior (turtle and snake) with winter.

Massings of the Five Planets, the Supernal Lord's "Minister-Regulators," solar and lunar eclipses, and other astronomical and atmospheric phenomena were seen as portents of imminent, usually ominous, events. There is only a smattering of astronomical records in the earliest written documents, the oracle bone divinations from the late Shang Dynasty (thirteenth to mid eleventh centuries BCE), though meteorological phenomena are abundantly represented. In the divinations a theory of astral–terrestrial reciprocity prefiguring later Chinese astrological thinking begins to appear. Because it was the abode of the Supernal Lord and the royal ancestors, what transpired in the heavens could and did profoundly influence human affairs, and, conversely, human behavior could and did provoke responses from the supra-visible realm beyond the limits of human perception. There was no real separation: the two "realms" were continuous. Shang divination was reactive and opportunistic and never focused on individuals beyond the royal person, his consort, and his officials. Of chief interest were affairs of state such as the sacrifices to the royal ancestors, the harvest, warfare, illness, and the like.

By the late Zhou Dynasty (1046–256 BCE), *tian wen*, "sky-pattern reading," had taken as its frame of reference the twenty-eight lunar lodges (later twelve equatorial hour-angle segments) into which the sky was divided. As Sima Qian would later say, "the Twenty-Eight Lodges govern the Twelve Provinces, and the handle of the *Dipper* seconds them; the origin [of these conceptions] is ancient." In classical "field-allocation" astral omenology of mid to late Zhou, these twenty-eight segments of uneven angular dimensions were correlated with terrestrial domains according to different schemes. Allocated among the astral fields for purposes of prognostication were either the Nine Provinces into which China proper was anciently thought to have been divided, or the twelve warring kingdoms of the late Zhou Dynasty whose successive annihilation by the most ruthless among them, the Qin, led to the establishment of the unified empire in 221 BCE.

The classical job description of the post of Astrologer Royal is found in the third-century BCE canonical text, *The Rites of Zhou*:

[The *Bao zhang shi*] concerns himself with the stars in the heavens, keeping a record of the changes and movements of the stars and planets, Sun and Moon, in order to discern [corresponding] trends in the terrestrial world, with the object of distinguishing (prognosticating) good and bad fortune. He divides the territories of the nine regions of the empire in accordance with their dependence on particular celestial bodies; all the fiefs and territories are connected with distinct stars, based on which their prosperity or misfortune can be ascertained. He makes prognostications, according to the twelve years [of the Jupiter cycle], of good and evil in the terrestrial world.[12]

In this scheme, movements of the Sun, Moon, and five visible planets formed the basis of prognostication, taking also into account their correlations with *yin* and *yang* and the Five Elemental Phases (Mercury–Water, Venus–Metal, Mars–Fire, Jupiter–Wood, Saturn–Earth). While sparsely documented in contemporary sources, in part due to the hermetic nature of astral prognostication, the available evidence shows that the influence of astral omenology was deep and pervasive. As a common aphorism put it, not long after the founding of the empire, "perspicacious though the Son of Heaven may be, one must still look to where Mars is located." Although a century ago it was claimed that Chinese astrology was influenced by Mesopotamia, research since then has shown that there is no basis for that claim.[13] Ancient Chinese cosmology and astrology are distinctive in so many respects that it is now clear that throughout its formative period Chinese astronomy–astrology developed in isolation from external influences. When it comes to China's immediate neighbors the transmission of ideas was mainly centrifugal.

In the early imperial period, Han Dynasty (206 BCE–220 CE) cosmologists melded field-allocation astral prognostication with hemerological concepts, *yin-yang* and Five Elemental Phases correlative cosmology, as well as the trigrams of the *Book of Changes*, to develop the systematic and highly complex method of divination embodied in the *shi* or mantic astrolabe so representative of that period. Examples of the latter excavated from Han tombs typically consist of a round heaven-plate with Ursa Major depicted in the center along with the twenty-eight lunar lodges, months of the year (solar chronograms) inscribed in bands around the circumference. The pivot of the heaven-plate is conventionally placed in or near the handle of the *Dipper* in recognition of

---

[12] Trans. Needham (1959, 190, modified).   [13] Pankenier (forthcoming).

its symbolic centrality and perceived numinous power, while the square earth-plate underneath is graduated around the perimeter in concentric bands showing the lodges, the ancient ordinal graphs marking the cardinal and inter-cardinal directions, the twenty-four fortnightly "*qi*-nodes" of the tropical year, and so on. Some examples substitute for the heaven-plate an actual ladle fashioned from magnetic lodestone, designed to revolve within a highly polished circular enclosure representing the circumpolar region.

As originally conceived, the lodges did not technically constitute a zodiac, since, with the exception of comets, novae and the like, the Sun, Moon, and planets did not actually appear among their constituent stars; many of the latter in ancient times actually lay closer to the celestial equator than to the ecliptic. Rather, astronomical phenomena occurring within the range in longitude of a given astral field were connected with noteworthy events in the terrestrial region identified with it. In terms of astral omen theory this was because the astral and terrestrial realms were continuous and composed of the same quasi-matter/quasi-*pneuma* called *qi*.[14] Theory held that disequilibrium at any point in the continuum could potentially provoke imbalance throughout by a mysterious process somewhat analogous to a disturbance in a magnetic field of force or sympathetic resonance. In case of disruption it was essential to identify the cause, based on *yin-yang* and Five Elemental Phases phenomenological correlations, and to take action to remedy the situation by restoring equilibrium (or, in physiological terms, homeostasis). Unlike the Ptolemaic scheme, which has aptly been dubbed "astrological ethnology," despite modifications designed to take account of historical changes in political boundaries and the relative balance of power between the empire and its non-Chinese neighbors, from the outset field-allocation astrology was resolutely sinocentric. For the most part the non-Chinese world remained unrepresented in the heavens and in astral omenology except as a reflex of Chinese concerns.

Though the casting of nativities (horoscopic astrology or genethliology) did not figure in ancient China's astrological repertoire, the increasing complexity of astromantic theory in Han times was accompanied by a proliferation of calendrical prohibitions and devotions directed toward quasi-astral deities and spirits. The ancient cult of Taiyi, "Supreme One," the numinous cosmic force resident at the Pole, was elevated to a prominent place in the imperial state sacrifices, being imaginatively linked in contemporary iconography with the image of the Supernal Lord driving his starry carriage (Ursa Major) around the Pole:

---

[14] This term *qi* is often left untranslated in view of the difficulty of devising a satisfactory English equivalent that captures its protean properties. I have adopted *materia vitalis* as an expedient in an attempt to express the combination of quasi-material yet *pneuma*-like animating qualities inherent in *qi*.

The *Dipper* is the Supernal Lord Di's carriage. It revolves about the center, visiting and regulating each of the four regions. It divides *yin* from *yang*, establishes the four seasons, equalizes the Five Elemental Phases, deploys the seasonal junctures and angular measures, and determines the various periodicities: all these are tied to the *Dipper*. (Sima Qian, "Treatise on the Celestial Offices")

The protection of Taiyi and lesser astral spirits was invoked both in local cults led by magicians and by imperial officials, in the latter case even before initiating major military campaigns, when

a banner decorated with images of the Sun, Moon, *Northern Dipper*, and rampant *Dragon* was mounted on a shaft made from the wood of the thorn tree, to symbolize the Supreme One and its three stars . . . The banner was called "Numinous Flag." When one prayed for military success, the Prefect Grand Scribe-Astrologer would hold it aloft and point in the direction of the country to be attacked.

Prognostication based on the appearance of the stars of the *Dipper* appeared, as well as that based on the color, brightness, movements, etc. of comets, guest stars (novae or supernovae), eclipses, occultations of planets by the Moon, cloud formations, and a variety of atmospheric phenomena.

Ancient precedent dating from the Three Dynasties (Xia, Shang, Zhou) of the Bronze Age in the second millennium BCE led to the establishment by Han times of certain astrological resonance periods, especially dense clusters of the five visible planets at roughly 500-year intervals, as the pre-eminent sign of Heaven's–Shangdi's conferral of the Mandate to rule on a new dynasty. Other alignments of the Five Planets, or simply their simultaneous appearance in the *yang* half of the sky, were later popularly held to be beneficial for China. Not surprisingly, given the close theoretical link in Han imperial ideology between anomalies, portents, and the conduct of state affairs, the popularization of prognostication by omens led to the politicization of astral portentology. It was exceedingly rare for observations of astronomical phenomena to be faked – that would have been suicidal since court rivalries virtually guaranteed that false reporting would be discovered – rather, after the fact, the prognostics were "spun" for effect, sometimes long after, when the lessons of history became clearer. Because of the connection between astrological omens and state security, only certain imperial officials were permitted to make observations and study the historical precedents in omen texts, and by imperial decree unauthorized dabbling in astrological or calendrical matters was proscribed, and at times made a capital offense.

Along with the pervasive spread of Buddhism in the centuries following the collapse of the Han Dynasty, efforts were made by Buddhist writers during the Six Dynasties period (CE 316–589) to integrate Indian Buddhist cosmological and astrological concepts and to reconcile incommensurable numerological categories, for example matching the Buddhist *mahābhūtas* (Four Elements) with the Chinese Five Elemental Phases. Subsequently, attempts were made to establish even more complex correspondences between Chinese and Indian astrological sets such as the twenty-eight lunar lodges with the twelve Indian zodiacal signs derived from Hellenistic astrology, the Nine Planets of Indian astronomy with the seven astral deities of the *Northern Dipper*, and so on. On the whole, however, these syncretic efforts had almost no influence on long-established Chinese astrological theory, especially given the drastic decline of Buddhism following the Tang Dynasty suppression in the mid ninth century and the subsequent resurgence of Neo-Confucianism. Assimilation was also hindered by the difficulty of rendering foreign concepts and terminology into Chinese, which was often accomplished by means of bizarre or idiosyncratic transliterations.

At the popular level, Chinese astrology continued to absorb influences (Iranian, Islamic, Sogdian) via the Central Asian and maritime trade routes. Certain Western numerological categories (e.g. the seven-day week) are represented in the enormously popular and widely circulated *lishu* or almanacs documented from the ninth century, and individualized horoscopic astrology appears in late horoscopes (from the fourteenth century). But in general Hellenistic concepts had no discernible impact on the practice of astral divination at the imperial court. Until modern times the most common popular forms of divination employed ancient prognostication techniques connected with lucky and unlucky denary and duodenary cyclical characters, paired to generate the sequence of sixty unique designations used to count the days since at least the Shang Dynasty, and fate calculation based on the eight characters (*bazi*) designating the exact day, hour, etc. of birth.

During the Song Dynasty (960–1279), astral portentology entered a period of routinization and gradual decline, in part as a result of over-exploitation by sycophants and careerists as a means of enhancing their status or prospects for advancement at court, and in part because of the resurgence of Neo-Confucianism and a return to a more rational and anthropocentric outlook. Along with an increasing emphasis on human affairs and moral self-cultivation, the archaic belief in an interventionist Heaven which communicated by means of signs in the heavens faded into the background, and *tian wen* or "sky-pattern reading" shifted focus from a genre of prediction fraught with risk to a safer and more manageable interpretive mode.

As a consequence, the objective status of natural phenomena declined, and the practice of astrology by imperial officials on the whole reverted to routine observing and recording of observations, focusing on the anomalous. The interpretation of "sky-patterns" was Confucianized – one might say domesticated – and only isolated instances of inductive generalization from observation are to be found, rather than interpretation more or less tendentiously based on historical precedent. Given its subservience to the state ideology, Chinese astrology was incapable of growing into an independent body of learning or science of the heavens, but remained throughout imperial history the handmaiden of politics when not dismissed as mere superstition, its status confirmed by the comparatively modest rank of the post of Prefect Grand Scribe-Astrologer.

* * *

The Han Dynasty Prefect Grand Scribe-Astrologer Sima Qian (c.145 or 135 BC–86 BC) said, "[looking] forward and backward a thousand years before and after [events], the continuity of the interactions between Heaven and Man will be fully in evidence."[15] In other words, this continuity constitutes a historical meta-narrative discernible within Chinese civilization from the earliest times to the historian's own day. In his "Treatise on the Celestial Offices," Sima Qian lists a number of historical examples to illustrate his theory of astrological history. In this book I offer a series of topical historical studies which argue the case not that astral influences are real in any physical sense, but that in the minds of the early Chinese they were thought to propagate through the *materia vitalis* (*qi*), and it is this notion that infuses nearly every aspect of ancient Chinese civilization.

Part One, "Astronomy and cosmology in the time of dragons," begins the story in the late Neolithic, as early polities were assuming precedent-setting shape. Chapter 1, "Astronomy begins at Taosi," starts with a discussion of the recent archaeological discoveries of altar platforms (later denoted "numinous terraces," *ling tai*) dedicated to ritualized celestial observation. Their discovery has pushed back the beginnings of Chinese astronomy and calendrical science into the third millennium BCE. Taking as its starting point the earliest account of the symbiotic relationship between dragons and human society, Chapter 2, "Watching for dragons," delves into dragon lore and symbolism in the earliest period. The aim is to study the cultural role of the vast *Cerulean Dragon* constellation of the East and revisit its significance as the most important harbinger of the seasons. In the process, however, we will also look into the possible origins of this most protean and iconic symbol of Chinese civilization and trace the history of its calendrical function back to the early second millennium BCE.

What this means for the genesis of astral myth, the calendar, record keeping, and cardinal alignment is the subject of Part Two. Chapter 3, "Looking to the Supernal Lord," pursues the ancient preoccupation with axial alignment back into the Neolithic, showing how the pivot of the heavens, the celestial Pole, was thought to be the abode of the Supernal Lord from an early date. Given the absence of a bright star at the Pole throughout the entire formative period of Chinese civilization, how the ancients achieved all-important cardinal

---

[15] *Shiji*, 27.1350.

alignment will be explored in practical terms. This in turn leads to a hypothesis about the origin of the graph used to write the word "Supernal Lord" (*Di*).

Chapter 4, "Bringing Heaven down to Earth," describes an original Chinese method for orienting on true north using the stars of the Great Square of Pegasus. This was the innovative technique in use during the Zhou Dynasty (1046–256 BCE), according to pre-imperial sources, and is potentially highly accurate. Although the ancient Chinese were already well familiar with alignment using solar shadows, the priority placed on using the stars to locate the Pole exemplifies the persistence of the preoccupation with the pivot of the heavens and the dominant role of the Supernal Lord.

In Chapter 5, "Astral revelation and the origins of writing," I argue that it was calendrical astronomy that lent impetus to the development of writing in China and prefigured its application to other forms of record keeping that emerged later, including the late second-millennium BCE Shang oracle bone divinations. I offer the conjecture that rhyming among the ordinal signs used for naming the days served as the notional inspiration for the full realization that the sounds of spoken words could be attached to conventional graphic signs and serve as analogs of speech in a new medium. Finally, I suggest that one of those original ordinal graphs, stem-sign *ding*, written as a small square, in all probability takes its inspiration from the Great Square of Pegasus and links the abode of the Supernal Lord above; the idea of a divinely revealed standard of what is right, true, and normative, both spatially and temporally; and the practice of representing the nexus of the above meanings graphically as a square from the beginning of the second millennium BCE.

Part Three, "Planetary omens and cosmic ideology," explores the ideological and religious implications of the cosmic ideology and forms a linchpin of the book since it provided the early inspiration for the rest. Chapter 6 sets out the evidence for the remarkably early co-ordination of celestial portents and watershed political and military events long remembered in the oral and textual traditions of the Chinese. Beyond this, there is an account of the ancient Chinese politico-religious imagination according to which macrocosmic/microcosmic correspondences legitimated the social order. Taking the idea of the conferral of dynastic legitimacy by the Supernal Lord (or Heaven, *Tian*) as their starting point, Chapter 7, "The rhetoric of the supernal," and Chapter 8, "Cosmology and the calendar," explore the cosmological, ideological, psychological, and practical dimensions of the conviction that political legitimacy is divinely sanctioned by the Supernal Lord by means of signs in the heavens.

Chapters 9 to 11 are devoted to the theory and practice of heavenly pattern reading as it is reconstructed from the ancient texts. Strictly speaking, astrology in the earliest period, particularly involving the planets and transient phenomena, is properly termed "astral omenology." The difference is this. when it is a matter of actual observation of celestial phenomena, the early practice was generally to infer the significance after the fact based on precedent, rather than to cast a horoscope or consider the implications of the positions of the Sun, Moon, or planets for a certain contemplated action. When I use the conventional terms "astrology" or "general astrology," it is this practice that is meant, although by the early imperial period judicial astrology as an aid to political and military decision making plays a larger role. Horoscopic astrology of the kind familiar to the West was not practiced in early China.

Chapter 9, "Astral prognostication and the Battle of Chengpu," explains the theory and practice of astral prognostication in the mid-Zhou Dynasty based on the scheme of correspondences between astral locations and terrestrial polities. The focus is on the role of planetary phenomena in conjunction with the watershed Battle of Chengpu in 632 BCE. Here the precedent-setting planetary massing of the mid eleventh century at the founding of the Zhou Dynasty will be seen as having played a significant role in military strategy and timing.

Chapter 10, "A new astrological paradigm," shows how Sima Qian's "Treatise on the Celestial Offices" attempted to fundamentally reform the traditional astral–terrestrial paradigm to adapt it to the new political reality of the universal empire. The case study, General Zhao Chongguo and the campaign of 61 BCE, documents the application of the new astrological paradigm in strategic decision making at the highest level. In addition, we will see how, by the Eastern Han Dynasty (26–220 CE), the ancient hermetic science of omen reading was transformed into popular astrology.

Part Five begins with Chapter 11, "Cosmic capitals," which shows concretely how the archaic preoccupation with orientation toward the celestial Pole came to be permanently established as a feature of the symbology of empire and was embodied in the actual layout of the imperial center. Chapter 12, "Temporality and the fabric of space–time," comes to grips with ancient Chinese notions of space and time and with how the spatio-temporal world of experience is constituted. Against this background the discussion returns to the theme of the ancient focus on the pivot of the heavens at the Pole and the principal structural metaphors adopted early on to conceptualize the cosmos. The goal, first, is to show how considerations of timing and the inherent potentiality of location in time and space, expressed symbolically in the *Book of Changes*, permeated many aspects of literate culture by the early imperial period; and second, to illustrate how the early adoption of the terminology of weaving and cordage provided the necessary structural metaphors for the conceptualization of the shape of the cosmos. Chapter 13, "The Sky River and cosmography," picks up the theme of the crucial importance of the Milky Way, after the celestial Pole the most important structural element in ancient Chinese cosmography. The Milky Way, China's *Sky River*, is crucially important in astrological and cosmological contexts, as will become evident in Chapters 9, 10, and 11, but the role of the *Sky River* has been little studied and is poorly understood despite its prominence in medieval poetry and folklore. This essay briefly explores the place of the *Sky River* in iconography and cosmography, in particular in relation to mythic progenitors of Chinese civilization, Fu Xi and Nü Wa.

Chapter 14, "Planetary portentology East and West," concludes the *tour d'horizon* of astrological theory and practice with a comparative study of planetary astrology East and West. The focus is on the cluster of the five visible planets in February 1524, which was anticipated by astrologers in both Reformation Europe and Ming Dynasty China. The goal is, first, to show how differently the theory and practice of planetary astrology and long-export prediction developed in China and the West; second, to show how, despite growing rational skepticism about astral omens, it became essential for China's imperial system to gain control over the dangerously subversive potential of Five-Planet clusters, long established as the paradigmatic sign of dynastic transition; third, to portray the utterly different historical contingencies at work in the contrasting sociopolitical environments of Ming China and Reformation Germany; and finally, to illustrate the end stage of China's three-millennia-long preoccupation with "bringing Heaven down to Earth."

The Appendix is a complete, annotated translation of Sima Tan's and Sima Qian's "Treatise on the Celestial Offices" (*c.*100 BCE), the first in a Western language since Edouard Chavannes's translation in his monumental *Mémoires historiques* over a century ago. This "Treatise" from the Simas' *The Grand Scribe's Records* is the summa of elite astronomy–astrology and omen reading from the early imperial period, whose form and content speak volumes about the theory and practice of official astrology in a crucial formative period.

My hope is that this book will show that "archaeoastronomy with Chinese characteristics" offers a promising approach to the study of profoundly important facets of ancient Chinese civilization. Much more remains to be done if we are to achieve an appreciation of the full range of the ancients' responses to their environment.

Part One

Astronomy and cosmology in the time of dragons

1   Astronomy begins at Taosi

Before the Three Dynasties, everyone knew the heavenly patterns.[1]

According to historical accounts and venerable traditions, the area from the Fen River in western Shanxi Province to where the Yellow River turns abruptly eastward and passes by Luoyang (Map 1.1) was the heartland of the first Chinese dynasty, the Xia, which dominated the western Central Plain from about 1953 to 1555 BCE.[2] Study of the large late Neolithic city at Erlitou and the traces of its extensive cultural influence across north China have demonstrated that state formation had already reached an advanced stage by the early second millennium BCE. Since 1959 several decades of study of Erlitou have revealed a highly stratified society characterized by stoutly walled urban centers, ruling elites, palatial complexes, luxury goods, elaborate religious ceremonies, and early bronze industry, all sustained by considerable agricultural surpluses and trade networks.[3]

Less well known until recently is Taosi, a late Neolithic Longshan city in Shanxi, located beside the Fen River at Pingyang (N 35° 52′ 55.9″ E 111° 29′ 54″). Taosi contains the ruins of at least four predynastic cities that successively occupied the site for 500 to 600 years, from about 2500 to 1900 BCE, ultimately expanding to enclose an area of some three square kilometers. Taosi was initially discovered at about the same time as Erlitou in the late 1950s, but not until two or three decades later were dwellings of commoners and an elite cemetery unearthed. More than 1,300 burials have since been found, including those of the chiefs of the early Taosi period, pointing to the emergence of a sizeable predynastic kingdom with far-reaching trade contacts. Dating of the site and its location are both consistent with the transition to the early dynastic period that followed.[4]

---

[1] Gu Yanwu (1613–82), *Ri zhi lu*, 30.1a.

[2] Dating of events in the second millennium BCE follows the astronomical benchmarks established in Pankenier (1981–82), subsequently elaborated in Pankenier (1995). A more detailed chronology that differs substantially in the first half of the second millennium is that of China's Xia-Shang-Zhou Chronology Project: Xia Shang Zhou duandai gongcheng zhuanjiazu (2000).

[3] *Yanshi Erlitou* (1999); Li Liu (2007).   [4] Xie (2007); Liu (2007, 109–11).

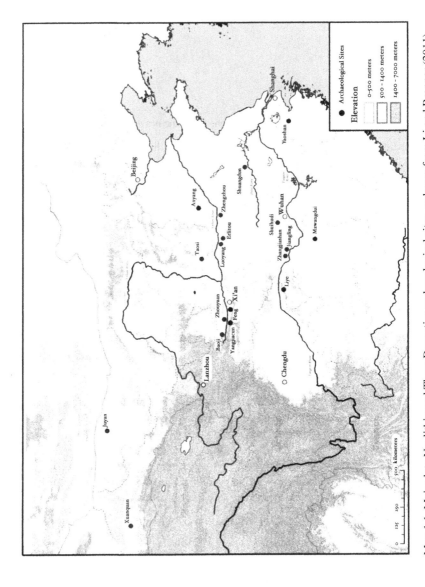

Map 1.1 Major late Neolithic and Three Dynasties archaeological sites, redrawn from Li and Branner (2011).

From 1999 to 2001, archaeologists excavated a large pounded-earth enclo-sure from the Middle Taosi period (2100–2000 BCE). Roughly rectangular in shape with a walled area of twenty-eight square kilometers, this discovery made Taosi the largest known walled town in prehistoric China. In addition to numerous pounded-earth building platforms, burials, storage pits, etc., the archaeological finds include a bronze bell, fine jades, polychrome pottery, and evidence of extreme social stratification. Even more sensationally, the discovery of brush-written glyphs reminiscent of the Shang Dynasty oracle bone script half a millennium later has led to speculation about the possible contempora-neous use of writing on perishable materials which have not survived.

By the time of its downfall, late-period Taosi was as advanced culturally as the earliest phase of Erlitou, with which it coexisted for about a century. Moreover, Taosi displays all the cultural characteristics of Erlitou enumerated above, if slightly less developed. Archaeological research has revealed that by Taosi's middle period, about 2100 BCE, major changes had occurred which had transformed the earlier small town into a large urban center five times its former size. Taosi became the largest Longshan predynastic city, with a sizeable elite enjoying palace culture on an unprecedented scale. A number of factors – date, location, cultural level – have led to Taosi's identification by some Chinese scholars as the capital of the predynastic culture hero, Emperor Yao, though this is much debated. Tradition and history have it that Yao conquered a local power identified as Taotang and established his capital at Pingyang.[5]

Having moved in and taken over what had been a minor regional power, Taosi's conquering elite brought with them an entirely new cultural assem-blage emblematic of Longshan cultures farther to the east – expansive palaces, rich burials, wheel-thrown pottery and piece-mold technology, new ceramic vessel shapes, polychrome decorative motifs, bronze making, scapulimancy, jade and lacquer work, and brush-written glyphs. Distinctive artifacts recov-ered from tombs give evidence of trade contacts with distant cultures such as Liangzhu 800 miles to the southeast. But Taosi people also engaged in some-thing else previously unknown in late Neolithic China – ritualized astronomy and timekeeping by solar observation – an innovation for which Yao's reign has been renowned since at least the mid first millennium BCE. Though unprece-dented in north China at this early date, such innovation need not surprise us. Fifty years ago Xu Fuguan wrote,

We need only think of the astronomical achievements of ancient Babylon for it to seem not the least surprising that in the age of Tang [Yao] and Yu there had already accumulated a certain knowledge of the order of the heavens, and that in government there were specific individuals who handed this knowledge down so that it was not affected by dynastic change, in that way being preserved for posterity.[6]

---

[5] *Zuozhuan* (Duke Xiang, twenty-fourth year); Cheng Pingshan (2005, 52).
[6] Xu Fuguan (1961, 14). Unless otherwise indicated, all translations are my own.

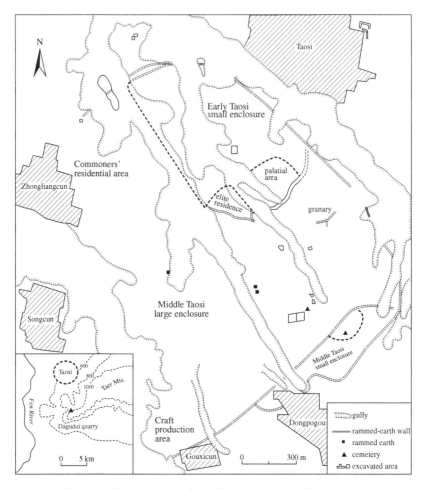

Figure 1.1 Plan of the Taosi middle period walled city with solar altar platform and elite cemetery, and location (inset) in relation to Mt. Taer and the
Fen River. Adapted from Li Liu (2007), © 2007 Li Liu. Reprinted with the
permission of Cambridge University Press.

In Parts Two and Three I will have more to say about cosmological orientation
in the prehistoric period and the Bronze Age, but for now we will focus on the
archaeological remains at Taosi (Figure 1.1).

Discovered in 2003 abutting the southeast wall of Taosi's large middle-period
city is a walled enclosure (IIFJT1) comprising three concentric pounded-earth
structures, covering an area of more than 1,400 square meters. The structure has

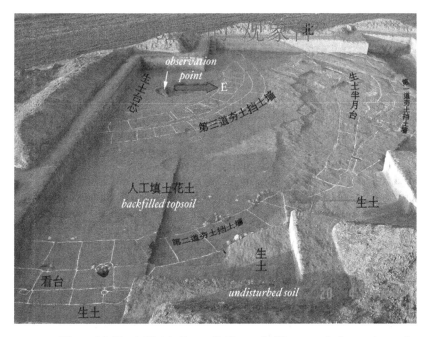

Figure 1.2 Taosi "Spirit Terrace" (*ling tai*). The central observation point is located at the base of the arrow at upper left. Pillars and apertures on the uppermost (third) platform are outlined in chalk. Adapted from photo courtesy of Wang Wei, Institute of Archaeology, Chinese Academy of Social Sciences.

been reconstructed as a three-level altar platform. The third, top level of the altar complex is a semicircular platform. That terrace is adjacent to an elite burial ground and was evidently used for sacrificial offerings. On it a curved pounded-earth wall faced south-southeast, surviving traces of its base still perforated by twelve regularly spaced grooves, about 1.4 meters apart and twenty-five centimeters wide, but now a mere four to seventeen centimeters deep. Analysis of these grooves indicates that the raised stumps of pounded-earth between them are probably the remains of more or less uniformly shaped earthen pillars, rectangular or trapezoidal in cross-section. Structurally, archaeologists surmise that a curved earthen wall was first built, the slots subsequently being cut through at regular intervals (Figure 1.2).[7]

---

[7] He Nu (2004); IACASS et al. (2004); IACASS et al. (2005); Liu Ciyuan et al. (2005, 129–30); Wu Jiabi and He Nu (2005); He Nu (2006); IACASS et al. (2007); Xu Fengxian and He Nu (2010).

Astronomy and the "Spirit Terrace"

Because the arc of the wall approximates the range in azimuth of sunrise along the eastern horizon throughout the year, archaeologists realized that the platform might have been used to observe the rising Sun, and that it may have had both ritual and astronomical functions. Standing at the center of the terrace platform and observing through simulated apertures placed in the slots of the wall, archaeologists found that most are oriented toward the Taer mountain ridge to the east-southeast. The range in azimuth defined by apertures E2–E12 matches the arc traced by the Sun as it moves along the horizon between the solstitial extremes. Initial analysis, based on calculation as well as on-site observations, suggested that the apertures may have been intended to permit observation of the rising Sun on the horizon on particular dates, including the solstices. In other words, the complex was thought to have been both a sacrificial altar and solar observation platform, in this way combining the attributes of the sacred structures called "spirit terraces" (*ling tai*) mentioned later in texts and bronze inscriptions from the early first millennium BCE.[8]

The archaeological team led by He Nu, which excavated the semicircular platform in 2003–4, proceeded on the assumption that the design of the pounded-earth foundation suggested a potential use in connection with sunrise observation or a solar cult. The team made preliminary measurements with transit and compass from the computed center of the arc, followed by simulated sunrise observations. The observations were accomplished by fabricating an iron frame with the same dimensions as the wall apertures, and then moving the frame from one slot to another to determine when the Sun rose in the bracketed space. It was only after these surveys and year-long observations had been completed that a circular pounded-earth pedestal was discovered under a previously unexcavated column of earth (Figure 1.3). At the very center of this small circular pedestal was a pounded-earth core twenty-five centimeters in diameter, which apparently marked the precise point from which the original observations were made. This observation point lay only four centimeters from the computed center of the structure, lending strong support to the presumptions about its function. Calculation of the Sun's rising points at Taosi on the solstices in 2100 BCE by Wu Jiabi and He Nu confirmed the hypothesis that the structure could have been used for solar observation. Further consultation with astronomers led the archaeologists to conclude that the preliminary measurements and observations should be repeated, this time from the newly discovered observation point using higher-precision equipment. This was accomplished in 2004–5.

---

[8] For the ode "Spirit Terrace" in the "Greater Elegantiae," one of the oldest sections of the canonical *Book of Odes*, see *infra*. Hwang (1996, 234–5).

Figure 1.3 The central observation point, located at point "0" on the plan in Figure 1.6. Adapted from Wu Jiabi, Chen Meidong, and Liu Ciyuan (2007, 2).

Structural features

Several salient facts are immediately apparent from the design of the structure. First, the Sun could not have risen in the southernmost aperture (E1) because it is oriented toward a point on the horizon several degrees south of winter solstice sunrise. This gave rise to speculation that slot E1 may have been intended to mark the Moon's major standstill, which occurs every 18.6 years. Second, for apertures E2–E12 to have served the intended purpose, the pillars which originally defined them would have to have been three to four meters tall in order to bracket the Sun's rising point on the mountain ridge some ten kilometers distant. Third, the two apertures E11 and E12 at the northern end of the arc are significantly offset from the curve of the main foundation and also deviate somewhat in their dimensions, suggesting the possibility of a different purpose and/or construction at a different time. Detailed analysis of the physical features of the structure indicates that it was crudely built in that the pounded-earth foundation conforms poorly to the arc of a circle.[9] In particular, several of the apertures (E1, E6, E9) do not precisely align with the original observation

[9] Wu Jiabi, Chen Meidong, and Liu Ciyuan (2007).

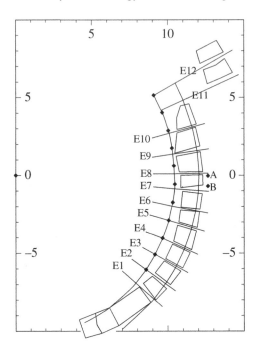

Figure 1.4 Scale drawing of the Taosi "pillars" and apertures. Redrawn from
Wu Jiabi, Chen Meidong, and Liu Ciyuan (2007, 2).

point (Figure 1.4). In some cases this has the effect of radically narrowing the
apparent size of the slot aperture as seen from the observation point. Fourth,
it is curious that the slots number exactly twelve and are regularly spaced,
since during the six months the Sun takes to travel along the horizon between
its northern and southern extremes it moves about six times faster around the
equinoxes than near the solstices. The intervals separating dates defined by
the twelve regularly spaced apertures, rather than demarcating the twenty-four
fortnightly periods ("*qi*-nodes" – *jieqi*) of later times, would necessarily have
varied greatly in duration.[10]

Although summer solstice sunrise cannot be observed in aperture E12 today,
when the change in obliquity of the ecliptic since 2100 BCE is taken into
account it is evident that sunrise on both summer and winter solstices could
have been observed through apertures E2 and E12 to within a few minutes of arc
(Table 1.1). The accuracy of the alignment in 2100 BCE was better than at
present, whether one defines sunrise as Sun centered on the horizon or upper

---

[10] Interestingly, in this the Taosi structure resembles the towers of Chankillo in Peru from two
thousand years later, which appear to have served a similar purpose. Ghezzi and Ruggles (2007).

Table 1.1 *Physical and astronomical features of the Taosi "Spirit Terrace."*

Backsight number	Midline azimuth	Horizon altitude	App. slot width	ΔAz	Midline δ GETDEC 4.0	Sun's δ 2000	JDN	Day (ss+n)	Δn days	Julian date in 2000 BCE (~Gregorian)
E1	131.07°	5.56°	0.36°	–	–28° 20′ 06″	–	–	–	–	–
					–28.33					
E2	125.06	5.81	1.23	6.02°	–23 53 51	–23.93 8:30LT	990928	179	–	Jan. 6, 1999 (~Dec. 20)
					23.90°					
E3	118.87	5.54	0.68	6.17	–19 32 16	–19.61	990894	145	34	Dec. 3 (~Nov. 16)
					–19.54	–19.59	990962	213	34	Feb. 9 (~Jan. 23)
E4	112.68	6.13	0.56	6.19	–14 26 48	–14.62	990877	128	17	Nov. 16 (~ Oct. 31)
					–14.45	–14.81	990979	230	17	Feb. 26 (~Feb. 9)
E5	106.00	7.20	0.70	6.68	–8 35 15	–8.40	990860	111	17	Oct. 30 (~ Oct. 13)
					–8.59	–8.96	990996	247	17	March 15 (~ Feb. 26)
E6	100.64	5.78	0.09	5.36	–5 14 41	–5.21	990852	103	8	Oct. 22 (~ Oct. 5)
					–5.23	–5.97	991004	255	8	March 23 (~March 6)
E7	94.46	4.27	0.76	6.17	–1 12 44	–1.51 6:27LT	990842	93	10	Oct. 12, 2000 (~Sept. 25)
					–1.21	–1.36 6:39LT	991016	267	12	April 4, 1999 (~March 18)
E8	89.11	3.32	1.02	5.35	2 32 37	2.59	990833	84	10	Oct. 3 (~Sept. 16)
					2.54	2.51	991026	277	11	April 14 (~March 28)
E9	82.30	2.26	0.53	6.80	7 23 41	7.41	990821	72	12	Sept. 21 (~Sept. 4)
					7.39	7.05	991038	292	12	April 26 (~April 9)
E10	74.59	1.91	0.61	7.71	13 23 15	13.66	990804	55	17	Sept. 4 (~Aug. 18)
					13.39	13.03	991055	309	17	May 13 (~ April 26)
E11	66.08	1.12	2.29	8.51	19 38 04	19.75	990783	34	21	Aug. 14 (~July 28)
					19.63	19.01	991076	330	21	June 3 (~May 17)
E12	60.35	1.27	1.42	5.73	24 12 04	23.92 5:19LT	990749			July 11, 2000 (~June 24)
					24.20°					

limb tangent. Thus the potential use of the complex to observe and conduct sacrificial rituals at sunrise on the solstices is confirmed by the astronomical analysis. The situation is more ambiguous with regard to aperture E7 and its possible association with the equinoxes. The degree of accuracy is far from that achieved on the solstices; indeed, aperture E7 does not capture equinoctial sunrises at all. In the absence of evidence of such an early focus on the equinoxes, the argument that E7 was used to determine the equinox is problematic. The earliest textual evidence of interest in the midpoint of the seasons does not appear until over a millennium later in the "Canon of Yao," though that text's astronomical content clearly does refer to a time considerably earlier. From inscriptional records of Sun sacrifices beginning in the late Shang oracle bones (*c*.1250–1046 BCE) it is clear that direct solar observation was already involved in determining when "the day is of medium length," according to the "Canon."[11] Apart from the date of the site and Taosi's location near the traditional location of Yao's capital, as suggestive as they are, there is no other evidence of a direct connection between Taosi and that culture hero.

Interpreting the table

Shown here in Table 1.1 are the features of the Taosi "Spirit Terrace" derived from the published survey data. The first six columns reproduce the physical features of the twelve apertures (E1–E12). Column four shows the approximately 5° to 6° difference in azimuth between aperture midlines, which is a reflection of the more or less regular size of the slots and the pillars separating them. As noted above, this regularity is not affected by the apparent slot width in those cases where the aperture does not accurately point toward the central observation point. However, that misalignment has the effect of compromising the apertures' usefulness, for example in the case of E6 whose apparent width is only 0.09°. Deviation from the norm is also apparent in the case of apertures E10 and E11 located at the transition between the main curve of the foundation and the significantly offset pillars that define apertures E11 and E12 on the north end of the array. This deviation also has the effect of altering the structural symmetry in terms of the number of days intervening between the Sun's appearance in successive apertures near the solstices (column 10, "Δn days" in

---

[11] Song Zhenhao (1993). On Shang knowledge of the quadrantal points identified in the "Canon," see also Needham and Wang (1959, 248). Henri Maspero (1932) had argued that nothing certain could be ascertained about the origins of Chinese astronomy before the mid first millennium BCE. Maspero was soon refuted by the publication of the astronomical records found in the thirteenth-century BCE Shang Dynasty oracle bone inscriptions; Needham and Wang (1959, 176, 242, 259, 460).

Table 1.1). For example, before and after winter solstice (WS), up to five weeks intervened between the Sun's appearance in apertures E2–E3, while near the summer solstice (SS) it took about half that time to move between E11 and E12.

Table 1.1, column 6, shows the conversion to declination of the midline azimuth of each aperture based on the survey data shown in columns 2–4. Column 7 shows the Sun's declination within the aperture (upper limb tangent) between January 6, 1999, and July 11, 2000 (Starry Night Pro 6.0 simulation), followed by the local time of sunrise in three cases: E2, E7, and E12. Columns 8–9 give the Julian day numbers and the number of days since the summer solstice. Column 11 shows those "best-fit" Julian calendar dates (Gregorian in parentheses) when the Sun could have been observed to rise in each of the apertures. Computed dates for the solstices (WS and SS) and equinoxes (VE and AE) using the complete VSOP theory for 2000 and 1999 are as follows:[*]

WS Jan. 6, 22h 26m TD JDN 990563.4347 (2000)
VE April 7, 17h 38m 990655.2417
SS July 11, 0h 40m 990749.5278
AE Oct. 9, 18h 59m 990840.2910
WS Jan. 6, 4h 26m 990928.6847 (1999)
VE–WS = 91.8070 days; SS–VE = 94.2861 days; AE–SS = 90.7632 days; WS–AE = 88.3937 days; year = 365.2500 days

From Table 1.1 it is apparent that apertures E2 (δ –23.46° to –24.50°) and E12 (δ 23.65° to 24.75°) precisely bracket winter and summer solstice sunrise respectively. In contrast, the Sun would have risen in aperture E7 (δ –0.91° to –1.52°) three days after autumnal equinox and three days before the vernal equinox. Curiously, this is the opposite result from what one would expect (as a consequence of the slightly elliptical shape of Earth's orbit) if the designers had determined the location of aperture E7 by counting the days between midwinter and midsummer and dividing the total in half. Aperture E1, whose northern edge was nearly 5° south of the Sun's southernmost declination, could not have been used to observe sunrise. The suggestion has been made that E1 could have been used to observe the Moon's major standstill limit. Below are data for a lunar standstill close to the epoch of Table 1.1, with declination as indicated by aperture E1 (δ –28.21° to –28.46°):

Moonrise: 1995.07.11 at 20:44LT
Age 14.8d (100% illumination)
Az. 132.85°, alt. 5.56° = δ –29.56° (geocentric lunar δ –28.86°)

This moonrise occurred at summer solstice on July 11, 1995 BCE, when the full Moon rose at least two lunar diameters south of aperture E1. Therefore aperture E1 does not precisely mark the southern major lunar standstill.

Although the co-occurrence of lunar maxima with the summer solstice may have been recognized at some point, to my knowledge there is no historical evidence of interest in marking the event.

* I am grateful to Prof. Salvo de Meis for providing this data.

Practical considerations

Examination of the plan of the observation platform (Figure 1.4) offers an indication of how the complex may have been built. The fact that the main arc of the pounded-earth wall foundation extended well south of the winter solstice position and yet fell short of the summer solstice position on the north end suggests that initially the builders may not have had a firm grasp of the range in azimuth that needed to be accommodated. The fact that the apertures nearest the winter solstice (E2–E5) are aimed most accurately (shown by the black dots marking their centerlines) suggests that the winter solstice alignment was fundamental. The remaining apertures appear to have been laid out at regular intervals, each about 1.2 meters (a Neolithic yard?) northeastward, with the result that by the time the halfway point was reached at E6, cumulative measurement error had already produced serious misalignment *vis-à-vis* the central observation point. Perhaps, when it was realized six months later that the structure failed to capture summer solstice sunrise in the northeast, an extension was added to create apertures E11–E12, but site constraints required that extension to be built as an outlier farther from the central observation point. The idiosyncratic location and dimensions of apertures E11–E12 would thus result from construction at a different time from E1–E10. Furthermore, the design of the complex, with its twelve uniformly spaced apertures, displays either a lack of understanding or a lack of concern with the variability in the Sun's daily progress along the horizon through the seasons. Interestingly, the curvature may reflect the existence of the concept of a circular heaven, since a straight-line array would have served equally well. Apart from fixing the solstices and the length of the tropical year, it is difficult to understand the function of the calendar generated by such alignments. The apparent trial-and-error approach and relative lack of sophistication displayed by the Spirit Terrace does, however, seem technologically "age-appropriate" with respect to the late Neolithic cultural level of Taosi.

It is worth noting that ritualized solar observation for the purpose of prediction may be said to be "doing science" in the broad sense. As Anthony Aveni said in reference to Stonehenge, "I find it curious that we feel the need to choose between 'ritual' and 'science' ... isn't there a bit of science in a ritual conducted at a time of year signaled by the passage of sunlight down an avenue or through a stone archway?"[12] More recently the scientific nature of the solar observations carried out at Taosi is supported by the discovery in a royal tomb there of a lacquered "shadow-rule" whose graduations appear to correspond to the length of the Sun's shadow cast by a gnomon of standard length on the dates indicated by the apertures in the terrace wall. Preliminary study suggests that this lacquered stick may have been used to track the Sun's progress through the year. Simulated measurement of the Sun's shadow with a modern replica is reported to confirm the correspondence between the lengths of the Sun's shadow on the dates marked by the apertures and the markings on the shadow rule.[13] Equally remarkable is the fact that to facilitate measurement the graduations on the lacquer rule are marked at intervals by differently colored bands – green, black, red – an exceptionally early instance of color coding in a technical context.[14]

The design and placement of the "observatory" and the discovery of the shadow rule indicates the monopolization of ritualized sunrise observation by the Taosi elite, no doubt for reasons of control and prestige; indeed, some of them are buried just inside the city wall adjacent to the Spirit Terrace. Exclusive access to the site was afforded by a passageway through the wall, and the high status and sacral function of the Spirit Terrace may explain why, during the violence that attended the decline and ultimate abandonment of the city, attributed to attacks by a rival polity to the south, the site was razed rather than being simply abandoned. Burials, palace walls, and other structures identified with the Taosi elite were similarly desecrated or demolished.[15]

Yaoshan and Huiguanshan solar altars

Of comparable interest with Taosi are two earlier altars from the middle period of the Liangzhu culture in Zhejiang Province to the south, dating from the middle of the third millennium BCE. First excavated in 1987 and 1991 near Hangzhou, the two altar platforms over twenty meters high were built only a few kilometers apart and share the same configuration as concentric squares in the shape of the Chinese character *hui* 回. Both are constructed of contrasting layers of differently colored soils (Figure 1.5). After their abandonment, both sites were subsequently used as royal cemeteries, with as many as fourteen elite burials in one case, the richly furnished tombs yielding many distinctively carved Liangzhu jades, weapons, and other artifacts.[16] In the case of Yaoshan,

---

[12] Aveni (2002, 67). For this "scientific" component of Neolithic religion, see Lewis-Williams and Pearce (2005, 232).
[13] He Nu (2009); Li Geng and Sun Xiaochun (2010).
[14] Li Geng and Sun Xiaochun (2010, 366, Figure 2, and color photograph following 372).
[15] Liu (2007, 111).
[16] On the association of graves with Liangzhu altars, see Keightley (1998, 789).

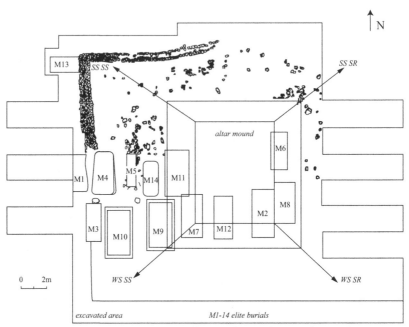

Figure 1.5 Yaoshan Liangzhu altar platform showing probable solstitial orientation. After Zhejiangsheng kaogu yanjiusuo bianzhu (2003, 6), reproduced by permission.

the burials did considerable damage to the pre-existing altar, so that although the hilltop clearly retained sacred significance the earlier altar platform must already have fallen out of use.

The original function of the two altar platforms is of particular interest because, although very different in design and cruder in construction than the solar observing terrace at Taosi, all three probably served a similar purpose. This is because both Liangzhu hilltop platforms share exactly the same orientation, with their four corners oriented toward the Sun's rising and setting directions on the solstices. According to the archaeological reports, the northeast corners of both platforms point toward azimuth 45°, the southeast corners toward 135°, the southwest corners toward 225°, and the northwest corners toward 305°.[17] Although the damage makes it difficult to establish the azimuths precisely, and detailed archaeoastronomical analysis has not yet been undertaken, the fact that the platforms share the same characteristics and were intentionally

---

[17] Liu Wu (2001).

positioned to afford a clear view of the horizon in those directions shows that the platforms could have been serviceable as functional horizon calendars using solar observations. The degree of precision achievable in fixing the length of the year would not have been as high as at Taosi, but the Liangzhu platforms are centuries older. Thus we have distinctive design solutions addressing the same scientific problem from two very different late Neolithic cultures located hundreds of miles apart. Both cultures obviously shared an intense interest in observing the sky, a preoccupation of great importance to their rulers, who were willing and able to commit considerable resources to the construction and maintenance of the sites over long periods. A classic interpretation which captures essential aspects of the sites' function is this:

Dumézil's interpretation of the "Great Time" [may be seen] not only as a wealth of exemplary events ready for enactment but also as a source of order and, accordingly, of power conferred to ordinary time. According to this view, the basic aim of the institution of the calendar appears to be the establishment of a division between time devoted to religious activities and time assigned to profane behavior. This alternation of times suggests that an important function of primordial rites may be the interruption as much as the abolition of profane time. Thanks to this ordering instituted by festal periods of time, new typical contrasts are introduced within profane time itself, which secures to it a certain amount of meaningfulness, such as the distinction between favorable and unfavorable times for doing or not doing this or that thing. This distinction may be cast in terms of days, lunar phases, months, seasons, or the magical reckoning of time-spans. We may indeed consider this ordering of profane time as an "infiltration," as it were, of sacred time into profane time; but, to the extent that it rules not merely the periodic return of specifically religious festivals but the works themselves, as in Hesiod's *Works and Days*, and besides works, all kinds of ordinary activities . . . it is profane time itself that is qualified.[18]

* * *

The Liangzhu and Taosi "spirit terraces" entail the use of horizon calendars, making their archaeological discovery unprecedented in China and lending strong support to historical accounts such as those in the "Canon of Yao" and the *Classic of Mountains and Seas (Shanhaijing)*, which document the use of such horizon methods by the second millennium BCE. The "Canon of Yao" famously records sacrifices to the Sun at the seasonal culminations of four cardinal asterisms, ritual observances datable to the late Shang Dynasty (thirteenth to eleventh centuries BCE):

[Yao] then commanded Xi and He, in reverence to August Heaven, to "calendar" (track) the [astral] signs, Sun, Moon, stars, and seasonal asterisms, so as to respectfully bestow the seasons on the people; to host the rising sun, and regularly arrange the initiation of affairs in the east; to take leave of the setting sun, and regularly arrange the completion

---

[18] Ricoeur (1985, 21).

Figure 1.6 Inscribed signs on a Dawenkou pottery jar. After Xu Fengxian (2010, 373), reproduced by permission.

of affairs in the west; to regularly attend to [the Sun's] change [of direction] in the north; to regularly attend to [the Sun's] southward displacement and reverently [mark] the solstice . . . The day being of medium length and the asterism being *Bird* [α Hya], he thereby determined mid-spring; the day being longest and the asterism being *Fire* [α Sco], he thereby determined mid-summer; the night being of medium length and the asterism being *Ruins* [β Aqr], he thereby determined mid-autumn; the day being shortest and the asterism being *Topknot* [7 Tau], he thereby determined mid-winter.[19]

Several scholars have argued that the striking glyphs (Figure 1.6) found etched on several pottery jars from Neolithic Dawenkou culture sites (*c.*3000 BCE) in Shandong are related to solar observations. The separate elements making up the compound pictographs bear a strong resemblance to the later graphs for Sun, Moon, and mountain. Some have suggested that the

---

[19] Hwang Ming-chorng (1996, 608–9). Cf Karlgren (1950a, 3); Needham and Wang (1959, 245). The best recent studies of the "Canon of Yao" are in Hwang Ming-chorng (1996, 607–10) and Liu Qiyu (2004). As Hwang (1996, 606 ff.) and Song Zhenhao (1985) have shown, the oracle bone inscriptions leave no doubt that the Shang were observing and sacrificing to the Sun on the solstices (and probably other dates as well).

pictograph on the left signifies ritualized observation of the Sun and Moon rising above the distinctive, five-peaked Sigushan ("Four Passes") Mountain ridge to the east of the site. There seems little doubt that the signs are somehow connected to the Sun and Moon, and Taosi and Yaoshan both now provide incontrovertible evidence for the existence of such observational techniques by mid third millennium BCE. This and an array of inscriptional and textual evidence led Hwang Ming-chorng to conclude, "The records of *ri-yue-suo-chu-ru-zhi-shan* ['mountains where the Sun comes out and goes in'] recorded in the *Classic of the Great Barrens* (*Dahuangjing*) [in the *Classic of Mountains and Seas*] seems [*sic*] to enable us to tie everything, including the meaning of this sign . . . the use of natural mountains as points of reference, and the textual records in the *Yaodian* and other classics, together."[20]

Other spirit terrace observations

What other celestial phenomena might those priestly skywatchers have observed, whether at Taosi or at other similar spirit terraces as yet undiscovered? It is hardly conceivable that they had not already identified eye-catching patterns in the night sky and learned to associate certain constellations with the seasons, as we will shortly see.[21] The very fact that skywatchers were attempting to determine the length of the solar year suggests that they were already motivated to devise a more accurate seasonal scheme than a simple count of the lunar months. In Parts Two and Three we will pursue this question in greater detail, but here it is appropriate to introduce a most significant near-contemporaneous observation, which appears to have played a seminal role in causing those early predynastic Chinese to focus even more intently on what transpired in the sky.

In a famous passage in the *Mozi* (fifth to fourth century BCE) there is a lengthy account of how dynastic transitions came about during the Three Dynasties of Xia, Shang, and Zhou in the second millennium BCE.[22] The first of these passages concerns the conferral of royal power on Yu the Great, putative founder of the Xia Dynasty, and deals with the earliest historical precedent for the direct intervention of the Supernal Lord, Shangdi (aka "Heaven," *Tian*) in human affairs:

---

[20] Hwang (1996, 612); see also Shao and Lu (1981); Wang Shumin (2006, 34–9, 58); Xu Fengxian (2010).

[21] This should not be surprising. Compare Ian Lilley's conclusion regarding the cognitive and linguistic abilities of the people who migrated out of Africa tens of millennia earlier; Lilley (2010, 24).

[22] "Against Aggressive Warfare," *Fei gong* (*xia*), *Mozi*, 1.11b; Enoki and Kimura, (1974, 33/19/44).

Anciently the Three Miao tribes brought about great disorder. Heaven ordered their destruction. The Sun rose at night and it rained blood for three days . . . ice came forth in summer, the earth cracked until water gushed forth. The Five Grains appeared mutated. At this the people were greatly shocked. Gao Yang ["Heaven" in some versions] then gave command [to Yu] in the Dark Palace (*xuan gong*). Yu held the auspicious jade command [scepter] in hand and set forth to conquer the Miao. Amidst thunder and lightning a god with the face of a man and the body of a bird was revealed to be waiting upon [Yu] with a jade *gui*-scepter in hand.[23]

Another later version recorded in the "Treatise on Tallies and Auspicious Signs" (*Fu rui zhi*) in the *History of the Song Dynasty* (*Song shu*, fifth century CE), adds further detail:

Yu showed himself at the [Yellow] River. A long being with a white face and the body of a fish emerged to say: "I am the Spirit of the River," then called upon Yu saying, "the Writing commands [you] to control the flood." When he finished speaking [the Spirit] presented Yu with the River Diagram [*Hetu*], which concerned the matter of controlling the waters, then withdrew into the deep. After Yu had finished controlling the waters, Heaven bestowed on him the Dark Scepter by way of announcing his success.[24]

We have now entered prehistoric mythic territory dealing with cosmogonic heroes like Yu, the tamer of the primordial flood; the Spirit of the Yellow River; and revealed esoteric diagrams. Despite the mythic tropes and fantastic subject matter, my contention is that this is a mythicized account of an identifiable astronomical event that took place in the twentieth century BCE. Later we will return to the *Mozi* account of analogous events accompanying the founding of the two succeeding dynasties, Shang and Zhou, for those refer explicitly to astronomical phenomena, but for now we will simply unpack some aspects of Heaven's charge to Yu to quell the flood and rule the kingdom.[25]

The *Mozi* reports the conferral of an auspicious jade scepter with accompanying charge on Yu the Great, whose claims to fame are quelling an all-encompassing flood, overcoming a troublesome enemy, defining the boundaries of the realm, and in consequence founding China's first hereditary dynasty, the

---

[23] Mei (1929, 222, trans. modified). Johnston (2010, 189). Despite the ambiguity of the term "Heaven" as a translation for Chinese *tian*, "sky; day; transcendent deity," and despite the non-Chinese theological baggage inevitably associated with it, for practical reasons I am disinclined to attempt a substitute rendering for the well-established "Heaven's Mandate." The context will, I hope, enable the informed reader to find his or her bearings among the range of historical meanings of the term *Tian*, from archaic sky divinity to anthropomorphized interventionist sky god, to abstract cosmic power, over the span of time encompassed by the book.

[24] *Song shu*, 27A.763.

[25] For different accounts of the etiology of ancient Chinese flood myths, see Major (1978, 4); Porter (1996); Lewis (2006b). The recent discovery of a highly unusual Western Zhou bronze vessel, the *Bin Gong xu*, has excited renewed interest in the exploits of the legendary Yu the Great, since it confirms the existence by the early ninth century BCE of accounts of his cosmogonic exploits and proves he was being held up as a paragon. For the inscription and its significance, see Li Ling (2002); Jao (2003); Li Xueqin (2002); Qiu (2004); Shaughnessy (2007).

Xia. In recognition of Yu's splendid achievements the Supernal Lord bestowed on him an inscribed dark jade *gui*-scepter signifying divine appointment to rule. Such scepters were customarily presented to royal officials and vassals as emblems of their delegated authority, certainly in the early Zhou Dynasty and no doubt earlier given their common occurrence in elite Shang burials. Whether inscribed or not they certainly comported implications of royal legitimacy, like the "diagram" mentioned in the "Testamentary Command" (*Gu ming*) chapter of the *Book of Documents* from the first half of the first millennium BCE.[26] The "Tribute of Yu" (*Yu gong*) chapter in that same text also explicitly identifies the object bestowed on Yu as a "dark jade scepter" *xuan gui*, given in recognition of the completion of his labors in hydraulic engineering. In the second account Yu had been previously presented with a "River Diagram" by the powerful spirit of the Yellow River, which apparently offered instruction on exactly how to tame the flood, not by attempting to confine the waters, but by channeling the water according to the natural contours of the land, as we learn from later tradition.

According to the *Mozi*, this auspicious token was bestowed on Yu in the "Dark Palace" (*xuan gong*), a location early commentators identify with a terrestrial temple. But by Mozi's time, "Dark Palace" conventionally referred to the winter or "watery" northern quadrant of the heavens, the abode of Zhuan Xu (the "Gao Yang" mentioned in the passage). A quotation from the fourth-century BCE astrologer Shi Shen's *Canon of Stars* (*Xing jing*) states that the term "Dark Palace" refers specifically to the lunar lodge in Pegasus known as *Align-the-Hall* (*Yingshi*), number thirteen among the twenty-eight lodges.[27] For as far back as we are able to trace the history of the system of determinative stars of the lodges (sixth century BCE), the brightest star in the constellation we know as the winged horse, α Pegasi, has always marked the location of this important asterism, which actually comprises the west side of the Great Square of Pegasus. In Chapter 4 we will see how the Square of Pegasus, known as *Determiner* (*Ding*) in the *Book of Odes* (Mao 50, *Ding zhi fang zhong*), played a crucial role in ritualized astronomy and calendrical contexts. For now, we need only take note of the fact, not coincidentally, that Pegasus also played host to a spectacular planetary massing.

---

[26] Mao ode 252 in the *Book of Odes*, in celebrating the Zhou Dynasty's receipt of Heaven's Mandate, likens the king to a jade scepter: "You are great and high, like a *gui*-scepter, like a *zhang*-scepter, with good fame ... fine to look at." In Mao ode 254, the simile reappears: "Heaven's guiding the people is like an ocarina, like a flute, like a *zhang*-jade, like a *gui*-jade; it is like taking hold of them, like leading them by the hand and nothing more." Bernhard Karlgren comments on the last stanza: "[Heaven] is mildly persuasive, like guiding people by the sound of mild music or by the sight of fine insignia of authority – not by violence or force." Trans. Karlgren (1950b, 210–14, modified).

[27] *History of the Jin Dynasty, Jin shu,* 11.301.

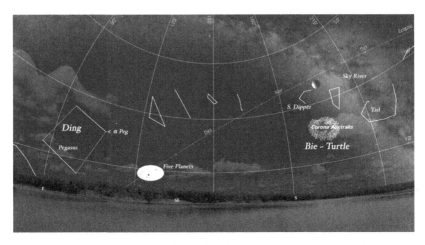

Figure 1.7 The spectacular conjunction of the five visible planets in February 1953 BCE. Mercury is so close to Mars that it cannot be shown here. The star α Pegasi became the determinative star marking the western boundary of lodge *Align-the-Hall*, *Yingshi* (Starry Night Pro 6.4.3).

In 1953 BCE, the five visible planets rose before dawn on twenty-seven consecutive mornings, from the heliacal [predawn] rising of Saturn on February 19 through the heliacal setting of Mercury on March 16. In the predawn hours of February 26, 1953 BCE, as the astrologer-priests of ancient China watched for the Sun to appear, they would have seen all five visible planets rise at very nearly the same point on the southeast horizon, precisely in the longitude of *Ding*, the Great Square of Pegasus (Figure 1.7). The planets were so close together that they could have been circumscribed by a circle a mere four degrees in diameter. To give an idea of the density of this cluster, it would be easily covered by a fist held at arm's length; at least four such four-degree circles could fit in the bowl of our Big Dipper.[28]

Here is how the scene was described by astronomer R.B. Weitzel, the first to draw attention to the phenomenon as an astronomical curiosity in 1945:

The five planets provided a magnificent spectacle. Mercury, Venus, and Mars approximated a triple star; Saturn was somewhat lower and to the left; while Jupiter, more apart, shone above and to the right of them. On the morning of February 26, 1953 B.C.,

---

[28] Weitzel (1945); Pankenier (1983, 85); Pang and Bangert (1993). Pang and Bangert put forward the proposition that Liu Xiang (77–6 BCE) knew the precise longitude in the late Neolithic (in degrees no less) where the five planets, Sun, and Moon were supposedly all stacked up on March 5, 1953 BCE, which date they claim marked the beginning of the Zhuan Xu astronomical system, a Warring States period (fourth century BCE) creation. Apart from the utterly ahistorical nature of their claim, by March 5 the planets were already separated by more than eleven degrees.

Mercury, Venus, Mars, Jupiter, and Saturn were clustered in a field of three and eight-tenth degrees – an exceptionally dense assemblage of planets.[29]

Weitzel understates the case. This was not merely an exceptionally dense assemblage – it was the most spectacular planetary massing in human history. In Parts Two and Three I will marshal evidence pointing to the conclusion not only that second-millennium BCE astrologer-priests witnessed and remembered this event as a sign from the Supernal Lord (Shangdi), but that in time they formulated an ideology of kingly legitimacy that ultimately made such celestial directives indispensable. First, however, it will be helpful to shed additional light on the other kinds of celestial phenomena those early observers were watching, and to what end.

[29] Weitzel (1945, 159).

2   Watching for dragons

It is difficult to think of any creature that has exercised an equal power over the imagination of any nation, let alone for so long.[1]

Dragons in the formative period

There is no more quintessentially Chinese symbol than the *long*, or dragon. Dragon images are virtually ubiquitous in the East Asian cultural sphere and have been in China since the Neolithic. Despite a superficial similarity with the fire-breathing reptilian familiar from Western mythology, by nature the Chinese dragon differs radically from the fearsome creatures battled by heroic figures in the West.[2] The Chinese *long*, while still "awful" in the true sense of the word, symbolizes instead the overwhelmingly powerful and protean, but ultimately beneficent, forces of nature. My aim here is to trace the iconic symbol of the Chinese dragon to its source in the life-world and imagination of the people of 4,000 years ago. Such study suggests that the symbolism of the *long* arose from close observation of nature and the stars in the Neolithic and was strongly conditioned by the implications of such observations for human adaptation to the environment. The *long* derives its dual nature – awesome power and shape-shifting ability – from its associations in both astral and terrestrial domains (Figure 2.1).

Here I will not attempt to survey all the primary sources and commentarial literature on the dragon in all its iconic manifestations. Such a catalogue has already been capably compiled by Jean-Pierre Diény.[3] Moreover, John Hay published a study of the metaphorical role of the *long* in later Chinese tradition in which he memorably characterized its symbolism this way:

[The dragon's] inseparable association with the flux of substance, seen in water and mist . . . its emergence from this flux and inevitable disappearance back into it, embodied the transformational processes of actualization itself. The transformations of the

---

[1] Hay (1994, 119).
[2] For a preliminary survey of the dragon as a cosmological symbol in a variety of cultures from the Hindu *makara* to the Andean *machácuay*, see Carlson (1982).
[3] Diény (1987). For other recent discussion of the dragon in Chinese tradition, particularly in the pre-classical period, see ibid., 119, n. 2; and, for the pre-imperial period, Yan Yunxiang (1987, 131–3).

38

Figure 2.1 Neolithic cosmo-priest's burial at Puyang, Xishuipo, c.3000 BCE. The clamshell mosaics depict the *Dragon* (right) and *Tiger* (left) in their correct positions according to the cosmology of two-and-a-half millennia later. After Loewe and Shaughnessy (1999, 51, Figure 1.5), © Cambridge University Press.

[dragon] along the axes and across the categories of existence were inherent in its nature.[4]

Clearly, we are dealing with something altogether different from the malevolent creature familiar from European mythology.[5] Neither Diény nor Hay concerned himself with the most ancient period, however. Indeed, for his part Diény dismissed out of hand the possibility of tracing the symbol to its origins. Undaunted, I propose to show that those origins are not quite as obscure as supposed. In the process I will refer to the later symbolic meaning of the *long*, but my principal aim will be to see what can be learned about the origins of the iconic symbol and its astronomical significance.

[4] Hay (1994, 149).   [5] Barber and Barber (2004, 231–44).

The first, extended account of how the fortunes of dragons and humans intersected is found in the fourth-century BCE annalistic narrative *Zuozhuan* (*Tradition of Zuo*), which chronicles the two and a half centuries of the Spring and Autumn period (722–481 BCE) during which the erstwhile vassals of the great Zhou Dynasty (1046–256 BCE) contended for hegemony.

Dramatis personae and glossary

Emperor Shun: a legendary predynastic sovereign, *c.* twenty-first century BCE

Kong Jia: (here) virtuous fourteenth king of the quasi-historical Xia Dynasty (1953–1555 BCE), in favor of whose founder, Yu the Great, Emperor Shun supposedly abdicated

Di: the high god dwelling in the celestial Pole, the "Supernal Lord" or Shangdi*

Shiwei: a chiefdom terminated by Kong Jia, later restored to service as dragon tamers during the Shang Dynasty (1554–1046 BCE); later also the name of an astral space, roughly Aqr-Psc

Tao Tang: lineage of the legendary Emperor Yao, who preceded Emperor Shun

Book of Changes: early first-millennium BCE (Zhou Dynasty) oracular text whose sixty-four hexagrams comprise all combinations of six *yang* (unbroken) and six *yin* (broken) lines and thus all possible configurations of space–time

qian: first hexagram in the *Book of Changes*, comprising six *yang* lines

kun: second hexagram in the *Book of Changes*, comprising six *yin* lines

* Throughout this book I have adopted Christopher Cullen's felicitous translation "Supernal Lord" for Shangdi.

Duke Zhao, twenty-ninth year (513 BCE)

In autumn, a *long* (dragon) was seen in the outskirts of Jiang. Wei Xianzi asked Cai Mo about this:[6] "I hear that there is no creature wiser than the *long* because they are never taken alive. Is it correct to call them wise?"

---

[6] Cai Mo (i.e. Scribe-Astrologer Mo of the state of Cai) is the most famous astrologer–diviner in the *Zuozhuan*, famed for the accuracy of his predictions. As Marc Kalinowski (2009, 370) remarks, "Cai Mo of the state of Jin (fl. 513–475 BC) . . . appears as a scholar of renown. His speeches are often the occasion of a vast display of knowledge of both astronomy and annalistic as well as ritual traditions. He is also involved in political debates. In one account, his opinions are considered on a par with the judgments of Confucius, and his clairvoyance earns him the privilege of being called a gentleman (*junzi*)"; Schaberg (2001, 7, 108).

[Cai Mo] replied, "The truth is that people are unwise, not that dragons are truly wise. The ancients raised dragons, and therefore the state had a clan of dragon breeders, the Huan Long, and a clan of dragon tamers [lit. 'drivers'], the Yu Long."

Xianzi said, "I have indeed heard of these two clans, but did not know their origin; why were they so called?"

[Cai Mo] said, "In ancient times, there was Liu Shu'an, whose descendant was Dong Fu. In truth, [Dong Fu] greatly loved dragons, and he managed to discover their tastes and appetites in order to feed and water them. Dragons flocked to him in great number. So he raised them and employed them in the service of Emperor Shun. [Shun] bestowed upon him the surname Dong, and his clan was called Huan Long, 'Breeders of Dragons.' He was enfoeffed at Zhong Chuan, and the Zhong Yi clan are his descendants. Thus, in the days of Emperor Shun and for generations afterward, there were domesticated dragons. Then it came to the days of Kong Jia of the Xia, who was obedient and faithful to Di.[7] Di bestowed upon him a team of dragons, two from the Yellow River and two from the Han River, each a pair – male and female. Kong Jia could not feed them, and he could not obtain [the help of] the Huan Long clan. The Tao Tang clan had already fallen into decline, but there was among their descendants one Liu Lei, who had studied the keeping of dragons with the Huan Long clan. In this way he came to serve Kong Jia and was able to feed and water the dragons. The Lord of Xia [Kong Jia] praised him, and granted him the clan name Yu Long, Driver of Dragons, employing him in place of the descendants of the Shiwei [clan].[8] One of the female dragons died, so [Liu Lei] secretly made it into mincemeat, which he served to the Lord of Xia. The Lord of Xia liked it, and when it was finished he sent for more. Liu Lei was frightened, and fled to the district of Lu, whose Fan clan are his descendants."

Xianzi asked, "Why is it that there are no dragons now?"

[Cai Mo] replied: "Now, as to the iconic creatures, each has its official who must adhere to the rules [of his office], bearing them in mind from dawn to dusk. If he should neglect his duty even for a single day, then death will come to him; he will forfeit his office and starve. If the official is diligent in his duties, then his iconic creature will come to him. If he abandons his duty, the creature will go into hiding, be cut off, and fail to flourish. Therefore, there were officials of the Five Elemental Phases called the Five Officials who in fact all received clan names and surnames; they were invested as Senior Grandees, and received sacrifices as noble spirits during the Five Sacrifices on

---

[7] The account of Kong Jia's reign in *The Grand Scribe's Records* (*Shiji*) closely follows *Zuozhuan* here, except that there Kong Jia is portrayed as the degenerate and dissipated late ruler of a dynastic house in decline. His misrule is said to have provoked rebellion. Nienhauser et al. (1994, 37).

[8] The Shiwei clan later came to be immortalized by being placed in the sky as the name of the central portion of the winter palace of the heavens.

the altars of grain and soil, being thus honored and revered. The Regulator of Wood was called Gou Mang, the Regulator of Fire was called Zhu Rong, the Regulator of Metal was called Ru Shou, the Regulator of Water was called Xuan Ming, and the Regulator of Earth was called Hou Tu. Dragons are creatures of water, but the Office of Water has been abandoned, so that dragons cannot be taken alive. Does not the *Changes* say of the first changing line of the *Qian* hexagram, *'the Dragon is hidden; do not act'*? Of the second changing line, *'the Dragon appears in the field'*? Of the fifth changing line, *'the soaring Dragon is in the sky'*? Of the sixth changing line, *'the recalcitrant Dragon; remorse'*? And of all the changing lines, *'there appears a flock of Dragons with no head; auspicious'*? [And furthermore], of the sixth changing line of the *Kun* hexagram, it says, *'Dragons battle in the wild.'* If Dragons had not been seen morning and night, who could have made them thus iconic?"[9]

Why is it, Wei Xianzi wonders, that dragons can no longer be caught? Is it because they are so much cleverer than men, so that they are able to elude capture? Cai Mo's somewhat cryptic reply, "the truth is that people are unwise, not that the dragons are truly wise," is key to the meaning of the passage.

This is one of a few early accounts in *Zuozhuan* to imply that "'there was a time that was blessed,'" in the sense that everything succeeded, that occasions were all favorable – this was the Golden Age. All succeeding times, and ours more than others, are, by comparison, unfortunate, untimely."[10] Like Confucius, Cai Mo "used a narrative myth or anecdote of an ancient past" for an ironic commentary on his own time, but, unlike Confucius, not yet used "in the service of a new ideal grounded in radically new insights into man's essential nature

---

[9] Trans. Legge (1972, 731, trans. modified). The curious expression "at the point Qian goes to hexagram X" (*zai qian zhi X*), rendered here for simplicity's sake as "changing line," is shorthand in the *Zuozhuan* for "the line of *Qian* which, on changing, yields hexagram X." Rather than being a genitive, *zhi* retains its archaic verbal sense "go to" as in the *Odes*, consistent with its dynamical significance here; cf Smith (1989, 445). This use of expert jargon obviously assumes that the listener is also perfectly conversant with the configuration of all sixty-four hexagrams. Reference to the individual lines in the form "nine in the first place, six in the second place, etc." is a later development. My rendering of verbal *wu zhi*, "make iconic," here is informed by the famous passage at Duke Xuan, third year (606 BCE), where King Zhuang of Chu (r. 613–591), showing no compunction in revealing his ambition to succeed the Zhou, asks a Zhou royal official about the condition of the Nine Cauldrons, the very symbol of royal power and legitimacy. Obliquely lecturing the king for his impertinence, the official says in part, "Anciently, when the Xia domain was virtuous, the distant quarters designed iconic figures [*wu*] and submitted nine ingots of bronze in tribute, [from which Nine] Cauldrons were cast representing the images. The plenitude of images were fully represented on them, enabling the people to distinguish the spiritually beneficial from the malevolent . . . By using the [Nine] Cauldrons [in the state sacrifices] superior and inferior could be harmonized and the beneficence of Heaven obtained thereby." Iconic dragon images in various forms are ubiquitous on early Shang and Zhou bronzes. As Wilhelm (1959, 275, n. 2) pointed out, the earliest stratum of the *Changes* did not include the judgments "do not act; remorse; auspicious." If those are bracketed then what remains of the middle four lines are perfectly rhymed four-character phrases no different from those found in the *Book of Odes*.

[10] Ricoeur (1985, 22); Schaberg (2001, 110); Levi (1977).

and powers." This "narrative myth... is that, but it is told of history and it clearly has a relationship to history as we know it."[11]

That relationship to history is also made clear in the excursus on the ancient history of the Five Offices (*wu guan*) prefaced by this set piece about dragons.[12] In it, however, the late Warring States term "Five Phases" (*wu xing*) is used only once, at the very beginning of the discussion. For the rest, as elsewhere in *Zuozhuan*, the terminology "Five Offices" and "Regulator" (*zheng*), as the office holder is titled, is used throughout, harking back to the usage of a millennium earlier, as we will see in Parts Two and Three below. The thrust of the passage is not to elaborate the system of Five Elemental Phases so much as to further amplify Cai Mo's perspective on the history of his profession. His subordinating latter-day theory to history is less a theoretical position than a consequence of the evolution of the Five Elemental Phases scheme from its archaic precedents.

The author of this passage obviously has a particular perspective on historical change. Brief as it is, perhaps one might (with Hayden White) tentatively categorize the rhetorical strategy prefiguring his historical narrative as *emplotment* in the form of ironic comedy: a protagonist unappreciated or rejected by society (hence lacking the usual comic reintegration) but portrayed as wiser than that society, his interlocutor an unknowing foil representing the powers that be or conventional wisdom, and disharmony rather than harmony (as in comedy) between the natural and the social; the *argument* as contextual: present circumstances explained by relationship to similar past events, tracing threads back to origins; the *ideological implication* as conservative: history evolves, the Golden Age is achievable, but change occurs slowly, and not necessarily in a desirable direction.[13]

On the surface it might seem as if Cai Mo's speech is a straightforward account of cultural tradition, but in fact the author of the passage is alerting the knowledgeable listener that there is a subtext to attend to. As Ronald Egan has remarked about the author of the *Zuozhuan* who "removes himself from overt control and manipulation of his material,"

---

[11] Fingarette (1998, 67–8), quoted in Bellah (2011, 415–16).

[12] This discussion has been characterized as "the most comprehensive elaboration of the system of the Five Phases in either *Zuozhuan* or *Guoyu*" – Schaberg (2001, 109). Schaberg goes on to observe, "Aetiology here ascribes to the Five Phases a priority that thinkers like Zou Yan would have found useful, but it subordinates them to history... Zou Yan's theory of ages keyed to the different phases, on the other hand, would subordinate history to Five Phases theory, something the *Zuozhuan* and *Guoyu* never do." As Wilhelm reassuringly notes in regard to the priority of the line texts from the *Changes* (1959, 276), "all the cases where the wording of the *Tso-chuan* differs from our present version are explained most readily if the *Tso-chuan* version is accepted as the earlier one." With the exception of three obvious phonetic substitutions in the Mawangdui silk manuscript version discovered in 1973 (e.g. "gate-latch" *jian* *jĕn for "heavenly; dry" *qian* *kan), the wording of the first-hexagram line texts here is identical; cf Shaughnessy (1996, 38–9).

[13] White (1975).

When an author chooses not to exercise his power to comment directly on events, it is to be expected that he will illustrate their significance in other ways. The narrative author whose narrator is silent is obliged to arrange his material, as the dramatist arranges his, so that the presentation itself elucidates the meaning.[14]

What that meaning is will presently become clear. First, in specific, what material facts do we learn about dragons?

(1) A certain individual initially acquired the secret knowledge of how to tend dragons, which knowledge was considered so valuable that he was called upon to serve the earliest rulers. He was rewarded with official duties, emoluments, and hereditary title, and his descendants carried on the same responsibilities generation after generation, down through the two dynasties of the second millennium BCE, Xia (1953–1555 BCE) and Shang (1554–1046 BCE).

(2) Anciently, *long*-dragons were bestowed by the Supernal Lord on rulers of conspicuous virtue who displayed reverence toward the spirits. The flourishing of dragons signified divine approval of the ruler.

(3) The dragon being the iconic "watery" creature (associated with clouds, thunder, rain, and the watery abyss beneath the earth), in course of time the function of "dragon tamer" became institutionalized as the hereditary royal office responsible for the elemental force of Water. Diligent performance of the duties of this office was of vital importance; indeed, it was a matter of life or death. If the responsible official should neglect his duties, the dragon would disappear (signifying the Supernal Lord's displeasure), with ominous consequences for the state.

(4) The reason why dragons can no longer be seen, much less tamed, is because the secret knowledge ("wisdom") of how to attract and tame them has fallen into disuse, government has become disorderly, royal virtue has declined, and the spirits no longer smile on human society.

(5) It is certainly not the case that dragons are fictional, for if that were true how could a canonical authority such as the *Book of Changes* possibly portray their behavior so vividly in its oracular imagery? Hence, in ancient times dragon were regularly seen (i.e. rational skepticism in Zuo Qiuming's day notwithstanding).

Astronomy and the harnessing of dragons

We will ignore the layers of metaphysical commentary in the *Book of Changes* that have accumulated through the ages around the first hexagram *qian* ☰

---

[14] Egan (1977, 325); also Schaberg (2001, 7).

(a)

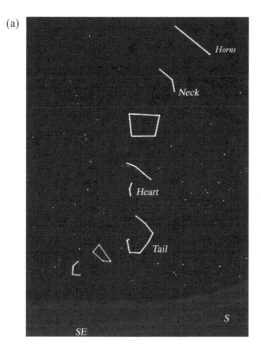

Figure 2.2 (a) The immense *Dragon* constellation, comprising stars from Vir–Sco. (b) Silk painting of a figure (immortal?) riding a dragon from the early Western Han tomb M1 at Mawangdui. After Loewe and Shaughnessy (1999, 743, Figure 10.41b), © Cambridge University Press.

and simply consider the earliest stratum of the text, the individual line texts associated with the lines of the hexagram:

(1) *hidden Dragon: do no work*
(2) *Dragon appearing in the fields*
(3) [unrelated]
(4) *perhaps bounding in the void*
(5) *soaring Dragon in the sky*
(6) *recalcitrant Dragon: there is regret*

And about the lines collectively, "*a flock of Dragons with no head: auspicious.*"

Now, though generally ignored in metaphysical commentaries, it was never a secret that these line texts of hexagram *qian* refer to the seasonal appearance of the *Cerulean Dragon*, a huge constellation comprising stars from Virgo through Scorpius (Figure 2.2a). (In what follows, "*Dragon*" in italics will refer to the constellation, and "dragon" to the iconic creature.) Wen Yiduo was the first modern Chinese scholar to elaborate on the connection in detail. As Wen says, "when ancient texts mention 'dragon,' most often they are referring

(b)

Figure 2.2 (*cont.*)

to the *Dragon* constellation"; numerous other scholars have also taken up the seasonally related astronomy of the hexagram's line texts.[15] My interpretation of the astronomical significance of the line texts differs from all the others in crucial respects, particularly in providing an account of seasonal role of both *Celestial Dragons* during the *yin* and *yang* portions of the year, which had been previously ignored.

Qian hexagram, consisting of all solid lines, symbolized the pure *yang* force: bright, warming, energetic, quickening all of nature, like the stirring of the

---

[15] Wen Yiduo (1993, Vol. 2, 231). Wen also proposes that *qian* was originally an astronym whose meaning was "to revolve" and that it referred specifically to the *Dipper*. Hsien-chi Tseng cites a colophon by Ouyang Yuan (1273–1357) on the lively *Nine Dragons* hand-scroll by Song Dynasty painter Chen Rong (fl. *c*.1244) in which Ouyang states explicitly that the *Nine Dragons* drew its inspiration from the lines of the *qian* hexagram and the first line of the "Commentary on the Image," *xiang zhuan*; Tseng (1957, 23). See http://scrolls.uchicago.edu/artists-short/chen-rong-陳容. The first Western scholar to explain the link between the constellation and hexagram was de Saussure (1930, 378). See also Li Jingchi (1978, 198; 1981, 1–4); Pankenier (1981; 1981–2, 29, n. 56); Xia Hanyi (1985); Kunst (1985, 380–419); Chen Jiujin (1987); Feng Shi (1990a, 113); Porter (1996, 46, 73); Shaughnessy (1997, 9, 197–219); Feng Shi (2007, 416–17).

dragon in the watery depths underground. China's first etymological dictionary, *Shuowen jiezi* (*Explicating Graphs and Analyzing Composite Characters*) by Xu Shen (*c*.58 CE–*c*.147 CE), in glossing the character *long*, says:

> Chief of the scaly creatures, it is able to be darkly obscure or brightly manifest, able to be minuscule or huge, able to be short or long; at autumn equinox it conceals [itself] in the watery void, at spring equinox it climbs into the sky.[16]

In the past, discussion of the seasonal appearance of the *Dragon* has focused almost exclusively on the *Dragon*'s changing appearance and orientation as it slowly makes its way across the southern sky throughout the spring and summer. According to this interpretation, during the winter, when the fields lie inert and farmwork is at a standstill, the *Dragon* supposedly hides in the watery void under the earth: "hidden Dragon, do no work." The first achronycal (evening) rising of the dragon's horn (α Vir, Spica) above the eastern horizon "in the fields," accompanied by the first full Moon of spring (the pearl the Dragon is often depicted chasing or mouthing), signaled the quickening of vegetative life and the approach of spring planting. When the *Dragon* rose it was time to conduct the great rain sacrifice (*da yu*), according to the *Zuozhuan* (Duke Huan, fifth year, 707 BCE).[17]

Here in Cai Mo's narrative, I believe we have the explanation for the account in the seventeenth year of Duke Zhao where, according to the *Zuozhuan*, "Tai Hao used *Dragon*(s) for recording, so he made a *Dragon* Master and named him for the *Dragon*." Tai Hao is none other than the mythic progenitor and benefactor of humanity Fu Xi. *Zuozhuan* is vague about exactly what was recorded (*ji*, lit. "threaded," "strung") and how dragons were implicated, but since traditional time reckoning typically involved keeping track of events by the days, months, and seasons (hence the title of the chronicle *Springs and Autumns, Chunqiu*),[18] it seems certain that the reference is to the scribe

---

[16] Even today the common expression "on the second of the second month (of the lunar calendar), the *Dragon* lifts its head" (*er yue er long tai tou*) is heard at "*Dragon*'s head festival" (*long tou jie*) in early spring. On that date "dragon scale cakes" and "dragon whisker noodles" are consumed, and other activities are performed to entice the *Dragon* to deliver mild spring breezes and timely rainfall; cf Chen Jiujin (1987, 208).

[17] Commentator Du Yu (222–85 CE) says this rite was held the month before the summer solstice. The Confucian *Analects* (XI, "*Xian jin*") indicate that this ceremony included ritual dancing. Michael Loewe (1987, 195) discerns in the rite traces of its ancient origins: "both the theory and the practice demonstrate a process that is seen in other aspects of China's cultural development; a comparatively late rationalisation and standardisation, based on philosophical principle, becomes imposed on an original act of faith that could well have been of a very early mythological origin."

[18] "Threaded" *ji* stands for *ji nian*, "sequentially arrange the years; reckon time," as in the story of the Old Man of Jiang District in *Zuozhuan*, Duke Zhao, thirtieth year. For more on the significance of *ji nian*, "reckon time," see Chapter 13 below.

(*tai shi*) in charge of astro-calendrical functions like seasonal observations of the *Dragon* and the Sun.

After his first auspicious appearance the *Dragon* leapt nearly vertically into the sky, leveling off only by summer solstice ("soaring dragon in the sky"), when its enormous 75° length extended horizontally across the entire southern sky.[19] By mid-August, as harvest season approached, the *Dragon*'s horns and head (Vir–Lib) had already disappeared below the horizon in the southwest, and the whole constellation was on the verge of sinking bodily into the depths once again. A new interpretation proposed here is that the *Dragon* constellation lingering in the sky at this season ("recalcitrant Dragon: there is regret") means that the count of the lunar months and the solar year were out of synchronization and in need of recalibration through intercalation.[20] Thus the sequential changes in the *Dragon*'s appearance indicated by the other lines of the hexagram, *when they conformed to the count of the months*, meant that nature and human affairs were in harmony: "there appears a flock of Dragons with no head: auspicious."[21] There, it would appear, the story ends, since we have run out of lines in the hexagram. As the "Appended Commentary" on the *Book of Changes* says, "Great indeed is *qian* . . . it responds to the season, driving six *Dragons* through the heavens!"

But hexagram *qian* ☰, the pure *yang* half of the complementary binary pair of *yin* and *yang*, is followed by the *kun* hexagram ☷ made up of six broken lines, the quintessential symbol of the *yin* force. *Kun* follows *qian* just as the withering of vegetation, the lengthening of shadows, and growing hours of darkness all signal the onset of winter, the season when the *yin* force reaches its peak. Hence the first line of the *kun* hexagram references mid-autumn when it says, "treading on frost [one knows] the hard freeze is coming." In fact, however, as any ancient farmer or sky-watcher certainly knew, the *Dragon*

---

[19] The third line has the *Dragon* "*bounding in the abyss*," suggesting that half the constellation has become visible while half is still immersed below the horizon. The text of line three, "*the perfected man is steadfast throughout the day; at night cautious*," does not mention the *Dragon*'s behavior.

[20] Twelve lunar months of 29.5 days amount to 354 days, more than eleven days less than the solar year. This means that a calendar based solely on a count of lunations will grow increasingly out of step with the seasons, the discrepancy amounting to more than a month after three years. Actual records of the intercalation of thirteenth months into the calendar to synchronize the lunar and solar calendars appear first in the thirteenth-century BCE Shang oracle bones.

[21] In other words, the tropical year and the lunar year are synchronized. The six lines of the hexagram are the monthly *Dragons*, of course. Another possible reading of "flock" *qun* is "nobleman, lord" *jun*. The third and the sixth lines of the hexagram are thought to signify tenuous circumstances so that their interpretations stress caution and *timeliness* – knowing when to act and when to refrain from acting. The sixth or uppermost line is especially vulnerable since it completes the hexagram, so it is not hard to grasp the analogy between inappropriate ("recalcitrant") seasonal behavior of the *Dragon* and arrogance on the part of the aspiring "lordly" man.

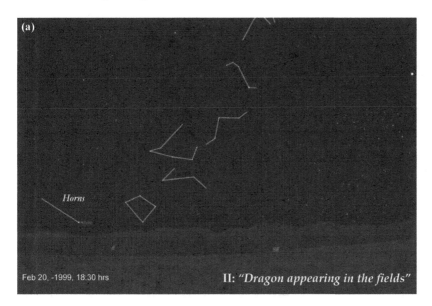

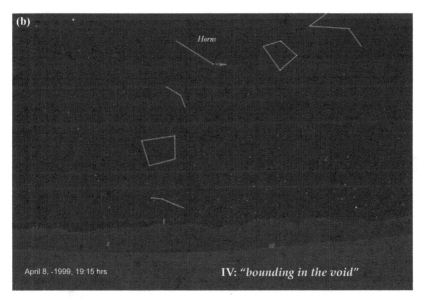

Figure 2.3(a–d) The correspondence between the line texts of the first hexa-gram *qian* and the changing posture of the evening *Dragon* constellation from the "Beginning of Spring" (*li chun*) to the "Beginning of Autumn" (*li qiu*); dates are Julian (Starry Night Pro 6.4.3).

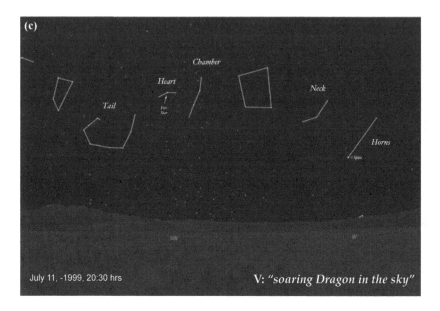

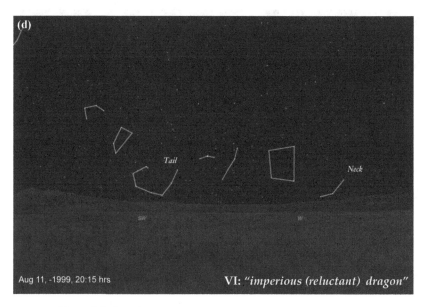

Figure 2.3 (cont.)

Oct 10, -1999, 04:00 hrs

Figure 2.4 (a) Predawn emergence of the horns of the *yin*-dragon at autumnal equinox; (b) The steeply climbing *Dragon* at winter solstice (Starry Night Pro 6.4.3).

never disappeared from the sky during the season of cold and darkness, much as the *yin* force never completely overcomes the *yang*.

Since the Sun advances one degree per day against the background of the stars, and the *Dragon* constellation is roughly 75° long, it stands to reason that the *Dragon* could not be obscured by the Sun for half the year. At the *Dragon*'s last evening setting in the west the Sun was approaching the *Horn* (Spica in Virgo) already below the horizon. While the Sun advanced eastward through the *Horn* one degree per day, the *Dragon*'s entire length quickly disappeared below the horizon because its body was nearly parallel to the horizon at this time of year (Figure 2.3d). After a month or so of invisibility, by mid-October the Sun would be a good bit of the way through the *Dragon* and once again the *Horns* would be peeking above the eastern horizon, only now in the *predawn* hours instead of in the evening. After this, the *Dragon*, rising almost vertically, would follow the same soaring path across the heavens as in spring and summer, only in half the time (Figure 2.4a and b). Numerous references to this "off-season" phenomenon appear in ancient texts, showing conclusively that the *Dragon*'s behavior served as a seasonal indicator throughout the entire year, not merely during the growing season.[22]

---

[22] Wen Yiduo (1993), Gao Heng (1973), and Gao Wence (1961) all display some degree of awareness of the contradictions among the classical passages that mention the appearances of

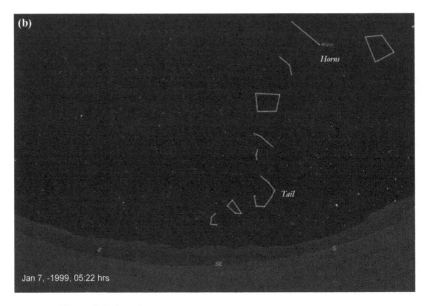

Figure 2.4 (*cont.*)

Consider a telling example. In *Discourses of the States* (*Guoyu*) there is the following series relating to the *Dragon*'s autumn and early winter indications:

When [the asterism *Dragon*'s] *Horn* [α Vir] appears [sc. before dawn], the rain stops. When [the asterism] *Heaven's Root* [Vir–Lib] appears, the rivers dry up. When [the asterism] *Base* [Libra] appears, the plants shed their leaves. When [*Heavenly*] *Quadriga* [i.e. *Chamber*, π Sco] appears, frost falls. When [*Great*] *Fire* [*Xin*, α Sco] appears, the clear wind forewarns of cold. Thus the teachings of the Former Kings say, "when the rains stop, clear the roads; when the rivers dry up, complete the bridges; when the plants shed their leaves, finish storing the harvest; when the frost falls, make ready the fur garments; when the clear wind comes, repair the inner and outer [city] walls, palaces and halls." Therefore, the *Ordinances of Xia* says: "in the ninth month, clear the roads; in the tenth month, finish the bridges." Its "Seasonal Admonitions" says: "store away your yield from the [threshing] floor; prepare baskets and carts, when *Align-the-Hall* culminates [it is time for] earthworks to begin, when the *Fire Star* first appears, [it is time to] assemble at the village overseer's." This is why the former kings, without employing rewards, were able to promulgate their virtue [*de*] widely throughout the world.[23]

the lodges comprising the *Dragon*, but without ever presenting a coherent explanation. Richard Kunst (1985, 409) mentions the conflicting indications only in passing; Shaughnessy (1983; 1997) not at all.

[23] *Guoyu*, "Zhouyu zhong," 2.9a. A striking parallel appears in the recently discovered Han Dynasty wall inscription, "Edict of Monthly Ordinances for the Four Seasons" (5 CE), from Xuanquanzhi near Dunhuang; Sanft (2008–9, 184).

There can be no doubt that this series is describing the appearance of the *Dragon*'s constituent parts as they rise before the Sun in the autumn. Following this, the all-important astral sign of the New Year and the arrival of spring some six weeks after winter solstice is said to be "when *Farmer's Auspice* [i.e. lodge *Chamber* in Sco, located at mid-*Dragon*] is upright on the meridian at dawn, and Sun and Moon are beneath the *Celestial Temple* [Great Square of Pegasus], the [*materia vitalis* of the] soil emerges in pulsations."[24] Here again, this cannot but be a predawn, late winter event – by definition, "Sun and Moon in the *Celestial Temple* [lodge *Align-the-Hall* in Pegasus]" is the first month of the Xia calendar. By this time, of course, the *yang* force is in the ascendant and *yin* is receding with the winter cold. Heretofore, all the autumnal and wintry indications have been ignored and all the focus with regard to both *qian* and *kun* hexagrams has exclusively been placed on the agricultural season:

The period of visibility of the Dragon constellation coincides so perfectly with the agricultural growing season in China that the progress of the dragon is equated with the maturation of the crops ... "Qian" was important insofar as it relates to the growing season; the birth of things in the spring, their growth in the summer, and final maturity in the fall ... This calendrical significance becomes explicit in "Kun" hexagram.[25]

Now, finally, it becomes clear why *Dragons* put in a final appearance in the sixth, or topmost, line of the *kun* hexagram long after its initial line mentioned the first freeze (the intervening lines give no indication of the month or the *Dragon*). As Cai Mo recited, "Dragons battle in the wild." After only four months the *yin* force is showing signs of exhaustion and the line suggests a contest for ascendancy, but what kind of contest? Again the answer lies in the behavior of the *Dragon* constellation, for at this time of year its performance is quite special. We saw above that during the winter months the *Dragon* re-enacts its spring and summer ballet, though *in half the time* – that is, it leaps steeply into the heavens until fully visible, then levels off and soars across the southern sky toward an oblique setting in the west. There the parallel ends, however, for

---

[24] *Guoyu*, "Zhouyu zhong," 2.9b.

[25] Shaughnessy (1997, 203–5). Following this Shaughnessy offers an extended discussion describing how *kun* hexagram must also relate to the harvest, even though its very first line begins by speaking about the frost warning of the hard freeze to come. "Frost Falls," the last fortnightly solar period of autumn, heralds "Lesser Snow," the beginning of winter. As we saw in *Guoyu*, the "*Ordinances of Xia* says: 'in the ninth month, clear the roads; in the tenth month, finish the bridges.'" In other words, the time for labor service is at hand. Surely the progression in time of the lines of *kun* hexagram must follow the same pattern as *qian* hexagram and others. Therefore, as *kun* hexagram begins the time of harvest is long past. It would be a very bad idea to wait for the frost to damage the crop before harvesting the grain. Unfortunately, Shaughnessy's discussion is also marred by the fact that all of his figures depicting the *Dragon* constellation (Figure 7.1–7.7) are reversed.

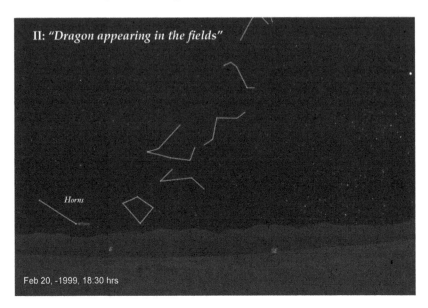

II: *"Dragon appearing in the fields"*

Horns

Feb 20, -1999, 18:30 hrs

Figure 2.5 The *yang-Dragon* reappears after sunset at the Beginning of Spring ready to begin the cycle again (Starry Night Pro 6.4.3).

just before the lunar New Year this magical creature performs an extraordinary feat.

Recall now the *Dragon's* size, some 75° from Spica in Virgo through its tail in Scorpius. After setting in the west in January–February, instead of disappearing completely from the sky as it did before the autumnal equinox, the *Dragon's* horn Spica (α Vir) would reappear above the eastern horizon at dusk (Figure 2.5) while a second *Dragon could still be seen in the western sky* on the very same day late at night until dawn (Figure 2.6). Given the appearance of differently postured *Dragons* in both predawn and evening skies at the same time at the start of the New Year, it follows that *two Dragons* would have been thought to coexist at the margins of the sky, one *yin* and one *yang*, contending with each other for supremacy. This phenomenon explains why the climactic line of the *kun* (*yin*) hexagram concludes with a combat: "Dragons battle in the wilds."[26]

---

[26] Cai Mo has truncated the line, which continues, *"their blood is reddish-brown."* Wen Yiduo (1993, 229–30), shows that translating *xuan huang* as "black and yellow" is mistaken. *Xuan* is dark red, bordering on black, the color of old coagulated blood, while *huang* (color of the loess soil, present-day "yellow") can shade all the way from cream colored into brown. Here, perhaps, we have the origin of the multiple personalities alluded to by Xu Shen in the *Shuowen jiezi*, "able to be darkly obscure or brightly manifest," and the different colors of their blood, dark (*yin*; water) and yellow (*yang*; earth). For differently colored *yin-* and *yang*-dragons (red and pale yellow) in the Mawangdui mortuary imagery (second century BCE), see Eugene Y. Wang

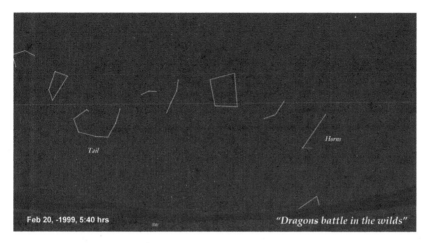

Feb 20, -1999, 5:40 hrs   *"Dragons battle in the wilds"*

Figure 2.6 The *yin-Dragon* lingering in the predawn sky at the Beginning of Spring on the very same day the *yang-Dragon* reappears in the evening (Starry Night Pro 6.4.3).

Their contest symbolizes the *yang* force stirring in the soil ready to succeed the *yin* force as the latter's *Dragon* sinks into the watery abyss. Of this juncture, the "Commentary on the Image" *Xiang zhuan* says, "Dragons battle in the wilds; their Way is at an end" – that is, the cycle is complete and renewal is at hand.

As an iconic creature the dragon is protean in nature. It is unique in embodying both *yin* and *yang* principles, if not in equal measure, at least sufficient for *yin* potentially to outweigh *yang*, as if the light and warmth might not return from the depths of winter. Dragons are quintessentially watery creatures, bringers of clouds and rain and denizens of the subterranean abyss, all *yin* attributes. At the same time, the Dragon symbolizes the nascent *yang* force as it stirs and rumbles under the earth before springing forth to take thundering flight in the spring, to serve as the dominant stellar image throughout the summer. (All the flight imagery relating to the *Dragon*, and later even the eccentric addition of wings, derive from the behavior of the constellation.) In the comprehensive interpretation presented here, the process embodied in the first two hexagrams of the *Changes*, *qian* and *kun*, evokes this dual nature, encapsulating the seasonal changes of an entire year.

Above, I alluded to the likelihood that Cai Mo knew more about the lost lore of *Dragons* than he was letting on – "the truth is that people are unwise, not that the *Dragons* are truly wise." As an astrologer and court diviner, it was

(2011, 53, 57). For a *yin*-dragon star in the Shang oracle bone divinations, see Jao Tsung-yi (1998, 44).

his business to read and interpret the patterns of the stars.[27] Recall now his comment on the significance of the line texts he quotes from the *Changes*: "if *Dragons* had not been seen morning and night who could have made them thus iconic?" Here, in a remark seemingly intended to deflect skepticism about the existence of *Dragons*, Cai Mo reveals his precise knowledge of the *Celestial Dragon*'s behavior – "if *Dragons* had not been seen *morning and night*." This expression can be read as a cliché, of course, but that does not preclude the author's ironic intent that it be taken literally. In this way the quotation from the *Changes* is transformed in the narrative from an oracular pronouncement into a self-referential sign – a wink and a nod to the implied reader.

Since we now know that already by 2100 BCE at Taosi, and probably centuries before at Yaoshan and Puyang (Figure 11.10), the Chinese were regularly watching for sunrise on the eastern horizon, naturally they would have observed the seasonal rising of the stars of the *Dragon* constellation.[28] There is strong evidence that rain-seeking rituals using dragon effigies occurred at least as early as the Shang.[29] Clearly Cai Mo (and, of course, Master Zuo) had a good understanding of the *Dragon* constellation's year-round function as seasonal indicator. By the Warring States period the increasingly rational and questioning spirit of the times – "mythospeculation," as Robert N. Bellah calls it – was overtaking the practice of divination. On this questioning spirit Marc Kalinowski remarks,

the role assigned in the *Commentary* [i.e. *Zuozhuan*] to the counsellors and scribes leaves no doubt of the existence of a deep crisis of belief in the traditional techniques of divination . . . It is, incidentally, in the *Commentary* that the first stirrings of a philosophical approach to the idea of individual fate may be found.[30]

John S. Major also reflected on the growing obscurity surrounding such cosmological myths, well under way by the time the *Zuozhuan* was composed:

It is clear that natural philosophers in the late Chou and early Han were involved in some very complex and abstract speculation linking schematic cosmography, numerology, and the operations of the Tao as expressed in the cyclical transformation of the Five Phases. At the same time . . . the myths in which the antecedents of the philosophical concepts had been conserved were being written down. They would no doubt have been intelligible as cosmological myths to educated men of the Warring States, but the frontier of scientific activity had long since moved elsewhere; the myths were understood increasingly in a religious sense only. Finally the scientific meaning of the language of

---

[27] As Donald Harper (1999, 823) notes, "Sima Tan and the *Han shu* bibliographic treatise attribute the emergence of *yin-yang* ideas . . . to men with knowledge of celestial and seasonal cycles, that is, astrological and calendrical knowledge."

[28] Pankenier (2008a, 141–48).

[29] Qiu Xigui (1983–85, 9–10); Jao Tsung-yi (1998, 36), who also reports sacrifices to a female (*yin*) dragon, or *long mu* "dagon mother."

[30] Kalinowski (2009, 394); cf. Bellah (2011, 275–6).

myth was lost almost beyond retrieval; by the Latter Han skeptics could comment that if heaven were round and earth square, the corners would not fit.[31]

Compare this now with a contemporaneous discussion of the by then mythicized Dragon's protean nature (here the iconic creature and not the constellation) from the *Book of Master Guan* (*Guanzi*):

> Those which, lying in obscurity, are able to preserve and extinguish [things], they are the turtle and the dragon. The turtle is born of water and expresses itself by fire, so it is prior to all things and the arbiter of misfortune and prosperity. The dragon is born of water and travels about mantled in the Five Colors; hence it is divine. When it wants to become small, it changes into a silkworm; when it wants to become large, it encompasses the entire sub-celestial realm; when it wishes to ascend it rises into the clouds; when it wishes to descend it enters the abyssal springs. Its transformations are not reckoned in days, nor its ascending and descending in seasons – hence they are called the divine turtle and dragon.[32]

Evidently, by the Warring States period (403–221 BCE) the *Dragon* constellation had already made the transition from being a central feature of Cai Mo's astro-calendrical science (by now diagrammatic and mathematized rather than observational) to its more familiar perennial role in the realm of myth and mystery. Indeed, the *Guanzi* even denies the creature's seasonality. Shortly after this, in Han times, the Dragon became the pre-eminent symbol of the ultimate authority and charisma of the emperor. This was inevitable because, according to the *Huainanzi*, "of the most honored spirits of Heaven none is more honored than *Cerulean Dragon* [*qing long*], who is sometimes called Heavenly One [*Tian yi*] and sometimes called Grand *Yin* [*Tai Yin*]."[33] In this way the Dragon became identified both as an astral deity and as a cosmological principle.

The celestial Dragon as astral indicator, *chen*

In 1978, Chinese archaeologists discovered the richly furnished tomb of a ruler of the state of Zeng dating from 433 BCE. Among the remarkable artifacts recovered from this famous tomb, identified as that of Marquis Yi of Zeng, was a lacquer hamper decorated with star patterns and astral motifs, including on its lid all the names of the twenty-eight lodges (Figure 2.7). In the center of the lid, surrounded by the names of the lodges, an outsized graph "*Dipper*" is written, representing the *Northern Dipper*. On either side of the *Dipper* are depicted a *Dragon* and a *Tiger*, representing the constellations of the eastern and western palaces of the heavens respectively. On the corresponding faces of

---

[31] Major (1978, 14–15). For the passage on the poor fit between round heaven and square earth from *The Record of Rites of the Dai the Elder*, see Needham and Wang (1959, 213).

[32] *Guanzi*, Chapter 14, "Shui di"; trans. Hay (1994, 132, trans. modified).

[33] Trans. Major (1993, 135); Lewis (1999, 92).

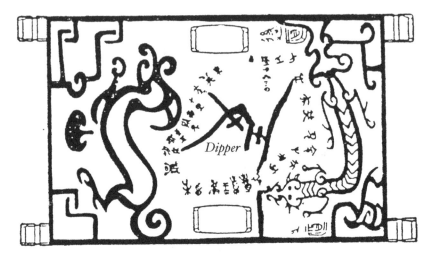

Figure 2.7 The astral–temporal diagram featuring the *Dragon* on the lacquer hamper lid from the tomb of Marquis Yi of Zeng, *c*.433 BCE. After Loewe and Shaughnessy (1999, 820, Figure 12.1), © Cambridge University Press.

the hamper are painted the principal asterisms of those two palaces, on the east the *Fire Star*, Antares in Scorpius and Orion on the west.[34] Used to store ritual regalia, the hamper was probably connected with seasonal ceremonies whose timing was determined by the stars and in conjunction with which the ruler "respectfully bestowed the seasons on the people," according to the "Canon of Yao." Numerous ancient texts allude to the function of those important seasonal asterisms.

In another fourth-century BCE narrative somewhat reminiscent of the *Zuozhuan*, the *Discourses of the States* ("Discourses of Jin," *Jinyu*), the *Fire Star* at the heart of the *Dragon* is referred to as Great *Chen*, a term that reappears elsewhere with some frequency. So the question arises, what does *chen* mean in astro-calendrical contexts? Better known as the fifth of the twelve earthly branches, *chen* is commonly also used to refer to time in general (*shichen*), or to the date (*richen*), or to the syzygy (conjunction of Sun and Moon or new Moon), or to the fifth of the twelve solar chronograms (roughly equivalent to the Western zodiacal signs used to designate 30° sections of the ecliptic).[35] Clearly *chen* has something to do with location in the sky and timing, essentially the same thing in terms of the Sun and Moon since their position gives the date.

---

[34] Feng Shi (1990a, 113).
[35] Liu Qiyu offers an illuminating comment, taken from the great Song Dynasty polymath Shen Kua's (1031–95) *Mengxi bitan*, on the accretion of meanings by the term *chen*; Liu (2004, 46).

The matter is clarified by a passage from the *Gongyang Tradition* (*Gongyang zhuan*), a third-century BCE catechetical commentary on the canonical *Spring and Autumn Chronicle*. There, glossing the chronicle's reference to "Great *Chen*," the *Gongyang* (Duke Zhao, seventeenth year) says, "What is a Great *Chen*? It is *Great Fire*; *Great Fire* [α Sco, Antares] is a Great *Chen*; *Attack* [Orion] is a Great *Chen*; and the *Northern chen* [*Dipper*] is also a Great *Chen*." Commentator He Xiu (129–82 CE) expands on this:

> *Great Fire* is called "*Heart*," "*Attack*" is called "*Triaster* and *Attack*" [Orion's belt and sword]. *Great Fire* and *Attack* are how Heaven displays to the people the earliness and lateness of the season. All in the sub-celestial realm derive the correct [time] from them, so they are the Great *Chen* – *chen* is "season [indicator]."[36]

From the *Gongyang* passage, Joseph Needham concluded that *chen* in context meant something like "celestial mark-point."[37] Liu Qiyu goes further and points out that the crucial celestial mark-points were seasonal indicators, so that their utility as temporal indicators was primary. Liu discusses the diachronic expansion of the meaning of *chen*, and concludes that it originally meant *xing*, similar to *aster* in the original nonspecific Latin sense of "celestial body." From the *Gongyang* commentary and the *Spring and Autumn Chronicle* passage, more or less contemporary with Marquis Yi, it is apparent that what is depicted on the lid and sides of his lacquer hamper are the three Great *Chen* or principal seasonal indicators.

Great Fire (Antares), the *Heart* of the *Celestial Dragon*, is the most prominent single star of that great constellation. Sometimes, because of their usefulness as adjunct indicators, the two lodges on either side of *Heart* are also included in a broader definition of Great *Chen*, as in the ancient glossary *Erya*, "Explicating Heaven" (*Shitian*): "'Great *Chen*' comprises *Chamber*, *Heart* and *Tail*; *Great Fire* [Antares, in *Heart*] is called Great *Chen*."[38] Guo Pu (276–324 CE) comments, "The bright ones [asterisms] of the *Dragon* constellation serve as harbingers of the seasons, so they are said to be the Great *Chen*. *Great Fire* or *Heart* is the brightest among them, so it is the chief seasonal indicator."[39] *Great Fire*'s dominant role is clearly shown on Marquis Yi's hamper by the outsized representation of the ancient graph for "fire" below the dragon on the face corresponding to the east.[40]

---

[36] *Chunqiu Gongyang zhuan He shi jiegu*, Duke Zhao, twenty-seventh year; cf. Feng Shi (1990a, 110).
[37] Needham and Wang (1959, 250).
[38] Similarly, in *The Grand Scribe's Records*, "Treatise on the Heavenly Offices," it simply says, "the Eastern Palace of the Cerulean Dragon is *Chamber* and *Heart*."
[39] See Pei Yin's *Suoyin* commentary in *The Grand Scribe's Records*; *Shiji*, 27.1296, n. 2.
[40] Feng Shi (1990c) analyzes all the asterisms depicted on the hamper in detail. See also Wu Jiabi (2010, 90–9; 2001, 90–4).

We saw above that the *Discourses of the States* preserves the details of the "ordinances" based on the sequential appearance of the *Dragon*'s components and designed to accurately time human activity between the harvest and the winter solstice. This is a perfect illustration of the autumnal–wintry *yin-Dragon*'s role as seasonal indicator once it appeared in the predawn sky after the autumnal equinox.[41] Finally, in *Discourses of the States* there is also a passage describing how at the beginning of spring, based on observations of natural phenomena, the Grand Scribe-Astrologer (our Cai Mo) would declare the time right for the initiation of farming activity.[42] A crucial seasonal indication was the culmination at dawn of the asterism known as *Farmer's Auspice*, another appellation for lunar lodge *Chamber* in Scorpius: "when *Farmer's Auspice* culminates at dawn and the Sun and Moon are beneath *Celestial Temple* [Pegasus], then the soil's *materia vitalis* (*qi*) emerges in pulses." Wei Zhao's (204–73 CE) comment explains the reference to the Sun and Moon: "*Celestial Temple* is *Align-the-Hall* [Pegasus]. In the first month of spring, the Sun and Moon are both in *Align-the-Hall*."[43] A check of the astronomical circumstances for February in the fourth century BCE shows this to be correct; at dawn when lodge *Chamber*'s four stars are due south and upright on the meridian, the Sun is in *Align-the-Hall* (which is, of course, invisible as a result).

Not only is the astronomy in *Discourses of the States* technically correct, the application of this calendrical maxim concerning *Farmer's Auspice* is confirmed by the inscription on still another lacquer box from the tomb of Marquis Yi of Zeng. This second box bears the inscription "it is *Chamber* to which the people sacrifice; when the syzygy [alt. 'sun's chronogram'] is at the inter-cardinal cord, the *Heavenly Quadriga* [i.e. *Chamber*] begins the year."[44] The meridian passage near dawn of *Farmer's Auspice* or *Heavenly Quadriga* (and, by implication, the new Moon marking the Beginning of Spring in the *Celestial Temple*) would have been serviceable as a harbinger of the arrival of

---

[41] For evidence of this from the Shang oracle bone inscriptions, see Jao (1998, 32–5, 37).

[42] *Guoyu*, 1.6b–7a.

[43] Ibid. Wei Zhao's gloss is supported by the *Shuowen jiezi*, which says "*chen* is *zhen* 'quake' (written with rain signific plus *chen* as phonetic); in the third month the *yang qi* (*materia vitalis*) stirs, thunder and lightning prod the people into activity for the farming season, and things all grow. Chen is the *Chamber* star, the season of Heaven." Similarly, the *Shuowen*, somewhat implausibly insisting on reading a semantic significance into the phonetic *chen*, glosses *ru*, "disgrace," as "to miss the season of plowing and be executed atop the boundary mound . . . *chen* is the farming season. Therefore, the star *Chamber* is *chen*, harbinger of field work."

[44] In an article discussing the imagery on the front face of the famous "twenty-eight lunar lodges" hamper, Wu Jiabi (2001) identifies the asterism depicted as lodge *Chamber* in its guise as a team of four; i.e., *Heavenly Quadriga*. Wu further conjectured that the hamper and inscribed box were both originally used in the *Farmer's Auspice* ceremony alluded to in *Discourses of the States* and documented in the inscription on the second box.

spring throughout the Xia, Shang, and Zhou dynasties of the Chinese Bronze Age.[45]

A naturalistic origin for the Chinese dragon

It is not surprising, therefore, that extraordinary dragon artifacts have appeared in quantity in burials of the Erlitou period and also at Taosi. From Xinzhai there are early dragon images inscribed on pottery sherds dating from between the late Longshan and early Erlitou periods. Beautifully fashioned bronze plaques inlaid with turquoise and representing the faces of dragons, as well as dragon sceptres or batons, have been excavated at Erlitou. Prefiguring the Erlitou drag-onitic images are those from Taosi, such as the primitive image on the famous "dragon basin," but the artistry and bronze workmanship at Erlitou shows great improvement over Taosi. Judging from the tombs in which the Erlitou dragon

---

[45] Richard S. Cook (1995) argues for an early representational and etymological linkage between *chen* and a scorpion, claiming a West Asian origin for the term and the asterism. Cook's conclusions are questionable. For example, quoting *The Grand Scribe's Records*, "Treatise on the Celestial Offices," which says, "*Align-the-Hall* is the *Celestial Temple* and *Detached Palace*. *Screened Causeway*'s four stars in the middle of the Milky Way are called *Heavenly Quadriga*," and missing the full stop in the middle of the line, Cook (at 25) misinterprets the line to say that asterism *Heavenly Quadriga* is to be identified with lodge 13, *Align-the-Hall* (Pegasus). Since this is the only passage in all the classical literature and scolia that appears to place *Heavenly Quadriga* in this part of the sky, either its mention here is an interpolation into Sima Qian's "Treatise" (see Pei Yin's note in his *Suoyin* commentary), or else the line has been transposed from the discussion of lodge *Chamber* earlier in the "Treatise." All other early sources, from *Discourses of the States* in the fourth century BCE on, identify *Heavenly Quadriga* with lodge *Chamber*, whose four stars do in fact straddle the Milky Way in Scorpius. As a result of the misreading, Cook divorces *chen* from *Heavenly Quadriga* and reads the mistake back into the Shang oracle bone inscriptions a millennium earlier. So, for example, in the inscription (HJ 28196c) 乙未卜瞆貞辰入史馬, 其 [X], Cook conjoins the two graphs *shi* 史 and *ma* 馬 and interprets the result to mean "*Quadriga*." Reading *chen* as "Scorpio," his rendering becomes: "the 20th day. A divination by the diviner Xu: At the setting of Scorpio the Great Square of Pegasus [has risen and we sacrifice to it]: A divination with regard to the ultimate fruits of the planting and reaping." An alternative translation would be, "crack-making on day *yiwei* [20], Xu divining: *Chen* sends reddish horses [in tribute]. It will perhaps be [unidentified graph: "advantageous"?]." *Shi* 史 here is a homophone substitution for 赤 *chi*, since 赤馬 *chi ma*, "red horse," recurs in the series of divination inscriptions on the same shell. In the common formula "X *ru* Y," X is invariably a toponym or ethnonym. Other Shang inscriptions confirm *chen* as a toponym, and Jao (1998) offers examples of *chen* as the name of the asterism in Scorpius. In any case, *Heavenly Quadriga* is certainly not Pegasus. For an epigraphic and linguistic analysis showing the Shang oracle bone graph *chen* to be a representation of a clam, *shen* (identified with the gibbous Moon), see Smith (2010–11). Richard Cook's conjectured diffusion of astronyms from Mesopotamia to China also does not bear scrutiny from the perspective of the history of Chinese astronomy. Joseph Needham had earlier concluded, "there was practically no parallel between the ancient names of the relevant constellations in China and the West," and "the number of cases in which any parallelism of symbolic nomenclature can be made out is remarkably small, and the same groups of stars were not seen in the same patterns." Needham and Wang (1959, 233, 271); see also, notably, Steele (2007).

implements and bronze bells were found, they were worn by high-status individuals with special skills, possibly the legendary dragon "tamers" or officials we encountered in Cai Mo's recounting of the history of the period. According to the archaeologists, the dragon accouterments were not the possessions of the king.[46]

The question now arises, how early is it possible to document this focus on the astronomical function of the *Dragon* constellation? Although dragon motifs of many kinds are nearly ubiquitous in early Chinese Bronze Age art, their abstract, multifaceted depictions have long discouraged speculation about a naturalistic origin. Some art historians have considered the often "dragonitic" mask-like *taotie* images fanciful, a pure product of human imagination. Others, on scant evidence, impute "shamanistic" significance to them (see the *Sidebar* on *taotie*).[47] A few scholars have alluded to a more down-to-earth connection of the dragon with the endangered Yangtze alligator, *Alligator sinensis*.[48] In fact, the climate of north China where the dragon motif was widespread by the early second millennium BCE was much warmer and wetter than at present. Abundant textual and archaeological evidence shows that the alligator, together with the Asian elephant, rhinoceros, and other subtropical flora and fauna, were common in north China, especially in marshlands and swampy areas in the east.[49] Archaeological finds of polished alligator scales in Neolithic burials at Dawenkou[50] and alligator skin drums containing alligator bones at Taosi, some five hundred kilometers from their customary range, attest to the presence of that creature in the eastern Yellow River drainage, and to the alligator's importance in elite ritual and trade.[51] Throughout Chinese history, drumming was an essential feature of rites intended to induce the Dragon Spirit to deliver rain and may have simulated the alligator's "thunderous" bellowing during the spring mating season.[52]

---

[46] Zhu Naicheng (2006, 15–21, 38).

[47] Interpretations run the gamut from "shamanistic" mystification to the outright denial of any iconic significance; Loehr (1968, 12–13); Paper (1978, 18–41); Kwang-chih Chang (1983, 56–80); Hay (1994, 125–6); Allan (1991, 124–70); Bagley (1993). According to Roderick Whitfield (1993), no scholarly consensus has been reached on the interpretation of the iconography. A cogent study is that by Kestner (1991).

[48] Porter (1993, 53–5); Feng Shi (1990a, 114); Yi Shitong (1996, 28).

[49] Needham and Wang (1959, 464, n. b); Keightley (1999a). Mencius' reference (3B/9) to a "watery" era in the "Central States," and to Yu the Great's banishing the snakes and "dragons" to the marshes, may represent the cultural memory of this very different climatic epoch.

[50] Porter (1993, 53); Li Liu (2007, 122–5).   [51] Li Liu (2007, 122–5).

[52] For an illustration of the alligator-skin-covered drums unearthed at Taosi, see Kwang-chih Chang et al. (2005, 94). The preferential use of alligator-skin drums persisted well into the Zhou Dynasty, as attested by the ode "Spirit Terrace" (*Ling tai*) in the *Book of Odes*: "The King was in the Spirit Park, where the does and stags lay [resting]; the does and stags were glossy, the white birds were glistening; the King was in the Spirit Pool; oh, the plentiful fishes leapt, [There were] uprights, crossbeams, and finials [forming the bell racks], there were panpipes, drums, and *yong*-bells; oh, assorted were the drums and bells; oh, musical was the [Hall in the] *Bi yong* Moat; the alligator-skin drums [resounded] *peng-peng*; the blind musicians gave

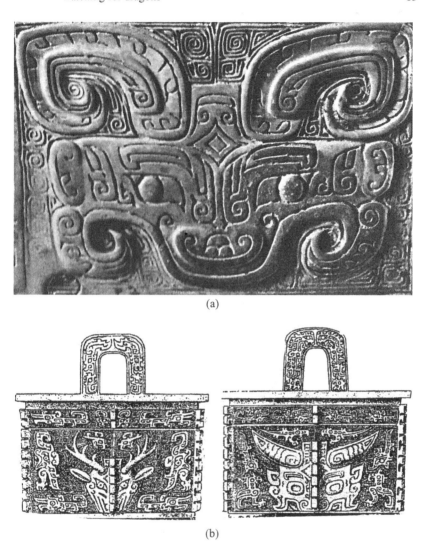

(a)

(b)

Figure 2.8 (a) *Taotie* ram motif from a Shang bronze. Detail from Pope et al. (1967, Vol. 1, Plate 36). (b) Late Shang square cauldrons with images of sacrificial animals shown *taotie*-style. Adapted from Chu Ge (2009, 31, Figure 17). (c–d) Pig faces on display at Mt. Qingcheng, Sichuan, 2005 (photo DWP).

their performance"; trans. Karlgren (1950b, 197, trans. modified). Especially noteworthy is that this ritual is being performed in the early Zhou version of the Luminous Hall (Mingtang), a ritual complex destined to become the very embodiment of astro-calendrical symbolism and ceremony (see Chapter 12 below). See also Loewe (1987); Porter (1993, 54).

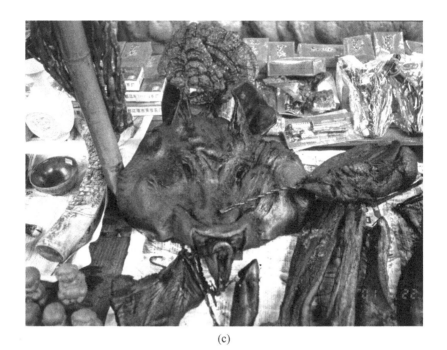

(c)

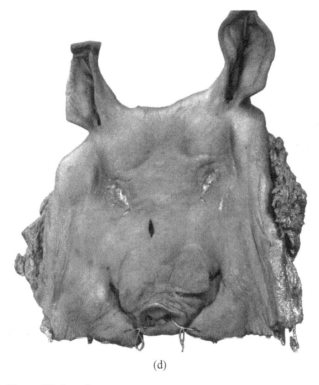

(d)

Figure 2.8 (*cont.*)

The *taotie* and the Dragon Festival

It is surprising that all the speculation about the meaning (or lack thereof) of the *taotie* virtually ignores the ethnographic evidence. Sacrifices to the ancestors and to the spirits of the landscape are the most enduring aspect of Chinese culture and are archaeologically documented deep into the Neolithic. The practice is ubiquitous of burying food, weapons, art objects, items of daily use, etc., with the deceased. Sheep, cattle, and deer are all represented as *taotie* on Shang and early Zhou bronzes (e.g. on the square cauldrons found in Fu Hao's tomb), along with innumerable stylized and abstract versions of the dragon motif, not simply as a mask-like face, but sometimes with the entire body splayed in a head-on view with both flanks visible. Over time artistic license led to these decorative motifs becoming increasingly imaginative (e.g. small dragons as horns on depictions of livestock), and then abstracted, dissolved, and made more fanciful, ultimately becoming almost unrecognizable.[a]

We know that specially bred livestock, in various combinations, were the most common sacrificial offerings in the Shang royal cult of the ancestors, as recorded in innumerable inscriptions. The *Book of Documents* records that a steer, a sheep, and a pig were sacrificed at the founding of the new Zhou capital of Luo in the mid tenth century BCE. Offerings were split open, dismembered, and arrayed in extravagant displays of disposable wealth prior to cooking and consumption in the context of worship and ritual feasting. Research has demonstrated the great political, economic, and religious importance of pigs earlier in the Neolithic: "ethnohistorical documents indicate that pigs provided the greatest source of animal protein for the ancient Chinese, that they were used in rituals, and that differential access to them was an ideological display of wealth and asymmetry."[b] Li Liu has drawn similar conclusions: "the interment of domesticated pig skulls, mandibles or even whole pigs in burials can be traced back to the early Neolithic period in the sixth millennium BCE . . . This practice became widespread during the Neolithic and continued during the Bronze Age in China . . . Most examples of pig skulls and mandibles are normally, but not exclusively, associated with rich burials in the late Neolithic period."[c] They are plentiful in burials at Taosi. The distinctive so-called "pig-dragon" jades from the Hongshan culture (fifth–third millennium BCE) are well known.

I suggest that the *taotie* motif may actually have originated in something as prosaic as the deboned and dried faces of sacrificial animals prepared for ritual consumption. These faces exhibit the same cleft-chinned death grin so characteristic of the *taotie* (Figure 2.8a), which also conspicuously lack chin and lower jaw. An origin for the *taotie* in this custom would hardly be surprising given the importance of domestic animals in sacrifices.[d] What could be more natural than for the faces of the buried skulls to have been displayed as part of the sacrificial rituals dedicated to the ancestors? Or for such faces to appear later, perhaps as an archaism, on the high-status bronzes used to make the actual offerings to the spirits? It is questionable whether the mask-like *taotie* ever represented anything so fanciful as the "spirits of the ancestors," or the "mystical transformation" of shamans into animal familiars, or "passage to the realm of the dead through the gaping maw of the monster," as has been speculated. Along with stylistic evolution of the bronze decor, the abstraction, dissolution, and fanciful recombination of features of the archaic motif gradually led to it becoming less and less recognizable, ultimately to be abandoned by late Zhou.[e]

To this day, however, one can see the dried faces of pigs looking exactly like *taotie* displayed for sale in rural market stalls, as I have in Hunan, Shaanxi, Shandong, Sichuan, and even Taipei, Taiwan, in 1974. These faces are carefully preserved to be consumed as a delicacy on special occasions, and a few moment's online search will turn up recipes and restaurants featuring the delicacy. The major feast day when pig face is consumed is the Dragon's Head Festival previously mentioned, when all manner of folk customs were performed to awaken the dragon and implore it to send the spring rains. Called "*La* pig face" (*La zhu lian*), or "stewed pig head" (*pa zhu tou*), in ancient times the pig's head was a sacrificial offering to the ancestors or to Heaven. In the north, on the "Second Day of the Second Month" (*er yue er*), the day "the *Dragon* raises its head" (*long tai tou*), peasant households would cook pig heads. Having already celebrated New Year's Day and the Fifteenth of the First Month, the Second Day of the Second Month is the last day of Spring Festival. Peasant families, having worked hard the entire year, would slaughter a pig on the twenty-third of the preceding *La* or last month of the year. Once New Year was past, the pork from the slaughtered pig would be finished, and all that was left would be the head. In the north the Second of the Second Month is called "Spring Dragon Festival." In the Zhou Dynasty the Second Day of the Second Month was a day of sacrifice, which by the Tang Dynasty had evolved into feast day among the common folk. Every food item consumed on that day would be dubbed "dragon-this" or "dragon-that"; so, for example, noodles become "dragon whiskers" (*long xu*), spring pancakes (*chun bing*) become "dragon scales" (*long lin*), fried crullers become "dragon bones" (*long gu*), and stewed pig face no doubt represented the "dragon's head" (*long tou*). This strikes me as an altogether more satisfyingly Chinese, if less mystical, origin for the *taotie* motif.[f]

[a] Chang (1983, 75–6).
[b] Kim (1994, 122).

c   Liu (2007, 123). Etched depictions of pigs or boars frequently appear on the bases of clay pots as early as the sixth millennium BCE; Xu Dali (2008, 78). See especially the clay figures of sheep and the face of a pig in Chang (1999, 45, Figure 1.3).

d   Dozens of early Western Zhou royal sacrificial pits have recently been unearthed containing the remains of sheep, horses, cattle, pigs, and dogs, some accompanied by humans, in one case including an eight-year-old child sacrificed together with a pig (*Zhongguo wenwu bao*, June 17, 2011, 4). For the possibility of shamanistic practices by "dragon-tamers" associated with these rites, see Chang (1983, 72–8). A pair of bronze *taotie* masks with pig-like snouts, presumably looted from a Western Zhou tomb, recently appeared on the antiquities market; see Rohleder (2011).

e   Chang (1983, 77, Figure 29). Consider especially the discussion of "iconic figures" (*wu*) above in connection with the decorative motifs on the Nine Cauldrons of state.

f   For the great variety of dragon motifs in Shang, including *taotie*, see Chen Zhongyu (1969).

One of the most striking artifacts discovered at the late Neolithic city site of Erlitou was the dragon-shaped scepter or mace pictured here (Figure 2.9) made of bronze with elaborate turquoise inlay.[53] A number of turquoise-inlaid bronze plaques have also been found, dating from late Longshan through Erlitou Period Three (early second millennium BCE). They have all been unearthed in elite burials (though not royal tombs), together with small bronze bells. Archaeologists have characterized the plaques as belonging to elite individuals who possessed highly regarded specialized skills and point to similarities with the depictions of dragons discovered at Taosi.[54] These may well be the "dragon officials" of legend referred to by Cai Mo. According to *Zuozhuan* (Duke Zhao, seventeenth year), "Lord Tai Hao kept track of [time] using the *Dragon*, so he created [the office of] Master of Dragons with dragon names." Tai Hao is, of course, Fu Xi, mythical culture hero and, in some versions, progenitor (with Nü Wa) of the Sinitic people, so this *Zuozhuan* passage is pushing the office of dragon watcher well back into the predynastic period, contemporaneous with Taosi.

Thus there is abundant evidence to suggest that the dragon ranked high in the pantheon of spirit entities revered by those early Bronze Age Chinese. The behavioral characteristics of the Chinese alligator suggest the reason why. Today, this Yangtze alligator is so called because it lives almost exclusively in or near the Yangtze River. They are denizens of swampland, which explains their gradual retreat southward as a result of the growing aridity of the north over the past three millennia. A noteworthy characteristic of their behavior

---

53   Liu and Xu (2007, 891, Figure 4); Allan (2007, 480, 483).

54   Zhu Naicheng (2006, 20–1); Zhang Tian'en (2002, 43–6). Jean M. James (1993, 100–1) argues that the "dragon" mosaic found in the Puyang cosmo-priest's grave is not a dragon but is actually intended to represent an alligator, though the length and curve of the neck would suggest otherwise. She does not mention a connection between the two.

1. 绿松石龙形器 (原物)

2. 仿制复原的绿松石龙形器

Figure 2.9 Erlitou dragon scepter, 70 cm, apparently grasped in the deceased's hand and laid diagonally across the chest. Adapted from *Zhongyuan wenwu* 4 (2006, Plates 1, 2, inside front cover).

is that during the winter they hibernate in underground burrows to conserve energy, emerging in spring to hunt during the warmth of the day. During the summer, in contrast, they switch to a nocturnal schedule. Like the peasantry, they store up caloric reserves from March through October to see them safely through winter hibernation. It can hardly be a coincidence that the seasonal behavior of this intimidating and sometimes aggressive creature is a perfect analog of the *Dragon* constellation's behavior and the seasonal activity of the late Neolithic and Three Dynasties farmers.[55]

---

[55] Li Xueqin (1989, 7). The discovery at Taosi of a pottery basin with a painted *long*-dragon with a stalk of grain emerging from its mouth has been interpreted to mean the dragon was a totemic creature. More likely the figure is indicative of the association of the dragon with fertility and farming; Li Xiusong (1995, 82–7); Porter (1993, 39). For a striking functional parallel with the Andean *amarus* (rainbow serpents) and their cosmic associations, based on their seasonal behavior, see Carlson (1982, 157); Urton (1981, 115). In Inka cosmology, virtually all the important Andean fauna had celestial counterparts: "in general, [the Inkas] believed that all the animals and birds on the earth had their likeness in the sky whose responsibility was their procreation and augmentation"; Urton (1981, 110), quoting Spanish chronicler Polo de Ondegardo (1571).

Figure 2.10 Shang Dynasty dragon basin (*pan*). Adapted from Pope et al. (1967, Vol. 1, Plate 3).

An eccentric dragon

The discovery of an extraordinary bronze vessel offers convincing evidence of the linkage between the alligator and the celestial *long*-dragon. In western Shanxi Province some archaeological finds are representative of a Northern Complex, so called because of the mix of stylistic influences that clearly distinguish the artifacts of this interaction sphere from the Central Plains style. Many share hybrid characteristics that reflect a mix of heartland and steppe cultures indicative of the complex archaeological picture of this area, where "northerners adopted into their own culture the manufacture and use of bronze vessels," some of which "have repeatedly been found together with vessels so eccentric that they must be local castings."[56]

---

[56] Bagley (1999, 225–6); Lin (1986).

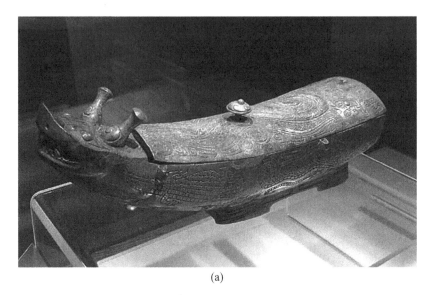

(a)

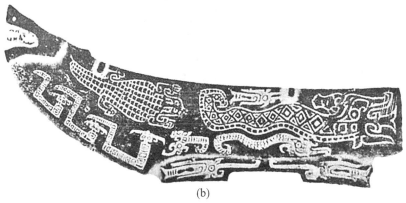

(b)

Figure 2.11 (a) Bottle-horned dragon *gong* wine vessel in the collection of Shanxi Provincial Museum. Adapted from *Shanxi Provincial Museum: Bronzes*. (b–c) Ink rubbings of the vessel's side and top. After Shanxi Provincial Museum available at www.art-and-archaeology.com/china/taiyuan/museum/pm05.html.

One such eccentric bronze is a boat-shaped zoomorphic wine vessel called a *gong* in the shape of a dragon (Figure 2.11a–c).[57] On the lid is a prominent raised knob, which serves as the handle. More than merely eccentric, this bronze

[57] In the collection of the Shanxi Provincial Museum, www.art-and-archaeology.com/china/taiyuan/museum/pm05.html; Xie and Yang (1960, 51–2); Feng Shi (1990a, 114); Yang Xiaoneng (1988, 177, Plate 170); Chang (1999, Figure 3.39g).

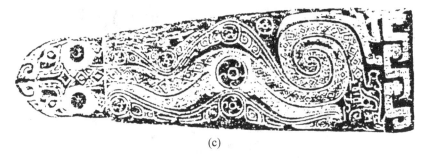

(c)

Figure 2.11 (*cont.*)

is exceptional in several respects. It is the only artifact known to juxtapose a realistic representation of an alligator (Figure 2.11b) with the iconic "bottle-horned" dragon motif familiar from the late Shang Dynasty bronzes of the Central Plains. Discovered at Shilou barely fifty kilometers north of Taosi in the heart of the ancient Xia homeland, this vessel's realistic representation of an alligator makes it unique. The juxtaposition of the naturalistic and imaginative depictions of the two creatures, alligator and dragon, is clearly intended to suggest their intimate association. The diamond-back dragons depicted on the vessel's sides are repeated on the lid, only now with their tails curling together, the much larger dragon giving shape to the vessel as a whole. Surrounding the bottle-horned dragons and alligators on the sides are various other scaly or serpentine creatures, suggesting common membership in the category of reptilians. In this respect, the iconography of this vessel is perfectly consistent with the tradition according to which the dragon is chief of the reptilian or "scaly" clan (compare Figure 2.10).

The most curious feature of this object, however, is the asymmetry of the design on the lid, where seven roundels are arranged atop the large dragon. Three straddle the dragon's midsection, the central and largest of the three forming the raised knob, while four others are arranged alongside the body, between the knob and the head. In addition, the two bottle-shaped horns on the head also bear roundels on their "mushroom-cap" tops. For anyone familiar with Chinese bronzes, what is immediately eye-catching about the roundels is their asymmetric arrangement and varying sizes, a highly idiosyncratic feature since strict bilateral symmetry is the norm in Shang bronzes. Such roundels (or "whorl circles," *wo wen*) are identified as "fire patterns" (*huo wen*) by Ma Chengyuan, who gives an account of their long history from the late Neolithic through the Warring States period. Later, in Western Zhou, the simple curlicues of early Shang evolve into more obvious flames.[58] "Fire patterns" are particularly common from the early Shang Dynasty on, especially on bronze ritual vessels

---

[58]   Ma Chengyuan (1992, 338).

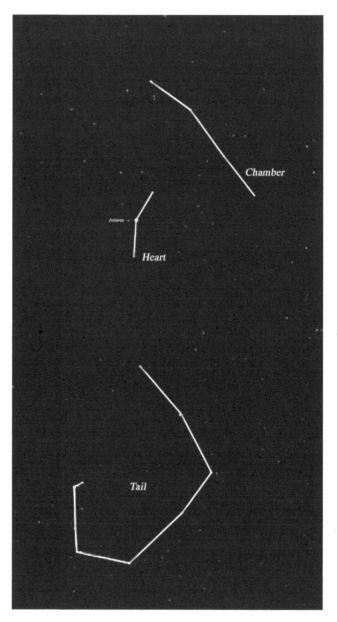

Figure 2.12 Chart of the *Celestial Dragon*'s midsection and tail (Starry Night Pro 6.4.3).

and in company with dragons and occasionally other creatures. Although the pattern regularly occurs as an ornamental motif, its presentation here in such a curious fashion in association with the iconic dragon demands an explanation. Do they have special meaning? The answer must be yes.

If we consider the depiction of the *long-Dragon* constellation in Figure 2.12, it will immediately be seen that the two asterisms comprising the *Dragon*'s midsection, lodges *Chamber* (π, ρ, δ, β_1 Sco) and *Heart* (σ, α, τ Sco), are made up of four and three stars respectively. The prominent middle star of *Heart* is first-magnitude α Sco or Antares, the *"Fire Star"* of Chinese tradition, so-called for its dull orange-red color. Antares is one of several stars mentioned in the earliest written documents, the thirteenth-century BCE Shang Dynasty oracle bone divinations.[59] As we saw above, the star served a very important function as a seasonal harbinger throughout the early period, both as the symbolic heart of the *Dragon* constellation and in its own right. Between *Chamber* and the two horns of the *Dragon* in lodge *Horns* (α Vir Spica, ζ Vir) lie the two somewhat nondescript lodges 3 *Base* (α^1, L, γ, β Lib) and *Neck* (κ, ι, φ, λ Vir). Neither is particularly bright or eye-catching, compared to *Chamber*, *Heart*, and *Tail* in Scorpius, a feature stressed in the quote above from the early glossary *Literary Expositor (Erya)*: "The Great *Chen* is *Chamber*, *Heart*, and *Tail*."[60]

Compare now the stone relief in Figure 2.13 from the Eastern Han Dynasty (second century BCE), where we also see the stars *Horns*, *Chamber*, *Heart*, and *Tail* represented, *Heart* once again straddling the dragon's midsection. It is apparent to any naked-eye observer that the stars in *Chamber* and *Heart* vary significantly in brightness, with Antares being especially prominent, not least because of its distinctive ruddy color. Here, I believe, we have the explanation for the varying sizes of the roundels depicted on the dragon vessel. Beyond simply representing stars on what must be an early star map, they are actually intended to suggest variations in the stars' apparent visual magnitude, with Antares rendered especially prominently (and cleverly) as the knob due to its brightness and importance in regulating the calendar.[61] In this curious bronze

---

[59] Jao Tsung-yi (1998).

[60] In his "Treatise on the Celestial Offices" in *The Grand Scribe's Records* (27.1295), Sima Qian lays particular emphasis on just this part of the dragon: "In the Eastern Palace of the *Cerulean Dragon*, [the principal asterisms are] *Chamber* (4) and *Heart* (5). *Heart* is the *Luminous Hall*. Its large star [Antares] is the *Celestial King*... *Chamber* is the *Heavenly Directorate*, called *Heavenly Quadriga*." Deborah L. Porter (1996, 45, n. 67) argues (following Qiu Xigui) that "Yu [the Great's] name is a pictographic representation of the lower half of the celestial dragon." Qiu Xigui (1992, 13). Deborah Porter argues (1996, 56) that, in his heroic cosmogonic role, "Yu emblematized the ideal of the sage ruler, the embodiment of the earliest human capacity to observe, measure, and predict celestial changes." Indeed, notes Porter (ibid., 202, n. 105, original emphasis), "Yu is actually described as the *incarnation of measures [du]* in *The Grand Scribe's Records (Shiji*, 1.2, 51)."

[61] Feng Shi (2007, 418). The identification of the raised knob as Antares was first proposed in Feng Shi (1990a, 114); cited in Porter (1993, 41).

(a)

(b)

Figure 2.13 (a) Han Dynasty stone relief of the *Celestial Dragon*, first to second century CE (photo DWP). (b) Ink rubbing of the stone relief, adapted from *Nanyang liang Han hua xiang shi*, 270.

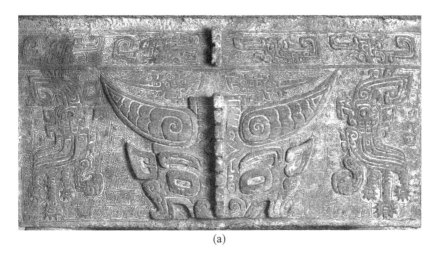
(a)

Figure 2.14 (a) *Niu fang ding* cauldron. Late Shang, excavated from tomb M1004 at Anyang in 1935. Enlargement of long side showing *taotie*-style mask flanked by parrots or owls. Available at www.ihp.sinica.edu.tw/~museum/tw/artifacts_detail.php?dc_id=9&class_plan=138. (b) Detail of bird image from the *Niu fang ding* (photos DWP).

from a zone of mixed cultural influences in the mid second millennium BCE we appear to have the earliest depiction of the *Dragon* constellation in any medium.[62]

* * *

Similar roundels are common on various styles of Shang bronzes, though not arranged so idiosyncratically as on the *gong* wine-pouring vessel. If this is an early depiction of the stars we may well ask; could similar motifs have a similar meaning on other Shang period artifacts, such as in the tail of the bird on the *Niu fang ding* (Figure 2.14a–b)? I think this is highly likely. Even if the appearance of the term "Bird Star," *niao xing*, in the Shang oracle bones is still disputed,[63] the prominent mention in the "Canon of Yao" of a *Bird Star* connected with the summer solstice in the second millennium BCE and its certain identification with the giant constellation later called the *Vermilion Bird* (lodges 23–8, roughly Cancer to Crater), make it fairly certain that the

---

[62] Except possibly for the famous clamshell mosaics from the 2500 BCE cosmo-priest's tomb in Puyang, Xishuipo, Henan, whose dragon and tiger may or may not represent actual constellations; but see Feng Shi (1990a, 108–11; 1996, 159; 2007, 374); Yi Shitong (1996, 22–31); and Figure 11.10 in Chapter 11 below.

[63] Li Xueqin (1999c; 2000).

(b)

Figure 2.14 (*cont.*)

constellation existed in Shang times.[64] Evidence in support of this comes from
the site of the predynastic Western Zhou capital of Fengjing. An eave tile (*wa
dang*) prominently displaying the name of the capital, *feng*, was discovered with
four iconic creatures deployed in their proper cardinal directions: the *Dragon* to
the east, a fish to the north, a long-necked bird to the south, and a tiger (bear?)

[64] Indeed, as Hwang Ming-chorng has shown, "almost all elements in the *Yaodian* can find an
earlier and more complete example in the *Dahuangjing*. This match between the *Dahuangjing*
and *Yaodian* has important implications in the understanding of the intellectual history in early
China. Assuming that our reading of the *Dahuangjing* as a Shang cosmology is a reasonable proposition, from the *Dahuangjing* to *Yaodian* there seems to be a transformation from
cosmology to history." Hwang (1996, 664).

to the west (Figure 2.15a).[65] Although the date and provenance are unknown, this appears to be an early depiction of the Four Iconic Images (*si xiang*) and only a slight variation on the later Warring States and Han configurations in which the Somber Warrior (*xuanwu*) replaces the Fish in the north.[66]

One problem, however, is that what could be a raptor depicted on the *Niu fang ding* bronze is utterly unlike either the long-necked *Vermilion Bird* constellation comprising lodges *Ghost in the Conveyance* through *Chariot Platform* (Cancer through Corvus) or the bird in Figure 2.15. Might there be another possibility? Indeed there is. Lying between the lodges *Hunting Net* (Taurus) and *Triaster* (Orion's belt), beneath *Topknot* (Pleiades), is the small triangle of stars called *Horned Owl* (aka *Beak*, lodge 20).[67] The identification of this asterism as an owl must belong to an early stratum in the development of the scheme. Unlike the *Dragon* of the eastern palace, none of the western asterisms except Orion are said to depict any part of a tiger's body, and variant names are already found in the *Book of Odes*. Figure 2.16 shows the *Owl* asterism as depicted on the early Han tomb ceiling from Xi'an. Here, corresponding to Taurus, a man is shown throwing a small-game net to catch a fleeing rabbit. Just behind him is the *Owl*, known as *Zuixi*, a term whose obscurity occasioned much discussion before the discovery of the Xi'an tomb ceiling in 1987. This is because in the "Treatise on the Celestial Offices" Sima Qian says that this "small triangle of three stars is the head of the *Tiger*."[68] Given the tiny asterism's small extent, scholars no doubt wondered how it could possibly represent the head of a huge constellation comprising the entire western quadrant of the sky. In the *Shuowen jiezi*, the first graph, *zui*, is defined as the feathery "horns" on the *Owl*'s head, as well as a western lodge, while a thousand years later the *Guangyun* (eleventh

---

[65] Yi Ding, Yu Lu, and Hong Yong (1996, 14, Figure 1-15); the reference is to the history of Chinese architecture by Itō Chūta (1938, 87), who reproduces the rubbing from *Jin shi suo* (Chapter 6) by Feng Yunpeng (1893). Western Zhou eave tiles were typically semicircular (*ban wa*), so both the identification of this unprovenanced artifact and Western Zhou dating are somewhat uncertain.

[66] A bear still survives as the spirit animal of the west in the "Artificer's Record" (*Kao gong ji*) in the *Rites of Zhou* (*Zhou li*, "Dongguan"). There, among the four cardinal asterisms, *Attack* (*Fa*, Orion's sword) is said to be represented by a flag with six pennants. In his commentary Zheng Xuan then associates the two main components of Orion, *Triaster*, *Shen* (belt), and *Attack* with the Tiger and Bear respectively; Yi Ding, Yu Lu, and Hong Yong (1996, 11–15). The Tiger may be the later addition, becoming conventional once the southlands were absorbed into the Hua–Xia cultural orbit, since the tiger motif is especially prominent in the iconography of the south; but see Jao Tsung-yi (1998, 39, 44) for a possible *Tiger Star* in the oracle bone inscriptions. Variants have a deer or *qilin* (unicorn) in place of the fish or turtle in the north; cf Feng Shi (2007, 427 and the Niya brocade on the cover).

[67] λ, φ1, φ2 Ori; *Shiji*, 27.1306. The discovery in 1987 of the Xi'an Western Han tomb ceiling with a horned owl representing this asterism resolved the confusion about the interpretation of the name *Zuixi* for this lodge; Hu Lin'gui (1989, 87). Its dual identity as both "Tiger's head" and *Horned Owl* reflects the late overlay of the iconic Tiger on the earlier individual asterisms of the western palace.

[68] *Shiji*, 27.1306.

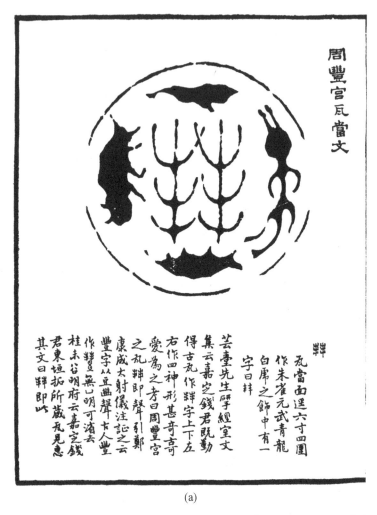

(a)

Figure 2.15 (a) Predynastic Western Zhou eave tile (?) from the capital, Fengjing. After Feng Yunpeng (1893), Chapter 6; cf Yi Ding Yu Lu, and Hong Yong (1996, 14, Figure 1-15). (b) Tracing of the *Vermilion Bird* constellation from the Western Han tomb mural discovered at Xi'an, Jiaotong University, in 1987. Detail redrawn from Rawson (2000, 178, Figure 9).

century) says it is a "beak." The second graph, *xi*, is defined as a horn or awl-shaped pendant worn on the belt to loosen knots, again a pointed or horn-shaped object. In some early literary contexts the compound *zuixi* is said to be a "large turtle" without further comment. About the only thing the two creatures, owl and turtle, have in common is the shape of their beak, so it is possible the compound alludes to the two attributes, feathery peak as well as hooked beak.

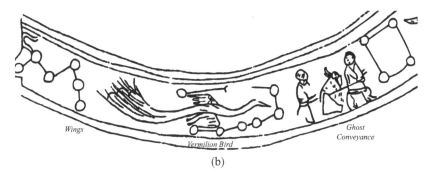

(b)

Figure 2.15 (*cont.*)

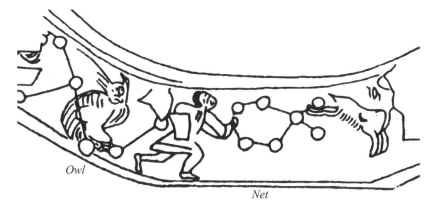

Figure 2.16 Tracing of western lodges *Hunting Net* (Bi) and *Owl* (*Zuixi*; λ, φ₁, φ₂ Ori) from the W. Han tomb ceiling mural at Xi'an Jiaotong University. Detail redrawn from Rawson (2000, 178, Figure 9).

In any case, the Xi'an mural conclusively confirms the existence of a tradition identifying this asterism as an owl, so that Sima Qian's characterization must simply refer to this part of Orion as marking the general location of the *Tiger*'s head. So here we have three stars anciently perceived to form a triangle and identified with the horned owl, a description that clearly fits the owl-like crested bird with a triangle of three stars in its tail depicted on the *Niu fang ding*. If this identification is correct, then we also have here in a classic late Shang bronze *ding* a representation of an asterism, this time exactly opposite the *Dragon* in the sky and therefore not to be identified with the giant *Vermilion Bird* constellation.[69]

---

[69] Strong support for this identification is found in the old astronomy of the Yizu minority of Sichuan, Guizhou, and Yunnan. Among the Yizu and Naxi people, the stars of lodge *Zuixi* are

Conclusion

In these two artifacts from the mid second millennium BCE – an eccentric zoomorphic bronze from a zone of mixed cultural influences and a classic Anyang period cauldron – we seem to have the earliest Chinese depictions in any medium of actual constellations as well as individual stars. These treasured bronze vessels provide the best evidence to date that by the Shang Dynasty at the latest, constellations familiar to us from late classical cosmology and cosmography had already been placed in the sky. The identification of the *long-Dragon*, in particular, ultimately derived from close observation of the activity of the alligator, whose seasonal behavior paralleled that of the constellation. This is not to say that one should understand the naturalistic explanation in terms of an apotheosis or an intellectualized process of elevation of a denizen of the environment to a cosmic place of honor in the sky. At the time "nature" as such was not differentiated from the human world. Rather, one should see this nexus of associations as an example of the "dynamic totalistic weaving of nature, society, myth and technology" characteristic of Neolithic and early Bronze Age thought.[70] We can further affirm that there was an elite class of priest-astrologers, Cai Mo's predecessors, who were responsible for maintenance of the calendar and management of ritual time. Their esoteric knowledge, given material form in the dragon-shaped bronze from the tomb of one of their number, was precisely the sort of "wisdom" alluded to in the passage from *Zuozhuan*, whose obsolescence astrologer Cai Mo obliquely laments in recounting the passing of the age when *Dragons* could still be domesticated.

identified as a *Parrot*, whose long tail feathers even more closely resemble the image on the *Niu fang ding*; Chen Jiujin, Lu Yang, and Liu Yaohan (1984, 90, 95). Both owls and parrots are amply represented in late Shang figurative art, with owl-shaped bronze wine pourers and beakers being particularly prevalent. The parrot was no longer a denizen of the Central Plains in Warring States and Han times, though in the late Shang it was, along with the elephant, tiger, and rhinoceros. Whether parrot or owl, the bird must have had a more auspicious significance in Shang than in late imperial times when the call of an owl in the night was considered an evil omen.

[70] Tambiah (1990, 106). As Nathan Sivin points out (Lloyd and Sivin 2002, 200), "before modern times, Chinese did not need a word that meant 'nature' (the physical or material universe)."

Part Two

Aligning with Heaven

3   Looking to the Supernal Lord

Chinese preoccupation with astronomical orientation has a very long history. Archaeological evidence from the fifth millennium BCE Neolithic cultures of north China shows that tombs and dwellings were already being oriented with particular attention to the cardinal directions and the seasonal variations in the Sun's location. The consistent pattern of cardinal orientation of graves among certain Neolithic cultures from the middle and lower course of the Yellow River to the lower course of the Yangtze River shows clearly that these peoples had already formed a concept of east and west based on the location of sunrise and sunset and had devised a method of determining the cardinal, and in some cases inter-cardinal, directions (Figure 3.1a–b). The entrances to early Yangshao dwellings at Banpo were oriented toward the location of the mid-afternoon winter Sun when at its warmest a month or so after the solstice.[1]

With the beginning of the Bronze Age and early state formation in the second millennium BCE, such concepts had progressed to the point where ritually and politically important structures were uniformly quadrilateral in shape and cardinally oriented, their longitudinal axes aligned with varying precision in a north–south direction.[2] Palatial structures and royal tombs from the earliest states in the second millennium BCE – Xia, Shang, and Zhou – consistently display such orientation (Figures 3.2, 4.1).

From the layout of the best-preserved of these city walls and palatial foundations, it is clear that the principal access was normally via a main gate in the south facade, with the inner sanctum located far from the entrance and against the rear wall. The Zhouyuan palace structure has the traditional screening wall in front of the entrance. This habitual, cosmologically significant architectural design remained essentially constant throughout the

---

[1] Some minority peoples of southwest China to this day call the month corresponding to this time "house-building month"; see Lu and Shao (1989). As David N. Keightley (1998, 794) has observed, "the great attention the early inhabitants of China paid to orientation and posture was a significant characteristic of both the Neolithic and Bronze Age cultures."

[2] Wheatley (1971); Wu Hung (1997). All Erlitou burials from Phases I–IV in which the head is oriented to the north are aligned to between zero and ten degress west of magnetic north. *Zhongguo shehui kexueyuan kaogu yanjiusuo* (1999, 141, 397–8).

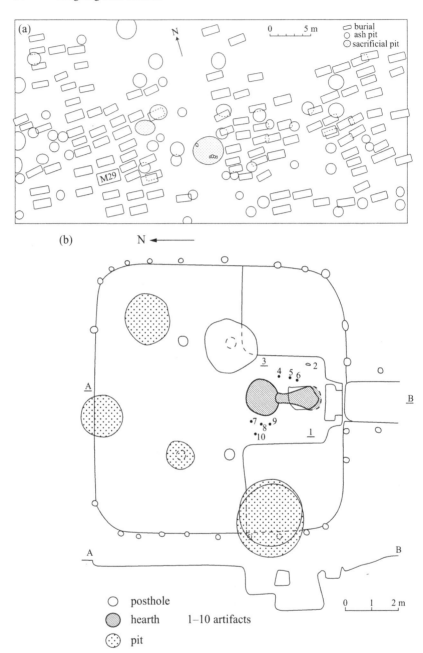

Figure 3.1 Neolithic cardinal alignments: (a) Peiligang culture burial ground, Henan (c.5300 BCE). (b) Yangshao house, Jiangzhai, Banpo phase (4800–4300 BCE), Shaanxi. After Li Liu (2007, 129, 37), © 2007 Li Liu. Reprinted with the permission of Cambridge University Press.

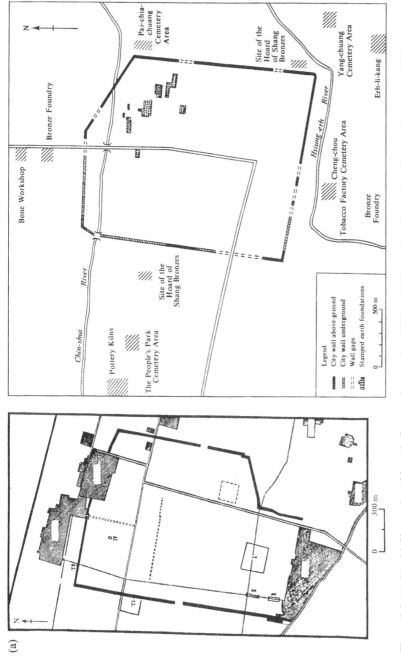

Figure 3.2 (a) Yanshi Shang city (fifteenth–fourteenth centuries BCE) and Zhengzhou Shang city (fourteenth century BCE) showing alignment slightly east of north. (b) Erlitou palace number 2 (seventeenth century BCE) oriented six degrees west of north. (c) Western Zhou predynastic palace, Fengchu, Shaanxi (early eleventh century BCE). Adapted from Wu Hung (1997, 84, 86–7, Figures 2.3, 2.6, 2.7). The arrows indicate magnetic north at the time of excavation (i.e. ignoring –3.5° magnetic deviation in the area). Of the four examples, the earliest, the Erlitou palace in (b), is most closely aligned on celestial north.

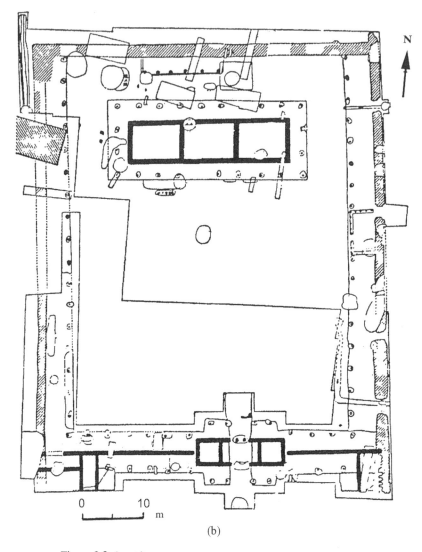

0   10
|___|___| m

(b)

Figure 3.2 (cont.)

entire history of China, most especially in edifices connected with royal prestige and power, as exemplified by the Ming period Forbidden City in Beijing,
which now houses the Palace Museum. This much is well known and has been
thoroughly documented.[3] What has been less well explored is the possible

<hr />

[3] Wheatley (1971), especially Chapter 5; Thorp (2006, 16, 37, 33, 69, 147, 131); Steinhardt (1999).

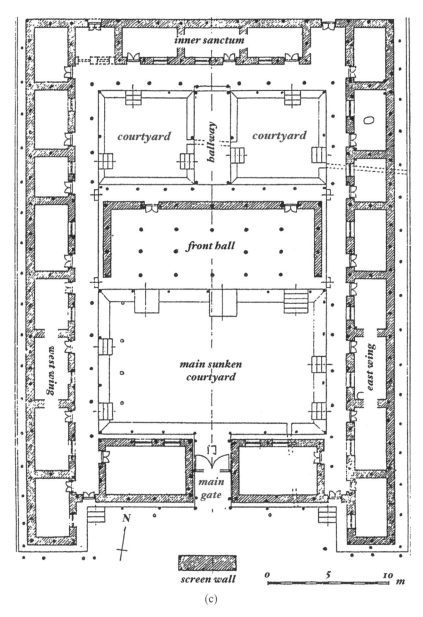

(c)

Figure 3.2 (*cont.*)

astral–terrestrial correspondence between the archaic kingship in the Bronze Age and the celestial Pole, or *Northern Culmen* (*beiji*), whose unique characteristics and powerful associations as the pivot of the heavens ultimately led to its becoming the celestial prototype of the cosmically empowered Chinese monarch. Already in the late Warring States period (403–221 BCE), well before the founding of the empire, Confucius famously drew on the metaphorical potency of the Pole to exemplify the charismatic virtue (*de*) of the sage ruler: "The Master said: 'To conduct government by virtue may be compared to the *Northern Asterism*: it occupies its place, while the myriad stars revolve around it.'"[4]

Taiyi and Northern Culmen

Recent studies of the cultic practices, ceremonial, and conceptual background of the numinous power "Supreme One" during the Warring States through Han periods draw on abundant textual and archaeological materials and underscore the identification of Supreme One with the celestial Pole. To cite a few examples:

> *Shiji*, "Treatise on the Celestial Offices": "The brightest star of the *Pole-star asterism* in the Central Palace is the constant abode of the Supreme One."[5]
>
> *Huainanzi*, "Heaven's Patterns": "The [*Palace of*] *Grand Tenuity* is the court of the Supreme One; the *Palace of Purple* [*Tenuity*] is the abode of the Supreme One."[6]
>
> *Heguanzi*, "Great Flood": "The center is the place of the Supreme One, the one hundred spirits look up to it and are controlled by it."[7]
>
> *Liji*, "Evolution of the Rites": "Now the Rites necessarily have their origin in the Supreme One, which divides to become Heaven and Earth, revolves to become *yin* and *yang*, and changes to become the four seasons."[8]

In the Warring States *Guodian* manuscripts from Chu, in the text "The Supreme One Springs from Water," *Taiyi sheng shui*: "Supreme One [*Taiyi*] is here

---

[4] *The Confucian Analects* (2/1); Brooks and Brooks (1998, 109).
[5] *Shiji*, 27.1289.
[6] Liu An (1974, 3.39). *Taiwei* here is a reference to the *Celestial Court*, *Taiweiyuan*, between Leo and Virgo; *Zigong* refers to *Ziweiyuan*, the *Palace of Purple Tenuity* at the Pole, already referred to above.
[7] *Heguanzi*, 10.71; photographically reproduced in Defoort (1997, 345).
[8] *Liji zhengyi, Shisanjing zhushu*, 22.1426; also *Lüshi chunqiu*, "Great Music" (Bi Yuan 1974), 5.46. In his commentary, Gao You glosses, "The Supreme One is the Dao"; Graham (1989a, 360).

an alternative appellation for the Dao . . . *Taiyi* in these texts is what is referred to as the Dao in the pre-Qin period."[9]

From *Zhuangzi*, "All under Heaven" (*Tianxia*):

Of the ancient traditions of the Way there is this: Guan Yin and Lao Dan heard about it and found pleasure in it. They established it on the constancy of nothing and based it on the Supreme One. They took weakness and submissiveness as its external manifestation, and emptiness and non-destructiveness to all things as its inner reality.[10]

And from "Lie Yukou": "The form of the Supreme One is emptiness."[11]

According to Li Ling,

The "Treatise on the *Feng* and *Shan* Sacrifices" in *The Grand Scribe's Records* [*Shiji*] and the "Monograph on the Suburban Sacrifices" in *The History of the Former Han Dynasty* [*Han shu*] record that in Han times the Supreme One was worshipped as the most revered. The abode of the Supreme One is the Pole of the *Dipper* about which the myriad stars revolve. The opening passage in the "Treatise on the Celestial Offices" in *Shiji* talks about just this, and the *Taiyi* method associated with the use of the mantic astrolabe (*shi*) also takes the Supreme One as central. All these serve to illustrate its importance.[12]

According to Li Jianmin,

The Supreme One dwells in the center, which is also the position of the Son of Heaven. There is a parallel between the position of the ruler among the people and that of the Supreme One among the heavenly bodies . . . *Heguanzi* speaks of "using the One," *Laozi* [Chapter 22] talks about "embracing the One," the Mawangdui silk manuscript "Essentials" [*Yao*] talks about "through the One," "obtain the One," and so on, so it appears as if abstract metaphysical thinking may indeed be the same as the study of computational techniques [*shushu zhi xue*] in this respect.[13]

---

[9] Quoted in Li Jianmin (1999, 51). My translation of the title of this work follows Chen Songchang, who argues, plausibly in my view, that the title should be understood as "Supreme One Springs from Water" *Taiyi sheng* [*yu*] *shui*; Chen (2000, 542–46). As Marc Kalinowski (2004, 104) points out, "les deux éléments présents dans la phase initiale de formation du monde dans les cosmogonies sont rarement dans un rapport d'engendrement mais plutôt de concomitance, et s'ils le sont parfois c'est en général la substance primordial (le chaos, l'océan, etc.) qui préexiste a l'apparition de l'instance créatrice." This would explain why, in the second part of the text where the cosmogonic process is traced in reverse, a phase in which water is supposedly engendered is not even mentioned. See *Guanzi* (Chapter 14, "Shuidi"): "Therefore, what is it that is complete unto itself? It is water – all things rely on it for their existence. Only those who understand this contingency can act as regulator [of all things]. That which is complete unto itself is water. Therefore it is said: What is water? It is the source of everything." Cf Kalinowski (2004, 105, n. 4).

[10] Wang Xianqian (1974), 33.461, 472.   [11] Ibid., 32.453.

[12] Li Ling (2000, 269); quoted in Li Jianmin (1999, 50).

[13] Li Jianmin (1999, 51). The Baoshan divination texts also mention Supreme One several times and there are Han accounts of ritualized invocation of the spirit in advance of a military campaign against Nanyue in the south. Graphic depictions of *Taiyi* exist (see Figure 3.10a) which make explicit the cosmic attributes of Supreme One. For discussion of this and Supreme One as the

These sources and opinions could be multiplied manyfold. From them we may conclude:

(1) Supreme One in Warring States and Han thinking is the supreme spiritual power residing at the center of the *Palace of Purple Tenuity* at the Pole and identifiable with the Dao. In metaphysical contexts it denotes the power of the ancient deity Shangdi, the Supernal Lord.[14]

(2) All other numinous influences are subordinate to the Supreme One; it is the ultimate source of all phenomena, invisibly animating and regulating the universe.

(3) An important attribute of the Supreme One is its protean nature; there is awareness that its nominal association with the bright star Kochab, β UMi (*tian di xing*), is an expedient, similar to the rationale for placing the heavenly pivot of the Han period mantic astrolabe *shi* in the handle of the *Dipper*.[15]

(4) As an inspirational focus, based on the testimony of *Shiji*, *Zhuangzi*, and others, Supreme One as *North Pole Asterism* has a history that reaches into the distant past where its attributes as celestial high god merge with those of the Supernal Lord.[16]

The virtue of nothing

The mysterious efficacy of charismatic virtue to which Confucius referred in the passage above in the alternative, Taoist vision of *Zhuangzi* becomes the constancy of inaction, or *wu wei*, the ultimate achievement of one who is in harmony with the invisible force of Dao (Supreme One) animating the universe. The aphoristic maxims of the *Laozi* repeatedly invoke the themes of nonaction, artlessness, and embracing "the One" through which non-intentional purposefulness achieves its objective. More to the point, however, may be this musing on the paradoxical virtue of nothing:

> Thirty spokes join at a single hub,
> But it is precisely where there is nothing
> That the utility of the wheel resides.

central deity in the Chu Silk Manuscript, see Harper (1999, 870). For the institution of the state cult of Taiyi by Emperor Wu in the Han, see Lewis (1999, 187 ff.).

[14] Twitchett and Loewe (1986, 661–8). Kalinowski (2004, 117).

[15] On the astronomical connections of the Supreme One, see Qian Baocong (1932). For the significance of the mantic astrolabe in ancient Chinese cosmology and mantic arts, see Li Ling (1991, 89–176); but see also Shi Yunli, Fang Lin, and Han Zhao (2012). As we saw in the discussion of dragons above, another spirit, *Tian yi*, or "Celestial One," is variously identified with the *Cerulean Dragon* of the East and with certain stars near the Pole, but appears not to have rivaled Supreme One in spiritual significance or astro-calendrical fate calculation; Harper (1999, 851). Michel Teboul (1985) attempts to sort out the confusion between polar deities *Supreme Unity* and *Celestial Unity*.

[16] See especially Li Ling (1995–96).

Clay is fired to make a pot,
But it is precisely where there is nothing
> That the utility of the clay pot resides.
Cut out door and windows [to finish a room],
> But it is precisely where there is nothing
> That the utility of the room resides.
Therefore,
> Having a thing is beneficial,
> But having nothing is useful.[17]

Then there is this from the *Springs and Autumns of Master Lü* (*Lüshi chunqiu*), "Great Music" chapter:

From the Supreme One emanate the Two Exemplars; the Two Exemplars give forth *yin* and *yang*; *yin* and *yang* change and transform, one arising one descending, they combine to form shapes. Confused and obscure in their separateness, they recombine, once combined they separate again; this is Heaven's constant rule. Heaven and Earth turn about like the wheel of a cart, ending then beginning anew; reaching their extremes they turn back again, and nothing is ever out of place. The Sun, Moon, stars and asterisms, some speed along, some move slowly; the Sun and the Moon differ, and thereby complete their movements. The four seasons arise in succession, now warm, now cold, now short, now long, now mild, now harsh. At the source from which all things emanate, they are initiated by the Supreme One, and transformed by *yin* and *yang*.[18]

We have twice seen the metaphor of a wheel appear with reference to rotational movement of the heavens, in the quotation from the *Spring and Autumn of Master Lü* and with reference to the utility of the empty space at the hub. In contrast to the *Analects* of Confucius, which was quite explicit about the celestial source of its evocative metaphor for the mysterious efficacy of charismatic virtue, the *Laozi* is indirect, allusive, yet down-to-earth in its choice of images. Nevertheless, it does not require a great imaginative leap to perceive the likelihood of a common inspirational source for their respective visions of ultimate attainment in the mysterious operations of the empty pivot of the heavens. Indeed, for his part, in the Postface to his monumental history Sima Qian (*c*.100 BCE) is quite explicit: "the twenty-eight lodges surround the Northern Asterism, thirty spokes turning ceaselessly around their hub, [like] supportive and indispensable officials attending it [sc. the Northern Asterism], loyally and faithfully carrying out the Way in service to their Master above."[19]

It can hardly be coincidental that during the two millennia while this mystical vision was taking shape *there was no star* located precisely at the Pole such as we have today, no obvious physical presence at the pivot of the heavens, so that

---

[17] Cf. Henricks (1989, 63).

[18] *Lüshi chunqiu* (Bi Yuan 1974, 5:46); Graham (1989a, 360).   [19] *Shiji*, 130.3319.

the marvel of an efficacious "nothing" at the center of the rotating dome of the sky was nightly on display, inviting wonder.

The Northern Dipper and the imperial power

Understandably, as the most distinctive stellar formation near the Pole, some of the mysterious aura of that location quite naturally attached to the *Dipper*. The rotation of the *Northern Dipper* around the mysterious pivot of the four quarters for centuries enabled it to serve as a celestial clock whose changing orientation marked the passing of the hours of the night as well as the seasons of the year. As the *Pheasant Cap Master* (*Heguanzi*) famously put it,

When the handle of the *Dipper* points to the east [at dawn], it is spring to all the world. When the handle of the *Dipper* points to the south it is summer to all the world. When the handle of the *Dipper* points to the west, it is autumn to all the world. When the handle of the *Dipper* points to the north, it is winter to all the world. As the handle of the *Dipper* rotates above, so affairs are set below . . . [20]

On the centrality of this polar imagery in the *Pheasant Cap Master*, Carine Defoort remarks,

Most attention in the *Pheasant Cap Master* is directed at the sagely ruler, the One Man, who is the unique fountainhead of order. This sage-ruler is the political Pole-star, surrounded by his ministers, impartially distributing responsibilities and penetrating to the smallest corners of his realm. The strength of this imagery, fully exploited in the *Pheasant Cap Master* . . . lies in its affirmation of the very different and independent position of the central pole in relation to the other political constellations.[21]

Explicit literary and graphic elaboration of the association of the North Pole and its attributes with the person of the Emperor, the Son of Heaven, came with the establishment of the universal empire, as in the famous Han Dynasty stone relief from a Wu Family tomb shrine dating from the Eastern Han

---

[20] Cf. Defoort (1997, 189, 320); *Heguanzi*, 5.21/1–4. The earliest such metaphorical use of stars or asterisms occurs in ode Mao 203 in the *Book of Odes*. There the brilliance of several asterisms, including the *Northern Dipper*, is ironically likened to the aristocratic elite who occupy positions of importance but do not apply their brilliance in ways that benefit the populace; Karlgren (1950b, 153–4). For an account of the *Dipper* in Han times, see Major (1993, 106 ff.).

[21] Defoort (1997, 120). Only the "Lesser Annuary of Xia" and *Heguanzi* among pre-Qin sources mention the use of the direction of the *Dipper*'s handle to determine the seasons. There is otherwise no trace of this feature in Shang and Zhou. Chen Jiujin, Lu Yang, and Liu Yaohan (1984, 65, 71, 107–8, 119, 214) show that both sources bear a strong resemblance to the calendrical astronomy of the Yi minority in the southwest, indicating that the use of the *Dipper* to determine the seasons was a cultural characteristic of the Qiang–Rong groups linked to the Xia and ancestral to the Yi people. Chang Hong (astrologer to the late Western Zhou King Li), Heguanzi himself, and Luoxia Hong, a central figure in the Grand Inception calendar reform of 104 BCE, were all from the Yi minority region of Ba in the southwest.

Figure 3.3 The Supernal Lord in his *Dipper* carriage surrounded by servitors and winged spirits (one holding Alcor, 80 UMa next to Mizar, ζ UMa). Ink rubbing of a carved stone panel from the Wu Liang shrine, Shandong, Eastern Han Dynasty (mid second century CE). After Feng Yunpeng and Feng Yunyuan (1893, Chapter 3).

(Figure 3.3).[22] Here we see the Supernal Lord dressed in the imperial garb and riding in the *Dipper*, accompanied by immortals, ministers, and spirit servitors. Like the mysterious *Northern Culmen* (*beiji*) at the center of the sky, the formal ritual pose of the terrestrial Emperor was to sit facing south, so that all his ministers, minions, generals, and subjects approached his exalted presence and prostrated themselves while facing north. As we will see in Chapter 11, "Cosmic capitals," historical accounts from the early Han Dynasty are quite explicit that the Qin Dynasty capital of Xianyang and the Han capital of Chang'an both were designed with this astral symbolism in mind, which emphasized the identification of the emperor with the celestial Pole.

With the inception of the imperial system the emperor also came to be titled *Di*, as in Shangdi or Supernal Lord. The *locus classicus* for the identification of his person with the cosmic functions of the *Northern Dipper* is in the "Treatise on the Celestial Offices" (*Tianguan shu*) in *The Grand Scribe's Records* (*c.*100 BCE). There, in the description of the astral correlates of the imperial court in the circumpolar region of the sky, Sima Qian says,

The *Dipper* is the Supernal Lord's carriage. It revolves about the center, visiting and regulating each of the four regions. It divides *yin* from *yang*, establishes the

---

[22] Compare the slightly foreshortened image reproduced in Needham and Wang (1959, 241, Figure 90) on which Needham remarks, "On the original relief, the spirit with the isolated star is in line with, and beyond the last of, the stars of the 'handle,' which is represented in a straighter line. The isolated star must therefore be Chao yao (γ Bootis)."

four seasons, equilibrates the Five Elemental Forces, deploys the seasonal junctures and angular measures, and determines the various periodicities – all are tied to the *Dipper*.[23]

As John S. Major pointed out, the Supernal Lord or "God of the Pivot" was in Han times associated with the star Kochab, which, although several degrees from the celestial Pole at the time, conventionally served as a Pole Star in astrological contexts.[24] This star thus became the pivot of the heaven-plate of Han mantic astrolabes (*shi*), around which the image of the *Dipper* inscribed on the plate rotated while serving as a pointer.[25] Beginning with the Han Dynasty the historical record amply reflects the crucial symbolic significance of the *Dipper* in the cosmo-magical imagery associated with the imperial office. For example, in the Grand Scribe's account of ritual procedures during the Western Han Dynasty there is the following passage:

That autumn [112 BCE], in preparation for a punitive expedition against Southern Yue, the attack was announced in prayers to the Supreme One. A banner decorated with images of the Sun, Moon, *Northern Dipper*, and rampant dragons was mounted on a shaft made from the wood of the thorn tree, to symbolize the Celestial Unity *Tianyi* and its three stars, vanguard of the Supreme One, *Taiyi*. [The banner] was called Numinous Flag. In praying for military success, the Grand Scribe-Astrologer would hold it aloft and point in the direction of the country to be attacked.[26]

And then there is this later account from the reign of the usurper Wang Mang (45 BCE–23 CE), first and only emperor of the Xin Dynasty, which intervened between the Former and Eastern Han dynasties. This occurred in 17 CE:

[fourth year of the *Tianfeng* reign period] in the eighth month, [Wang] Mang went in person to the place for the suburban sacrifice south of the capital to superintend the casting and making of the Ladle of Majesty. It was prepared from minerals of five colors and from copper. In shape it was like the *Northern Dipper*, measuring two feet five inches in length. [Wang] Mang intended [to use it] to conquer all rebel forces by means of conjurations and incantations. After the Ladle of Majesty was finished, he ordered the Directors of Mandates [of the Five Elemental Phases] to carry it solemnly on their shoulders in front of him whenever he went out, and when he entered the palace, they waited upon him at his side.[27]

Six years later, during the rebellion of 23 CE, when the burning palace was invaded and Wang Mang and his retinue were about to be killed by Han Dynasty loyalists, the following scene ensued:

---

[23] *Shiji*, 27.1289. This was clearly the common perception throughout the Han Dynasty. In glossing a reference to the Supernal Lord, in the *Gongyang Commentary* (Duke Xuan, third year), He Xiu (129–82 CE) says, "[this is] *Di* – the Great Lord of August Heaven of the *Northern Asterism*, overall ruler of Heaven and Earth, the Five Emperors [of the Five Elemental-Phases], and all the spirits"; *Chunqiu Gongyang zhuan He shi jiegu*, Sibu beiyao ed., 15.4b.

[24] The actual Pole star of Han time was 4339 Cam; Needham and Wang (1959, 261).

[25] Major (1993, 107).   [26] Cf. Nienhauser et al. (2002, 239).

[27] Trans. Needham and Wang (1962, 272, trans. modified).

Meanwhile, [Wang] Mang, dressed all in deep purple and wearing a silk belt with the imperial seals attached to it, held in his hand the spoon-headed dagger of the Emperor Shun. An astrological official placed a mantic astrolabe [*shi*] in front of him, adjusting it to correspond to the day and hour. The Emperor turned his sitting position, following the handle of the *Dipper*, and so sat. Then he said, "Heaven has given the [imperial] virtue to me; how can the Han armies take it away?"[28]

Imperturbable in his faith in the protection of Heaven and the *Dipper*, in this pose Wang Mang and his Xin Dynasty met their end.

Seasonal timekeeping and the Northern Dipper

As in Homeric Greece, where the orientation of the *Dipper*'s handle was used to time sentries' watches at Troy, the *Dipper* served as a celestial clock hand, indicating the passage of the seasons and the hours of the night. This enormously useful function of the *Dipper* is certainly very ancient.[29] In a more uniquely Chinese application of the usefulness of the circumpolar stars, it was evidently also common practice to key their meridian transits to the positions of other asterisms by means of alignments, and in this way by indirect means determine the location of the Sun, Moon and other heavenly bodies among the stellar lodges not currently visible in the night sky.[30] Given the resolutely polar-equatorial orientation of Chinese astronomy from the outset, this technique is likely to have arisen quite early. It is clearly reflected in the astronomical decor painted on the lid of the famous lacquer hamper from the tomb of Marquis Yi of Zeng, dating from 433 BCE, which also for the first time deploys the entire sequence of twenty-eight lodges. In the diagram of the heavens painted on the lid the oversized central graph representing the *Northern Dipper* is exaggerated in specific ways to point to certain lodges, indicating the date.[31]

Bronze Age antecedents

We have seen how the microcosmic–macrocosmic analogy between the emperor and *Di*, the Supernal Lord at the apex of the heavens, is abundantly

---

[28] Ibid.

[29] Divinations concerning sacrifices to the *Dipper* appear already in the thirteenth century BCE oracle bones; Jao Tsung-yi (1998, 42). There is evidence that the bright star Arcturus (α Boo) or *Great Horn* participated in this function of the *Dipper*'s handle until precession caused the ecliptic to move away from its vicinity, so that Arcturus's *Left* and *Right Assistant Conductors* "could no longer serve as indicators" as Sima Qian put it. Needham and Wang (1959, 252); *Shiji*, 26.1257.

[30] According to Joseph Needham (Needham and Wang 1959, 232, 239), "it was J. B. Biot [*c.*1835] who was the first to realise that the choice of the *hsiu*-determinatives had depended on their having the same (or approximately the same) right ascensions as the constantly visible circumpolars." It will shortly be seen that this principle was applied in locating the celestial Pole itself.

[31] Feng Shi (2007, 320 ff.); Sun and Kistemaker (1997, 20); Harper (1999, 820, 833 ff.).

documented for the early imperial period. In addition, as Paul Wheatley and others have shown for the Shang and Zhou dynasties, each of the basic modes of symbolism displayed by the ideal-type city throughout much of the ancient world is evident in the planning of ancient Chinese capitals. These aspects of traditional symbolism have been succinctly formulated by Mircea Eliade:

(1) Reality is a function of the Imitation of a Celestial Archetype.
(2) The parallelism between the macrocosmos and the microcosmos necessitates the practice of ritual ceremonies to maintain harmony between the world of the gods and the world of men.
(3) Reality is achieved through participation in the symbolism of the center, as expressed by some form of *axis mundi*.
(4) The techniques of orientation necessary to define sacred territory within the continuum of profane space involve an emphasis on cardinal directions.[32]

In addition to the cosmo-magical physical layout of dynastic capitals and ritual centers discussed in Chapter 1 above and below in Chapter 11, abundant inscriptional evidence attests to the ancient Chinese preoccupation with each of these basic modes of symbolism. For example, the late Shang oracle bone inscriptions integral to the Shang royal cult contain numerous examples of divinations motivated by propitiatory impulses. Besides those concerning the Supernal Lord *Di*'s manipulation of powerful natural forces such as wind, rain, and the like, numerous other divinations relate to the bestowal of good harvests and to relief from natural disasters, belligerent neighbors, etc. by the spirits of the Four Quarters, the *Dipper*,[33] and so on. These and other divinations also embody the conceptualization of Shang as the symbolic center from which the royal charisma radiates in all directions. From the inscriptions it is also clear that the supra-sensible pantheon, with *Di* at its apex, mirrored the hierarchy of the temporal Shang state.[34] Though we have no explicit contemporary statement to that effect in the oracle bone divinations, it seems clear that in the Shang conception the Supernal Lord's abode, as in Warring States and Han times, was at the center of the heavens from which cosmic control appeared to emanate, a location made all the more mysterious by the lack of a Pole Star. The absence of a distinctive Pole Star at this epoch is an issue that warrants further exploration, both because of the axial orientation of monumental architecture in the landscape, and to better understand the role of the Supernal Lord and the significance of the character used to denote him – *Di*. First, however, a brief account of the career of the celestial Pole from the Neolithic through the Bronze Age is in order.

---

[32] Wheatley (1971, 418).
[33] This is a reflection of the protean role of the *Dipper* asterism whose most archaic association (*c*.1300 BCE) was already with a ladle (*dou*), or peck-measure.
[34] Hsu and Linduff (1988, 98).

Migration of the Pole

Precession of the equinoxes may seem a rather sophisticated concept to be of concern to the Chinese in the mid second millennium BCE, but for a civilization with an emphatically polar-equatorial astronomy and cosmo-political culture, the inconstancy of the Pole's location ought to have been somewhat problematical. Although the explanation of precession *as a phenomenon* was not put forward until the third century, long after the founding of the empire, its effects were certainly noticed and accommodated centuries earlier, most obviously in the debates leading up to the well-documented Grand Inception calendar reform of 104 BCE.[35] Given the focus on the celestial Pole in early Chinese astronomy, it would be surprising if the inconstancy of such a highly symbolic location did not eventually register with observers in the Chinese Bronze Age.

The perceivable effect of the precession of Earth's axis around the Pole is that it gradually causes a given star or constellation to rise later than usual. The rate of change amounts to about one day's delay every seventy-two years. In order to observe precession over two to three generations, some sort of reference point is necessary. Here is Harald A.T. Reiche's explanation:

Consider the slow eastward slippage, past a fixed and ancient horizon marker, of the familiar constellations marking solstices or equinoxes, clearly noticeable after but a few generations and entailing the gradual obsolescence of any given polar star . . . All one needed to notice this (in Professor Philip Morrison's apt phrase) was an old tree and faith in the veracity of one's grandfather. Obviously, a set of stone markers of the sort amply documented by Professor Alexander Thom's work would have done even better. Given these naked-eye horizon phenomena, there is no empirical reason for down-dating the recognition of the slippage in question to the time (150 B.C.) when Hipparchus correctly traced the visible phenomenon at issue to the westward "precession" of the invisible intersection of the invisible celestial equator with the equally invisible ecliptic. After all, the correct explanation of a phenomenon commonly comes long after its routine observation.[36]

Imagine that your grandfather told you, as his grandfather had told him, that on the morning of the winter solstice, a particular star rose exactly over a certain old tree or mountain peak on the horizon. Then suppose that years later on the solstice you stood at the designated place and did not see the

---

[35] Cullen (1993). As J. Norman Lockyer (1964, 300) remarked on the continuity of Egyptian observation, "The great point, however, is that in Egypt the change of cult might depend upon astronomical change – upon the precession of the equinoxes, as well as upon different schools of religious or astronomical thought. We gather from this an idea of the wonderfully continuous observations which were made by the Egyptians of the risings and settings of stars, because, if the work had not been absolutely continuous, they would certainly never have got the very sharp idea of the precession of the equinoxes which they undoubtedly possessed." See also Magli (2004).

[36] Reiche (1979, 157).

star rising at that point. A day or two later, when you did spot the star's first predawn appearance, if you believed your grandfather and the astral myth he had recounted for you, you might think something was amiss. Of course, you might suspect your grandfather misremembered the details, but after some generations the conclusion would have been inescapable – either time or the stars were out of joint.

This is all the more true when it comes to the celestial Pole, since in the Neolithic there had been a comparatively bright star, α Draconis (magnitude +3.65), ideally located precisely at celestial north. Indeed, in about 2775 BCE, Thuban (i.e. α Dra, dec. 89°53') was even closer to the Pole then than our own Polaris is now. In 1894 the British astronomer and "father of archaeoastronomy," Sir Norman Lockyer (1836–1920), demonstrated in *The Dawn of Astronomy* (1894) that the ancient Egyptians aligned their temples on the rising points of stars in the east, and over time rebuilt temples to realign them with changes in those rising points caused by precession. Those results prompted Giorgio de Santillana to conclude,

When a stellar temple is oriented so accurately that it requires several reconstructions at intervals of a few centuries, which involved each time the rebuilding of its narrow alignment on a star, and the wrecking of the main symmetry that goes with it; when Zodiacs, like that of Denderah, are deliberately depicted in the appearance they would have had centuries before, as if to date the changes, then it is not reasonable to suppose the Egyptians were unaware of the Precession of the equinoxes, even if their mathematics was unable to predict it numerically. Lockyer let the facts speak for themselves, but it is he who has given the proof. Actually, the Egyptians do describe the Precession, but in a language usually written off as mythological or religious.[37]

A handoff of pole stars

Beginning in the late third millennium BCE and throughout the entire Chinese Bronze Age, the track of the North Pole did not bring it anywhere near as close to a comparably bright object as Thuban had been (Figure 3.4). As we saw above, however, the lack of a precisely located Pole star did not deter the Shang royal architects in the latter half of the second millennium BCE from aligning their built environment with true north. This is especially true in the Shang ritual center of Yinxu at Anyang, where palatial foundations, city walls, royal tombs, and other structures were laid out with their longitudinal axes aligned on the Pole. To cite just one well-known example, the consistent displacement from the present true north by some five to twelve degrees east displayed by late

---

[37] Giorgio de Santillana in Lockyer (1964, vii). See also ibid., 162. For a graphic explanation of precession, see http://robertbauval.co.uk/articles/articles/anchor.html. Above we noted how the ancient Chinese, too, were well aware of the impermanence of the seasonal rising points of certain "fixed" stars; Cullen (1993).

Palace of Purple Tenuity

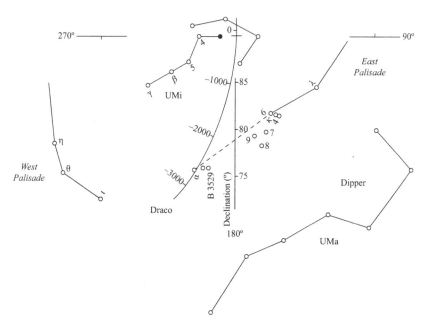

Figure 3.4 The trajectory of the north celestial Pole among the circumpolar asterisms from –3000 through –100. In 2775 BCE it was located at the star α Draconis or Thuban. Redrawn from Maeyama (2002, 7, Figure 4).

Shang tombs is, as Joseph Needham observed, "not far from what we should expect if the Shang people had taken care to site their tombs in accordance with the astronomical north of their time."[38] Needham thought that Kochab (β UMi), about 6.5° from the Pole (dec 83° 26′ 54″) in 1200 BCE, may have served as the Pole star for alignment purposes. This is a reasonable enough conjecture, but there are other possibilities that might account equally well for the archaeological facts observed on the ground.

[38] Needham and Wang (1962, 313). On Kochab as the Pole star of about 1000 BCE, see 261.

Table 3.1 *Declinations of Thuban, the first Pole star, and Kochab at intervals from –2000 to –600.*

Epoch	Star	Declination	Polar distance
–2000	Kochab	82° 01′ 54″	$c.8°$
	Thuban	85° 32′ 36″	$c.4.5°$
–1600	Kochab	82° 58′ 11″	$c.7°$
	Thuban	83° 18′ 00″	$c.6.6°$
–1200	Kochab	83° 26′ 54″	$c.6.5°$
	Thuban	81° 04′ 20″	$c.9°$
–1000	Kochab	83° 28′ 46″	$c.6.5°$
	Thuban	79° 57′ 56″	$c.10°$
–600	Kochab	83° 07′ 05″	$c.7°$
	Thuban	77° 46′ 13″	$c.12.25°$

Note the changing polar distances of the two stars in Table 3.1. Thuban, located precisely at the Pole in 2775 BCE, grew ever more distant from ninety degrees north over the course of the next two millennia, all the while Kochab (mag. +2.0, about the same as Polaris), the brightest star within ten degrees of the Pole at that epoch, grew progressively closer. Unlike Thuban, however, Kochab never approached closer than 6.5° to the Pole, and by mid first millennium, during Confucius' time, Kochab was already withdrawing from the vicinity of the Pole. What is striking, however, is that the crossover point during the sixteenth century BCE, when both stars were equidistant from the Pole, coincides with the date of Shang ascendancy. It is a curious fact that the earlier palatial foundations at Erlitou, which are attributed by many archaeologists to a distinctly different Xia polity, depart radically from subsequent Shang period structures in being oriented roughly the same average number of degrees west of north, rather than east of north (Figure 3.2).[39] Perhaps a transition from the obsolete Pole star, Thuban, to the upstart, Kochab, to the east, might provide an explanation for this phenomenon. It is possible, therefore, that along with

[39] For example, Erlitou Phase III, Palace 1 is oriented eight degrees west of north, while Palace 2 is oriented six degrees west of north. All Erlitou burials from Phases I–IV in which the head is oriented to the north are aligned to between zero and ten degrees west of north; *Zhongguo shehui kexueyuan kaogu yanjiusuo* (1999, 141, 397–8). Recently, Tang Jigen and others have echoed my observation (Pankenier, 1995) that this contrast in orientation is an index of cultural difference: "a new approach, though, has been to argue that the different cultural character of the Erlitou site and the Zhengzhou and Yanshi Shang cities stems from their having different sources. The following traits are used to support this: at Erlitou, palaces and other major architecture and graves are all oriented west of north, but the orientation of the Yanshi Shang city and its palaces is east of north, the same as Zhengzhou and the culturally similar and contemporaneous Shang city at Panlongcheng 500 km further south in Hubei province, all of which are oriented within 20° east of north." Tang (2001, 39).

the political hegemony and the new cult of the royal ancestors, the Shang also introduced certain other ritual and cultural innovations, including using a different Pole star or method of alignment. Or, conceivably, Kochab was used by both Xia and Shang, but in the case of Xia the timing of observation had been consistently at a time when the star's circular course around the Pole took it to its greatest elongation to the west. Whatever its source, it is noteworthy that this same alignment west of true north is also found in important predynastic Zhou palatial foundations in the Zhou homeland at Fufeng (Figure 3.2c), implying that the Zhou claim to be perpetuating the legacy of the Xia may be more than merely rhetorical. But there is still another possible reason for the discrepancy, more conjectural, to which we now turn.

On the origin of the character *Di* "(Supernal) Lord"

In what follows we pursue the question how this northerly alignment might have been accomplished in practical terms, given the absence of a true Pole star, which will also lead to a new interpretation of the notoriously obscure shape of the oracle bone character *Di* used to write the name of the Shang Supernal Lord. In view of the inconspicuousness of the stars near ninety degrees north when the late Shang royal tombs were being constructed, some technique must have been devised to locate true north in order to orient such structures in the landscape. In speaking of a technique, we should have in mind something akin to the method still employed today to conveniently locate a naked-eye object among the vast array of stars. First, one identifies an unmistakable constellation nearby – for example, to find Polaris one uses Ursa Major. By sighting along the line formed by the two bright stars Mirak and Dubhe (β and α Ursa Major, forming the outer edge of the bowl of the *Dipper*), the eye is easily guided toward Polaris. This is a simple and effective device, both serviceable and still necessary in naked-eye astronomy, even in an age when we have an excellent Pole Star. How much more indispensable would such a technique have been in an age when there was no star located at the Pole? Indeed, we know that similar techniques were commonly used later.[40]

Some might argue that the objective of locating astronomical north could have been accomplished more conveniently had structures been oriented in conformity with a north–south axis determined by the bisection of the angle between the directions of the rising and setting Sun. Indeed, such procedures are alluded to in an ancient ode from the early Zhou Dynasty.[41] They are explicitly recommended much later in the *Rites of Zhou* (*Zhou li*).[42] However, precise

---

[40] Feng Shi (2007, 320 ff.).
[41] See Chapter Four below. Wheatley (1971, 426); also Chang (1980, 160).
[42] Wheatley (1971, 426).

north–south alignments established by bisecting the angle between the directions of the rising and setting Sun are difficult to achieve unless the horizons to the east and west are level. Otherwise a more sophisticated geometric method must be employed. More importantly, had alignments been established using a gnomon and the Sun's shadow they should still be accurate today and would not have produced the skewing exhibited by Shang architectural remains.[43]

More likely, in my view, at a time when the Supernal Lord's intentions vis-à-vis the Shang state were very much a "national security" concern, taking direction, literally, from the ultimate source of supra-sensible power probably called for a more direct, polar method. As Robert Thorp comments,

the people responsible for a hall like the one pictured at Panlongcheng mastered a variety of specialized tasks. Someone had to locate the building site. Its environment had to be evaluated, including conditions underground. A high water table would make digging any deep trench impossible. Aligning and laying out such a trench required ability to sight on the Pole-star or another natural beacon and drawing straight lines on the ground with right-angle corners.[44]

The role of the Supernal Lord

Assume that at some point, after the ancient Pole star Thuban had departed very noticeably from the north celestial pivot point, say by about five degrees, the need arose for a convenient means to locate true north. Now, consider the curiously obscure shape of the Shang oracle bone (OBI) graph for *Di* "(Supernal) Lord" (Figure 3.5). Over the centuries lexicographers have puzzled over the origin of this word and the graph used to write it, with no single etymology winning general acceptance.[45] In the thirteenth century BCE the same character was used in the inscriptions to denote both the Supernal Lord and the ritual sacrifice dedicated to him. Later the two would be disambiguated by the addition of an "altar" signific to the sacrificial term (second form from the top in the figure). From later accounts, especially in the early imperial period, we know that this sacrifice was conducted on an open-air platform like the one described in the "Treatise on the Feng and Shan Sacrifices" in *The Grand Scribe's Records*. Performed in person by the emperor, this rite involved a role reversal, in that after the prayer of supplication the emperor, now enacting the

---

[43] Interestingly, the foundation of a small shrine that originally stood atop the underground tomb of Fuhao, consort of the late thirteenth-century BCE Shang king, Wu Ding, was more accurately aligned to the cardinal directions than the tomb beneath. Zhang and Zhou (1995, 117, 120); Wu Hung (1997, 111); Thorp (2006, 187). At the other extreme, the largest and most southerly outlier of the early Shang Erligang culture, Panlongcheng, is oriented fully twenty degrees east of north.

[44] Thorp (2006, 73, 83).   [45] Pankenier (2004a).

Figure 3.5 Commonest oracle bone script variants of the character *Di*, Supernal Lord. Redrawn from Yu Xingwu (1996, Vol. 2, 1082). Forms 2–4 are commonly used to denote the *di*-sacrifice to the Four Quarters, later disambiguated through the addition of the "altar" signific.

role of devoted subject, faced north and performed ritual prostrations like those prescribed for subjects entering his own presence.

In the Han Dynasty, in *Explicating Graphs and Explaining Composite Characters* (*Shuowen jiezi*), Xu Shen (*c*.55–*c*.149 CE) claimed that "Lord" *Di* is equivalent to its homophone *di* "to look into, examine," which adds only the "speech" signific on the left. In a similar vein, later interpreters, influenced by the existence of the character for "stem" *di*, written with the "grass" signific, identified *Di* as a pictograph of the peduncle or footstalk of a plant.[46] Yu Xingwu's dissection of the graphic components of the character led him to propose (with Yan Yiping) "burnt sacrifice" as the basic meaning, implying

---

[46] For graphical analysis, see Yu Xingwu (1996, Vol. 2, 1082).

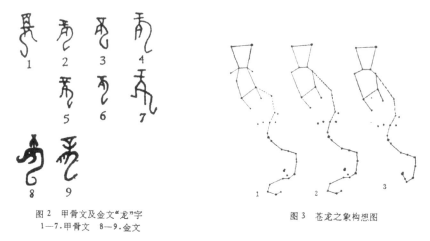

图 2   甲骨文及金文"龙"字
1—7.甲骨文   8—9.金文

图 3   苍龙之象构想图

Figure 3.6 Shang oracle bone script (left 1–7) and Shang bronze variants (8–9) of the character for *long*, "dragon"; (right) impression of the *Dragon* constellation formed by linking together stars from Virgo through Scorpius. Adapted from Feng Shi (1990c, 112).

that "*di*-sacrifice" is used metonymically to stand for an otherwise "graph-less" Supernal Lord. Of the oracle bone variants he illustrates (Figure 3.5), the first two predominate in the inscriptions from the reign of King Wu Ding (r. *c*.1239– 1181). The shape of the character subsequently remained comparatively stable throughout the Anyang period inscriptions of late Shang.

Let me suggest that the character *Di* is in fact representational. There is precedent for this kind of representation in the oracle bone inscriptions. One obvious example is the OBI character for *dou* 斗, "ladle, peck-measure, dipper," which clearly resembles a ladle and refers in some contexts to the *Dipper*. An even more striking example is the resemblance between the bone form of the character for dragon, *long*, and the stellar configuration to which the iconic Cerulean Dragon actually corresponds (Figure 3.6). As we saw in Chapter 1, the *Dragon* corresponds to constellations from Virgo through Scorpius, its horn marked by Spica and its tail by Scorpius, roughly seventy-five degrees distant. It is hardly surprising that the hooked array the Babylonians saw as the tail of a scorpion should have been identified by early Chinese observers as the tail of a dragon, particularly in view of the seasonal significance of the *Celestial Dragon*, and the *Fire Star*, Antares (α Sco), at its heart. Indeed, that star appears in the oracle bones as the *Shang* star (*Shang xing*), and is one of only two or three stars mentioned by name in the formulaic divinations.

What, then does the character *Di* for the Supernal Lord actually depict? Given the practical necessity of locating true north, here is a possible method.

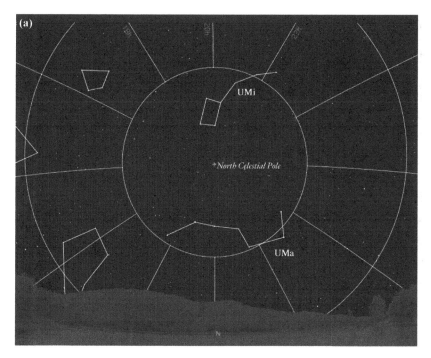

Figure 3.7 (a) Star chart showing the location of the north celestial Pole in 2100 BCE, midway between the two dippers. (b) Star chart at the same date and time with character *Di* superimposed and linking stars on opposite sides of the celestial Pole. The intersection of the three lines connecting the principal stars in UMa and UMi marks the exact location of the Pole in 2100 BCE. Kochab (β UMi), the middle star at the top of the configuration, later came to be denoted the August Emperor star, possibly an echo of its archaic role (Starry Night Pro 6.4.3).

Figure 3.7a shows the circumpolar region in 2100 BCE. Superimposed on this same chart in Figure 3.7b is the Shang graph for the high god *Di* drawn to scale and connecting the bright stars in the handle of Ursa Major (below: ζ, ε, δ UMa) and scoop of Ursa Minor (above: γ, β, 5 UMi). By comparing the two charts it is evident that the three longitudinal lines common to all forms of the graph *Di* triangulate on the actual location of true north in 2100 BCE. A plumb line held up so that it passed through any pair of stars on opposite sides of the Pole would intersect at celestial north at that date and permit one to lay out a true north–south line on the ground. We do not know the epoch of creation of the character *Di*, except that it had to have occurred well before the

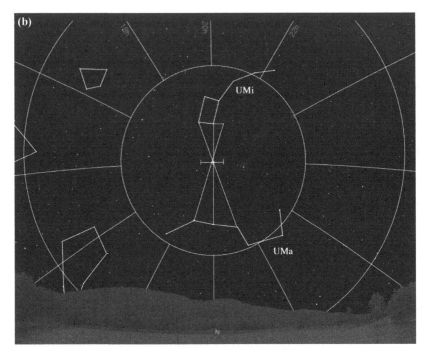

Figure 3.7 (*cont.*)

thirteenth-century BCE date of the earliest Shang inscriptions, which already give evidence of a script and written language in fully developed form.[47]

John Didier (2009) speculates, based on subjective interpretations of Neolithic images culled from an assortment of cultures, that there was a huge pan-Asian "square" constellation centered on the north Pole which was specifically worshipped by the Shang. (Needless to say, Didier's subtitle notwithstanding, the northern circumpolar stars only impress inhabitants of the northern hemisphere.) Didier's thesis is an illustration of what David N. Keightley has called "ethno-artistic" arguments which amount to no more than "scholarly Rorschach tests."[a] I know of no precedent anywhere of a quadrilateral constellation bracketing the Pole such as Didier proposes,

---

[47] It seems the OBI character for *Di* could even be written upside down (the inscription is *Hou* 1.26.5). This is only surprising until one realizes that this stellar configuration is constantly rotating about the Pole (though it must be admitted this variant is rare). Liu Xinglong (1986, 178).

whose dimensions would have been in the decidedly un-square ratio of one to five, the short sides being separated by twenty degrees of empty space with no intervening stars to link them together. In fact, of course, the circumpolar stars describe obvious circles around the Pole, which is reflected in the early "wheel" metaphors we saw above.[b] Unlike the *Dipper, Dragon, Fire Star*, and Orion, no trace of a polar "quadrilateral" can be found in any ancient Chinese inscription, text, or graphic representation of the circumpolar sky. The stars of UMa (ζ, ε, δ UMa) which Didier thinks formed one side of the imaginary square in fact *already comprised the handle of the Dipper*, the most ancient and recognizable constellation in the northern sky. The proposition that the same stars could have done double duty as part of a giant constellation of which no trace has survived is simply not credible. References to and representations of the *Dipper* are ubiquitous in China, beginning with the Puyang, Xishuipo shaman's burial from about 3000 BCE.[c] The *Dipper* was perceived as a unitary constellation by the Shang Chinese, just as it was by the Egyptians, Babylonians, Greeks, Lakotas, and others, even if it was depicted differently – as a ladle, wagon, plow, leg of a bull, bear, etc.[d] No less an authority than Owen Gingerich inclines to the view that the *Dipper* has been recognized as a unitary constellation since the Paleolithic.[e]

[a] Keightley (1998, 766).
[b] For ancient "axial" imagery, see Major (1978, 5 ff.).
[c] Feng Shi (1990a, 52–60, 69). See Chapter 11 below.
[d] Feng Shi (1990a, 137, 144); Jao (1998, 42).
[e] Gingerich (1984, 218–20).

I am not suggesting that this configuration was actually perceived to be a constellation at the time. What I am suggesting is that the bone graph for *Di* depicts the shape of a device designed to permit one to establish true north in the absence of a Pole star. As a practical means of establishing the location of the Pole in order to lay out a true north–south line, the kind of device or method posited here could conceivably have been employed; indeed, it or something like it would have been indispensable. Thinking in purely practical terms, some sort of sightline to the north had to have been established, and right angles to it would probably need to be laid down in order to determine east–west. We know of course that the Four Quarters, *si fang*, figured importantly in Shang sacrificial rites. A sighting device or template such as that drawn on the star chart above, placed atop a standard and used with a plumb line, could have been designed to serve the purpose. One can imagine a graphic representation of this implement subsequently being used to denote the supreme supra-sensible power resident at that mysteriously empty spot in the sky. The alignment technique involved is also entirely consistent with documented later alignment practices using the stars. Consider, for example, the following example from the early imperial technical manual, *Zhou bi suan jing*:

> To fix the pivot of the north pole, the centre of the *xuan ji* [i.e., *Dipper*], to fix the centre of north heaven, to fix the excursions of the [north] Pole [star]. At the winter solstice, at the time when the Sun is at *you* [LT 17–19], set up an eight-*chi* gnomon, tie a cord to its top and sight [along the cord] on the large star in the middle of the North Pole [constellation = ζ UMa]. Lead the cord down to the ground and note [its position]. Again, as it comes to the light of dawn, at the time when the Sun is at *mao* [LT 5–7], stretch out another cord and take a sighting with your head against the cord. Take it down to the ground and note [the positions] of the two ends . . . The [line of] separation of the two ends fixes east and west, and if one splits [the distance] between them in the middle and points to the gnomon it fixes south and north.[48]

"Setting up a gnomon" would necessarily have entailed the use of a plumb bob to ensure it was perfectly vertical (even at Taosi in the twenty-first century BCE). Although the *Zhou bi suan jing* is describing a technique practiced in the late Warring States period, there is every reason to assume there was interest in developing a similar method earlier, as we shall see.

An Egyptian parallel

As early as the mid third millennium BCE it is evident that the ancient Egyptians were capable of aligning monumental structures like the great pyramids of Giza on true north with an accuracy of a fraction of a degree. So much was established by J. Norman Lockyer's pioneering research in *The Dawn of Astronomy*.[49] How they might have done it can shed light on our discussion. On the right in Figure 3.8 is a representation of the ancient Egyptian goddess Seshat (*c*.3000 BCE) in her role as patroness of the "stretching of the cord" ceremony, which was the essential first step in laying out temples, pyramids, and other high-value structures.[50] On her head, there is a star-shaped headdress ⚹ (Figure 3.8b), which is also depicted atop a standard in the procession illustrated in (Figure 3.8a) at the left. It has been argued that this device is the very instrument

---

[48] Trans. Cullen (1996, 191, trans. modified).

[49] The degree of exactitude achieved in aligning the three greatest pyramids in Egypt, all at Giza, has recently been reconfirmed by Clive Ruggles: "The sides of each of the Giza pyramids were carefully aligned upon the cardinal directions (north–south or east–west). This alignment followed established practice, but the accuracy with which it was achieved at Giza is truly impressive, particularly in the case of Khufu's pyramid [the greatest one]. Each of its sides is cardinally aligned to within six arc minutes, or one-tenth of a degree. This is equivalent to no more than one-fifth of the apparent diameter of the sun or moon. The other pyramids are only slightly less well aligned, Khafre's to within about eight arc minutes and Menkaure's to within sixteen." Ruggles (2005, 353).

[50] Lockyer (1964, 173–5).

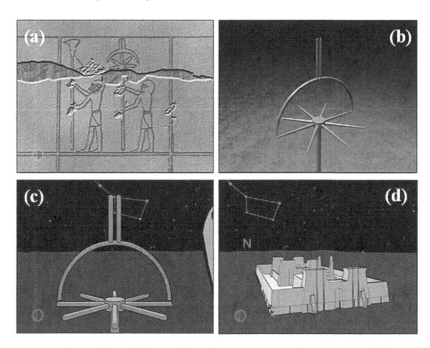

Figure 3.8 (a) Seshat's device shown atop an attendant's standard in the procession for the Stretching of the Cord ritual. (b–d) Artist's rendering of the use of the device. Image courtesy of Juan Antonio Belmonte, reproduced by permission. (e) A Fifth Dynasty (c.2450 BCE) depiction of the goddess Seshat with her headdress penetrating the frieze of stars above. After Miranda, Belmonte, and Molinero (2008, 58), reproduced by permission.

that was used to establish north–south alignment, which subsequently came to be the hieroglyph for the goddess Seshat herself. In the Egyptian case, when the two stars Megrez and Phecda (Figure 3.8c) in the bowl of the *Dipper* were precisely aligned between the two vertical tines of the device, the eight horizontal spokes on top of the standard accurately pointed in the cardinal and inter-cardinal directions. At the epoch in question the alignment formed by those two stars, when centered between the tines, would have pointed straight toward the celestial Pole. This Egyptian innovation was in all probability the ancestor of the device known as a *groma* or "surveyor's cross" (Figure 3.9), used by ancient Etruscan and Roman engineers to accomplish essentially the same engineering task:

In founding new cities or Roman colonies, *gromatici* (surveyors) emphatically refer to upholding the ancient Etruscan practice of astronomically orienting towns. For example, the first-century AD historian Hyginus Gromaticus says: "The *limits* were established

(e)

Figure 3.8 (*cont.*)

not without consideration for the celestial system, since the *decumani* (east–west streets) were laid out according to the sun and the *cardines* (north–south streets) according to the celestial axis. This system of measurement for the first time established the teachings of the Etruscans . . . and from these terms the boundaries of the temples also came to be described."[51]

[51] Aveni (2008b, 147–8); Miranda, Belmonte, and Molinero (2008, 57–61).

(a)

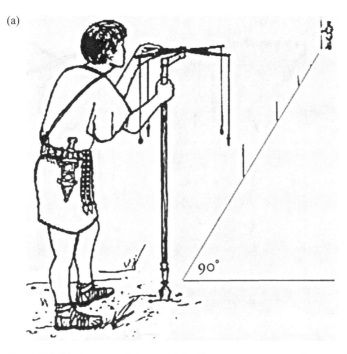

Figure 3.9 (a) A *groma* being used by Roman surveyors to lay out right angles. Adapted from http://blog.edidablog.it/edidablog/tuautem/2009/11/25/introduzione-alla-torino-romana-3/groma. (b) Reconstruction (right) of the *groma* repesented on the tomb of a surveyor at Pompeii. After Miranda, Belmonte, and Molinero (2008, 59, Figure 3), reproduced by permission. The word *groma* is thought to derive from Greek *gnomon* (γνόμων), "indicator," possibly via Etruscan.

Kate Spence (2000) proposed that the Egyptians used a "simultaneous transit" method, strikingly similar to the technique I proposed, relying on the simultaneous transits of the meridian by a pair of circumpolar stars above and below the Pole. A vertical between them established by means of a plumbline would have passed exactly thorough the celestial Pole in 2500 BCE.[52]

[52] See Spence (2000, 230–4). Gingerich (2000, 297–8); Rawlins and Pickering (2001, 699); Spence (2001, 700). Unaware of Kate Spence's posited "simultaneous transit" hypothesis, in 2003 I proposed that the Chinese used an analogous method to locate true north. Pankenier (2004a, 287–308). Needham and Wang take note of a parallel passage in the *Rites of Zhou* similar to the one just quoted, noting that "Han commentators took this to mean that cords of equal length were fixed one to each corner of the base, but Chia Kung-yen in the Thang supposed that four suspended plumblines were used; if this was so, the instrument was very like the *groma* of the Roman surveyors." Needham and Wang (1959, 286).

(b)

 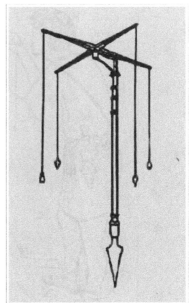

Figure 3.9 (cont.)

Paleographic and linguistic evidence

Above we deferred analysis of the possible significance of the crossbar element ⊢⊣ in the bone graph for *Di*. Now, perhaps, we are in a position to establish its meaning. The "I-beam" form of this element depicts the earliest form (and function) of the tool we know as a set square or carpenter's square, used to make accurate right angles. Written in the Shang bone inscriptions like this 𠂤 (modern *ju*), the graph for this tool is an unmistakable depiction of a man holding an "I-square" in his left hand.[53] Not surprisingly, this element is found in the character for "square" or "quarter; quadrant" itself, *fang* 方, the term used by the Shang to refer to the Four Quarters of their world. This graph, whose basic meaning is "square," is also represented in some diviner groups by the variant 中.[54] Similarly, the oracle bone graph *yang* 央 incorporates the same I-square and means "center."[55] Insofar as that element emphasizes "centrality," the parallel

---

[53] Li Xiaoding (1970, Vol. 5, 1593).

[54] Compare *Tahuantinsuyu*, "Land of the *Four Quarters*," the name the *Inka* gave to their empire.

[55] Li Xiaoding (1970, Vol. 5, 1825); Yu Xingwu (1996, Vol. 1, 223); Ban Dawei (2011). Sarah Allan (2009, 22–4) discusses this I-square element, but without explaining what its precise semantic contribution might be to the meaning of the graph *Di*.

Table 3.2 *The* Di *word family.*

The di 帝 family of root meanings: ~ match; mate; fit; conform with; join
di 敵   be a match for > opponent > enemy [~ be a match for in strength, power, wits, etc.] 抵擋:「寡不敵眾」; 相當, 匹敵:「勢均力敵」:《孫子兵法 · 謀攻》:「敵則能戰 之」
di 蒂   peduncle; footstalk; i.e., part joining the flower, melon, etc. to the main stem [~mate with; join]
di 締   join together as one [~connect; conjoin] 締結:賈誼《過泰論》:「合從締交,相與為 一」
di 嫡   primary consort; principle wife > legitimate heir [~mate; match]
shi 適   be fitting; suitable; match [~match; conform to] 適合, 適宜, 恰好《商君書 · 畫策》: 「然其名尊者,以適 于時也」
di 禘   principle sacrifice to Shang di (Supernal Lord) and/or the ancestors collectively
di 帝   Supernal Lord; sky god resident at the Celestial Pole

with the graph for *Di* 帝 is unmistakable. Not only does the I-square contribute importantly to the meaning of the graph *Di*, it seems to signify that, just as in the case of the instrument identified with Seshat, the analogous Chinese device also had one or more horizontal crosspieces on the top at right angles to the shaft and to each other, to indicate the cardinal directions. Like both the Egyptian and Roman implements, it would necessarily also have incorporated plumb bobs to establish verticality.[56]

As noted above, in order to accomplish the practical objective of achieving north–south axial alignment of structural foundations beginning as early as Erlitou, it would certainly have been necessary to use a sighting technique of some kind. As we saw in Chapter 1, the Taosi calendar priests may already have invented a shadow-measuring rule, so there is no doubt they were familiar with the concept of establishing a sight line using a vertical marker like a gnomon. Subsequently, just as in the case of the Egyptian goddess Seshat, the symbol for this device could have been adopted as the glyph for the abstract concept of the Supernal Lord resident at the empty Pole. As we saw above, this notion of a sky god at the Pole persisted throughout the two millennia from the late Neolithic through the Han Dynasty, and was ultimately revivified and transformed into a cult devoted to the metaphysical locus of power at the center of the sky, Supreme One. But, unbeknownst to its ancient designers, equally persistent ritual use of the routine alignment method described throughout the second millennium BCE, over time, would imperceptibly have led to substantial error in locating true north, producing misalignments as significant as those observable in high-value Shang structures.

Finally, Table 3.2 reproduces some of the phono-semantic series to which *Di* belongs.[57] If we try to isolate the root meaning of those words which appear

[56] Needham and Wang (1959, 286).   [57] Karlgren (1964b), GSR877: k–l.

to be intrinsically related or cognate, leaving aside those which appear to be simply homophones, the result is as shown in the table. This series implies a constellation of meanings similar to "match, mate, fit, conform with, join" and so on, which would seem apt if the original meaning of the character derived from the process of locating true north by means of a device designed to be congruent somehow with the stellar configuration in the sky.

This ancient ritual preoccupation with taking divine direction from the spirit of the celestial Pole literally entailed bringing a normative aspect of mysterious Heaven down to Earth. Like the practical use of the *Dipper*'s handle as a clock hand, the idea survived to inspire seminal metaphors about charismatic potency until the late first millennium BCE when the Supernal Lord experienced a resurgence of devotional interest, both in his own right and in the guise of a state cult devoted to Supreme One, his latter-day alter ego. Compare the striking resemblance between the form of the early OBI character for *Di* and the spread-eagle posture of Supreme One (*Taiyi*) on the Warring States period "Weapon Deterring" dagger-axe (Figure 3.10a). Another classic example of this heroic posture is seen in an early depiction of the mythological cosmic miscreant *Chi You*, "The Wounder," shown on the right in Figure 3.10b.[58]

Conclusion

By the end of the Shang Dynasty in the mid eleventh century BCE, the evolution of the Shang ancestral cult had culminated not merely in the identification

---

[58] "Chi You's Banner," a token of the miscreant's defeat placed in the sky by the Yellow Emperor, was a comet with curving tail, first mentioned in the third century BCE *Springs and Autumns of Master Lü*, "Ming li" chapter; Lü Buwei (1966, 6.9b). The comet appeared twice in the Western Han dynasty, between July 3–August 1 and August 31–September 29, 135 BCE; *Han shu*, 6.160, 27.1517. The second of these records reads, "Sixth year of the Jianyuan reign period of Emperor Wu of the Han Dynasty, eighth month; a long star appeared in the east, so long that it stretched across the sky; after thirty days it departed. The prognostication said, 'this is Chi You's Banner; when seen the ruler will attack the four quarters.' After this the army punished the Four Yi [barbarians] for several decades." Pankenier et al. (2008, 19). This was the "long star," *chang xing*, that stretched across the sky throughout September which prompted Emperor Wu just a few weeks later to inaugurate a new reign period beginning in the tenth month that year (then the first month of the civil calendar). The new reign period was aptly dubbed *yuanguang*, "Primal Brilliance," to commemorate the celestial sign. For a comparison of the Chinese and Roman accounts of the comet of 135, see Ramsey (1999) and Kronk (1997). Chi You's Banner is one of the comets illustrated in the Mawangdui cometary atlas from two generations before the 135 BCE apparition. *Zhongguo shehui kexueyuan kaogu yanjiusuo* (1980, 22, Figure 20). For the myths about Chi You and Huang Di, see Bodde (1975); Lewis (1990); Puett (1998). See also Pankenier, "*Huainanzi's* 'Heavenly Patterns' and *Shiji's* 'Treatise on the Celestial Offices': What's the Difference?" (in press). A depiction of Chi You in his role as the more peaceful rain god on a late second-century wellhead is found in Liu et al. (2005, 298). See also the image of the deity Supreme One and discussion by Donald Harper (1999, 851, 870). For a highly suggestive auroral explanation for the worldwide depiction of this god-like posture in ancient times, see Peratt (2003, 1192–1214).

(a)

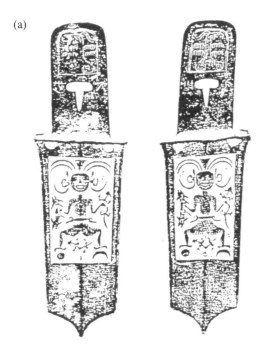

Figure 3.10 (a) Taiyi depicted with cosmogonic attributes on the "weapon repelling" Taisui dagger-axe from the Warring States period. Adapted from Harper (1999, 872). (b) Chi You, "The Wounder." Late Eastern Han stone engraving, Yi'nan, Shandong. Detail from *Zhongguo huaxiangshi quanji bianji weiyuanhui* (2000, 143, no 194).

of the royal Shang ancestors with the Supernal Lord *Di*, but even in the posthumous adoption of the title *Di* for the ruler. With this, it seems, the precedent was assuredly established for the assertion of a genetic relationship between the Shang temporal authority and the Supernal Lord, which was to be made still more explicit in the term "Son of Heaven" later applied to Zhou kings, and ultimately to all Chinese emperors. Quite apart from metaphorical associations between the temporal power and the potency of the numinous void, whether in Confucian ideology or in Taoist mysticism, one is struck by a remarkably strong and persistent preoccupation with quite specific astral–terrestrial correspondences, especially concerning true north as the locus of supra-sensible power. Although well documented in the classical and imperial period for which abundant literary sources are available, the history of this preoccupation with the pivot of the heavens and its potent symbolism have not previously been elucidated as early as the second millennium BCE. By bringing together a variety of evidence from archaeology, the paleographic record, and astronomy,

(b)

Figure 3.10 (*cont.*)

it is possible to piece together a coherent picture that identifies the Supernal Lord with the potency of the pivot of the heavens at least by the founding of dynastic Shang in the early Bronze Age, and probably well before.[59]

---

[59] In a recent article, Sarah Allan (2009, 4) misconstrues my earlier argument, averring, "as Pankenier [2004b] discusses . . . there was always a Pole-star by which buildings and tombs were oriented. This is clear from Pankenier's discovery that when the nearest star to the pole changed in the late Neolithic and early Bronze Age . . . the north–south orientation of buildings also changed." Allan fails to distinguish between the celestial Pole (i.e. the pivot of the heavens) and the Pole Star, attributing to me a confusion about the existence of a bright star located at true north during the Chinese Bronze Age ("there was always a Pole-star"), when in fact there was none. The major thrust of my 2004 article, as here, had to do with a method of aligning on true north *in the absence of a true Pole Star*. In an effort to distinguish her thesis from my own, Allan asserts that Pankenier (2004b) "has argued that the potency of Shang Di derives from its association with the pivot of the heavens at the North Pole, whereas I am arguing that the Pole-star – that is, the star which was used to determine a northward orientation – was called 'Shang Di.'" If there was such a star Sarah Allan does not identify it, nor reveal how it was used to achieve "northward orientation." As I explained, there had indeed been a bright star, Thuban (the Pole star of contemporary Egyptians), located at ninety degrees north in the early third millennium BCE. The memory of that Pole star, coupled with the obvious wheeling of the stars about the pivot of the heavens, explains the hub's numinous significance; that is, the metaphysical preoccupation with the unmistakable void at the apex of the sky. As we saw, the nearest bright star, Kochab (β UMi), was never closer than 6.5° from the true Pole (more than

Joseph Needham conjectured that Kochab might have been taken to be the Pole Star in Shang times, and while that is not inconceivable, the fact that at the time Kochab actually described quite a large circle around the Pole should not have gone unnoticed. Since some Xia and Shang structures were aligned on true north with a higher degree of accuracy, it is unlikely that they were using Kochab as their Pole star. Then, too, we saw how by mid-Zhou the "virtue of emptiness" and the immobility of the Pole had become an established trope in the philosophical literature. In view of all this, it is difficult to accept without explanation that the ancient Egyptians of the mid third millennium BCE were capable of achieving alignments on true north accurate to within a tenth of a degree, fifty to one-hundred times more precise than their counterparts in East Asia a millennium later.

the width of a fist held at arm's length) – quite a substantial margin. Had the Zhou Chinese *not* distinguished between celestial north and Kochab's location, it is highly unlikely a metaphysical focus on the efficacy of the mysterious *void* at the "hub of the wheel" could have arisen.

4 Bringing Heaven down to Earth

> This is how the former kings, without employing rewards, were able to promulgate their virtue (*de*) widely throughout the world. (*Discourses of the States*)

We have seen how ritual specialists in the late Neolithic and early Bronze Age, like their counterparts in ancient Egypt, could have used the circumpolar stars to find true north, a task complicated by the absence of a comparatively bright star near the Pole. Archaeological discoveries from the Xia, Shang, and Zhou clearly show that it had become crucially important to achieve accurate cardinal orientation of the built environment – walls, palaces, temples, tombs, common burials, and even storage pits give evidence of a preoccupation with north–south axial alignment.[1] It has long been understood that cardinality is an index of the paradigmatic roles of "the Center" (*zhong*) and "the Four Quarters" (*si fang*) – both core organizing principles of early Chinese cosmological thinking.[2] Here, however, our concern will be less with cosmological perspectives than with how, in practical terms, cardinal orientation was achieved after the obsolescence of the archaic method put forward in Chapter 3, and what this tells us about a fundamental mindset that figured importantly in the formation of early Chinese civilization.

There are, of course, a variety of methods for achieving cardinality described in the literature on archaeoastronomy, most involving observations of the Sun's shadow using a gnomon. Variations on these methods documented in the Warring States period (403–221 BCE) have been known for some time and will be described below, but here we are not primarily concerned with comparison of Chinese methods with those of other ancient cultures. Instead, we will continue

---

[1] As David N. Keightley (1998, 794) has concluded on more than one occasion, "The great attention the early inhabitants of China paid to orientation and posture was a significant characteristic of both the Neolithic and Bronze Age cultures." Examples have repeatedly been identified from different Neolithic cultures, although axial orientations are sometimes inconsistent even within a single site. See the examples from Liangzhai, Banpo phase, illustrated in Liu (2007, 37, Figure 3.3, and 81, Figure 4.5).

[2] See Wang (2000); Keightley (2000); Allan (1991). The Bronze Age Chinese were hardly unique in this respect; see Wheatley (1971); Selin (2001).

to focus on recovering traces of uniquely Chinese solutions in the late Bronze Age to the problem of orienting structures in the landscape using the stars. Not surprisingly, this method underscores the distinctive polar-equatorial focus of Chinese astronomy.[3] Until now, researchers have overlooked the technique, which takes advantage of the unique orientation of the Great Square of Pegasus known as *"Determiner," Ding*, in pre-Han China.

Alignments and misalignments

We have seen that important structures in the Three Dynasties period (Xia, Shang, Zhou) were more or less accurately oriented to the cardinal directions, demonstrating that painstaking efforts were being made before construction to achieve this. It is still something of a conundrum why the Shang produced many high-value structures whose central axis is consistently oriented several degrees east of true north (Figure 4.1). David N. Keightley has commented on this phenomenon:

By late Shang the cardinality of the great royal tombs at Xibeigang was notable, the axis of their long ramps generally being north-by-east/south-by-west. The generally north–south orientation of Shang or Shang-style elite burials is confirmed by a variety of finds from such sites as Dasikong *cun*, north of Xiaotun, Panlongcheng in Hubei, Luoshan in Henan, and Sufutun in Shandong. The north–south orientation of the tombs in all these sites was similar to that of their rammed-earth ramparts (if present) or building foundations. The so-called "commoner" burials at Yinxu West were also oriented to the cardinal directions, with a clear bias, generally of ten degrees "to the right." The same ten-degree bias is found at other Shang sites. The intense concern with orientation is further suggested by the cardinality of the burials of Shang sacrifice victims.[4]

The ten-degree bias Keightley reports is based on a total of 938 burials at Yinxu West, which led him to observe further that "the remarkable consistency of the deviation strongly suggests that the Shang tomb builders, at least in the 'commoner' cemetery, had a definite directional scheme in mind."[5] A few scholars, including Keightley, suspect that the Shang "venerated, or at least gave priority to, the northeast."[6] But it is not at all clear what the rationale might have been, and the question where the Shang elite came from is still

---

[3] Needham and Wang (1959, 239).
[4] Keightley (2000, 82). The eastward bias of the temple palace foundations at Xiaotun, Anyang, is somewhat less than ten degrees. Keightley (1999b, 259, Figure 4.7). See Zou Heng (1979, 69), who notes that the majority of house foundations at Xiaotun were rectangular, with comparatively more facing east–west than north–south, unlike the later palaces, which Zou says faced south. When surveyed, the long axis of the foundations of the latter were all close to magnetic north. Kwang-chih Chang (1980, 112, 121) mentions alignments only in passing. See also Thote (2009, 108, 120).
[5] Keightley (2000, 83, n. 8).   [6] Ibid., 87; Zhu Yanmin (2003, 27–33).

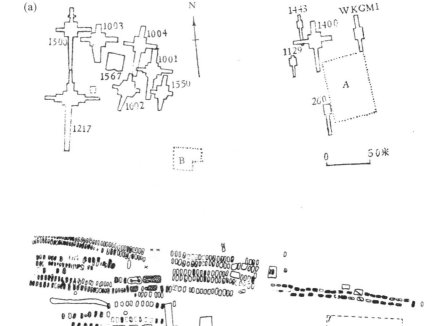

Figure 4.1 (a) Site plan of the late Shang royal tombs at Xibeigang, Anyang.
(b) Burials of petty elite nearby. Adapted from Loewe and Shaughnessy (1999,
187, Figure 3.20), © Cambridge University Press, reprinted by permission.
The arrow shows magnetic north at the time of excavation. True north at the
time was just over three degrees west, which means that the eastward offset
of these tombs was less than depicted on the archaeological plan.

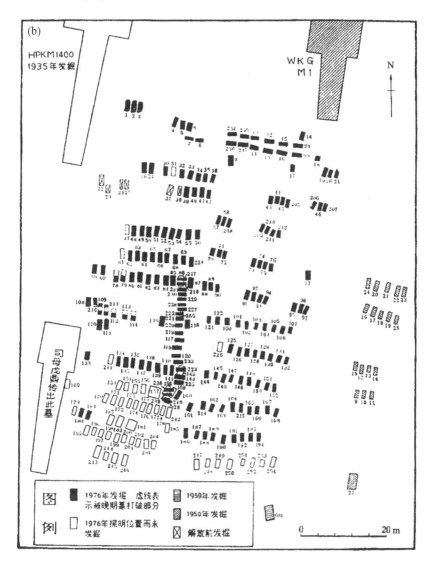

Figure 4.1 (cont.)

far from settled.[7] Sacrifices to the four quarters recorded in the oracle bones reveal no particular preference for a given direction. The axial orientation of the rectangular tombs, access ramps, royal skeletons, and petty elite burials strongly

[7] Zhu Yanmin (2005).

favors north–south alignment, as even a cursory perusal of archaeological studies will confirm (Figure 4.1b).[8] Had the Sun's shadow been employed for the purpose, then those highly symbolic Shang edifices – palaces, royal tombs, city walls – should all still be correctly oriented today. If the alignments of houses at Xiaotun had been based on sunrise observations they would seem to point to a generalized focus on the east-southeast, and a date shortly after the autumnal equinox, rather than the northeast. Joseph Needham, too, noted that Shang sites such as the royal tombs at Xibeigang display consistent misalignment in the order of five to twelve degrees (two to nine degrees true).[9]

By contrast, the foundations of pre-Shang (i.e. Xia) walls and palaces in Erlitou Phases I–II display a consistent misalignment *west* of true north by about six to eight degrees, which when corrected for magnetic declination yields an actual offset of some nine to eleven degrees.[10] Misalignments this large are very difficult to explain as a result of measurement error, even in the mid second millennium BCE, when, as we saw above, the Egyptians were capable of aligning the Great Pyramid of Khufu with an accuracy of 0.1°. More to the point, however, the contrast between the westward Xia bias and the eastward Shang bias suggests that contrasting methods or motives must have been involved.[11] What is especially intriguing is that the alignment of a predynastic Zhou palace foundation at Zhouyuan, homeland of the Zhou

---

[8] *Zhongguo shehui kexueyuan kaogu yanjiusuo* (1994, 100–35).

[9] Needham and Wang (1962, 313), based on the report by Shi Zhangru (1948, 21). It should be noted that the bearings given in Chinese archaeological reports were measured using transit and magnetic compass and do not take magnetic declination into account. On average, the magnetic declination of the core area of north China in question at the time of archaeological excavation was slightly more than three degress west of true north, so that the corrected figure for the average bias east of true north noted by Keightley would actually be around seven degrees (about a fist-and-a-half held at arm's length). Needham provides data on similar misalignments from the Tang through Qing dynasties, but those are clearly attributable to the use of a magnetic compass, which was unknown in the mid second millennium BCE. Needham and Wang (1962, 310, Table 52).

[10] With few exceptions, burials at *Erlitou* from Phases I–IV tabulated in Zhongguo shehui kexueyuan kaogu yanjiusuo (1999, Appendix 1) display a deliberate north–south alignment, with the head of the tomb occupant oriented to the north in nineteen out of twenty-five cases where skeletal remains survived. In six cases the heads pointed south, with the long axis of those tombs oriented east to west. Alignment errors of the tombs oriented to the north vary between three and seventeen degrees west (magnetic), with one outlier displaying a twenty-four-degree eastward bias.

[11] Von Falkenhausen (2006, 215). There is strong evidence that the magnetic declination in the north China plain fluctuated widely in the 2000–1000 BCE period. A recent study of paleomagnetism in China found that "during the last 4500 yr the direction of the geomagnetic field underwent obvious changes . . . [T]he declination also changed over a wide range: from ~38° E at –4600 and 4000 yr BP to 37° west at ~1900 and 2900 yr BP." Cong and Wei (1989, 71), Wei et al. (1983, 138–50). The change in the direction of the magnetic field is certainly intriguing. Nevertheless, it must be stressed that, although the attractive quality of lodestone had been recognized by the mid first millennium BCE, there is no evidence that the ancient Chinese made use of the north-pointing properties of magnetite before the south-pointing lodestone

rulers prior to the conquest of Shang, follows the west-of-north bias of Xia orientations centuries earlier, possibly suggesting recovery or reinstatement of a method or cultural bias that pre-dates Shang.[12] The priority given to the north continued into Western Zhou and is clearly reflected in the bronze inscriptions, where the king invariably faces south in ceremonial contexts, and, significantly, "north" is given pride of place in naming the directions.[13] Facing south came to symbolize the posture uniquely reserved for the "Son of Heaven."

City building in the earliest sources

There are few undertakings of greater symbolic significance to dynasts than the siting and construction of a new capital. With King Wen's receipt of Heaven's Mandate (see Chapter 6), explicit references to the procedures involved in laying out the new capitals emblematic of the changing political circumstances make their appearance in the literature. Almost immediately after receiving the Mandate, the dynastic founder King Wen aggressively pursued the prerogatives signified by his appointment as the Shang court's Earl of the West, which included campaigns to subjugate the states of Mi and Chong in the eastern Wei River valley. Chong surrendered to Zhou in the thirty-fourth year of Di Xin of Shang (1053 BCE), the sixth year of King Wen's "Mandate," according to the Bamboo Annals.[14] Immediately after this, construction was begun on a new capital of Feng southeast of present-day Xi'an, from which the Zhou would continue their efforts to consolidate power on the Shang western flank. The seat of royal Zhou power was soon relocated to Feng and in 1051 BCE the heir apparent, Fa (King Wu), was ordered to create a new ritual center at Haojing, adjacent to the new capital. Haojing, which would also become known as Zongzhou, the "spiritual" or ceremonial capital, demonstratively incorporated a so-called "Spirit Terrace" already mentioned in Part One, as well as major state temples, and it was here that regular dynastic rituals of renewal, appointment, and tribute were subsequently conducted. After the successful conquest of Shang still another more centrally located administrative center and royal seat, Chengzhou, was built near present-day Luoyang nearly 250 miles to the east.

Two passages from the earliest texts in the Book of Documents, the "Announcement at Luo" (Luo gao) and the "Announcement of Duke Shao"

spoons of the Han dynasty, and even their application was mantic and not geomagnetic. See Needham and Wang (1962, 229–30); also Wheatley (1971, 461); cf Carlson (1982).

[12] A preliminary survey of predynastic Zhou and early Western Zhou foundations reveals a range of orientations and it is not yet clear if there was a consistent preference for any one of them. Interestingly, the Qin kingdom, which occupied the former Zhou homeland from the eighth century BCE on, also oriented their capital of Yongcheng fourteen degrees (magnetic) west of north.

[13] For example, Li Xueqin (1999a, 122) adduces the Wu Hu ding and Wu si Wei ding cauldrons.

[14] Loewe (1993, 39–47).

(*Shao gao*) deal with the founding of the new capital, Chengzhou, in 1036 BCE, a decade after the Zhou conquest of Shang (1046 BCE). In fulfillment of King Wu's (r. 1049–1045) wishes, the texts record that for symbolic as well as strategic reasons the Zhou founders deemed it appropriate to relocate the royal administration, including the residence of King Cheng (r. 1042–1021 BCE), to a more central location on the north China plain. After divining about the auspiciousness of different sites, a suitable location was identified beside the Luo River near its confluence with the Yellow River. Of the Grand Protector (*Tai bao*, i.e. Duke of Shao) it is said in "The Announcement at Luo" that "he arrived at Luo at daybreak and divined about dwelling [there]. Having obtained the oracle's [response], he laid out and aligned [the city]." The next day, the Duke of Zhou, in his capacity as regent, "arrived at daybreak, whereupon he came through and inspected the layout of the new city."[15] Then, after a lengthy exposition of the two dukes' merits in supervising the multiple divinations which led to the siting of the new city, King Cheng is recorded as saying, "The Duke [of Shao], having fixed the site, sent a messenger for me to come; he has come to show me the grace and constant auspiciousness of the oracles."[16]

Many of the accounts of the events of the conquest period and of the founding of Zhou had become hallowed traditions by the time the *Book of Documents* was compiled some centuries later, hence they were prone to hagiographic excess. Nevertheless, the historicity of this particular episode was conclusively confirmed in 1963 with the discovery of a large bronze beaker known as the *He zun*, whose dedicatory inscription dates its casting to the fifth year of King Cheng, when "the King initially removed to dwell in Chengzhou" ("Accomplished Zhou"; that is, the new city of Luo). Then the inscription, recounting the meritorious service of the maker's ancestor at the founding of the dynasty, recalls that it was King Cheng's father, King Wu, who had decreed the move, though he died before giving effect to his plan: "after King Wu had defeated the Great City Shang, he made a ritual announcement to Heaven saying, 'Let me reside in this central territory, and from here govern the people.'"[17] The language of these early accounts is suggestive, but still somewhat vague, so that discussion of its precise import must wait until we have looked more closely at the earliest texts that explicitly mention the techniques involved in laying out high-value structures.

A number of poems from the *Book of Odes* (*c*. tenth to seventh centuries BCE) also celebrate the elaborate undertaking of city building in both Shang and Zhou times. These events are commemorated in numerous early Zhou

---

[15] Trans. Karlgren (1950b, 47, trans. modified). Arthur Wright (1977, 35 ff.) provides an account of the founding of the early Zhou capitals.
[16] Karlgren (1950b, 50).
[17] For a photograph of the vessel and a rubbing of the inscription, see Chang et al. (2005, 183).

texts, though perhaps none more memorably than in the *Odes.* "Wen Wang's Renown" (*Wen Wang you sheng,* Mao 244) in the "Greater Elegantiae" (*Da ya*) section, is a paean to the founding rulers of the dynasty and recounts the events:

King Wen was renowned, he made great his fame; he sought their [sc. the people's] tranquility; he saw his work achieved; King Wen was splendid.

When King Wen received the Mandate, he had these martial achievements; after having attacked Chong, he raised the city at Feng; King Wen was splendid.

The wall he built was moated, the city Feng matched it; he did not alter his intentions, mindful of his predecessors he was filial; the royal ruler was splendid.

The King's works were brilliant, the walls of Feng were where [the principals of] the Four Quarters came together; the royal ruler was their pillar; the royal ruler was splendid.

The River Feng flowed to the east, being a vestige of Yu; here was where [the principals of] the Four Quarters came together; the August King was a true sovereign; the August King was splendid.

The capital Hao had its *biyong*-moat; from west, east, south, north, there were none who gave a thought to not submitting; the August King was splendid.

The King himself examined the oracle on the siting of Haojing; the turtle settled it, and King Wu accomplished it; King Wu was splendid . . . [18]

In the Feng River there are *qi*-reeds; did King Wen not strive? He handed down his plans to his descendants, in order to make tranquil and assist his son, King Wu [who] was splendid.[19]

The ode "Spirit Terrace" (*Ling tai,* Mao 242), quoted in Part One above, provides additional details:

He laid out and commenced the Spirit Terrace, he aligned and delimited it; the people worked at it, in less than a day they achieved it.

He planned and commenced it without urging them on; but the people diligently came [to work].[20]

Notice here the reappearance of the same terminology that appears in the *Book of Documents* in conjunction with the building of Chengzhou: the site is determined by divination, whereupon the new city (or ritual complex) is "aligned" (*jing,* i.e. laid out with cords), and again "aligned and delimited"

---

[18] *Zhai* here is not "to reside," but refers to the process of "settling" on a location. Hwang (1996, 227).

[19] Trans. Karlgren (1950b, 199, trans. modified).

[20] Ibid., 197 (trans. modified). I follow Hwang (1996, 234–5) in taking *yu yue bi yong* in the final stanzas to refer to the music emanating from the ritual performance in the Luminous Hall (*Mingtang*).

(*jing zhi ying zhi*) prior to construction.[21] Fundamentally, *jing* refers to the warp threads which, when tied on the loom, establish the structure of the woven goods. Here, in the context of city building, *jing* and *ying* refer to the use of stretched cords to establish the orientation, alignment, and *enceinte* of the pounded-earth footings for the walls and buildings to come.[22] There will be much more to say below about the concrete meaning of these terms, but it is worth pointing out here that the same terms derived from the delimiting of physical space are also used figuratively in the broader sense of "rectifying and encompassing" the kingdom as a whole and even in cosmogony. So, for example, in the ode "What Plant Is Not Yellow" (*He cao bu huang*, Mao 234) we read, "What plant is not yellow; what day do we not march; what man is not going [to help in] *aligning and encompassing* [*jing ying*] the Four Quarters?"[23] Centuries later, in the *Huainanzi* chapter "Quintessential Spirit" (*Jing shen*), speaking of the emergence of *yin* and *yang*, the text says, "two spirits sprang from the obscurity, *aligning Heaven and delimiting Earth.*"

The early Zhou ode "Jiang Han" (Mao 262) again records how during the reign of King Cheng (r. 1042?–1021) the Duke of Shao led a pacification campaign to the Huai River area to suppress restive erstwhile allies of Shang:

The Jiang and the Han rivers were surging, the warriors formed a rushing flood; they rectified and encompassed [*jing ying*] [the regions of] the Four Quarters, reporting the achievement to the king; [the regions of] the Four Quarters were pacified, the king's state began to be settled; then there was no strife and the king's mind was at peace.[24]

In this ode, which reproduces remarkably closely the form and content of early commemorative bronze inscriptions, the theme is the pacifying and rendering exploitable of the vast Huai territory not previously under royal Zhou control. Deploying the "order and delimit" figure to describe the political process of incorporation of new territories into the kingdom – pacification, establishment of boundaries and divisions, promulgation of the royal order, initiation of taxation – is particularly apt, since it recapitulates on a macro level the functional use of the binome *jing ying*, to delineate "consecrated" space on formerly profane ground.

---

[21] The parallel with the Egyptian ritual known as the "stretching of the cord" ceremony (Miranda, Belmonte, and Molinero, 2008) and the (*c*. fifth-century BCE) Hindu Vedic text *Sulva Sutra* (Ritual Book on Ropes) is striking. For a mathematical approach to the analysis of ancient methods of "squaring space" in ancient contexts, see Ranieri (1997, 209–44). Arthur Wright (1977, 37) discusses the ritualized activities in connection with the founding of capitals.

[22] This is precisely the terminology used, for example, in Han Dynasty sources dealing with the layout of Chang'an. Wei Hong's (fl. *c*.25 CE) *Han jiu yi* states of the walls of Chang'an, "the 'warp and the woof' were each 15 *li* long"; *Han jiu yi*, 2.14. Xu Shen (*c*.55–*c*.149 CE) defines *ying* as *za ju* "circumscribe and dwell in," which Duan Yucai glosses as "*za ju* means 'to encircle and dwell in.'"

[23] Trans. Karlgren (1950a, 184, trans. modified).   [24] Ibid., 233 (trans. modified).

A detailed celebration of settlement founding is the ode "Silk Floss" (*Mian*, Mao 237), from the "Greater Elegantiae," this time a paean to Venerable Ancestor Tan Fu (fl. *c*.1150 BCE), the illustrious ancestor responsible for settling the Zhou people in the Plains of Zhou (Zhouyuan) at the foot of Mt. Qi:

> And so he called the Master of Works, he called the Master of Multitudes,
> he made them build houses, their plumb lines were straight;
> they lashed the boards and thus erected the building frames;
> they made the temple straight.
> In long rows they collected it [sc. the earth for the buildings],
> in great crowds they measured it out, they pounded it, [the walls] rising high;
> they scraped and [repeated=] went over them again, [so they became] solid;
> one hundred *du* [measures] of walls all rose, the [rhythmic drums] could not
>   keep pace.
> And so he raised the outer gate, the outer gate was high;
> he raised the principal gate, the principal gate was grand;
> he raised the grand earth-altar, from which the great armies marched.[25]

Here we have a vivid description of massed labor gangs under the direction of their foremen, erecting wooden forms that were then lashed together to contain the pounded earth. As foundations and walls rise, a plumb line is applied to keep the walls vertical, and basket brigades maintain a constant supply of fill, all to the rhythmic cadence of drums. What techniques of siting or orientation in the landscape might be involved?

"When *Ding* Had Just Culminated"

The answer to this question is provided by the ode "When *Ding* Had Just Culminated" (*Ding zhi fang zhong*, Mao 50),[26] in the "Airs of Yong" (*Yong feng*) section of the *Odes*. The theme of this poem is the correctness with which Duke Wen of Wei carried out the rebuilding of his destroyed capital in 658 BCE, and it speaks to the rectitude and uprightness of the duke's character:[27]

---

[25] Ibid., 190. The *she* or altar mound dedicated to the spirits of the earth (somewhat comparable to the Roman *lares*) is the only other sacred structure specifically mentioned as being constructed at the founding of a capital. By definition, a city with a *she* and an ancestral temple is a capital, *du*. Wright (1977, 39).

[26] In what follows, given the occurrence of several homophones, *Ding* is capitalized and in italics whenever it denotes the asterism.

[27] The ode celebrates the restoration of Duke Wen of Wey at Chuqiu in 658 BCE after Wey had been destroyed by an invasion of the *Di*, "barbarians." Resettlement of Duke Wen and the remnant population of Wey was brought about through the intervention of the Hegemon, Duke Huan of Qi, who drove the *Di* out of the area. Legge (1972, 128). Much later, in his "Way of the Lord" (*Jundao*) chapter, Xunzi chooses a strikingly parallel metaphor to characterize the "superior man or lord" (*junzi*): "The lord is the gnomon; if the gnomon is straight, the shadow is straight." See Wang Xianqian (1975, Vol. 3, 154); Goldin (2005, 45).

When [the asterism] *Ding* just culminated, he started work on the Chu Palace;

when he had measured it by the Sun, he started work on the Chu Hall . . .

He ascended the tell in order to look out over Chu; he looked out over Chu and Tang; he measured hills and mounds by their shadow;

He descended and inspected the mulberry grounds; the turtle-shell oracle was auspicious, all through it was truly good . . . [28]

Once again, pride of place among the activities described is the correct orientation in the landscape of the main temple.[29] Commentators agree that the time to commence work, "just when *Ding* was centered" (*ding zhi fang zhong*) refers to the moment when the asterism *Ding* transited the local meridian (culminated) due south in the evening. The next, parallel line alludes to a collateral astronomical technique: "when he measured it by the Sun" (*kui zhi yi ri*) he started work on the hall.[30] The Chu Hall is then surveyed (*wang*) from atop a hill, from which the location (presumably on the south-facing slope of an elevation) is gauged by means of shadows. The "Mao Commentary" on the first of these lines reads as follows:

Ding is *Yingshi* [lodge 13 *Align-the-Hall*, in Pegasus]. "Just culminated" [*fang zhong*] [means] at dusk to rectify [*zheng*] the four directions. "Chu Palace" is the hall at Chuqiu . . . "to gauge" [*kui*] is to measure – to measure sunrise and sunset in order to ascertain east and west. Watching to the south [he] observes *Ding*, and to the north he aligns on the Pole, in order to rectify [*zheng*] south and north. A hall [*shi*] is the same as a palace [*gong*].[31]

Since the Sun and shadows are explicitly mentioned, we can guess at the method: a gnomon was used to measure the Sun's shadow at sunrise and sunset to lay out a proper east–west line. The ode "Gong Liu" in the "Greater Elegantiae" (Mao 250) is a paean to the distant predynastic Zhou ancestor Gong

---

[28] Trans. Karlgren (1950b, 33, trans. modified). Wheatley (1971, 461), mistranslates *fang zhong*, "just centered [on the meridian]," as "attained the zenith." Arthur Wright (1977, 38) merely alludes to the passage in passing.

[29] Indeed, Hwang Ming-chorng (1996, 346) adduces evidence to show that the structure being built is, in fact, the highly symbolic ritual center of the state, the Luminous Hall or *Mingtang*: "[The Preface to the *Odes*] says this poem is about the rebuilding of the capital city of the state of Wei in 658 BCE. There is no dispute about this interpretation . . . however, we believe what is described in this poem is the entire process of rebuilding a sacred architecture – a *ming-tang*."

[30] The second mention of a "raised ceremonial hall" here is *tang* (**daŋ*), but it was probably the **-aŋ* end-rhyme that dictated the substitution of *tang* for *shi*, "chamber." The parallelism suggests we are dealing with the same structure, rather than a distinction between "temple" and "residence," as some commentators would have it. For the rhyme scheme of the ode "*Ding* zhi fang zhong," see Baxter (1992, 601); the Old Chinese phonetic reconstructions given here and below are from *Thesaurus Linguae Sericae* (*TLS*), http://tls.uni-hd.de/home_en.lasso.

[31] Ruan Yuan (1970, Vol. 1, 59).

Liu, who led the Zhou people to settle at Bin. There Gong Liu is said to have "measured by the shadow, making use of the ridge, inspecting its north-facing and south-facing slopes." As we saw in Chapter 1, recent discoveries at Taosi and Shang Dynasty references to the solstice show that the shadow method could certainly have been in use among Gong Liu's people in the mid to late second millennium BCE.[32] Had the shadow method been used to align palace foundations, tombs, and walls, however, they would still be accurately aligned today. That they are not shows that a different technique was used; stars cast no shadows, of course.

Where, then, did Mao Heng's comment come from – "watching to the south [he] observed *Ding*, and to the north he aligned on the Pole, in order to rectify south and north"? Zheng Xuan (127–200 CE) expands on the Mao commentary:

"Chu Palace" means the ancestral temple. When asterism *Ding* culminates on the meridian at dusk it is upright, so that one can use *Ying*[-*shi* = Pegasus] to construct temples and halls. That is why it is called *Align-the-Hall*. "When *Ding* culminates on the meridian at dusk [and is upright]" means that at the time of Lesser Snow[33] its [*Ding*'s] shape and [that of] *Dongbi* [*Eastern Wall*34] join in rectifying the [directions of the] Four Quarters.[35]

Zheng Xuan has added some important clarification, which will become more meaningful shortly, but for now it is apparent that in this deceptively straightforward, terse comment there is much more of a technical nature going on than meets the eye.

Now, we know in some detail from a passage in the "Artificers' Manual" (*Kao gong ji*) section of the canonical *Rites of Zhou* what such alignment procedures entailed in the late Warring States period:

When the builders construct a walled city, they [use] water [to level] the ground with a plumb line [?]; they set up a gnomon using a plumb line; then they watch and use the shadow to inscribe a circle; then they note the [length of the] shadow at sunrise and sunset. By day they align them with the Sun's noon shadow and by night they check against the Pole star, in order to rectify daybreak and dusk [i.e. east and west, north and south].[36]

The technique being used to level the ground is obscure, as the sentence appears garbled, but is of no great consequence here. The use of suspended cords is

---

[32] Li and Sun (2010); Needham and Wang (1959, 293).

[33] "Lesser Snow" is the fortnightly solar period that begins thirty days before winter solstice.

[34] *Eastern Wall*, comprising the two stars Alpheratz and Algenib in Pegasus, is lodge 14 immediately to the east of *Align-the-Hall*, about which more below.

[35] *Shisanjing zhushu*, Vol. 1, 59.

[36] Wheatley (1971, 426); Biot (1851, Vol. 1, 555). Needham and Wang (1959, 231) mistranslate "check against the Pole star" as "investigated the culminations of stars." Cf. also Wright (1977, 47).

said by some commentators to refer to a procedure whereby cords of identical length are stretched to the ground in all directions from the top of the gnomon to ensure its perpendicularity. More likely, the reference is to a plumb line. A cord is, however, used to measure when the morning and afternoon shadows are of equal length, whereupon these are connected by means of an inscribed circle with the gnomon at its center. In this way it is not strictly necessary to measure the Sun's shadow precisely at sunrise or sunset, and one can avoid error arising from different horizon elevations of the rising and setting Sun. Connecting the two points where the shadows intersect the arc will establish a true east–west line. Bisecting that line and connecting its midpoint with the gnomon will establish a true north–south line. When the midday gnomon shadow is superimposed on this line it is noon and the Sun is on the local meridian. A similar geometrical procedure is prescribed in *Huainanzi*, "Heavenly Patterns" (*Tian wen*) chapter:

To fix [the directions of] sunrise and sunset, first set up a gnomon in the east. Grasp another gnomon and retreat ten paces from the former gnomon, sighting on the sun as it first leaves the northern edge. Exactly at sunset plant a further gnomon in the east, sighting in conjunction with the western gnomon on the setting sun as it is about to enter the northern edge. Thus [the positions are] fixed. The midpoint of the two eastern gnomons and the western gnomon define [a line running] due east–west.[37]

Bringing down the Pole

Notice, however, that in the "Artificer's Record," just as in the ode "*Ding zhi fang zhong*" above, it was deemed inadequate merely to use the Sun to lay out a true east–west line; it was essential to bring down a true north–south line using the Pole. The question is, how was this done? We have seen from commentary on this ode and elsewhere that the asterism *Ding* is none other than *Yingshi* in Pegasus, the thirteenth lodge.[38] We saw above that *Yingshi* means "lay out/align the hall," and we have seen the *ying* of *Yingshi* used in

[37] Trans. Major (1993, 272). Note in particular Major's comments on the paraphrases "morning and evening twilight" (*zhao xi*) for east–west, and especially *can wang*: "the usage *canwang* refers to situations where two objects define a line of sight running in the direction of a third. It is interesting therefore that Karlgren notes early uses of *can* as both 'a triad' and 'straight.' Pronounced *shen*, it is, of course, also used for the line of three stars that forms Orion's belt. The fifty-seventh definition of the Mohist Canon has *zhi can ye*. As the basic meaning of *zhi* is, of course, 'straight,' the 'alignment' sense of *can* is confirmed."

[38] The third-century BCE glossary *Literary Expositor* (*Erya*), in the section "Heaven Explicated," says, "*Yingshi* is called *Ding*." Guo Pu (276–324 CE) comments, "*Ding* is *zheng* 'correct; straight.' In building temples and halls, all take *Yingshi*'s culmination [on the meridian] to be straight and true." *Shisanjing zhushu*, Vol. 2, 2609. The *Shuowen jiezi* glosses *zheng*: "*zheng* 'correct' is *shi* 'be right'; *shi* is *zhi* 'straight; square'; from *ri* 'Sun' and *zheng* 'correct.'" Duan Yucai (1735–1815) comments, "to take the Sun as correct; to pattern on the Sun. The meaning is compounded from 'Sun' and 'correct.'"

precisely this sense in the context of city building in both the *Book of Documents* and the *Book of Odes*. So the asterism's function is actually embodied in its name, *Align-the-Hall*, which had begun to supplant the name *Ding* by Han times. The two bright stars of *Align-the-Hall* (*Yingshi*) on the north and south are β and α Pegasi (Scheat and Markab), which form the western side of the prominent Square of Pegasus (in the West the square forms the trunk of the winged horse). Immediately to the east is *Eastern Wall* or *Dongbi*, comprising the two stars δ and γ Pegasi (Alpheratz and Algenib) in lodge fourteen.

Looking at Figure 4.2, one can see why the nearly parallel orientation of *Eastern Wall* also implicates *Dongbi* in the alignment function ascribed to *Yingshi*, as suggested by Zheng Xuan in his comments above: "[*Ding*'s] shape and [that of] *Dongbi* [*Eastern Wall*] combine in rectifying the [directions of the] Four Quarters." The accuracy of this statement is proven by careful examination of the earliest depiction of the entire scheme of twenty-eight lodges on the lid of the famous lacquer hamper (Figure 4.3) from the 433 BCE tomb of the Marquis Yi of Zeng. Rather than representing the asterisms using the "dots and bars" method so familiar from later star maps, the relative positions of the lodges are shown using the actual names of the asterisms written in seal script. All are recognizable, despite a few minor stroke variations and homophone substitutions. But in the case of *Yingshi* and *Dongbi*, in lieu of those later conventional names we find *Xiying* and *Dongying* or "*West Aligner*" and "*East Aligner*."[39] So here we have not only confirmation of our postulated parallelism between the two lodges, but also an indication that the separation of a single ancient asterism *Ding* into two had already occurred well before the system of twenty-eight lodges was systematized in the late fifth century BCE.[40] The astronomy of the Yi minority of southwest China, much of which pre-dates the Warring States period, underscores the antiquity of the perception of lodges thirteen and fourteen as a single square asterism. In the Yi scheme of lunar

---

[39] Qiu (1979, 25–32); Wang Jianmin et al. (1979, 40–5). Luo Qikun (1991, 242) points out that by the time the Qin dynasty *Shuihudi* "day books" (*rishu*) were composed in the late third century BCE, the later conventional names *Yingshi* and *Dongbi* had already appeared. *Yunmeng Shuihudi Qinmu bianxiezu* (1981, Plate 151, slip nos 987, 988).

[40] Zhu Kezhen (1979, 234, 237) already pointed out in 1944 that *Yingshi* and *Dongbi* originally comprised a single asterism – *Ding*. As Zhu also points out, an earlier scheme of twenty-seven lodges would have provided a better approximation of the Moon's nightly progress against the backdrop of stars, since the lunar sidereal period is 27.32 days, versus the synodic period (phase to phase, or lunation) of 29.53 days. But twenty-seven is not divisible by four (seasons) and so was incommensurate with the numerological systematization that became dominant in the Warring States period. Luo Qikun (1991, 242) concurs that before the Spring and Autumn period (i.e. before 722 BCE) the two were a single asterism called *Ding*. See also H. J. Zhong et al. (1983, esp. 11–12); Waley (1937, 164); Needham and Wang (1959, 244). None of the above mentions the polar-alignment function of asterism *Ding*. Chang Zhengguang (1989a, 177) merely alludes to the possibility in passing, as does Li Qin (1991, 36).

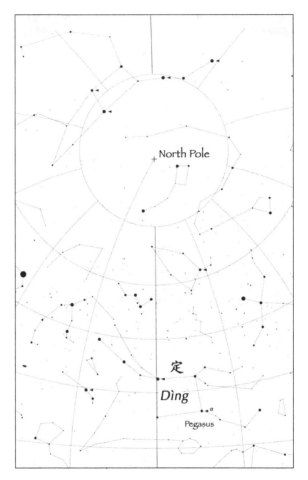

Figure 4.2 The orientation of the two sides of the Square of Pegasus (known as *Ding*) formed by the two pairs of stars in *Align-the-Hall* (right) and *Eastern Wall* (left) (Starry Night Pro 6.4.3).

lodges, the name of the asterism corresponding to *Yingshi* and *Dongbi* means "a big square of four stars" and is none other than *Ding*.[41]

In *Erya*, "Heaven Explicated," where *Align-the-Hall* (*Yingshi*) is identified as *Ding*, the text goes on to say, "*Juzi*'s mouth is *Yingshi* and *Dongbi*." *Juzi* is the late Warring States and Han designation for the astral space corresponding to chronogram *hai* (roughly Aqr–Psc), which includes lunar lodges *Yingshi* and *Dongbi*. "Mouth" (*kou*) in the gloss, of course, refers to the shape of the

---

[41] Chen Jiujin, Lu Yang, and Liu Yaohan (1984, 95–6).

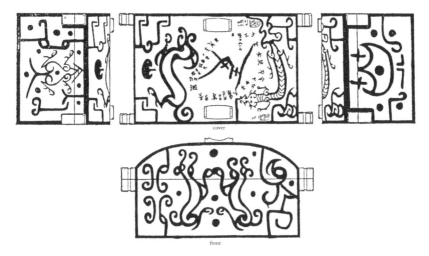

Figure 4.3 Imagery on the lacquer hamper from the tomb of the Marquis Yi of Zeng with the twenty-eight lodges depicted on the cover. *Yingshi* and *Dongbi*, here called *Xiying* and *Dongying* are just in front of the dragon's snout. Adapted from Loewe and Shaughnessy (1999, 820, Figure 12.1), © Cambridge University Press, reproduced by permission.

Great Square of Pegasus. Guo Pu's comment reads, "The four sides of asterisms *Yingshi* and *Dongbi* resemble a mouth [*kou* 口], hence the name."[42] The *Erya* gloss is no doubt an allusion to the record in *Zuozhuan*, Duke Gong, thirtieth year (543 BCE), where Jupiter's position is identified as "Jupiter was in the mouth of *Juzi*," and the commentary equates the "mouth" with *Yingshi* and *Dongbi*. The memory of the early history of *Yingshi–Dongbi* as a single asterism still lingered into the Tang Dynasty (618–907), since Chapter 61 of the eighth-century *Prognostication Classic of the Kaiyuan Reign Period* (*Kaiyuan zhanjing*) preserves a comment by the Eastern Han astronomer Xi Meng (fl. *c.*100 CE) stating that "the two stars of *Yingshi* are the west wall, and together with the two stars of *Dongbi* they combine to form a foursome, their shape an open square resembling a mouth."[43]

---

[42] *Shisanjing zhushu*, Vol. 2, 2609.
[43] John C. Didier (2009, Vol. 2, 169) alleges that "Pankenier's idea to project *ding* 丁 astronomically could only have originated in these very chapters," referring to his work *In and Outside the Square*. In fact, however, the ancient Mesopotamians, Greeks, Chinese, and even the Lakota of North America all perceived the four stars in Pegasus as forming a square and all represented it as such – they were not inspired by Didier either. Unbeknownst to Didier, the identification of asterism *Ding* with the Great Square of Pegasus is made explicit in standard early texts and commentaries. The Sumerians and Akkadians identified it as the square celestial "field," *ikû*. Even in the traditional astronomy of China's southwest the Yi minority name for the asterism means "big square of stars." Only after a formal presentation of mine of an early version of the

(a)

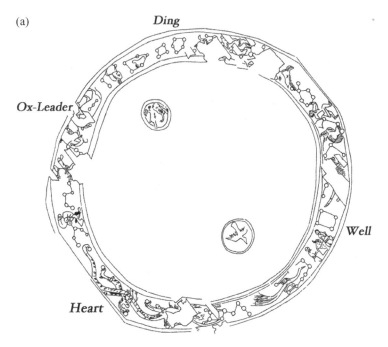

Figure 4.4 (a) The circular band of lunar lodges depicted on the ceiling of a Han Dynasty tomb at Xi'an Jiaotong University. *Ding* is in the ten o'clock position. The fifth star forming the peak of the square is a curious idiosyncracy. Redrawn from Li Qin (1991, 25). (b) Fu Xi and Nü Wa on the silk sky banner from Gaochang, Turfan (*c.*500 CE). *Ding* is below the elbow of Nü Wa on the left. Adapted from *Zhongguo shehui kexueyuan kaogu yanjiusuo* (1980, 58, Figure 56, and 9, Plate 7), reproduced by permission.

In addition to the textual evidence, archaeological discoveries also graphically confirm that *Yingshi* and *Dongbi* were once a single asterism. In 1987 a late Western Han tomb was discovered on the campus of Jiaotong University in Xi'an, its walls brilliantly decorated with colorful wall paintings. On the ceiling of the tomb archaeologists discovered a map of the sky depicting Sun and Moon surrounded by the twenty-eight lodges arranged in a circular band, their constituent stars shown in the familiar "dots and bars" style (Figure 4.4a).

present chapter in February 2009 did Didier add an appendix to his *In and Outside the Square* vigorously attacking my identification of *Ding* with the Square of Pegasus and stem-sign *ding*. Prior to my public presentation Didier had never discussed asterism *Ding* or Pegasus, had never mentioned the ode "*Ding* zhi fang zhong," and had never broached the phonological and semantic roots of the **t–ng* word family (see Chapter 5, "Finding inspiration in the sky," below), topics I first raised in Ban Dawei (2008), about whose existence Didier also displayed no awareness: see www.pankenier.net/Refute_Didier-Mair.html.

(b)

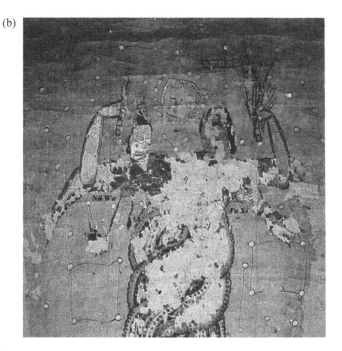

Figure 4.4 (*cont.*)

This is the earliest star map of its kind so far discovered and detailed study of the asterisms it depicts reveals that the painter was not only a talented artist, but also well familiar with the five heavenly palaces, the twenty-eight lodges, their correct orientation, and the astral lore associated with them.[44] In other words, unlike cruder such ceiling paintings previously discovered which are more evocative than representational, the painter of the Xi'an murals displays expert knowledge of the stars and astral lore.[45] Study of the arrangement and configuration of the lodges indicates that the small square with a peaked roof on the upper left in about the ten o'clock position should be *Ding*, i.e. the four stars of *Align-the-Hall* and *Eastern Wall*.

We have seen ample evidence that it was still well understood in Han times that *Align-the-Hall* and *Eastern Wall* were anciently a single asterism. Perhaps more surprising, however, is that the square of *Ding* is still depicted as such in the iconic representation of cosmic demiurges Fu Xi and Nü Wa on the large silk painting of the sky from Gaochang (Turfan), dating from about 500 CE

[44] Tseng (2001, 202 ff.).
[45] Compare the amateurish rendering of the same motif in the Five Dynasties (907–60) tomb of Wang Chuzhi; Rawson (2000, 182).

(Figure 4.4b).[46] It too was discovered in a tomb, having originally hung from the ceiling of the burial chamber above the coffin. Originally, therefore, *Yingshi* and *Dongbi* were not perceived as two parallel lines comprising two stars each, as they conventionally came to be represented later, but were linked in the form of a square, representing not a geometric abstraction but, in fact, a building foundation.

Yingshi–Dongbi as "Heavenly Temple"

In the fourth-century BCE narrative history *Discourses of the States*, "Discourses of Zhou," there is a passage describing how anciently at the beginning of spring the Grand Scribe-Astrologer *Tai shi*, based on observations of natural phenomena, would declare the time was right for the initiation of farming activity.[47] One of the crucial seasonal indications was the culmination at dawn of the asterism known as *Farmer's Auspice* (*nong xiang*), a variant name for lunar lodge *Chamber*, in Scorpius: "when *Farmer's Auspice* culminates at dawn and the Sun and Moon reach Celestial Temple, the soil's *materia vitalis* [*qi*] emerges in pulsations."[48] Wei Zhao's (204–73 CE) comment identifies *Farmer's Auspice* as lodge *Chamber*, and he glosses "culminates at dawn" as follows: "that is to say, on the day Beginning of Spring [*li chun*] [i.e. forty-five days after the solstice], at dawn [*Chamber*] is centered on the meridian." Wei Zhao then explains the reference to the Sun and Moon: "The *Celestial Temple* is *Yingshi* [*Align-the-Hall*]. In the first month of spring, the Sun and Moon are both in *Yingshi*." A check of the astronomical circumstances for February in the late Warring States period shows this to be precisely correct; at dawn when lodge *Chamber* is due south on the meridian, the Sun is in *Align-the-Hall*, which is, of course, invisible as a result. The passage in *Discourses of the States* is the earliest to identify *Align-the-Hall* as the *Celestial Temple*.[49]

In *The Grand Scribe's Records*, "Treatise on the Celestial Offices" (*Tianguan shu*), the first alternative appellation given for *Yingshi* is *Pure Temple* (*qing miao*),[50] an allusion to the Grand Ancestral Temple for sacrifices to the Zhou founder, King Wen, whose august solemnity is celebrated in the "Hymns in Praise of Zhou" (*Zhou song*) section of the *Book of Odes*. Furthermore, in

---

[46] Zhongguo shehui kexueyuan kaogu yanjiusuo (1980, 9, Plate 7); Sun and Kistemaker (1997, opposite 112).

[47] *Guoyu*, 1.7a.

[48] Cf Wu Jiabi (2001, 90–4). For the Shang Dynasty precedent, see Jao Tsung-yi (1998, 35). *Farmer's Auspice* was introduced above, in the section "The celestial Dragon as astral indicator, *chen*."

[49] *Taiping yulan*, 20.5b, quotes a gloss on the passage by Tang Gu (d. 225 CE) affirming this reading.

[50] *Shiji*, 27,1309. Another variant designation given here by Sima Qian is "Separate Palace" (*li gong*), which later star maps generally show as an appendage connected to the northernmost star of *Yingshi*. Sun and Kistemaker (1997, 73, 158).

glossing *Yingshi* in *The Grand Scribe's Records*, "Treatise on Harmonics and the Pitchpipes" (*Lü shu*), Sima Zhen's (fl. eighth century) *Suoyin* commentary says, "*Yingshi* – the asterism *Ding*. At *Ding*'s culmination one can build halls, hence it is called *Align-the-Hall*. Its stars have the shape of a hall, so the 'Treatise on the Celestial Offices' makes it the [astral] governor of temples."[51] Clearly, then, what we have in the two asterisms *Yingshi* and *Dongbi* are the eastern and western walls of an archetypal celestial temple, so that prior to the complete elaboration of the scheme of twenty-eight lodges sometime after the late fifth century BCE the Zhou Chinese also recognized the square in Pegasus as a single asterism.[52]

The alignment function of asterism *Ding*

We saw above that the eastern and western walls of *Ding* are typically depicted as two more or less parallel lines comprising two stars apiece. But they share an even more important characteristic in common. If one looks carefully at the longitudinal meridian lines in the chart in Figure 4.2, which shows asterism *Ding* due south in 658 BCE (the date of the event alluded to in the ode), one immediately sees that the eastern and western walls of the *Celestial Temple* align perfectly with the meridians converging on the Pole over seventy degrees to the north. Actually, the alignment of *Dongbi*, the eastern wall (i.e. Algenib to Alpheratz), is even more precise than that of *Yingshi* on the west. Calculation shows that in 1105 BCE the alignment of the *Eastern Wall* would have been exact, the deviation from true north being a minuscule 0.001′ of arc.[53] At most, throughout the Shang and Zhou periods the deviation from true polar alignment of the two stars of *Eastern Wall* never exceeded about two minutes of arc, roughly one-fifteenth the width of a forefinger held at arm's length. In the case of the west wall, *Align-the-Hall*, the deviation did not exceed about thirteen minutes of arc, six times as much, but still less than half the Moon's apparent diameter. Such tiny discrepancies would certainly have been overwhelmed by measurement errors elsewhere in the sighting process, so that by early Zhou at the latest the Chinese certainly possessed a technique capable of precisely locating true north in the absence of a bright star at the Pole.[54]

---

[51] *Shiji*, 25.1244.

[52] It follows, of course, that there were not twenty-eight lodges when the usage *Ding* was current. Hence the epoch of *Ding* may also mark the *terminus post quem* for the elaboration of the twenty-eight lodges, making the full scheme an early Warring States period (fifth century–221 BCE) innovation.

[53] I am grateful to Dr. Salvo de Meis, Istituto Italiano per l'Africa e l'Oriente, Milan, Italy, for assistance with calculating the secular change in polar alignment of the Square of Pegasus during the first millennium BCE. For another perspective on the role of Pegasus going back as far as the Neolithic, see Galdieri and Ranieri (1995, 155–71).

[54] It is instructive that in the Han period a slightly different mechanism was employed to locate the Pole by means of the so-called "excursions" of the "North Pole Star" as it revolved around the

Now, the large distance from the *Celestial Temple* to the Pole means that it was not possible to observe the circumpolar sky in the north while facing the *Ding* asterism in the south. In addition, the diurnal and annual revolutions of Pegasus mean that the *Celestial Temple* would only have been useful for the purpose of aligning on the Pole at a particular time – on transiting the meridian in the evening when the two parallel sides of *Ding* would have been perpendicular to the horizon and pointing overhead through the zenith to the Pole at one's back. At other times of the year when *Ding* was either invisible or oriented at some oblique angle to the horizon, it could not have served the stated purpose. Here, then, we have the true meaning of Mao Heng's obscure comment above whose discussion we deferred: "*Ding* is *Yingshi*; just culminated [*fang zhong*] [means] at dusk to rectify [*zheng*] the [directions of the] Four Quarters . . . Watching to the south [he] observes *Ding*, and to the north he aligns on the Pole, in order to rectify (*zheng*) south and north."[55]

Investigation reveals that the optimal time for such alignment activity in late Shang and Western Zhou would have been in early evening in late autumn. The precise date would have varied depending on the time of observation. In mid-November, *Ding* would have been optimally positioned at nightfall right after sunset. Various sources confirm that it was in autumn after the end of the agricultural season that this activity would have taken place. In the *Spring and Autumn of Master Lü*, "Monthly Ordinances" (*Yue ling*), concerning the activities appropriate to mid-autumn, it says, "in this month one may construct inner and outer walls and build capitals and cities."[56] Moreover, the "Canon of Yao" in the *Book of Documents* states, "separately he charged He Zhong to reside in the West [at the place] called Dark Ravine to respectfully bid farewell to the setting Sun, and to arrange and regulate the achievements of the west. The night being of medium length and the asterism being *Ruins* [Xu; lodge 11, β Aqr], he thereby determined mid-autumn."[57]

pole. The star used for the purpose, β UMi or Kochab, lay some seven degrees from the Pole and so traveled in a small circle. Cullen (1996, 191). In *Zhou bi suan jing*, β UMi is the circumpolar star whose culmination is correlated with *Eastern Wall*, so the procedure described in *Zhou bi* is essentially the reflex of the method using *Eastern Wall*, with the procedure performed facing north rather than south. For an example of alignment of the royal city of Wangcheng in Eastern Zhou (*c*.550 BCE), see Von Falkenhausen (2006, 172) and www1.lit.edu.cn/heluo/Article_Show.asp?ArticleID=1892.

55 In the Qin bamboo slip texts from Fangmatan (269 BCE), in an astronomical context *zheng* carries the implication "to face south to make an observation." Zhong Shouhua (2005, 93); also Chang Zhengguang (1989a, 180).

56 Bi Yuan (1974, Vol. 7, 76). A striking parallel is the recently discovered Han Dynasty wall inscription, "Edict of Monthly Ordinances for the Four Seasons" (5 CE), from Xuanquanzhi near Dunhuang. The parallel passage reads, "[in the second month of autumn] it is permitted to build walls, construct cities and towns, and dig storage [cellars]. This means one may undertake large-scale earthworks." Sanft (2008–9, 184, trans. modified).

57 *Ruins* is the central lodge in the winter quadrant of the sky, about one hour west of *Yingshi*. In the "Canon of Yao" there are references to separate sacrifices to the rising and setting sun, scheduled

In *Discourses of the States*, "Discourses of Zhou," we read, "When *Ying Palace* is centered [on the meridian], the work of building may begin."[58] Similarly, in the *Zuozhuan* (Duke Zhuang, twenty-ninth year), it says,

As to the work of building, when the *Dragon* [asterism] appears [farming] labors end, for [the *Dragon*] alerts to the undertakings [to come]. When the *Fire Star* [Antares in Scorpius] appears, [the laborers] are put to work. *When Water culminates at dusk the foundations are built; at winter solstice [the work is] finished.*[59]

As we saw above in Cai Mo's account from *Zuozhuan* (Duke Zhao, twenty-ninth year), it was the hereditary office of Regulator of Water that had charge of the "care and feeding of dragons." It can hardly be coincidental that the lineage that was restored to office as dragon tamers in the Shang Dynasty (1554–1046 BCE) was the Shiwei. Like the great Shang official, Fu Yue, for whom a star was named, the Shiwei too were memorialized by elevation to celestial dignity, in their case as the name of the astral space corresponding to winter, *Water*, and significantly, *Ding* – the Square of Pegasus.[60]

---

to occur in mid-spring and mid-autumn respectively. Chang Zhengguang (1989a) argues that in the sectioning of the officer Xi He charged with the observances into four individuals (Xi Zhong, Xi Shu, He Zhong, He Shu) who are separately charged with the affairs of morning and evening, East and West, the text also preserves the memory of rites dedicated to the rising and setting sun on the equinox in spring and fall; that is, the same rising and setting sun sacrifices documented in the oracle bone inscriptions. For a reinterpretation of seemingly mythical locations in the "Canon of Yao," see Liu Qiyu (2004).

[58] *Guoyu*, 2.9b. Recall now how that passage begins (2.9a): "This is how the former kings, without employing rewards, were able to promulgate their virtue [*de*] widely throughout the world."

[59] Ruan Yuan (1970, Vol. 2, 1782). Cf. Legge (1972, Vol. 5, 116). Legge identifies "Water" in the passage as Mercury, the "watery" star, and translates *hun zheng* as "culminates at dusk." This is problematical for several reasons. Mercury is a denizen of dusk and dawn twilight and, consequently, hard to observe since its elongation from the Sun can never exceed twenty-nine degrees. It follows also that Mercury can never be observed to cross the meridian. If Legge is merely using "culminate" in the astronomical sense of "to reach the highest point above an observer's horizon," then in Mercury's case this occurs three to four times a year, with the planet most readily observable in evening twilight in spring and morning twilight in fall. This is impossible to reconcile with the plain meaning of the text. "*Water*" is instead an allusion to the Five Elemental Phases scheme in which the three northern or "watery" chronograms *chou, zi, hai* were correlated with lodges *Southern Dipper* through *Eastern Wall*, which "culminate" in autumn. Kong Yingda's (574–648 CE) *Zhengyi* commentary confirms this interpretation, as does the passage just cited from the "Discourses of Zhou."

[60] *Shiwei* is thus an alternate name for Jupiter station *Ju zi. Guoyu*, "Discourses of Jin," provides the same genealogy of dragon tamers/regulators of water as Cai Mo in Part One above. Shiwei was the ruling clan of the state of Peng in Shang time. Ying Shao (153–96), in his *Feng su tong*, "August Hegemons – Five Lords Protector" section, remarks, "coming to the decline of Yin [Shang], the Da Peng and Shiwei lineages continued the line, [exemplifying] the saying, 'when the Kingly Way was abandoned, the career of Hegemon arose.'" In *Guangya*, Chapter 8, "Zhang Ji zhuan" (*Siku quanshu* digital ed., 14a), *Yingshi* is specifically identified with *Shiwei*. Unlike Shang Dynasty paragons there is no trace of the Zhou founders in later stellar nomenclature, notwithstanding Sarah Allan's (2009) speculation that the Zhou founders were immortalized as stars.

Here, then, we have the full explanation of Mao Heng's commentary on *"Ding zhi fang zhong"* as well as Zheng Xuan's amplification – implicit in the reference to the culmination of *Ding* (*Guoyu*'s *"Ying Palace"*) is the identity of Pegasus as the prototypical *Celestial Temple* and the square temple's specialized function as an accurate guide to align structures on the Pole. The *Heavenly Temple*'s evening culmination precisely marks the season reserved for laying out walls for sacred structures whose construction is to follow.[61] It is worth recalling that two millennia earlier the Egyptian ceremony of "stretching the cord," in which a sighting instrument is employed to bring down the polar alignment of the two innermost stars of the bowl of the *Dipper*, employs essentially the same technique identified here using *Ding*. The main difference is that the Egyptian method was performed while facing the Pole and could presumably be used nightly, while the Chinese faced south and could only make use of *Ding*'s alignment on the Pole in autumn during the season for undertaking public works.

Ding "right and true"

For the heavens, the ruling principle is to be regular. For the earth, the ruling principle is to be level. For human beings the ruling principle is to be tranquil . . . If you can be regular and tranquil, only then can you be stable.[62]

In all references to these alignment procedures above the word *zheng* (**tɕieŋ **tjeŋs*) "right ~ straight ~ correct ~ true" characterizes both the observations integral to, and the outcome of, the specific alignment procedures.[63] Similarly, *jing* (**keŋ **keeŋ*), "arrange in order," and *ying* (**jieŋ **ɢʷleŋ*), "delimit ~ delineate ~ lay out," which we saw in both the "Announcement at Luo" and the "Announcement of Duke Shao" in the *Book of Documents*, refer to the large-scale arrangement of walled settlements or temple compounds, as well as the "Four Quarters" of the kingdom. All three words share a common rhyme, as well as a close semantic relationship – "be or make straight ~ make right ~ put in order." More than that, however, as the phonologically attuned reader will no doubt already have noticed, they share a rhyme with the name of the asterism actually used to accomplish the task, *Ding* (**deŋ **deeŋs*) "fix

---

[61] Chang Zhengguang (1989a, 180) paraphrased Mao Heng: "to accurately delineate the four cardinal directions, besides observing the solar shadow, in order to correctly establish north–south one must also use *Ding*'s alignment on the Pole." Chang draws important parallels between the use of solar shadows alluded to in *"Ding zhi fang zhong"* and Shang sacrifices to the Four Quarters (*si fang*) and to the rising and setting sun. Ibid., 177.

[62] *Guanzi*, "Nei ye," vii–viii; trans. Roth (1999, 58–61). For the significance of *zheng*, "to square up ~ center ~ align," in the *Nei ye*, see ibid., 109 ff.

[63] Reconstructed phonological profiles for Old Chinese (*) and Middle Chinese (**) are taken from *Thesaurus Linguae Sericae* (http://tls.uni-hd.de/home_en.lasso).

[in true orientation]."[64] When, therefore, the "Announcement of Duke Shao" represents King Cheng as saying, "when the Duke had fixed the site (*gong ji ding zhai*)," *ding zhai* may mean more than merely to "settle on" a location. It could actually connote making the layout conform to the celestial standard using the *Ding* asterism. In the "respectfully bestow the seasons on the people" (*jing shou ren shi*) passage in the "Canon of Yao" already mentioned, *zheng* (**tɕieŋ **tjeŋs*) is used in the sense of "determine correctly," for example in "to fix correctly mid-winter," *yi zheng zhong dong*. It seems clear that *Ding* and *zheng* are essentially the same word in such contexts, so that alignment procedures such as those described (i.e. meridian transit) draw on the root meaning of *zheng*, "be straight, erect, correct, right."[65]

To divine (*zhen* 貞) and cauldron (*ding* 鼎)

It is worth noting that the cognate "to divine" (*zhen *ʈieŋ **teŋ*) is used to introduce the charge to be confirmed by divination in the Shang oracle bone inscriptions. The meaning of this word is often rendered by means of functional circumlocutions such as "testing the proposition" ("inquire by divining," *bu wen*, according to the *Shuowen* dictionary), which does not capture the root meaning. The character "cauldron" (*ding *teŋ **teeŋ?*) is used interchangeably with "to divine" (*zhen *ʈieŋ **teŋ*) in the oracle bone divinations, in some instances even in the same sentence.[66] In Chapter 3, I discussed the politico-religious imperative behind the impulse to correctly align sacred precincts and structures on the Pole using the circumpolar stars or asterisms, arguing that "at a time when the Supernal Lord's intentions vis-à-vis the Shang state were very much a 'national security' concern, 'taking direction,' *literally*, from the ultimate source of supra-sensible power may well have called for a more direct polar method."[67]

---

[64] Takashima (1987, 408–9) gives the root meaning of this well-established word family as "fixed ~ stable ~ settled ~ secure ~ certain"; see also Boltz (1990, 1–8).

[65] Yu Xingwu (1996, Vol. 1, 790 ff.); Zhang Yujin (2004, 38–44). Relevant too in this context is Starostin's compilation of cognate words in Sino-Tibetan/Chinese, which confirm the postulated root meaning of *zheng* and its antiquity: 正 **tɕieŋ **tjeŋs* straight, correct; 貞 **ʈieŋ **teŋ* divination, straight and proper; Tibetan: *draŋ* straight; Burmese: *tanʔ* be straightward [*sic*], direct from one point to another; Kachin: *diŋ¹* be straight, rectilinear; Lushai: *diŋ* right, right-handed (cf. also *dīŋ* go straight or direct, as arrow); Lepcha: *diŋ* (1) to be erect, to be high, to be perpendicular; the highest point or degree; (2) to stand, to remain, to exist. Sergey S. Starostin, The Tower of Babel: Evolution of Human Language Project (1998–2003); http://starling.rinet.ru/intrab.php?lan=en – bases.

[66] Yu Xingwu, *Jiagu wenzi gulin*, Vol. 3, 2718 ff.; Karlgren (1964b, 834). Boltz (1990, 2) characterizes the graphic interchange between "cauldron" (*ding*) and "correct, right" (*zheng*) as "long settled."

[67] Following Stanley J. Tambiah (1990, 85), I prefer "supra-sensible" to "supernatural" since (quoting Lévy-Bruhl) "the 'savage' . . . made no demarcation between a domain of nature as

We have seen above how the method attested in "When *Ding* Had Just Culminated" could have produced accurate alignment on the Pole in Shang and Western Zhou. More important, the intentionality this technique discloses is surely revealing. Given the clear connection between the concrete meaning of "fix ~ true up ~ make straight ~ rectify" at the root of the *ding–zheng* series, to which *zhen–ding* "establish ~ fix ~ settle ~ confirm" also clearly belong, then one can discern in the use of *zhen–ding* in the oracle bones the analogous noetic impulse to "verify congruence with" the supra-sensible forces, which lies at the heart of the divination phenomenon.[68] As Joseph Needham so memorably remarked, the *Book of Changes* too exemplifies

the need for at least classing phenomena, and placing them in some sort of relation with one another, in order to conquer the ever-recurring fear and dread which must have weighed so terribly on early men. Any hypothesis which would take some of the terror out of disease and calamity there must at all costs be.[69]

"Making right," *ding–zheng*, the delimiting of physical space by aligning on the locus of celestial power has its psychological counterpart in the exercise in

opposed to the supernatural, it was better to describe his view of certain beings, forces or powers as 'supra-sensible' rather than as beliefs in "supernatural beings.'"

[68] See David S. Nivison's (1989, 125) gloss of "to divine ~ straight and proper" (*zhen*) in the oracle bone inscriptions, "officially verify the correctness of the results of a divination about"; and that of Paul Serruys: "If we try to explain 貞 of the introductory formula of divination, not in the light of *Shuowen* and later, rare usages, but of a majority of usages, we can only think of a verbal sense 'to test, to try out, to make true, correct' in the sense of 'find out the right (course of action)' parallel with 'tried, tested, reliable, correct, good' already points to a good morphological pair: *zhèng*/*tjings* 正 'to be right, correct' and *zhēn*/*trjing* 貞 'to test,'" quoted in Takashima and Serruys (2010, 23). See also Karlgren (1970, 76, 1752), glossing the line in the "Announcement at Luo": "We two men have both verified [*gong zhen*; sc. the reading of the oracles]." Zheng Xuan's gloss, quoted in *Shuowen jiezi zhu*, is "*zhen* as 'to ask' [means] to inquire as to correctness; first one must rectify it [i.e. confirm the propriety of the query] and then inquire [of the oracle] about it." It is not perfectly clear what Zheng Xuan means by "first one must rectify it." Presumably, he means one must first establish what is right before looking to the oracle for confirmation. This general point with regard to divination is made more than once in *Zuozhuan* (e.g. Duke Zhao, twelfth year: "the *Changes* may not be used to divine insincerely"; following Karlgren's (1970, 1428) gloss of *xian*). As Kidder Smith (1989, 440) observed, "it is nonsense to divine about something that is ritually incorrect." But one thing is clear, the equivalency between "to divine" *zhen* and "correct" *zheng* was uppermost in Zheng Xuan's mind. Paul R. Goldin (personal communication) points out that my assertion that *Ding* 定 and *zheng* 正 are interchangeable "is borne out by the fact that 正 (*tengs*, or *tjengs* in Baxter-Sagart) is merely the B-syllable version of 定 *ttengs*. People rage over what the whole A/B syllable distinction represents, both phonologically and semantically, but it's clear that such words must be cognate." For discussion of word families in Chinese and further examples of the *t-ng* series, see Karlgren (1933, 65, 70).

   In the specialized language of the "Inward Training," *Nei ye*, the practice of *zheng*, "aligning," means "adjusting or lining up something with an existing pattern or form," though in *Nei ye* the focus is the physical alignment of the body. Roth (1999, 109). This, of course, calls to mind that in the *Analects* it was said of the Sage, "if the mat was not straight, he did not sit." *The Confucian Analects*, 10/7. There may have been more to Confucius' fastidiousness than previously suspected.

[69] Needham (1969, 336).

mental space of establishing the correctness of a proposition through oracular communication *zhen* with those supra-sensible entities. Given the great antiquity of the preoccupation with orientation, reaching deep into the Neolithic, perhaps the early impulse to "make right" in terms of physical alignment (e.g. to face dwellings south to capture the light and warmth of the winter sun) over time was transformed into the psychological intentionality manifested in the context of divination and sacred alignment.

There may be more to the magical use of language in divination and the logic of orientation than first meets the eye, however. Divinatory "charges" to the turtle shell or bone in the form of "weighted" declarative statements were freighted with the psychological desire for a particular outcome – that is, the performance itself embodies the message.[70] This being the case, a rhetorical intentionality also enters the picture. As Kenneth Burke has shown, "the realistic use of addressed language to induce action in people became the magical use of addressed language to induce motion in things (things by nature alien to purely linguistic orders of motivation)." Therefore, Burke says, "magic is 'primitive rhetoric,' it is rooted in an essential function of language itself, a function that is wholly realistic, and is actually born anew; the use of language as a symbolic means of inducing cooperation in beings that by nature respond to symbols."[71] It follows that the "declarative" intentionality signaled by correctly aligning works and acts in time and space is likewise a symbolic means of inducing benevolence on the part of the supra-sensible forces of nature.

* * *

We have seen that the ancient Chinese were intensely interested in the circumpolar region, and especially in the mysterious Pole itself, from the very beginning of Chinese civilization. We have analyzed one method that could have been used throughout the first millennium BCE to precisely locate true north in the absence of a Pole Star. This method is well documented beginning in the early Western Zhou Dynasty, though it could potentially have been exploited centuries earlier. Therefore a final question to consider is: how early can we trace the focus on asterism *Ding*, the Square of Pegasus, and its special attributes? Earlier reference was made to the passage in the *Discourses of the States* where the calendrical function of the *Farmer's Auspice*, lodge *Chamber* in Scorpius, was mentioned: "when *Farmer's Auspice* is right on the meridian at dawn, the Sun and Moon reach the Celestial Temple." Wei Zhao's commentary provides a detailed explanation:

Farmer's Auspice is asterism *Chamber*. "Right" at dawn means to say, on the day Spring Begins, at dawn [*Chamber*] is on the meridian. [*Chamber*] is the harbinger of the agricultural season, so it is called *Farmer's Auspice* . . . the *Celestial Temple*

---

[70] Tambiah (1990, 54, 82); Kern (2009).   [71] Burke (1969, 42–3).

is *Yingshi* [*Align-the-Hall*]. In the first month of spring, Sun and Moon are both in *Align-the-Hall*.[72]

Not only is the astronomy in *Discourses of the States* technically correct, but the application of this calendrical maxim in Warring States times is confirmed by the inscription on a second lacquer box from the tomb of Marquis Yi of Zeng (*c*.433 BCE), the same tomb which yielded the famous lacquer hamper with a depiction of the entire scheme of twenty-eight lodges on its lid. This second box bears the inscription "it is *Chamber* to which the people sacrifice; when the syzygy [alt. "Sun's chronogram"] is at the [inter-cardinal] cord, the *Heavenly Quadriga* [*Chamber*] begins the year." *Heavenly Quadriga* is another name for the line of four stars resembling a chariot team that comprise lodge *Chamber*.[73] The meridian transit near dawn of *Farmer's Auspice* or *Heavenly Quadriga* would have been serviceable as a harbinger of the arrival of spring throughout the Xia, Shang, and Zhou dynasties. Not to be overlooked, of course, is the stipulation that the location of the Sun in *Ding* marks the first or "correct" *zheng* month of the year.[74]

In Chapter 1, we saw that ancient Chinese calendar priests from Taosi in Shanxi were observing the sunrise daily at least as early as 2100 BCE. Needless to say, they and their successors would also have paid attention to the sequence of asterisms rising in regular succession just prior to sunrise during each month of the year. They could not have failed to notice the correlation of the stars of the *Cerulean Dragon* constellation (with lunar lodge *Chamber* at its center) with the arrival of spring and the all-important initiation of farming activity. As we have seen, this is a principal reason why the *Dragon* came to figure so prominently in myth and iconography and as a seasonal indicator in folk lifeways, as well as in the line texts of hexagram *qian* in the *Book of Changes*. Ancient skywatchers awaiting sunrise in the twentieth century BCE could not have failed to notice still another dawn phenomenon. In Chapter 1, I drew attention to the earliest and most impressive instance of a rare phenomenon, the extraordinary cluster of all five visible planets that occurred in late February of 1953 BCE in the longitude of the star α Pegasi (Figure 4.5).[75] Now, α Pegasi is none other than Markab, the determinative star of lunar lodge *Align-the-Hall*, the *Celestial*

---

[72] *Guoyu*, 1.6b–7a.

[73] In a recent article discussing previously unidentified imagery on the front of the famous lodge hamper from the tomb, Wu Jiabi (2001, 90–4) identified the asterism depicted as lodge *Chamber* in its guise as *Heavenly Quadriga*. Wu further conjectured that the hamper and inscribed box were both originally used in the very *Farmer's Auspice* ritual alluded to in *Discourses of the States* and documented in the inscription on the second box. See also Hubeisheng bowuguan (1989).

[74] Here we have the rationale for Sima Qian's gloss on *Yingshi* in the "Treatise on the Heavenly Offices": "*Yingshi* governs the engendering of *yang qi* and gives birth to it." *Shiji* 25.1242 and 1244, n. 2.

[75] Pankenier (1983–85).

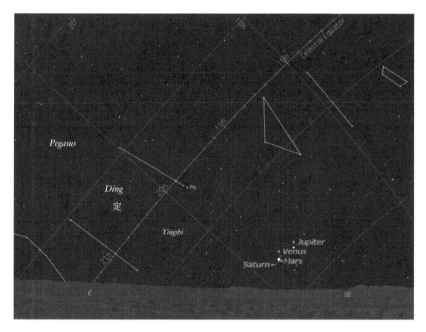

Figure 4.5 The cluster of the Five Planets in *Yingshi* at dawn on February 26, 1953 BCE (Mars is obscured by the disk of Venus and Mercury cannot be seen at this resolution). Markab or α Pegasi is near the center (Starry Night Pro 6.4.3).

Temple. Clearly, we have here a persuasive explanation for why the ancients' attention might have been powerfully drawn to asterism *Ding* as early as the twentieth century BCE. This sanctioning by means of a spectacular celestial phenomenon, together with *Ding*'s unique polar alignment, probably explains that asterism's later function as the standard in architectural, calendrical, and ritual contexts.

Conclusion

We have marshaled evidence for why the ancients' attention would have been powerfully drawn to the square asterism *Ding* as early as the twentieth century BCE. This "sanctification" by means of a spectacular celestial phenomenon, together with *Ding*'s precise polar alignment, could well explain its later function as the standard in architectural, calendrical, and ritual contexts. It also accounts for the iconic linkage of axial orientation on the Pole of terrestrial temples and palaces as materialized in the Luminous Hall, *Mingtang*, symbol of celestially derived power and legitimacy. Indeed, it is probably no coincidence that it is the early second millennium BCE that marks the transition from organic growth and unplanned towns to planned capital cities, square in shape, with street grid, rectilinear palace buildings with rows of columns, multiple courtyards, and pounded-earth foundations all arranged around a longitudinal axis (compare Figures 1.1 and 3.2b).

Summing up insights into the layout of Erlitou deriving from the excavations that disclosed those characteristics, archaeologist Xu Hong concluded,

Before Erlitou, even at exceptionally large-scale agglomerations like Xiangfen, Taosi and Xinmi, Xinzhai, without exception the construction of walls follows the lay of the land and does not attempt a square shape. To date, no [pre-Erlitou] palace enclosure has been found to exhibit a group of orderly, pounded-earth foundations with a regularly shaped, square wall surrounding them. If we compare the large-scale walled cities of Taosi and Xinzhai with the pattern of the agglomeration at the Erlitou site, a vast change has occurred, while comparison with the Shang cities at Zhengzhou and Yanshi, as well as later Chinese capitals, shows even greater continuity. Accordingly, Erlitou is the earliest capital discovered to date which can be acknowledged to clearly display planning, one which is in the direct line of succession to later Chinese capitals in planning and construction.[76]

Such planning implies a prefiguring of spatial arrangement informed by theoretical reasoning, and in the case of early Chinese civilization, that theory was cosmological. As Eliade noted, axial alignment is fundamentally the manifestation of a cosmological worldview. As Merlin Donald has observed, "The earliest evidence of an elementary form of theory formation is found in ancient astronomy. Astronomical knowledge, like writing, was a powerful device of social control." Further,

Astronomy was probably the earliest example of widespread, socially important theoretic development in human history. Astronomical observations and predictions could not have been achieved without some form of external storage of data and could not have been modeled without some form of computation. By combining a simple token system of counting with various analog measurement and computational devices, humans were able to improve their mental models of time and space in significant ways, while using these calculations to run their growing agricultural society. Analog measurement devices were thus interwoven with the evolution of the theoretic process. The resulting visual "models" reflected the state of theory, as they were a direct product of the theorizing

[76] Xu Hong (2004); Thorp (2006, 29). As yet there is no archaeological evidence conclusively identifying Erlitou with the Xia Dynasty. However, Li Liu and Xingcan Chen (2003, 148) clearly establish the existence of "a complex political and economic system centered at Erlitou, whose operation required a level of administration well beyond that of a chiefdom society." Kwang-chih Chang (1999, 73) concluded that "present evidence suggests that there indeed was a Xia dynasty."

process. Theory had not yet become as reflective and detached as it later would; but the symbolic modeling of a larger universe had begun.[77]

In early Bronze Age China, then, along with urbanization and state formation we find clear evidence for a cultural nexus of celestially inspired cosmic religion, calendrical science, and writing as crucial formative elements of emergent Chinese civilization.

Sometime during the late second millennium BCE it was realized that the Great Square of Pegasus ~ *Ding* offered an accurate method of locating celestial north, which could supplant an increasingly inaccurate method of alignment using the circumpolar stars. Exactly when this occurred is unclear though the *Ding* method was ritually prescribed by early Zhou.

It remains a conundrum why the ancient Chinese did not take full advantage of the potential for high precision that their methods afforded. No large-scale structures have so far been found to be aligned on true north with the precision approaching that of the Egyptian pyramids until the early imperial period (Chapter 11, "Cosmic capitals"). In view of the high level of technological prowess exhibited in other respects (e.g. metallurgy, ceramics, construction, mensuration), and the evident fixation on north–south axial alignment, we can rule out a lackadaisical approach to orientation and measurement. The misalignments, if that is what they are, are generally too large to be attributable to measurement error. Other factors must have been at play, since in the earliest period, as we have seen, consistent alignment east or west of true north appears to be systemic and radical changes appear to be associated with political transitions.[78]

A strikingly obvious case of such a transition is the fully ninety-degree shift in the orientation of burials that occurred when the state of Qin assumed control over the former Zhou homeland in the eighth century BCE, but the

---

[77] Donald (1991, 335, 340). Erik A. Havelock (1987, 44), too, speaking of the same theoretical modeling of the larger universe in the Greek context, characterized it this way: "By an act of cosmic projection, they translated the human mind into the cosmos . . . It was left to Parmenides clearly to grasp the truth that the dimensions of this mind lie in the human thought processes."

[78] It is a curious fact that pre-Columbian buildings of Central Mexico generally display a north–south axial alignment slightly east of astronomical north strikingly similar to the Shang case. In the Mesoamerican context, however, the emphasis was mainly on equinox sunrise, and as Aveni and Gibbs (1976, 516) show, "by varying the elevation of the observer relative to the observed event, the position of equinox sunrise can be shifted horizontally to match many of the orientations . . . (especially those in the range of 0° to 10° E of N)." The authors also emphasize (ibid., 515) "the importance of considering dimensions in any study of the astronomical relations of building orientations." Unfortunately, in the Chinese Bronze Age we have no monumental stone structures of known dimensions to work with but only tomb alignments and the remains of walls and rammed-earth platforms. Even in the case of Central Mexico the authors' extensive survey of structures led them to conclude that "no single explanation, astronomical or otherwise, can be advanced to explain the peculiar orientation of all pre-Columbian buildings in Central Mexico." Ibid., 517.

transition from Xia to Shang is also distinctive. From the early Zhou on, the picture becomes surprisingly inconsistent. We saw above (Figure 3.2c) how the predynastic Zhou palace unearthed at Fengchu village in the central Wei valley mimics the Erlitou alignment west of north. On the other hand, the walls of the early Western Zhou administrative capital of Chengzhou near present-day Luoyang appear consistent with the Shang preference for alignment slightly east of north. Complicating the picture still further is that site plans show that the axis of the large palace (5) in King Wen's predynastic Western Zhou capital of Haojing was aligned about twenty-four degrees east of north (magnetic, ~20.5° true). In this case, however, it may be that the east facade was actually oriented to face sunrise on the winter solstice at an azimuth of about 119°, a more satisfactory explanation than an inexplicably large twenty-four-degree error.[79]

Thus it is clear that cultural or political factors must have influenced what is observed on the ground. Regional variation, too, may have played an important role; since the Neolithic, eastward orientation seems to have been preferred among east coast cultures and westward orientation in the western hinterland. The situation from the Western Zhou through the Spring and Autumn period is complex, certainly compared to the consistency exhibited during several centuries of Shang hegemony. It would not be surprising if, during the Zhou, high-value structures were oriented differently depending on their occupational use or ritual function, such that, for example, polar alignment dominated in the case of ancestral shrines or cosmologically symbolic edifices like the Luminous Hall. Until a complete survey of alignments is undertaken and early capitals are reliably mapped and surveyed we can only speculate.

---

[79] Shaanxisheng kaogu yanjiusuo (1995, Figure 3). Only an accurate survey of the local horizon to the east will settle the question.

5   Astral revelation and the origins of writing

> Heaven suspends images, to manifest the propitious and the inauspicious, and the Sage makes of himself their semblance . . . Anciently, in ruling all under Heaven, Paoxi looked up to observe the images in Heaven, and looked down to observe the patterns of Earth
>
> *Appended Commentary* to the *Book of Changes, Xici zhuan*

It has long been thought that the *tiangan* and *di zhi* or "heavenly stems" and "earthly branches" may provide a clue to the origins of the Chinese writing system.[1] Indeed, it is probable that the ten stems and twelve branches are the most archaic remnant of a very early stage of written Chinese. Even though some appear originally to have had concrete referents or to bear a resemblance to Shang graphs whose meaning is known, the one and only application of the binary stem–branch combinations is as ordinals and, uniquely in the case of the ten stems, as cultic appellations for the royal ancestors. As E.G. Pulleyblank remarked,

The curious thing about these twenty-two signs is that neither the graphs nor the names attached to them have any separate meaning. Their meaning is simply the order in which they occur in the series to which they belong. It is true that a few of the characters are also used to write other homophonous words, but these are a small minority and such words have no apparent relation to the cyclical signs as such.[2]

These unique characteristics suggest that by the Yinxu period in late Shang (thirteenth–mid eleventh centuries BCE) the semantic origins of these cyclical signs were obscure. Indeed, if traditional historiography is any guide the ten stems were already being used as (posthumous?) royal appellations by the rulers of Xia who preceded the Shang. This would mean that their invention pre-dates by several hundred years the first appearance of the oracle bone script in the archaeological record.

It has been suggested that the origin of the stems and branches may be traced to their use in the late Shang ancestral cult, but this is a minority opinion.[3] As

---

[1] Boltz (2011, 73, n. 41).   [2] Pulleyblank (1991, 39–80); Kryukov (1986, 107–13).
[3] Postgate et al. (1995, 463). In William G. Boltz's (1999, 108) view, "we cannot assume writing to have arisen in an exclusively religious context." David N. Keightley (1989, 197) has pointedly noted that "it is a mere accident of archaeological preservation and discovery that the earliest

we shall see, the calendrical use of the two sets of cyclical signs is considerably more archaic and may have originated in a pre-Shang culture. Moreover, it is difficult to understand why, given an imperative to devise ordinal designations for the deceased ancestors, signs lacking separate meanings like the ten stems would have been adopted unless they possessed a special significance, either by virtue of their very archaism, or because of their supposed numinous origins, or because of a connection with temporal power and authority, like the calendar.[4] Arbitrariness in the initial choice of signs to represent numbers is well documented (e.g. in cuneiform) and illustrates the essential independence of writing in being able to represent ideas directly – as it were, semasiographically or ideographically. Writing was obviously not initially a "graphic echo of speech." As Merlin Donald says, "writing was, after all, an attempt to represent the message visually, not the sounds associated with a narrative version of the same message."[5]

The conventional interpretation of the ten stems as denoting the mythical solar progenitors of royal Shang clans elides the question of their origin as written signs. There is still no scholarly consensus on how the Shang kings' temple names were chosen or why they were selectively adopted, though perhaps the original ordinal significance of the signs was being invoked in some way, even if the information thus encoded is now obscure. It is also true that some stems were thought more auspicious than others, which was certainly the case later.[6]

corpus of Chinese writing, the Shang oracle bone inscriptions, was religious in nature; one cannot, on this basis, argue that early Chinese writing was developed to communicate with dead ancestors." After a critical review of the issue Robert W. Bagley (2004, 226) also concluded, "The idea that writing in China was confined to the ritual context in which we first encounter it, though firmly embedded in the literature, has no basis"; but see Boltz (2011, 68). Speculation that the emergence of writing in China came about through diffusion from western Asia is equally unfounded. Houston (2004b, 233); citing Boltz (1986; 1994; 1999).

[4] On this point, David N. Keightley (2000, 51) says, "The Shang ritualists were . . . certainly calendar, day, and sun watchers, whose temporal and jurisdictional concerns were sanctified by profound religious assumptions."

[5] Donald (1991, 294); Samson (1994); Harbsmeier (1998, 33, 40). Moreover, it is obvious that the oracle bone script was a mature writing system which had long since made what philosopher David Deutsch (2011, 125 ff.) calls the "jump to universality."

[6] Keightley (2000, 33). Kwang-chih Chang's (1980, 169) analysis ruled out the possibility that the heavenly stems in posthumous royal appellations were assigned on the basis of birth or death dates, because the sequence of posthumous temple names is anything but random. Instead, Chang proposed (ibid., 172) that "the Shang royal lineages were organized into ten ritual units, named after the ten *gan*-signs (day-signs). Kings were selected from various units and were named posthumously according to their day-sign units, which also regulated the rituals performed to them." David N. Keightley (2000, 35) has offered an alternative hypothesis. K. C. Chang's (1980, 169–70) tabulation of 1,295 bronze inscriptions with ancestral names containing heavenly stems showed that the even-numbered stems (*yi, ding, ji, xin, gui*) were far and away the most preferred, and of these the first two outstripped the others in frequency by a wide margin. Li Xueqin (1989, 4–11) provides conclusive evidence of this preference of *rou* (even) stem days over *gang* (odd) stem days from Shang all the way through Han. This feature may be almost as old as the stems themselves.

On ideographs

My use of the disputed term "ideograph" is informed by Merlin Donald's discussion of the reading ability of the congenitally deaf and the neuropsychological evidence for the independence of neural pathways for visual and phonological reading. As Donald points out, "The more important issues are (1) does an alphabetic writing system really *depend* on the link to speech for its expressive power, and (2) is an alphabetic system necessarily any less 'visual' than an ideographic system?"[a] Moreover, "this assumption, that alphabetic writing consists only of phonetic cues, appears to be the chief source of difficulty. It is not a solid assumption; an alphabetically written word may be a phonogram, but it can also simultaneously serve as an ideogram or logogram. Frequently used words, in particular, are recognized so rapidly that there appears to be no time to perform grapheme-to-phoneme mapping; highly trained speed-readers can take in whole phrases and short sentences as fast as single words. Alphabetic reading thus utilizes rapid, direct links between visual words and their conceptual referents . . . Unlike ideographic writing, however, a parallel phonetic path would allow a reader to simultaneously reconstruct an accurate spoken version of the same message . . . This is more than speculation; it is the only possible interpretation that can be placed on the reading skills of the congenitally deaf. Most deaf readers have no phonetic training, and many have no skill at signing, yet their reading abilities are considerable."[b] Walter Ong makes a similar point about ideographs: "the meaning is a concept not directly represented by the picture but established by code."[c] David N. Keightley says much the same in reference to certain elite Neolithic art motifs: "such art functioned like Chinese characters: you had to know the conventional code before you could read the meaning."[d]

More recent brain research strongly reinforces the point: "until recently, researchers who study reading abilities focused mainly on Western alphabets. English and 218 other languages, from Alsatian to Zulu, share variations of the same Latin character set. But that set is only one of 60 writing systems used among the world's remaining 6,912 spoken languages. Even so, those studies convinced many scientists and educators that the brain's response to the written word, regardless of the language, is universal. The new research suggests they're wrong. The schooling required to read English or Chinese may fine-tune neural circuits in distinctive ways. To learn the ABCs of English, we essentially harness our listening skills to a phonetic code. To become literate in Chinese, however, we must make much heavier use of memory, motor-control and visual-perception circuits located toward the front of the brain . . . 'we have to recognize that the writing system in China is different, the demands on the brain are different and the characteristics of dyslexia are different' . . . 'once you have different writing systems in place . . . they may reinforce the perceptual and cognitive trends that preceded the invention of writing. They may go hand in glove.'"[e]

My contention is not that written Chinese is a purely ideographic medium, but rather that an exclusive focus on glottography may distort our understanding of how early writing first emerged and which cultural factors contributed to the path-dependency that prevented a transition in China to a more economical writing system. As Malcolm D. Hyman has argued, "the typological model of pure glottographic or non-glottographic systems is unhelpful. Rather, we may conceive of writing as a *system of systems* . . . Additional subsystems are present in texts that deal with specialized domains . . . well-known examples from numerous cultures are calendars and schemes for recording astronomical observations . . . we should perhaps better view glottography not as a *type* of writing, but as a *function* of one *subsystem* within the system of writing."[f]

---

[a] Donald (1991, 298). On this point, see also John S. Robertson (2004, 18, 19): "visual interpretation of signs is more immediate than acoustic interpretation . . . Writing includes both the holistic characteristics of visual perception, and at the same time, without contradiction, the sequential character of auditory perception."

[b] Donald (1991, 300, original emphasis). For the evidence that "speech itself retains a strong visual component," see Corballis (2011, 68).

[c] Ong (2002, 85).

[d] Keightley (1996, 84).

[e] Hotz (2008); see especially Tan et al. (2005).

[f] Hyman (2006, 245–6, original emphasis). From the perspective of the science of informatics, Paul Beyton-Davies (2007, 313, original emphasis) argues, "Part of the reason for considering the Inca quipu is to demonstrate the need to broaden the view of 'writing' as simply *the graphic representation of the spoken word*" (for discussion of the *khipu/quipu*, see below). Houston (2004b, 226) also considers the concept of "bundling": "Notations of essentially different character do not replace one another in sequence but often occur together, in bundles or as separate marks."

Calendrical notation as a cultural imperative

In Chapter 1 we considered the design and function of the Taosi solar observing terrace and altar. Astronomical analysis has shown that this structure would have permitted its users to devise a calendar based on the movements of the rising Sun along the horizon as it oscillated between the solstitial extremes. Such a horizon calendar could have yielded an approximation of the length of the solar year to within a week or so. This degree of attention paid to the solar year clearly shows that Taosi's designers were interested in correlating the tropical year with the lunar months, which effort eventuated in a luni-solar calendar of the type that had become conventional by late Shang, as demonstrated by the Shang use of intercalary thirteenth months to maintain synchronization between solar and lunar cycles. Some have suggested, based on the number of viewing apertures at Taosi, that the observing platform represents an early effort to create a fortnightly scheme of twenty-four solar nodes (*jieqi*) like that familiar from much later times, though this suggestion is problematical, since the twelve observing slots are evenly spaced.

It is immediately apparent from the design and layout of the viewing platform and graduated lacquer shadow-rule that those early calendar priests (and priest-astronomers they most certainly would have been) must have possessed a number of crucial concepts and related specialized terminology.[7] Whether in the construction or use of the facility, those concepts and terms would have included Sun, Moon, stars, shadow, horizon, rise, set, direction, location, north, south, east, west, height, aperture, curve, straight line, to measure, units of measure (inch, foot, yard), length, color, shadow, and so forth. More apropos the present discussion, their technical vocabulary must also have included temporal concepts like day, night, month, dawn, twilight, midday, sunrise/sunset, moonrise/moonset, solstice, and possibly even achronychal (evening) and heliacal rising in reference to the stars. The implications of this are momentous. It was Otto Neugebauer who called astronomy the first of the exact sciences, and as Merlin Donald has said,

The earliest evidence of an elementary form of theory formation is found in ancient astronomy. Astronomical knowledge, like writing, was a powerful device of social control; the measurement of time in terms of astronomical cycles was probably the ultimate controlling activity in early agricultural societies, setting dates for planting, harvesting, storage, and distribution of grain for religious observations, as well as a number of cyclical social functions . . . Quite early in the history of visuographic symbolism, analog devices were invented that served both a measurement and predictive function in representing time. These devices eventually allowed humans to track celestial events, construct accurate calendars, and keep time on a daily basis.[8]

---

[7] Cf. Campion (2012, 62).

[8] Donald (1991, 335). Joseph Needham cites the great Song polymath Su Song (1020–1101), who in a memorial to the emperor in 1092 said, "We know that if the calendar is always well adjusted the work of the farmers will keep perfect time with the seasons, and so (apart from special calamities) bring the best harvests." Quoted in Needham and Wang (1959, 361); see also Kalinowski (2004, 87–8). Tony Aveni (2002, 91) noted that "the *calends*, from which we derive our word calendar, were the first days of the month, traditionally the time when religious leaders called people together to outline the festal and sacred days to be kept during that month." Aveni also wonders whether the two distinctively different calendars in *Hesiod*, in the "Works" and "Days," might not indicate the use of dual "fiscal and seasonal" calendars employed for different purposes, reflecting "a dialogue continually going on between nature and culture" (ibid., 43, 44). For his part, David Brown (2006, 113) questioned whether "the motivation behind the appearance of astronomy in Mesopotamia and elsewhere was calendar control" and argues that systematic reconciliation of the months and seasons began "only after the accurate prediction of celestial phenomena had begun," and that the "motivation behind prediction came from the advantages it gave to the celestial diviner." It is not clear to me how one distinguishes observation for the purpose of keeping track of time from prediction – they are two sides of the same coin. As

Doubts are often raised about whether a calendar would have been required in an agricultural society whose farmers were thoroughly familiar with the seasonal indications (*wu hou*) and stars, like the signs recorded in the "Lesser Annuary of Xia" (*Xia xiao zheng*), the *Book of Odes*, and other later texts. Merlin Donald makes a strong case that "all early agricultural societies, out of necessity, had calendars based on astronomical science."[9] Two further arguments come to mind. First, the discovery of large numbers of stone knives in storage pits at Anyang (3,600 in a single pit in one instance) points to centralized control of farming in the Shang Dynasty, as Robert L. Thorp has noted.[10] Second, if, as David N. Keightley and Tony Aveni have suggested, two calendars, one liturgical (or fiscal) and one agricultural, were simultaneously in use, and if the months of the latter were named according to the farming activity, feast day, or seasonal signs in nature which defined that month (*wu hou*), it would have been crucial for rulers to keep the two calendars synchronized with each other and with the seasons in order to maintain "cosmic" legitimacy.[11] Clear evidence of just such a second calendar in the Shang will be presented below. Tony Aveni has pointed to some non-Chinese parallels:

The Athenian months were named after gods and festivals. In this the calendar differed from the Mesopotamian models that lie behind all Greek lunar calendars. In the Sumerian and Babylonian prototypes, for instance, the months were named after the main agricultural activity practiced in that month. Many Athenian festivals did have links with different stages of the agricultural cycle, such as festivals of planting or harvest. This perhaps added to the need to keep lunar and solar calendars roughly aligned, though this was not always achieved. The year of farmers, however, was not the primary focus of the calendar.[12]

Merlin Donald (1991, 340) points out, "for the most part, astronomical record keeping was similar to commercial record keeping and involved mainly the construction of lists of observations." This then led to "systematic and selective observation, and the collection, coding, and eventually the visual storage of data; the analysis of stored data for regularities and cohesive structures; and the formulation of predictions on the basis of these regularities" (ibid., 339). See also Bellah (2011, 274). Early calendars did not *begin* with "reconciliation" of lunations with the seasons, as Brown implies, but were the result of millennia of observation. It was important even in the Paleolithic for hunter-gatherers to know in advance when there would be sufficient moonlight to move about after dark, so that observation of the Moon's phases and keeping track of the passage of "sleeps" and the Moon's location among the stars had simultaneous calendrical and predictive motives. A hunting party certainly needed to keep track of how many days' walk they were from home and when to meet up with the tribe.

9 Donald (1991, 339).   10 Thorp (2006, 159).
11 Keightley (2000, 44). On the time dimension in particular, Keightley observes, "to the Shang diviner, time was as portentous as place and direction; observed, shaped, and regulated, time was, like space, an indispensable dimension of religious cosmology, an integral part of all religious observance and divinatory prognostication" (ibid., 17).
12 Aveni (2002, 91). As an example, in the Tofa language in Siberia, which now has fewer than thirty speakers, months are named as follows: *teshkileer ay* – roughly February, or "hunting animals on skis month," *ytalaar ay* – March, "hunting with dogs month," *eki tozaar ay* – April, "good birch-bark collecting month," *aynaar ay* – August, "digging edible lily bulbs

With the emergence of more complex societies, conformity with the natural rhythms became a structural premise of social organization: "There were certain verifiable events in the upper realm of the sky that happened regularly – again and again, year after year. The continuity to which these events pointed concerned not just seasonal time, but also social structure because society dovetailed with the cosmos."[13]

Stems and branches in the calendar

In what follows, I will put forward the hypothesis that the set of cyclical stem–branch signs was a mental tool initially devised in response to the conceptual demands outlined above, that their origin is crucially related to the origin of the calendar, and that it was calendrical astronomy that lent impetus to the invention of writing in China.[14]

Calendar tables from among the Shang inscriptional materials provide important insights. They are clearly not divination texts, nor do they all represent calligraphy practice.[15] In their orderly arrangement they clearly display a scribal facility with visuographic presentation of written information that is certainly not spontaneous, but conventional. The neat graphic arrangement or charting of written elements like this could not possibly represent an early stage of writing but testifies rather to representational and graphic usage born of established convention, deep internalization, and regular use.[16] In 1929 Guo Moruo

month," *chary eter ay* – October, "round up castrated male reindeer month." Though they are non-agriculturalists, the naming pattern in Tofa time-keeping illustrates the point; cf also ibid., 96.

[13] Lewis-Williams and Pearce (2005, 232).

[14] Calendrical notation and list making were early steps on the path. Stephen D. Houston (2004a, 11) makes the point that "writing is a sequence of step-like inventions" and that "most early script did not expand to fulfill every conceivable function – an anachronistic fallacy – but served, at least initially, very limited needs." See also Ong (2002, 82).

[15] An example of a practice inscription would be HJ 18946, on which essentially the same sequence of fewer than ten characters is repeated in five separate lines. A number of such tables of cyclical signs may be found following HJ 38044. For practice inscriptions as an index of literacy in late Shang, see Smith (2011). David N. Keightley (2000, 39) concluded that many examples such as those identified by Guo are, in fact, written calendars used for reference. In a similar context, Qiu Xigui (1996, 41, Figure 6; 2000, 62, Figure 6) cites the *Xiaochen Qiang* bone – the longest non-oracular Shang inscription so far discovered – one side of recorded events with the reverse displaying a table of cyclical signs, suggesting a connection between historical record and reference calendar.

[16] One need only compare the irregular arrangement of examples of early writing from Mesopotamia to recognize this. "Cuneiforms acquired grammatical conventions about the same time that they became partially phonetic. They also became linear: whereas the earliest scripts had been read in loosely clustered boxes of rectangles, later cuneiforms were turned around ninety degrees, and written from left to right in straight lines, starting to imitate the spoken order of words. The progression was thus from a primarily visual medium, inventing completely new representations like lists of numbers, to a medium which, increasingly, tried to map the narrative products of the language system." Donald (1991, 289). On this point, see also Ong

analyzed the two examples described below in his pioneering monograph on the origins of the cyclical signs, "Explicating the Cyclical Signs," *Shi zhigan* [*sic*].[17] Guo points out that examples which repeat only the first three ten-day weeks (*xun*) several times in succession are actually about as numerous as those that reproduce the whole series of sixty cyclical signs. Guo inferred that these thirty-day tables are an indication that the Shang months originally comprised three *xun* of thirty days, which means that every month would have begun with day *jiazi* (1) and ended with *guisi* (30). This is an entirely reasonable proposition, since alternation of long and short months must have appeared as a corrective some time after the invention of the twelve-month tropical calendar, when it was realized that twelve nominal months of thirty days are actually slightly more than five days longer than twelve lunations of 354 days.

The arrangement of some of the tabulations cited by Guo proves that they are calendars. In one (HJ 21783), the cyclical signs from *jiazi* (1) through *guihai* (60) are arranged in four registers, the first two registers together comprising twenty-nine days and the second two comprising thirty-one days. Furthermore, the distribution of the days among the four registers is 14–15–17–14, reproducing a count of days for two successive months, the first short and the second long, divided at the full moon. This arrangement could hardly be accidental, nor could this be intended as a tabulation of cyclical signs designed purely for reference or scribal practice, since the irregular layout and the month of thirty-one days are both highly unusual. Conclusively, however, in Figure 5.1a–b (HJ 24440) a scribe has again reproduced the sequences of stem–branch signs, but in this unique inscription the names of the months are supplied – "Month One Regular is called 'Eat Wheat' [*shi mai*]" and "Second month 'Father 秋.'"[18] In addition to showing that this table is indisputably a fragment of a calendar, the thirty days of two successive long months are enumerated using the cyclical signs one through sixty, with one fortnight per column. Since adequate space was available it is curious that the scribe felt no compunction in splitting *jisi* (5) at the bottom of column one and *gengxu* (47) at the bottom of column six, which appears to suggest that their pairing was less firm than 3,500 years of subsequent usage has led us to expect.

(2002, 99, 122); Goody (1977). In the case of Mesopotamia the progression Donald describes took half a millennium. There is no reason to presume that it proceeded any faster in China. See Boltz (1986, 424, 429).

[17] Guo (1982b). Many of Guo's philological analyses and his hypotheses concerning the astral correlates of the stems and branches and their supposed Babylonian origins have not stood the test of time; cf., e.g., Wang Ning (1997), Smith (2010–11).

[18] As is often the case in inscriptions, the oracle bone graph for *shi*, "eat," was left unfinished, with lateral strokes left undone, as also with numerous other graphs on this famous bone. Guo (1982a, 161); Yang (1992, 121). The reading of the graph in the name of the second month is not known. The transcription and graphic are Ken-ichi Takashima's, who interprets the meaning as "cut, mow" (e-mail communication of August 13, 2011).

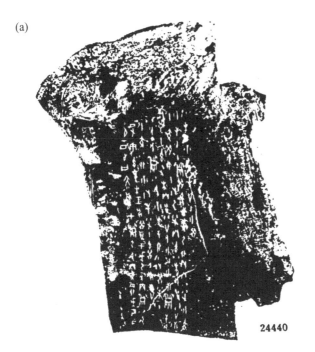

(a)

24440

(b) ⊞

月	巳	戊	丁	秋	壬	戌	己
一	庚	寅	亥	甲	寅	辛	未
正	午	己	戊	午	癸	亥	庚
日	辛	卯	子	乙	卯	壬	申
食	未	庚	己	未	甲	子	辛
麦	壬	辰	丑	丙	辰	癸	酉
甲	申	辛	庚	申	乙	丑	壬
子	癸	巳	寅	丁	巳	甲	戌
乙	酉	壬	辛	酉	丙	寅	癸
丑	甲	午	卯	戊	午	乙	亥
丙	戌	癸	壬	戌	丁	卯	
寅	乙	未	辰	己	未	丙	
丁	亥	甲	癸	亥	戊	辰	
卯	丙	申	巳	庚	申	丁	
戊	子	乙	二	子	己	巳	
辰	丁	酉	月	辛	酉	戊	
己	丑	丙	父	丑	庚	午	
		戌					

Figure 5.1 (a) Ink rubbing of oracle bone calendar and (b) transcription, HJ 24440. Adapted and retranscribed from Hu and Guo (1979–82).

Remarkably, the inscription records what must have been the conventional names for the first two months of the year, the first of which, "Eat Wheat" (*shi mai*), is corroborated by later textual evidence, e.g. from the "Monthly Ordinances" chapter of the *Lost Books of Zhou* (*Yi Zhou shu*).[19] Guo Moruo called this inscription "China's earliest calendar."[20] It certainly is the earliest discovered to date. It must reflect early calendrical usage, since the noteworthy features of this calendar do not bear much resemblance to the dating formulas of the actual Shang divination texts, which typically alternate twenty-nine- and thirty-day months and invariably enumerate the months rather than naming them.[21] Pondering the implications of these idiosyncratic tabulations, it seems appropriate to look into the reconstructed Old Chinese readings of the twenty-two signs to see what patterns might emerge from different arrangements. Table 5.1 reproduces the Old Chinese (OC) reconstructions of the cyclical signs. To the right of the OC reconstructions the rhymes are labeled: D, A, B, C, a, and X (X signifying no obvious rhyme with the other signs in the set or with each other).

Several features of the second set, the earthly branches, are immediately apparent; these occupy what one would expect to be the stressed, rhyming position when the binomial series was recited. First, apart from *wu* and *xu*, the other ten signs all share four rhymes, only about twelve percent of those available, one of which, "a," is in assonance with "A." Second, the "A" rhymes divide the twelve signs roughly in two. Third, remarkably, among the codas there are no labials, only a single nasal, and no velars (excluding the seemingly overrepresented, undetermined coda, $*$-q). Compare these features with those of the ten stems. Rhyme pairs are almost entirely lacking, but the stems display a full range of codas. The contrasting features of the twelve earthly branches are certainly eye-catching and appear prima facie to suggest that, by comparison with the ten stems, some deliberate process of selection must have

---

[19] The activities prescribed in the "Monthly Ordinances" for the first month of spring "when the Sun is in *Align-the-Hall* (*Yingshi* ~ Pegasus)" include the admonition to "eat wheat and mutton." Winter wheat is harvested in late spring, so some have argued on this basis that if the Shang month was named for the first fruits of that harvest, their first month should have fallen near the summer solstice. Yang (1992, 121). Conversely, the association of wheat and mutton with the first month of spring in the "Monthly Ordinances" seems incongruous. Another rare example of a named month in an inscription occurs in the *Chen Ni gui* from the late Spring and Autumn period (475 BCE), in which the eleventh month is called "Ice Month" (*bing yue*). Guo Moruo (1971, Vol. 3, 215b).

[20] Guo Moruo (1982a, 161); also Guo Moruo (1982b, 216); Chen Mengjia (1988, 219). Liu Xueshun (2009, 24–8) adduces HJ24440 as proof that the Shang were already using a prescriptive calendar.

[21] David N. Keightley (2000, 44) observed, "I suspect, in fact, that 'the start of the year' could have involved more than one kind of year. The Shang diviners might have pegged the first moon of their luni-solar calendar to the first lunation after the winter solstice, while the peasants might have tied their agricultural calendar to the observation of stars and constellations. It would have been the first, liturgical system, not the second, agricultural system, that gave rise to the numbered moons recorded in the divination inscriptions." See also Chen Mengjia (1956, 228–37); Liu Qiyu (2004, 47).

Table 5.1 *Old Chinese phonetic reconstructions of the stems & branches.*[22]

Stem		OC	Rhyme
甲	[八部]	*kkrap*	
乙	[十二部]	*qrik*	
丙	[十部]	*prang*	D
丁	[十一部]	*tteng*	
戊	[三部]	*mu-s*	
己	[一部]	*kə-q*	
庚	[十部]	*kkrang*	D
辛	[十三部]	*sing*	
壬	[七部]	*nəm*	
癸	[十五部]	*k^wij-q*	

Branch		OC	Rhyme
子	[一部]	*tsə-q*	A
丑	[三部]	*hnru-q*	B
寅	[十二部]	*lin*	C
卯	[三部]	*mmru-q*	B
辰	[十三部]	*dər*	a
巳	[一部]	*s-lə-q*	A
午	[五部]	*ngnga-q*	X
未	[十五部]	*mət-s*	a
申	[十二部]	*hlin*	C
酉	[三部]	*lu-q*	B
戌	[十二部]	*s-mit*	X
亥	[一部]	*ggə-q*	A

been operative at the time the twelve branches were created. In other words, the choice of rhymes and perhaps even the sequence of signs may not be random.[23]

The implications of this become apparent when we examine a thirty-day tabulation of the cyclical signs, considering only the rhyme of the second element in each binary combination:

First column:　　A B C, B a A, x a C, B
Second column:　x A ‖ A B C, B a A, x a
Third column:　　C, B x A ‖ A B C, B a A ...

---

[22] I am grateful to Paul R. Goldin, Wolfgang Behr, and David Prager Branner for comments and corrections with regard to Old Chinese rhyming and phonetics. The OC reconstruction given here (the "Baxter–Sagart" system, before its last revision), is that used by Behr and Gassmann: *-q* represents a final of as yet undetermined quality (notionally a glottal stop), doubled initials represent "type A" syllables.

[23] E. G. Pulleyblank's (1991) thesis about the deliberate selection of the twenty-two stems and branches to serve as phonograms as early as the second millennium BCE seems too self-conscious and sophisticated a linguistic analysis to impute to such an early stage in the development of Chinese writing.

Double vertical lines show where the sequence begins to repeat, so that here we have two and a half repetitions of a sequence of twelve ordinals, with alternation between two rhymes in the third position as in $- - C/ - - A/ - - C/ - - A$. Even without speculating about sources of uncertainty in the reconstructions, the features of the earthly branches suggest that a pattern such as this, even if based only on vocalic assonance or generic rhyming, may have played a role in the arrangement.

To this it will be objected that the recursive pattern is merely an artifact of the pairing of twelve branches with ten stems to produce the set of sixty signs, since a recursive pattern must necessarily emerge. This is true, of course, but one cannot ignore the stark contrast between the two sets of terms – the ten stems with their random selection of rhymes and codas and the twelve branches with their prominent rhymes and series of codas that conspicuously avoids labials and velars (except for the special case of coda *-q).[24] Some will also object that extrapolating Old Chinese reconstructions back more than a thousand years before the *Book of Odes* is a risky proposition, but while the criticism may apply to the precise details of the reconstructed pronunciation, phonetic change does follow more or less regular patterns, so that the same rules should apply to all members of a given set of words. Thus it is probable that, while the phonological complexion of the individual members of the two sets in the early to mid second millennium BCE may not have been exactly as represented, the fundamental contrast between the linguistic features of the two sets is unlikely to have changed much. A further objection might be that there is no unequivocal evidence of rhyming of any kind earlier than the Western Zhou bronze inscriptions. This is also true, but here I am not arguing for self-conscious use of rhyme as literary embellishment but merely as a simple device that may have been useful in remembering the repetitive sequence of binary cyclical signs whose recitation would naturally have tended to be rhythmic.[25] Even more significant than mnemonic functionality, perhaps, linguistic embodiment of the rhythm inherent in the passage of time may have been intentional. As Paul Ricoeur remarked in a discussion of the plurivocity of cyclical time,

We have to do here with a new important concept, one which should not be confused with that of ritual repetition. I mean the rhythmic structure of time, which relies on the succession of strong and weak intervals and the recurrence of the same patterns on the same occasions . . . A rhythmic sense of ordinary time may or may not be linked to rites of regeneration; it may or may not generate a sense of boredom, or of the meaninglessness of ordinary time. The notion of rhythm deserves to be held as a type of its own in the typology of sacred and profane temporality.[26]

---

[24] This latter feature was pointed out to me by Paul R. Goldin (email communication of 7 June 2008), who stressed the unlikelihood of this being a random occurrence. By contrast, fully half the ten stems have labial or velar finals, and nasals are well represented too.

[25] Ong (2002, 34, 40).   [26] Ricoeur (1985, 22).

Keeping in mind the likely calendrical origin of the ten stems, one might infer that the two series were created at different times. Initially the ten stems were invented to enumerate the days of the ten-day week in a repetitive cycle, and only later were they complemented by the twelve branches that may once have denoted the months.[27] Originally the days would have been named using just the ten stems, an arbitrary set of signs which perhaps named items of daily use easily recalled. But this meant that each stem had to repeat three times a month, once each week. At some point, possibly to help resolve ambiguity in scheduling ritual events, the series of twelve branches was paired with the ten stems in sequential fashion by matching successive branches with one of the original ten stems. Proceeding in this fashion for six ten-day weeks until the reappearance of the first pair, *jiazi*, produced the familiar series of sixty unique signs (in fact, only half the 120 possible combinations). But now each combination of signs would repeat six times a year, in different months sixty days apart, in contrast to thirty-six appearances spaced ten days apart for the unpaired stems. This meant, of course, that the number of unique combinations requiring memorization would have increased by a factor of six, so that at this point rhythmic repetition and rhyming might conceivably have been very helpful as an aid to memorization. The sequence of rhymes illustrated above, minimally – – C/ – – A/ – – C/ – – A, would repeat five times within the sequence of sixty cyclical signs, providing the basic rhythmic cues.

It seems, therefore, that the two sets of cyclical signs – stems and branches – may initially have been devised to respond to the conceptual and record-keeping demands of the calendar, so that the origins of the two are crucially related. It is likely that it was calendrical astronomy that lent impetus to the development of writing in China and prefigured its application to other forms of record keeping that emerged later, including the Shang divinations in which we see a mature written language fully formed.[28]

This is consistent with Merlin Donald's view:

The idea that writing produced scientific and technological development probably reverses the real order of things; it is just as likely that the invention of writing and other notational devices was driven by the conceptual needs of emerging theoretic culture. Writing, and graphic symbolism in general, arrived on the scene long after a number of major conceptual developments had already taken place. The historical

---

[27] The twelve "earthly branches" are generally thought to derive from a different source than the ten stems, about which there has been much speculation, including a putative origin in Babylonian astronomy (Guo Moruo) for which there is no persuasive evidence. For a stimulating new hypothesis, see Smith (2010–11).

[28] It appears that the same sequential process of evolution occurred in ancient Egypt: "The Egyptians viewed their moon god, Thoth . . . as the one in charge of counting and mathematical notation, the inventor of writing, and the one who set the dates of the festivals. Thus they believed that notation originated in counting celestial cycles and that it preceded writing." Barber and Barber (2004, 178).

order is important; if all theoretic development had followed writing, and especially if it had followed the invention of the alphabet, an argument could be made that it somehow depended on the latter. A long list of technological and protoscientific inventions preceded writing.[29]

Rhyme may have provided a crucial connection between orality and functional notation, linking the practical use of the two sets of signs and the idea of writing spoken words. That is, rhyming may have served as the notional stimulus prompting the realization that the sounds of individual spoken words could be attached to specific conventional graphic signs and thus serve as analogs of speech, in effect inventing a new medium – true writing. So far this is conjectural, of course, so it remains for us to establish if possible a direct connection between the early calendar, astronomy, and the inspiration leading to the invention of the cyclical signs.

Visuographic recording before writing

> By these knots they counted the successions of the times and when each Inca ruled, the children he had, if he was good or bad, valiant or cowardly, with whom he was married, what lands he conquered, the buildings he constructed, the service and riches he received, how many years he lived, where he died, what he was fond of; in sum, everything that books teach and show us was got from there. Martín de Murúa (1615)[30]

There is no denying that, whatever other cultic or ritual purpose Taosi might have served, the observing platform and lacquer shadow-rule were certainly analog devices for measuring and predicting time in the form of the Sun's progress along the horizon. We know from traditional accounts that other analog methods, likely inspired by the art of weaving, relied on knotting cords to record information. As the "Appended Commentary" to the *Book of Changes* says, "In high antiquity they knotted strings and brought order; the Sages of later generations switched to writing with inscribed graphs." In Warring States and Han accounts, Fu Xi is specifically identified as the inventor of this tool, along with the trigrams of the *Changes*. In *Zhuangzi*, "Qu qie" chapter, record keeping using knotted strings is attributed to the time of twelve prehistoric chiefdoms (including Fu Xi, Shen Nong, etc.), which preceded Yao, Shun, and Yu the Great.[31] No examples of such "tools of governance" have survived in

---

[29] Donald (1991, 333). Robert N. Bellah (2011, 273) makes the same point with more specificity: "early writing is clearly a significant step beyond painting in the amount of cognitive information that could be stored, but the unwieldy early writing systems and the limited number of people who could use them meant that they were precursors to, rather than full realization of, the possibilities of theoretic culture."

[30] Quoted in Brokaw (2003, 111).

[31] Lai (1972, 432). For the great antiquity of knotting strings to keep records in China, see Needham (1969, 100, 327, 556); Needham and Wang (1959, 69, 95). The Shang oracle bone form of *yue*,

Figure 5.2 (a) Inka *khipu* in the collection of the Archaeological Museum of Puruchuco, Lima. (b) Varieties of *khipu* knots with different meanings (photos DWP); compare Conklin (1982, 265, Figure 4). (c) Lumberjack's *khipu* from the Ryukyu Islands (early twentieth century) encoding type of tree to cut, length, diameter, quantity, etc. Adapted from Simon (1924, 666, Figure 10).

China proper – the invention of writing is too ancient for that – but analogous devices have appeared elsewhere in the form of the *khipu* for recording accounts, calendars, tribute, head tax, crop yields, etc., across the Pacific from the Marquesas to Hawaii, and famously among the Andean peoples. Figure 5.2c illustrates one of several Ryukyu *khipu* interpreted by their owners for Simon, and the kinds of information it contains.[32] The use of knotted fibers in the Ryukyu Islands, off China's east coast, survived into the early twentieth century, so that

"contract, pact," appears to show a hand tying a string. The later form of the character has a silk signific and in ancient times the word signified a form of contract in which duplicate sets of knotted strings, documents, or tallies were retained as proof of the agreement, one of which, according to the *Rites of Zhou*, was deposited with an officer called the supervisor of contracts (*si yue*).

[32] Simon (1924, 660). The use of knotted fibers in the Ryukyu Islands survived into the early twentieth century, so that Edmund Simon's 1924 study is an invaluable guide to the actual use of several kinds of *khipu* for recording and transmitting information there. For *khipu* use in Polynesia, see Handy (1923); Best (1921); Jacobsen (1983). According to an account by non-native observers, "the last time the *Kumulipo* [Hawaiian Creation Chant] was recited in its entirety for one of the Kings sometime during the 1800s . . . orators took turns in the recitation and . . . referred to the knotted string devices they held in their hands." Martha Noyes (private communication).

(b)

Figure 5.2 (*cont.*)

Edmund Simon's 1924 study is an invaluable guide to the actual use of several kinds of *khipu* for recording and transmitting information there.[33]

Figure 5.2a–b is an illustration of knots of several kinds used in Andean *khipu*, each conveying different information. Although in the Inka context the precise significance of each knot is lost, by means of their orientation, configuration, color, number, spacing, etc. they conveyed specific information about date, place, quantity, category, event, associated action, and the like.[34] It is now clear

---

[33] Confucius' famous adage springs to mind: "When the Rites are lost, seek them in the country-side" (*Li shi er qiu zhu ye*), attributed to Confucius by Ban Gu in the *Han shu*, "Bibliographical Monograph," *Yiwen zhi*, preface to "Zhuzi lue." In *Zuozhuan* (Duke Zhao, seventeenth year) the Sage is actually quoted as saying, "I have heard that, 'when the Son of Heaven misgoverns [lit. "loses his officials," *shi guan*], the knowledge [of good governance still] resides among the Four Barbarians.'"

[34] "*Inka*" and "*khipu*" are the preferred Aymara and Quechua spellings of these words. Other common spellings are "Inca" and "Quipu." "*Khipu*" (lit. "knot" in Quechua) has come to be used as a generic term for all information-recoding devices of this type. In the Ryukyu Islands the

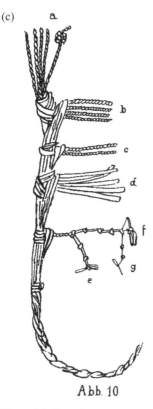

Abb. 10

Figure 5.2 (*cont.*)

that there are many varieties of *khipu*, and not all were accounting devices simply tallying numerical quantities. Galen Brokaw makes these crucial points with reference to our approach to understanding the semiotic capacity of such devices:

In anthropology, writing has been a benchmark used to measure the level or nature of human civilizations. Cultures have often been classified as those with (historic) and those without (pre-historic) writing. This scheme creates an opposition between writing and orality that has no room for alternative forms of representation. Faced with an other mentality, an other "literacy" based on a technology that establishes a different

aerial roots of banyan trees and other fibers were commonly used to make them, and the ancient Chinese term *jie sheng* was used to name the device (no doubt assigned by the Japanese). The following discussion is concerned with the actual emergence of writing, not as it was imagined to have happened by Warring States and Han authors. Lewis (1999, 199–208). For studies of the coding methods employed in the *khipu* and the practice of locally archiving duplicates of *khipu* submitted to the Inka capital, see Ascher and Ascher (1975); Conklin (1982); Urton (2003); Mann (2003).

relationship between medium and discourse, the European colonial episteme can only understand this other in terms of an already known, in-between category: mnemonics. Indeed, the most common analogue used to describe the *khipu* is the mnemonic rosary. An understanding of the *khipu* requires a deconstruction of this writing–(mnemonics)– orality opposition . . . there are several dimensions of the *khipu* that converge in such a way as to suggest that this medium of knotted, colored string is much more than a mnemonic device. Undeniably the *khipu* employed a set of highly complex conventions capable of encoding semasiographic or even phonographic information (cord configuration, numeric quantities, extra-numeric knots, colors and color patterns, ethnocategories, etc.), and these features would only have developed in response to a semiotic desire or need. Abundant testimony from the colonial period claims that the *khipu* was a narrative encoding device and transcriptions of *khipu* historiography . . . attest to the existence of highly stable genres of discourse. Furthermore, the structural features of this discourse exhibit a close correlation to known semiotic conventions of *khipu*.[35]

In Peru, many *khipu* survived the holocaust of the Spanish Conquest, some in the tombs of their owners, but examples of calendar *khipu* are rare.[36] A unique illustration of one kind of calendar *khipu* for the Conquest year 1532 has survived, however, in the form of a reproduction in an early seventeenth-century Spanish text (Figure 5.3a). Maintained by the specialists who were the schedulers of the religious rites of the Inka and guardians of their cultural astronomy and cosmology, *khipu* were zealously sought out as repositories of "pagan devil worship" by Spanish missionaries and consigned to the bonfires of pre-Columbian cultural artifacts. The connection between the keeper of the *khipu* and astronomy is clear in the original caption to the depiction of the astrologer (Figure 5.3b):

> And our Astrologer-Poet!
> who knows the revolution of the Sun,
> and of the eclipse of the Moon,
> of the stars and comets,
> of the hours, the Sundays, the months and years
> and of the four winds of the world
> and of the planting time for seeds for the food,
> – since time immemorial.[37]

---

[35] Brokaw (2003, 141). See also Beyton-Davies (2007, 310–12).

[36] For diverse forms of Andean *khipu*, see Conklin (1982). Recent research raises the possibility that the technology's circumpacific distribution could be a result of transmission. According to Geoffrey Irwin, "In recent years there has been increasing evidence and arguments for prehistoric Polynesian contacts with parts of the American coast stretching from Chile north to California. Certainly the sweet potato and probably the gourd were taken from America to East Polynesia. Various items travelling the other way, from Polynesia to America, could have included the coconut, the domestic fowl, the technique of building boats with stitched planking, and certain distinctive kinds of portable artefacts and sometimes even the Polynesian names for them' (Jones et al. in press [2011]) Although there is continuing debate about the details, we can be satisfied that the general case for contact has been made." Irwin (2010, 60, also 65, 67); Allen (2010, 147, 150, 163–4). But see Jones et al. (2011, 42, 44).

[37] Trans. Conklin (1982, 262).

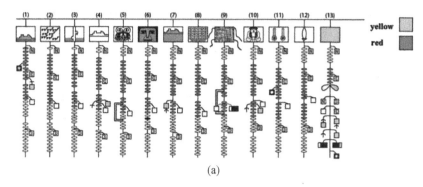

(a)

Figure 5.3 (a) Inka calendar *khipu* for the year 1532. Adapted from Laurencich-Minelli and Magli (2008), reproduced by permission. (b) Inka knot-reader/astrologer, carrying in his right hand a forked astronomical sighting rod and in his left a *khipu*. Horizon cairns or *sucancas* appear on the mountain ridges. After Felipe Guaman Poma de Ayala, *El primer nueva corónica y buen gobierno* (1615/1616) (Copenhagen: Det Kongelige Bibliotek, GKS 2232 4°), available at www.kb.dk/permalink/2006/poma/info/en/frontpage.htm. For a line drawing of the image, see Conklin (1982, 262).

Given its unique characteristics, this early Spanish illustration of a *khipu* is worth examining in some detail for what it might tell us about this method of record keeping in preliterate societies, or at least how it was thought to work by early chroniclers based on the testimony of native informants.[38] The thirteen square cartouches arrayed along the top cord are ideograms representing the noteworthy agricultural activities or ceremonials for which each month of the year was named or with which it was identified.[39] Suspended from these are pendants on which groups of red and yellow knots mark the days. These are

---

[38] Laurencich-Minelli and Magli (2008). The authenticity of the historical account of the Spanish conquest of Peru that the work contains has been questioned. Similarly, the text's discussion of literary *khipus* and how they are read is unprecedented. Domenici and Domenici (1996, 56). It may be, however, that the *khipu* calendar is a solar calendar of the kind alluded to in the sources. An authority on the Inka calendar, R. Tom Zuidema, mentions "references in Inca culture that allude to another calendar that also existed . . . of 12 solar months." Zuidema found evidence for that calendar's origin in the pre-Inka cultures of Huari and Tiahuanaco. Zuidema believes that the calendar *khipu* he analyzed was probably a variation on a regular solar month calendar such as described by the chronicler Polo de Ondegardo (d. 1575). The more complex "*ceque* calendar" was used "for describing exactly astronomical periods that are not dependent on, and cannot be described only in terms of, sequences of months." Zuidema (1989, 334). In such *khipus*, "the knots are a symbolic representation and do not represent a one-to-one correspondence between objects (e.g., days) and knots." Ascher and Ascher (1975, 337); see also Conklin (1982).

[39] The three cartouche images resembling Bactrian camel humps represent particular *sucancas*, or pairs of pillars or cairns bracketing crucial astronomical rising or setting points on the high mountain ridges surrounding Cuzco. Similar images appear in the drawing of a so-called "literary" *khipu* elsewhere in the same early seventeenth-century work. Domenici and Domenici (1996, 53).

(b)

Figure 5.3 (*cont.*)

grouped into ten-day weeks by spacing the knots, and further into groups of fifteen to define mid-month. Tags or sub-pendants attached to the cords at particular days denote events of significance, like astronomical phenomena – full Moons, the rising of the Pleiades, eclipses, and so forth. Seven long months of thirty days and five short months of twenty-nine days occur in irregular sequence, and a thirteenth "month" of ten epagomenal days is appended, bringing the total number of days represented to 365, matching the number of days in the solar year.

Contemporaneous accounts of Inka *khipu* and their use as recording devices attest to their impressive capacity to preserve complex information, including periodic tribute, barter and exchange agreements, contracts, and narratives, which information was read out as required by elite officials known as "knot-readers," *khipukamayuq* (Figure 5.3b). Their testimony was even accepted by the Spanish colonial authorities in legal cases.[40] According to Laura Laurencich-Minelli, professor of pre-Columbian studies at the University of Bologna, the early seventeenth-century document she analyzed, *Historia et Rudimenta Linguae Piruanorum*, contains a description of various kinds of *khipu* and provides clues to how literary *khipus* may have been read:

> Quechua . . . is a language similar to music and has several keys: a language for everyone; a holy language [which] was handed [down] only by knots; and another language that was handed [down] by means of woven textiles and by pictures on monuments and in jewels and small objects. I will tell you . . . about the *quipu*, which is a complicated device composed of colored knots . . . There is a general *quipu* used by everyone for numbering and daily communication and another *quipu* for keeping all religious and caste secrets, known only to the Kings, the Priests, and the Philosophers . . . I visited . . . archives for those *quipus* that tell the true story of the Inka people and that are hidden from commoners. These *quipus* differ from those used for calculations as they have elaborate symbols . . . which hang down from the main string . . . The scarceness of the words and the possibility of changing the same term using particles and suffixes to obtain different meanings allow them to realize a spelling book with neither paper, nor ink, nor pens . . . [The] *curaca*[41] emphasized that this *quipu* is based by its nature on the scarceness of words, and its composition key and its reading key lie in its syllabic division . . . [The] *curaca* explained, "If you divide the word *Pachacamac* [the Inka deity of earth and time] into syllables *Pa-cha-ca-mac*, you have four syllables. If you . . . want to indicate the word 'time,' *pacha* in Quechua, it will be necessary to make two symbols [in the *quipu*] representing *Pachacamac* – one of them with a little knot to indicate the first syllable, the other with two knots to indicate the second syllable" . . . [The *curaca*] listed the main key words with an explanation of how to realize them in *quipus*.[42]

The *khipu* is thus an example of a kind of composite analog device that would have preceded the invention of writing (in contrast to the mercantile tokens familiar from the Mesopotamian context). It illustrates the kinds of

---

[40] Urton (1998, 409–38). Accounts of the process indicate that the specialist reading an accounting *khipu* "parsed the knots by inspecting them visually, and by running their fingers along them Braille-style, sometimes accompanying this by manipulating stones." Mann (2003, 1650).

[41] A member of the Inka nobility responsible for the affairs of his lineage.

[42] Trans. Domenici and Domenici (1996, 50, 52). If this is an accurate description of how *khipus* recorded spoken language using a conventional syllabary as the key, then it is clear that the technology had crossed a crucial threshold and achieved a form of universal application; that is, it had become capable of representing every word in the language, which David Deutsch (2011, 125 ff.) calls the "jump to universality." Nevertheless, as Deutsch points out (ibid., 127): "it seems to be a recurring theme in the early history of many fields that universality, when it was achieved, was not the primary objective, if it was an objective at all. A small change in a system to meet a parochial purpose just happened to make the system universal as well."

information that ultimately had to be reproduced in written form – a number set to count the days, units of weight and measure, the ten-day week, festivals, terms for the phases of the Moon, solstice, zenith passage, eclipse, comet, meteor shower, stars, names of the months, colors, rituals, festivals, social classes, tribute items, seasons, a variety of action verbs, etc., and a technique for reading out the encoded information. This information is analogous in many respects to the vocabulary that the designers of the Taosi "observatory" would have had to employ.[43]

In other words, once the mental leap was made, say, from the ideographic representation of the months on a *khipu*-like device to zodiographs,[44] in order to accomplish the transition to a written calendar a substantial repertoire of contextually related signs would have to have been invented, or appropriated from existing religious symbolism, textile motifs, and other visuographic symbols or "iconic images" (*xiang*) such as those abundantly documented from the Chinese Neolithic. The application of writing in this specialized way could have come about in fairly short order. To be functional, such a written calendar, once conceived, would need to embody from the outset the kinds of elements and concepts itemized above. Short of pointing to the corresponding *tocapu* (see below) or *khipu* pendant, a glyph or ideograph for "corn-planting month," say, would have to be named to be spoken of, and all that is required to produce a zodiograph is for the spoken word to stick to the conventional glyphic representation. In the case of the depictions of *sucancas* (pillars on the horizon marking the Sun's location on critical dates) in cartouches 1, 4, and 7, for example, because of their similarity the reading of each would have to have been distinctive and quite specific.

The ethnographic evidence is too limited for us to know whether *khipu* "cartouches" like those depicted may already have been more than merely incipient zodiographs. It is not so difficult to imagine how the non-glottographic representation of a concept – say the graphic for "corn-planting month" or an Inka king's *tocapu* glyph – could imperceptibly be transformed in conventional usage into the glottographic denotation of the name. Once the technique of representing words graphically emerged, as it independently did in different contexts and cultures, the conversion of a notational system into writing and the spread of the process to other contexts could occur relatively rapidly.[45]

---

[43] In Chapter 12, we will see how the ancient Chinese equivalent of the *khipu* provided the structural metaphors used to conceptualize the cosmos, suggesting again the ancient use of such an analog device by the very individuals with a need to do the conceptualizing.

[44] In terms of the developmental stages of writing: "when a graph is primarily a depictive representation of a thing, it is a *pictograph* and is not writing. When the same graph, or a modified version of it, represents primarily the name of the thing, that is, the word for the thing, and stands for the thing itself only as information conveyed by the word, we call it a *zodiograph* and define it as writing." Boltz (1999, 110, original emphasis).

[45] Houston (2004b, 238).

Whether such a development was imminent in the Inka context is unknown, although there are highly suggestive indications of incipient writing. Inka *tocapu*, for example, are distinctive geometric designs found on textiles and ceramics composed of individual square blocks filled with geometric or pictographic designs painted or woven in bright colors. A large assortment of intricately patterned and uniformly square cartouches can fill an entire piece of woven goods like a patchwork. They also feature prominently in decorative bands on ceremonial libation cups (*queros*). The patterned designs are now understood to be a conventional visuographic code conveying significant information, including the names of Inka rulers, toponyms, notable victories, etc. They resemble the cartouches labeling the months in the calendar *khipu* in Figure 5.3a.[46]

Calendrical use of the cyclical signs

Visual symbols had immediate advantages over speech. Lists of transactions and numbers were much better expressed in writing than in speech. Lists of genealogies, and other historical sequences, were also much clearer in written form, and *devices such as astronomical almanacs . . . simply could not be formulated or expressed in spoken language.*[47]

For the purposes of counting the days between full Moons at Taosi, not to mention solstices, harvest festivals, and so on, a primitive number system and reliance on memory alone simply would not do. At a bare minimum one either devised a scheme to represent 1, 2, 3 . . . 10, 20, 30, etc., or maintaining a horizon calendar over time would have been impossible. A rudimentary number

---

[46] Use of *tocapu* was apparently an exclusive prerogative of members of the Inka imperial clan and other elite individuals. The highly varied patterns in the squares may also have encoded information conveying rank, lineage, the privileges of the individuals possessing the items, and perhaps the ethnic groups over which they ruled. Stone-Miller (1995 210); Ziółkowski (2009). An early Italian commentator on the *Historia et Rudimenta Linguae Piruanorum*, Joan Anello Oliva (c.1637–8), added to the document, in ciphered Italian, information provided by an Indian *khipu* reader, together with "definitions for symbols known as *tocapu* that appear in many Inka weavings." Domenici and Domenici (1996, 52). These have not yet been published, nor are they mentioned in Ziółkowski (2009). Similar cartouches appear alongside Mayan hieroglyphic writing as well. For example, on a fresco from the Temple of the Murals at Bonampak a frieze above a scene depicting royal ceremonial represents the sky and contains similar symbols denoting certain constellations. These are completely unlike the hieroglyphic writing that appears on the same mural and they presumably comprise zodiographs. Baudez and Picasso (1992, 119).

[47] Donald (1991, 290, italics mine). Walter Ong expressed much the same view: "Goody [1977, 52–111] has examined in detail the poetic significance of tables and lists, of which the calendar is one example. Writing makes such apparatus possible. Indeed, writing was in a sense invented largely to make something like lists . . . Primary oral cultures commonly situate their equivalent of lists in narrative . . . not an objective tally but an operational display in a story." Ong (2002, 97).

set consisting of "1, 2, 3, many" coupled with an oral narrative listing regularly observed astronomical or meteorological events like full Moons would also be inadequate to the task – biological memory is far too limited. Elaborating on the advantages of lists, Merlin Donald writes,

> Part of the gain was in the transportability and permanence of records; but another important part was in the ability to arrange virtually endless lists of items. The *list* is a peculiarly visual institution. The usefulness of oral listing is very limited, owing to memory limitations; orally memorized lists tend to tie up working memory, preventing further processing of the list. In contrast, visual lists can be arranged in various ways, and juxtaposed to simplify the later treatment of the information they contain. List arrangement can facilitate the sorting, summarizing, and classifying of items and can reveal patterns otherwise not discernible. With the invention of visual lists, the newly created state could acquire, analyze, and digest the information it needed to function.[48]

Given the conceptual toolkit of the elite users of the Taosi observing platform during the two centuries or so it was in use, they must have possessed some sort of external recording device like a *khipu*, if not a system of written signs, and this some eight centuries before the definitive appearance of writing under Shang king Wu Ding. It is worth recalling here that the prehistoric culture hero Fu Xi is credited with the invention of *both* the calendar and record keeping by means of knotted strings.[49] As we saw above, at a minimum a *khipu*-like device would lend itself perfectly to recording the recursive series of cyclical signs of the calendar, perhaps even with the help of color coding for which there is already evidence from Taosi.

Furthermore, individual glyphs have been found at Taosi written with brush and red pigment on the sherds of a large jar. One of these (Figure 5.4) bears a startling resemblance to the Shang OBI graph *wen* 文 "patterned; cultured" in appearance, stroke order, and execution. Needless to say, the discovery has provoked heated debate about whether these unique traces represent true writing, a question beyond the scope of our discussion here.[50] However, a detailed firsthand analysis of this early example of brush-written signs by one of China's pre-eminent paleographers, Li Xueqin, may be of interest:

> The color of the graphs on the jar is bright and the brush strokes are clear; it is not difficult to see that they were written with a large, pointed, and soft writing brush. Because the clay surface is relatively coarse and absorbent, the brush tip had to be able

---

[48] Ong (2002, 288), original emphasis.   [49] Needham and Wang (1954, 164).

[50] Walter Ong (2002, 82) characterized the meaning and significance of true writing this way: "The critical and unique breakthrough into new worlds of knowledge was achieved within human consciousness not when simple semiotic marking was devised but when a coded system of visible marks was invented whereby a writer could determine the exact words that the reader would generate from the text. This is what we usually mean today by writing in its sharply focused sense." Unfortunately, one cannot prove the existence of a writing system based on the appearance of individual (or even several) glyphs.

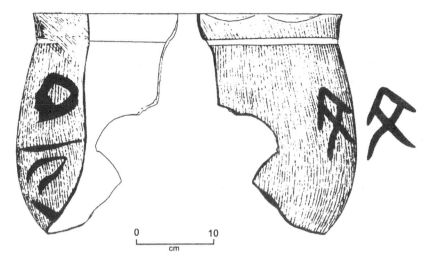

Figure 5.4 Large pottery jar from Taosi with brush-written glyph "pattern" (*wen*). Redrawn from Li Jianmin (2007, 620).

to hold the rich red ink. The graphs on both convex and flat surfaces of the jar are exactly the same color, confirming that they were written at the same time by the same hand. In addition, traces of red ink on the broken edges of the shards shows that the red writing was added after the jar had been broken. This also explains why the graphs on both faces of the jar are centered. The graph on the convex surface is obviously *wen*. It consists of four strokes, without the heart in the middle; the same *wen* can be found on Yinxu oracle bones. By mimicking the way the *wen* was written on the jar, one can see that the writing sequence of the four strokes is the same as we all use today, and also that the graph was written with the right hand. The stroke on the upper left is short, straight, and strong, while the right-hand stroke has been written with a little pressure, so the middle of the stroke is thicker than the end. The bottom left stroke is drawn out with a lifting of the brush, causing the bottom half to narrow and curve, while the right-hand stroke is written with more practiced force, coming to a fine point as the brush was lifted off. Doesn't expressiveness like this deserve to be called calligraphy?[51]

Finding inspiration in the sky

In Chapter 4 I concluded that the nexus of phonetic and semantic linkages within the *Ding – zheng – zhen – ding* word family reveals the use of Pegasus to have been the method designed to bring the normative celestial images (*xiang*) down to earth.[52] Now we also begin to discern a plausible celestial inspiration

---

[51] Li Xueqin (2005c); Li Jianmin (2007); Xu Fengxian (2010).
[52] The nexus I am attempting to describe in this chapter signals the formulation of a new theoretic culture, whose manifestation in Erlitou urban planning was described by Xu Hong above. As

for the "exchanging of knotted cords for written signs" alluded to in the passage from the "Appended Commentary" to the *Book of Changes*. We know that the asterism *Ding* is the Celestial Temple, the Square of Pegasus. In the oracle bone inscriptions celestial stem *ding* (丁) is written as a square 口, in all likelihood taking its inspiration from Pegasus. We have also established a connection between early astronomy, orientation, celestial simulacra, and word family of the stem-sign *ding* 口 (丁). This one graph may now be seen to provide a touchstone linking the abode of the Supernal Lord above; time management in the form of the first or "correct," *zheng*, month of the calendar; the idea of a divinely revealed standard of what is *zheng*, "right" and "true," both spatially and conceptually; and, I would argue, the practice of representing the nexus of the above meanings graphically as a square. *Ding/ding*, in other words, is none other than the root sememe in the pre-Shang language "square; be straight ~ be square; make straight ~ make square ~ fourth."[53]

The epigram beginning this chapter contains two statements ascribing the origin of writing to the sages of prehistory:

Heaven suspends images, to manifest the propitious and the inauspicious, and the Sage makes of himself their semblance . . . Anciently, in ruling all under Heaven, Paoxi [Fu Xi] looked up to observe the images in Heaven, and looked down to observe the patterns of Earth.

These observations led Fu Xi to devise the eight trigrams, from which the all-encompassing system of the *Changes* then arose. According to the tradition, the imagistic hexagrams of the *Changes* materialize all phenomenal nature and civilizational invention. Note, however, that the role of the Sage is that

---

Deborah L. Porter (1996, 71) shows, "the pointed use of the term *hsiang* to refer to astral entities displayed by heaven, as well as the process of their symbolization (that is, 'the sages symbolized them [the astral entities] in the *Book of Changes*') suggests that the process of symbolization itself is closely connected to heavenly bodies." As Merlin Donald (1991, 275) put it, "The critical innovation underlying theoretic culture is visuographic invention, or the symbolic use of graphic devices."

53 In non-calendrical contexts, and when not used alone as shorthand for "*ding*" ancestors like Wu Ding, a word written 口, graphically indistinguishable from stem-sign *ding*, can have distinctly different referents as the recipient of sacrifices: e.g. high ancestor Shang Jia; the Sun; even *fang*, "quadrant/direction." Shima (1979, 403; also 1969) argued that in Shang oracle bone inscriptions, when it occurs in the contexts like "*di*-sacrifice to 口," the square graph should actually be read as standing for the Supernal Lord, *Shangdi*. Keightley (1997, 517–24), in contrast, demonstrates that in a majority of cases 口 actually stands for 日, "the Sun." Didier (2009, Vol. 2, 166–232), although he does not cite Shima, rehearses the same evidence, claiming instead that the square should be read *ding* 丁, and that it refers to his speculative, decidedly un-square constellation bracketing the Pole, as well as to a supposedly sacred geometrical abstraction whose influence he thinks permeated Shang religious life. Keightley's study and the obvious parallels, where the object of "*di*-sacrifice" may be *fang*, "quarter, quadrant," account for the seemingly problematical inscriptions. If there had been a celestial square venerated by the Shang, surely it would have been the Great Square of Pegasus – *Ding* – identified by the Babylonians as the square irrigated *Field* (*ikû*), which for them doubled as a unit of area.

of enlightened observer and recipient of revelation. His invention of the trigrams is actually one step removed from the true source of inspiration – the patterned images of Heaven themselves. This is a crucial distinction, whose significance has not been fully appreciated, or has at best been assumed to be simply metaphorical. If, however, we hope ultimately to vicariously achieve a holographic comprehension of the culture, then in studying the history of the "textualization" of the cosmos investigators can no longer legitimately ignore those early Bronze Age skywatchers' perception of the heavenly patterns.[54]

Visuographic depiction of astronomical phenomena: *heshu* and *luotu*

I proposed that the two sets of cyclical signs, stems and branches, were initially devised to respond to the conceptual and record-keeping demands of the calendar, and that the origins of the two are crucially related. I suggested further that it was calendrical astronomy that lent impetus to the development of writing in China and prefigured its application to other forms of record keeping that emerged later, including the Shang divinations in which we see a mature written language fully formed and capable of expressing virtually anything.[55] Rhyme may have provided a link between orality and notational convention, acting as a bridge between the use of a non-glottographic notational scheme and the *idea* of true glottographic writing. In other words, rhyming served as the notional precursor to the full realization that the *sounds* of spoken words could be attached to conventional graphic signs and serve as analogs of speech in a new medium.

That profoundly important cultural innovation – true writing – was acknowledged in the canonical tradition to have been Heaven-bestowed. If the *Lesser Annuary of Xia* (*Xia xiao zheng*), the "Canon of Yao," and other Zhou texts can still preserve seasonal stellar correlations and seasonal indications from the second millennium BCE,[56] then it is not merely a rhetorical flourish when the "Appended Commentary" also claims that

---

[54] To say that "[w]hen hexagrams were imagined as the origins of script, reading natural phenomena as signs became the prototype of reading graphs" (Lewis (1999, 263) would seem to attribute to the ancient Chinese a misapprehension about the ultimate origin of writing. To paraphrase the Bard, however, "the credit lies not in ourselves but in the stars."

[55] On this point, Robert Bagley (2004, 225) commented, "without the pressure of new needs, or the lure of new possibilities, full writing would never have come into being. Comparison with these well-charted developments in the Near East argues that the writing system we encounter in the Wu Ding oracle texts is the end product of a gradual spread to a broad range of applications."

[56] For the antiquity of certain seasonal indicators (*wu hou*) in the *Lesser Annuary of Xia* in particular, see Hu Tiezhu (2000); Li Xueqin (1989, 4–11); Chen Jiujin, Lu Yang, and Liu Yaohan (1984, 63).

Heaven suspends images, to manifest the propitious and the inauspicious, and the Sage makes of himself their semblance. Out of the River there emerged a Diagram, and from the Luo [River] there emerged a Writing; the Sage models himself on them.[57]

Chapter 9 will offer a detailed account of the identity between astral fields and terrestrial polities in which the Yellow River's astral correlate is the Milky Way, but by now we should be alert to the possibility that the reference to a River in the *Changes* is to the *Silvery River* (*yin he*) in the sky no less than to the terrestrial Yellow River. The above passage could thus be read as an explicit claim about the celestial origins of writing. It is noteworthy that the prefaces to the "Treatise on Astrology" in the *History of the Jin Dynasty* (265–420 CE) and the *History of the Later Han Dynasty* explicitly identify the River Diagram *Hetu* as a text in which was recorded revealed wisdom concerning the heavenly bodies. For example, the latter says, "Xuan Yuan [i.e. the legendary Huang Di or Yellow Emperor] first received the *Hetu doubao shou* which plotted out the images [formed by] the Sun, Moon, planets, and constellations. Therefore, books concerning the celestial offices begin with the Yellow Emperor."[58] Similarly, the great Eastern Han commentator Zheng Xuan (127–200) asserted that the Luo Writing revealed on the carapace of the Mantic Turtle reproduced a list of the "coordinates of the *Dipper* and the lunar lodges, calculations needed to keep track of the rise and fall of kingdoms."[59] A River Diagram was listed among the royal regalia in the *Book of Documents*,[60] and it is clear from accounts in the Han Dynasty apocrypha that a principal significance of such talismans was to symbolize the conferral of the right to rule on its possessor. They were, in one definition, "associated with the *pao* or precious objects which a ruler keeps as a sign of his investiture by the spiritual powers of Heaven."[61] Such diagrams, typically delivered by a Sacred Turtle

---

[57] Lai (1972, 428).

[58] *Hou Han shu*, 10.3214; also *Jin shu*, 11.277. The standard account of the transmission of esoteric knowledge in the form of "River Diagrams" or "Luo Writings" is found in *Han shu*, 27.1315. In Chapter Six we will examine the account from *Sui shu* (27A.763) of the River God's bestowal on Yu the Great of both the River Diagram and the Dark Scepter. For a discussion of the military application of revealed "texts" in the Warring States period, "in imitation of divine patterns that inform the cosmos," see Lewis (1990, especially pp. 98 ff. and 137–63). Ignoring the "sky pattern reading" (*tian wen*) motivating early astral omenology and cosmology, Mark E. Lewis argues for a Warring States date for this "textualization" of the cosmos, contending that the notion of a cosmic kingship, wherein the legitimate authority of the king is derived from his ability to "read" the hieroglyphics of the cosmos, is a Warring States invention. In view of the evidence presented here regarding the privileged role of sky pattern reading, cosmology, and cosmic legitimation in the Bronze Age, Lewis's late dating of such "textualization" is hardly tenable.

[59] Quoted in Porter (1996, 36). See also the apocryphal *Chunqiu kao ling yao* quoted at same location. What is noteworthy is not the precise details of Zheng Xuan's testimony but the obvious astronomical and numerological associations of the revelatory diagram in his mind.

[60] Dull (1966, 7); Saso (1978, 405–6); Ho Peng-yoke (1966, 42, 58).   [61] Saso (1978, 411).

(*shen gui*) or dragon-horse (*long ma*) and sometimes by a Phoenix (*feng huang*), are generally described not "as a piece of jade but as a chart written on jade."[62] Moreover, "the presence of phoenixes, the number five, and the red script are therefore symbols essentially related to the *Ho-t'u*," which are understood to symbolize a contract "not between the visible lord and his subject . . . [but] rather connected with the ruling of the hills, streams, and the spiritual forces of nature."[63]

There is strong agreement about the cosmological content and context of those revelations. In one Han source, the *Images of the River Diagram Bestowed from Above* (*Hetu jiang xiang*), the River Diagram is described as "red writings depicting the heavenly stars that control nature."[64] Another Han text, the *Investigation into the Numinous Lights of the River Diagram* (*Hetu kao ling yao*), says, "[every] 500 years [there appears] the talisman of the Sagely Epoch."[65] The same text continues, "the River Diagram is the periodic record [*ji*] of the Mandate; [it] charts the periods of ending and beginning, of preservation and extinction of Heaven and Earth and of Kings, and the rule of dynastic succession of the Register [of Heavenly Legitimation]."[66] Much earlier these associations were made even more explicit in the *Guanzi*, "Wang yan." There Guanzi explains to his lord, the Hegemon Duke Huan of Qi (d. 643 BCE), that the Mantic Dragon and Turtle, River Diagram, and Luo Writing were the unambiguous signs of the conferral of the Mandate on the Three Dynasties (Xia, Shang, Zhou) of the second millennium BCE, and that their not having appeared to Duke Huan means he cannot legitimately claim to possess the Mandate.

Recall now Arthur Waley's prescient observation that "an omen is regarded as in itself a momentary, evanescent thing. Like silverprints, it requires 'fixing.' Otherwise, it will refer only to the moment at which it was secured." Commenting, David N. Keightley says, "the carving of the [Shang oracle bone] inscriptions, too, presumably involved the same desire to 'fix' the charge."[67] Seen in this light, the evidence adduced thus far fully justifies the conjecture that the original esoteric diagrams were, in fact, depictions of the five visible planets in a dense cluster, the first and most spectacular of which (1953 BCE) took place in Pegasus in winter, with the *Turtle* asterism (*Bie*, or *Tian yuan*, our Corona Australis) in plain view beside the *Sky River* and the tail

---

[62] Ibid., 408.   [63] Ibid., 409.   [64] Ibid., 410.
[65] "*Wu bai zai sheng ji fu*" identified as a Diagram, according to the commentators. *Yuhan shanfang ji yishu*, Vol. 4, 1999.
[66] "*Hetu ming ji ye; tian di diwang zhongshi cunwang zhi qi, lu dai zhi ju.*" Ibid.
[67] Keightley (1975, 22). The "Cinnabar Writing" bestowed on King Wen and subsequently inherited by King Wu could well have been such a graphic depiction like a Shang display inscription, presumably on jade, of the planetary omen taken as the sign of Heaven's conferral of the Mandate on the Zhou (Chapter 6 below in this volume).

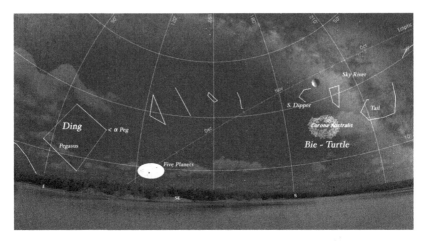

Figure 5.5 The unique configuration of the *Sky River*, Pegasus, and *Celestial Turtle* constellation at the time of the planetary massing of 1953 BCE in *Ding*, the *Celestial Temple* (Square of Pegasus). Note the location of the *Heavenly Turtle Tian bie* (Corona Australis) above whose back the Sun, Moon, and Five Planets pass just after fording the *Sky River*. Note also the tail of the *Dragon* as it "leaps [heavenward] from the watery void" in the densest region of the Milky Way (Starry Night Pro 6.4.3).

of the *Dragon* (Figure 5.5),[68] the second (1576 BCE) more complex back and forth movement at the ford on the *Sky River* between the *Dragon* and *Turtle* constellations, and the third (1059 BCE) in the *Beak* of the *Vermilion Bird*, or *Phoenix*.[69] Graphic representations of such omens would certainly have

---

[68] In some early traditions the iconic creature representing the north was not the *Somber Warrior*, *Xuanwu*, or turtle-serpent combination of Qin–Han time, but either a fish or a deer, the latter perhaps reminiscent of the *qilin* or "unicorn," thought to have been a now extinct species of deer. Karlgren (1950a, 7); Feng Shi (1990a, 117); Feng Shi (2007, 427–30). The Warring States astrologer Shi Shen considered the asterism *Southern Dipper* (in Sgr) to be the head of the *Primal Turtle* (*dou, yuan gui zhi shou*), though he may simply have meant that the *Turtle*'s head lay in the same longitude. Perhaps this suggests that for Shi Shen the entire winter quadrant of the heavens constituted the *Turtle*. In any event, *Southern Dipper* lies just above the crown of stars known as Corona Australis (Figure 5.5), which came to be identified as a celestial soft-shelled fresh-water turtle (*bie*). Following publication of the planetary grouping of 1953 BCE (Pankenier 1983–85), Pang and Bangert (1993) loosely speculated about a possible link between the conjunction in Pegasus and the Luo Writing based on a vague reference in the "Hong Fan" chapter of the *Book of Documents*, but without offering any supporting evidence.

[69] Pankenier (1983–85; 1995). Accounts of the omen in Han apocrypha identify the bird as a "peacock," *kong que*, or simply *que*, or as a phoenix, *fenghuang*, e.g. *Chunqiu yuan ming bao*, in Ma Guohan (n.d., Vol. 4, 2113). The Liang Dynasty (502–57) astrological treatise *Diagrams of Auspicious Signs (Ruiying tu)*, says, "as for the Red Peacock, when a True King moves Heaven to respond, then it comes clasping a Writing in its beak," quoted in *Kaiyuan zhan jing*, 115.2a. In the *Bamboo Annals* (Fang Shiming and Wang Xiuling 1981, 232), King Wu received the Cinnabar Writing from his tutor, Lü Shang, upon succeeding his father as ruler of Zhou. This

qualified them for inclusion among the royal regalia symbolizing the legitimacy of the ruling dynasty. Though perhaps unaware of the role of the planets in the conferral of the Mandate, Confucius was familiar enough with the implications of such River Diagrams to lament, "The Phoenix does not come; the River sends forth no Diagram. It is over!"[70] The Sage knew full well that revelation of the Mandate of Heaven was a fact of history and not merely the invention of early Zhou propagandists.

* * *

We saw how *Align-the-Hall* or *Ding* is optimally positioned south of the *Sky River*, the Milky Way. As we will see in Chapter 11, this special relationship was deliberately replicated in the layout of the Qin capital, Xianyang. At the time of the planetary massing of 1953 BCE in *Align-the-Hall* some eighteen centuries earlier, the *Sky River* would also have been on brilliant display, arching across the sky from northeast to southwest midway between *Align-the-Hall* (*Celestial Temple*) and the abode of the Supernal Lord at the Pole (Figure 5.5). It seems, therefore, that the "Diagram" or "Writing" that emerged "out of the River" was none other than the Great Square of Pegasus or *Ding* ($\Box \approx ding$ 丁) with the compact cluster of the Five Planets adjacent to it in the same longitude.[71] Of course, the famous representation of the River Diagram as a "magic square" with the number five at its center (Figure 5.6a) appears quite late. Nevertheless, the key elements which are all mentioned in Warring States and Han references to the revelation – square, symbolic number five, Mantic Turtle (the "dark" *xuan* iconic creature representing the winter quadrant of the sky) – are highlighted as of central importance. Note especially, therefore, the position of the *Celestial Turtle* in Figure 5.5 (*Bie*, aka *Tian yuan*, our Corona Australis), at precisely

same item is called the "Phoenix Writing" in *Chunqiu yuan ming bao*. According to a comment by Zheng Xuan on the *Yuan ming bao* (Ruan Yuan 1970, Vol. 1, 503.2, subcommentary to ode "Wen Wang"), the Cinnabar Writing and the famous Luo Writing are one and the same. The Cinnabar Writing and the revelatory connotations of such mysterious documents are particularly interesting since they call to mind both the Shang practice of applying red pigment to display inscriptions and the Warring States custom of smearing the inscribed texts of solemn oaths (*meng shu*) with the blood of sacrificial victims. This suggests that all such writings partake of the same contractual character involving the participation of supra-sensible powers; cf. Keightley (1975, 13–17, esp. 16).

[70] *Analects* 9/9.

[71] It is significant that, besides ode 50, "When *Ding* just culminated," the only other occurrence of *Ding* as a noun in the *Book of Odes* is in the ode "Feet of the *Lin*" (Mao 11), where *ding* refers to the crown of the head (定 *dey > 頂*ding* *teŋ) of the *qilin* (unicorn), traditionally held to be square. In some regional variants, like the remarkable third-century brocade remnant from far-off Niya in the Tarim Basin, the iconic creature identified with the north was a *lin* (Chapter 10, Figure 10.5, below). Thus there may be a link between the square, the mythical *lin*, and Pegasus as well. A connection, if any, between *anzû*, the Babylonian constellation of the fabulous winged horse (or eagle), Pegasus, and the *lin* is a subject worthy of further study. Horowitz (2011, 35, n. 20). Although there may be a Babylonian connection, the origin of the Greek winged horse is obscure, so any trans-Asiatic link between it and the auspicious Chinese quadruped is purely speculative. Rogers (1998b, 88).

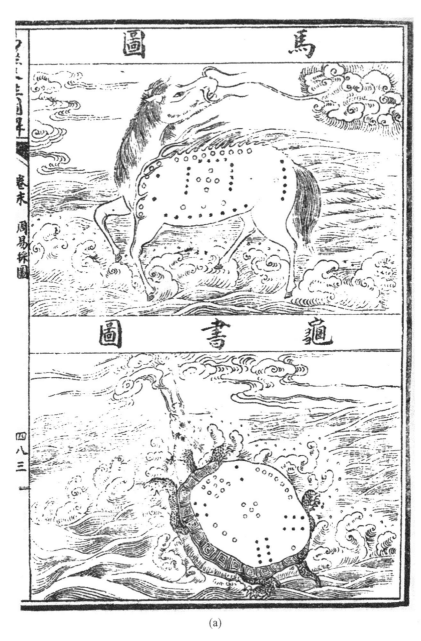

(a)

Figure 5.6 (a) Classic depictions of the River Diagram (top) and Luo Writing (bottom), revealed by their respective sacred creatures. Adapted from Lai (1972, 483) (b) Star ceiling from a Tang Dynasty tomb in Astana, Xinjiang, in which the central configuration of five discs represents the Five Planets. Adapted from Zhongguo shehui kexueyuan kaogu yanjiusuo (1980, 69, Figure 66), reproduced by permission. (c) Page from the Ming Dynasty astral prognostication manual *Tianyuan yu li xiang yi fu* (*c*.1600) showing the gathering of the Five Planets thought to foretell the conferral of Heaven's Mandate on the virtuous. Courtesy Library of Congress, East Asian Special Collections.

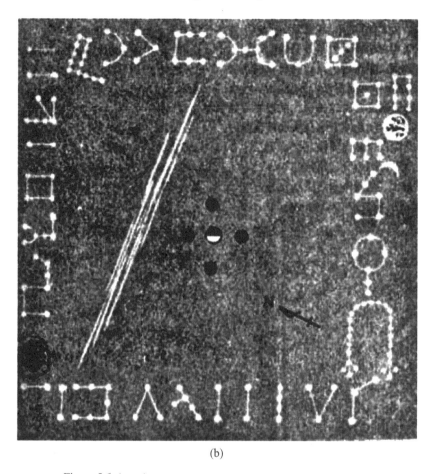

(b)

Figure 5.6 (*cont.*)

the location where the Five Planets forded the *Sky River* (Milky Way) on their way to their gathering in Pegasus. This configuration of "five" on the Mantic Turtle's back remained the standard pictorial representation of the Five Planets throughout history (Figure 5.6b–c).

Noteworthy too is that the two-dimensional arrays represented by the patterns of dots on some diagrams are curiously reminiscent of the visual representation of numerical relations by the strung knots on Inka *khipus*.[72] Thus, if a

---

[72] Ascher and Ascher (1975). The same association occurred to John B. Henderson (1995, 214): "the standard diagram of the *Luo shu*, that favored particularly by the Neo-Confucian cosmologists of the Song era, both dispenses with the squared outline and represents each of the

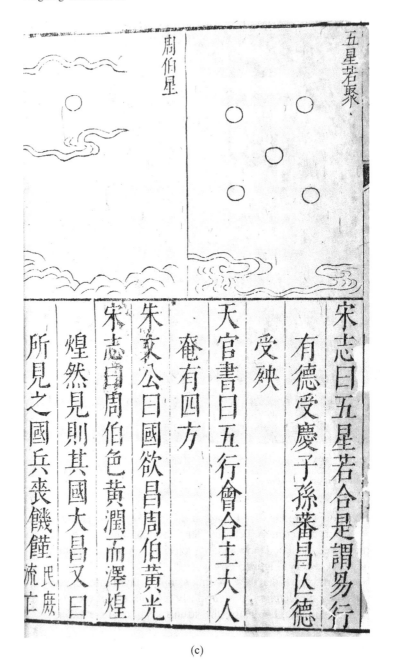

(c)

Figure 5.6 (*cont.*)

diagrammatic depiction of what was observed in 1953 BCE had been produced at the time, encoding shape, number, location among the stars in the sky, etc., this may well reflect a visuographic convention of the transitional period to a quasi-literate society.

Then too there is this account of Zhou king Wen's receipt of the Mandate found in Huan Tan's (d. 28 CE) "New Discourses" (*Xin lun*):

> The Phoenix soars [down] on spreading wings;
> Clasping a "Writing" it comes gamboling, thereby to command Chang [King Wen].
> I gazed up at Heaven and examined the Diagram;
> Yin [Shang] is about to expire [it portended].
> Great Heaven is azure, azure;
> First there [sc. in the heavens] was a presage.
> The linked essences of the Five Spirits [the planets]
> met in lunar mansion *Chamber* [Sco] to deliberate.[73]

Here, the intimate connection between the ancient astral portent, its graphic representation, and the implications for dynastic change are self-evident.[74] Significantly, the *Great Commentary* to the *Book of Documents* (*Shangshu dazhuan*) states, "Heaven's Great Command to King Wen was not [conveyed] sonorously, by means of sound," which is as one would expect if it had been a revelation in the sky.[75]

We now know from recent discoveries that by the Han Dynasty there had appeared abstract cosmological charts, such as the example below from the *Xingde* silk MS from a mid-second-century BCE aristocratic tomb at Changsha, Mawangdui (Figure 5.7).[76] In the *Xingde* diagram we see the central square

---

magic-square numbers by figures that resemble knotted cords rather than numerals." As Walter Ong (2002, 97–99) remarked, "Orality knows no lists or charts or figures . . . orally presented sequences are always occurrences in time, impossible to "examine," because they are not presented visually . . . charts, which range elements of thought not simply in one line of rank but simultaneously in horizontal and various criss-cross orders, represent a frame of thought even farther removed than lists are from the oral noetic processes which such charts are supposed to represent."

[73] Quoted in *Taiping yulan*, 84.5b.

[74] Needham and Wang (1959, 57) discuss the evidence, including an account in *The Grand Scribe's Records* (*c.*100 BCE), for the presentation of a River Diagram to the First Emperor of Qin in 230 BCE, and offer the opinion that "this account . . . refers in all probability to the beginning of the process whereby the ancient diagrams became a nucleus of crystallization for the magical–divinatory material incorporated during the Han in the Chhan-Wei apocryphal treatises." It is in these same treatises, as we just saw, that one finds the earliest written accounts of planetary massings in the second millennium BCE (Chapter Six).

[75] "*Tian zhi da ming Wen Wang, fei hengheng ran, you shengyin ye.*" *Shangshu dazhuan*, 4.5b.

[76] For interpretation and analysis of the *Xingde* text and its accompanying illustrative diagrams, see Kalinowski (1998–99, 195–202). From the cosmological importance given to the word *fang* in such astromantic contexts, and in the term *fang shi* used to denote the practitioners, *fang* may refer to numerology or mathematics. It may have little or nothing to do with compounding "recipes" as it might in medical contexts.

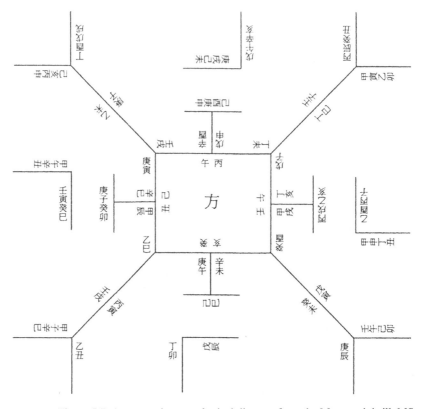

Figure 5.7 Astromantic cosmological diagram from the Mawangdui silk MS *Xingde*. Adapted from Kalinowski (1998–9, 144), reproduced by permission.

explicitly identified as such by the character "square/formula/method" *fang*. As in the dots at the center of the diagrams in Figure 5.6a–c, the four cardinal hooks (corner "Vs") in the Mawangdui diagram, together with the central square, make up the same archetypal configuration – *Five*.[77] Of course, by this date the Five Elemental Phases *wu xing* were firmly established as a principal heuristic numerological category, along with *yin* and *yang*, into which all manner of phenomena were classified and according to which their mutual relations were interpreted. But the Han through Tang examples of cosmographic diagrams represent only a high point in the development of visuographic representations of cosmological conceptions.

The role of the turtle in Chinese culture is worthy of a monograph in its own right. It is clear that the Mantic Turtle figured importantly in religious

---

[77] See the identical late Western Han divination gameboard from Jiangsu, in Liu et al. (2005, 376).

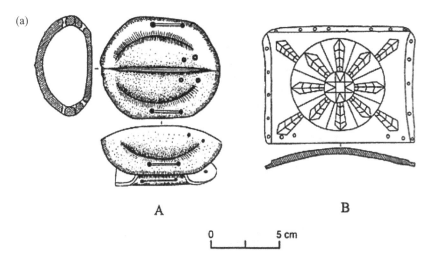

A   B

0   5 cm

Figure 5.8 (a) Jade model of the cosmos as a turtle, Hanshan, Lingjiatan. After Li Liu (2007, 66), © 2007 Li Liu, reprinted with the permission of Cambridge University Press; (b) typical late Shang Dynasty turtle plastron with divinatory inscription. Shown is *Bing bian* no 8, unearthed at Anyang. (c) Early Western Zhou Dynasty (King Cheng, mid eleventh century BCE) clan sign "Great (or Heavenly) Turtle" *Da/Tian gui*; after Tang Lan (1986, 85).

and cosmological thinking from the Neolithic through the imperial period. A startlingly sophisticated model exists from the late Neolithic when one would hardly have expected to find its like. The jade turtle shell and plaque found at Hanshan, Lingjiatan, in Anhui (Figure 5.8a), belong to a culture strongly influenced by the nearby Liangzhu jade-working culture of comparable date (*c.*2500 BCE).[78] As we saw above, the turtle has a very long history in China as a sacred animal and cosmological model, its carapace having been thought to represent the dome of the heavens and, one would think it follows, the belly plastron the Earth. It can hardly be coincidental that the configuration of the turtle plastron preferred for divination purposes in the Shang is so suggestive of the terrestrial topography with its paradigmatic Nine Provinces separated by the sinuous courses of major rivers. Numerous uninscribed plastrons and rare examples inscribed with single glyphs (some resembling the Shang oracle

[78] Li Xueqin (1992–3, 1–8); Feng Shi (2007, 503–4). Feng Shi (ibid., 514) also discusses the connection between the common Neolithic eight-pointed "star" (solar?) image and the River Diagram. Li Xueqin (1992–3, 7) studies the "star" image and its occurrence in association with other glyphs inscribed on pottery of this and neighboring cultures, as well as the connection of the jade plaque with later traditions of cosmological representation.

(b)

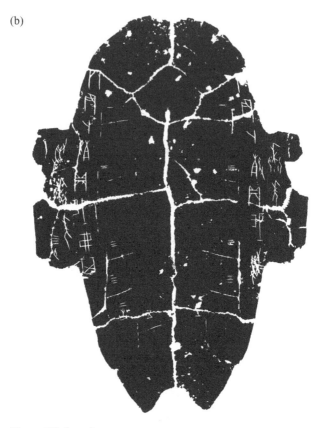

Figure 5.8 (*cont.*)

bone graph for "eye" or "Sun") have been discovered at the Peiligang site of Jiahu dating from 6000 BCE, as have Neolithic turtle-shell rattles and drums whose function suggests a connection with ritual use.[79] Well over one hundred thousand inscribed plastron shells have been unearthed from the late Shang and early Zhou periods in the late second millennium BCE, when they were buried in pit caches after being used in divinations connected to the rituals and sacrifices associated with the cult of the royal ancestors (Figure 5.8b).[80] In addition, judging from Shang and Zhou iconography, the *Heavenly Turtle* figured importantly in other contexts as a potent symbol, since it also occurs as

---

[79] Li Xueqin et al. (2003); Liu (2007, 124, Figure 5.5A); Keightley (1996, 87).
[80] Allan (1991); Keightley (1985).

(c)

Figure 5.8 (*cont.*)

a "clan sign," which may have possessed a kind of heraldic significance (Figure 5.8c) in addition to being an astronym and possibly a toponym.[81]

The inscribed jade plaque from Hanshan was originally sandwiched between the two halves of the turtle shell. The holes drilled through the three parts are thought to indicate that they were tied together in that configuration when buried. Nothing is known about the jade turtle's function (it probably also contained small pebbles so it could have been a rattle), but the fact that it was made of jade and appears to have had cosmological significance means that it was undoubtedly a treasured object belonging to a member of the elite. Of course, there is no possibility of a direct connection with the later representations such as the *Xingde* diagram, and it is unknown whether the plaque has true cosmographic significance rather than simply being an elaborate representation of the Sun's rays (the frequently depicted eight-pointed cross at the center is taken to be a solar symbol), albeit with remarkably regular indications of cardinal and inter-cardinal directions.

Whichever the case may be, there is no denying the existence of a venerable tradition of symbolic representation centuries before both Taosi and the spectacular planetary grouping of 1953 BCE. There is no reason to assume a priori that skywatchers in 1953 BCE would not have been capable of representing what they witnessed in a graphic or symbolic form. As David N. Keightley observed,

Neolithic ornament, Shang graph form, and Han iconography belong to the same tradition of representation, a further instance of the extraordinary persistence of some of these early Neolithic shapes . . . [It] seems likely that there was both a semantic and phonetic similarity between 虹 *g'ung/hong*, "rainbow," and 龍 *l'iung/long*, "dragon," and that both words derived from a still earlier word, which may be reconstructed as close to *kliung*, which had the basic meaning of "arched, vaulted." Once again we see how the early Chinese were using the shapes, first employed in elite Neolithic art, to record sounds and thus to record words. The Shang graphs for elephant [Figure 3.6], for example, which incorporated the same C-shaped profile, were not just *drawings of* an elephant; they recorded the *word for* elephant.[82]

Among the earliest and most striking Neolithic pictographs so far unearthed are those discovered in 1992 at a site near the village of Shuangdun in Bengbu, Anhui, dating from as early as 5330–4900 BCE. Numerous depictions of fish, deer, pigs, and goats, as well as stilt huts, the Sun, woven patterns, and the like, were found etched on the bases of clay pots, and this three millennia before

---

[81] The clan-sign (heraldic emblem?) of the Great/Heavenly Turtle commonly appears on Shang and Zhou ritual bronze vessels, beginning at Zhengzhou in early Shang. Tang Lan (1986, 86, n. 6) thinks may have begun as a place name.

[82] Keightley (1996, 86, original emphasis).

Figure 5.9 Shang oracle bone graph for "elephant" (*xiang*) with the extended meaning of "image, iconic figure."

Taosi.[83] During the many intervening centuries it is impossible to imagine that those early peoples did not also look up and imagine distinctive patterns in the sky as pictures of the domesticated and wild creatures, landforms, implements, etc., on which their lives depended. Indeed, the very graph which later came to denote such iconic images is *xiang* – in its earliest oracle bone form itself a pictograph of an elephant (Figure 5.9).[84]

---

[83] Xu Dali (2008, 75–9); Cheung (1983). What is particularly interesting is that even at this early date some of the same symbols are reported to have been found at other sites of similar date as much as sixty kilometers from Shuangdun. Comparable later signs from Dawenkou display an even greater geographical distribution. For an illustration of Neolithic pottery figurines from Shijiahe of sheep, pig, elephant, tiger, etc., see Chang et al. (2005, 107).

[84] For a comparison of the Shang graphs in Figure 3.6 with the actual configuration of the stars in the *Dragon* constellation, see Feng Shi (1990a, 112).

Part Three

Planetary omens and cosmic ideology

6   The cosmo-political mandate

Astronomy was a science of cardinal importance for the Chinese since it
arose naturally out of that cosmic "religion," that sense of the unity and even
"ethical solidarity" of the universe.[1]

As we have seen in the preceding chapters, in China by the early second millennium BCE there was already firmly established a mindset characterized by
a self-conscious dependence on regularly scrutinizing the sky for guidance.
Not only the calendar, but the correct orientation of any consecrated space,
the scheduling of religious ceremonies, and the proper conduct of seasonal
occupations all depended on the ruler's competent performance of his cosmo-
magical role of theocrat. The ability to comprehend the celestial rhythms and
to maintain conformity between their changes and human activity was a fundamental qualification of early Chinese kingship.[2] This suggests why, during
the second millennium BCE, the correlation of unusually dense clusters of all
five naked-eye planets with major political transitions was thought to signify
the conferral of legitimacy on the rising dynastic power.[3]

Here in Part Three, I set out the evidence for the remarkably early co-
ordination of celestial portents and watershed political and military events long
remembered in the oral and textual traditions of the Chinese. Beyond this, there
is an account of the ancient Chinese politico-religious imagination according
to which macrocosmic/microcosmic correspondences legitimated the social
order. The argument will be developed in three parts: first, a discussion of the
state of our knowledge of correlations between celestial and political events.
This is followed by a study of the formative role played by such knowledge

---

[1] Needham and Wang (1959, 171). More recently, Robert N. Bellah (2011, 266) made the same
point: "both tribal and archaic religions are 'cosmological,' in that supernature, nature, and
society were all fused into a single cosmos. The early state greatly extended the understanding
of the cosmos in time and space, but as Thorkild Jacobsen argued, the cosmos was still viewed
as a state – the homology between sociopolitical reality and religious reality was unbroken."
[2] For a recent, general discussion of the cross-fertilization between astronomy and governing in
ancient China, see Jiang Xiaoyuan (2004).
[3] This hypothesis was first put forward in Pankenier (1981–82) and subsequently refined and
developed in Pankenier (1995).

in the emergence of astral omenology, cosmology, and the management of the calendar as central concerns of the bureaucratic state. Finally, I take up the connections between ancient Chinese notions of universal sovereignty and the Supernal Lord, Shangdi (aka "Sky" or "Heaven," *Tian*), especially as manifested in the concept of Heaven's Mandate (*tian ming*) and in the rhetoric of early Chinese politico-religious discourse.

Astronomical phenomena and their correlation with political dynasties

In Part One we first saw how in the predawn hours of late February 1953 BCE the densest gathering of planets in human history took place. Almost a millennium later, at dusk in late May 1059 BCE, another occurred. These exceptionally close encounters of the "Five Pacers," *wu bu* (to use a later expression that recalls the Greek πλανήτης, an alternative form of πλάνης, "wanderer"), would have captivated even casual observers for days, if only because of the extreme rarity of such spectacular conjunctions of those unblinking stars.[4] Both clusters were certainly witnessed, and, more importantly, remembered, by the ancient Chinese, who must have gazed in amazement as they strove to comprehend their significance (Figures 4.5, 5.5).

An abundance of other literary and chronological evidence drawn from numerous Zhou Dynasty (1046–256 BCE) and Han Dynasty (206 BCE– 220 CE) sources suggests that these celestial events were taken from the start to signal the Supernal Lord's recognition of the legitimacy of a new regime, first Xia in 1953 BCE, followed by Shang in 1576 BCE, then Zhou in 1059 BCE, and finally Han in 205 BCE.[5] The three celestial events to which I previously drew

---

[4] The phrase "the *Five Planets* gathered in one lodge," *wu xing ju yu yi she*, generally used to denote a massing of planets (*Shiji*, 27.1312, "Treatise on the Celestial Offices," Appendix 1), should be understood to refer to a clustering of planets within a span of at most fifteen degrees in longitude. The Mawangdui MS "Prognostications of the Five Planets," *Wu xing zhan*, clearly implies this definition of "one lodge" by describing Venus's fifteen-degrees of retrograde motion as "travel in the opposite direction for one lodge" (*fan xing yi she*); Xi Zezong (1989b, 49). Similarly, in the *Han shu* "Monograph on Astrology" (26.1286) it says, "whichever lodge the *Five Planets* gather in, that state will rule the sub-celestial realm" (*fan wuxing suo ju xiu, qi guo wang tianxia*). In contrast, the critic Huang Yi-long (1990, 97) assumes a definition of thirty degrees for *yi she* and then generates an imposing number of supposedly qualifying planetary events. On this basis he questions the rarity and significance of observable planetary massings. When the narrower fifteen-degree definition is applied instead, only *four* of twenty-four planetary clusters Huang lists during the last two millennia BCE are actually found to qualify, for an average of one every 500 years. In fact, the spectacular massings of 1953 and 1059 BCE were much denser, spanning four and seven degrees respectively.

[5] Pankenier (1981–82; 1995). The connection between the founding of Xia and the spectacular planetary cluster of 1953 BCE was first pointed out in Pankenier (1983–85, 175–83). Subsequently, astronomers Pang (1987) and Pang and Bangert (1993) claimed the same discovery without acknowledging Pankenier (1983–85). Criticism of my identification of the 1576 BCE

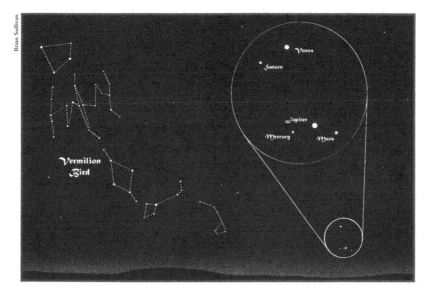

Figure 6.1 The cluster of the five visible planets in May 1059 BCE just ahead of the "*Beak*" of the *Vermilion Bird*. The enclosing circle is about seven degrees in diameter, making this the second-densest grouping in the past 5,000 years. After Pankenier (1998b, 34).

attention, in February 1953 BCE in lodge *Align-the-Hall* (*Yingshi* in Pegasus) (Figure 5.5), in November–December 1576 BCE in *Tail* and *Winnowing Basket* (Sco/Sgr), in May 1059 in *Ghost in the Conveyance* (Cancer) (Figure 6.1), are the earliest verifiable planetary events which the Chinese demonstrably witnessed, remembered, and interpreted as signs of the Supernal Lord's intentions. These massings of planets, which persisted for days and even weeks, would surely have impressed observers throughout the ancient world, although no other ancient records of their sighting from either Egypt or Mesopotamia have so far been found.[6]

phenomenon by Keenan (2002), following Baillie (2000, 223 ff.), is refuted in Pankenier (2007, 138). Both critiques result from a misreading of the discussion in Pankenier (1981–82, 18–19). The phenomenon recorded in the *Bamboo Annals* as "the Five Planets criss-crossed" describes a process that unfolded over more than a month and not a specific point in time. Baillie is not a historian and because he uncritically accepts Pang's uninformed views his otherwise pathbreaking study is marred by misstatements and ahistorical assumptions about Shang and Zhou history and chronology. The alignment of planets in May 205 BCE, which was taken at the time as the sign of the Mandate's conferral on the Han, will be dealt with in Part Four, Chapter Nine, "Astral prognostication and the Battle of Chengpu."

[6] Though they were certainly looking. Cuneiform texts exist which preserve a two-decade-long sequence of precise observations of the planet Venus during the reign of the penultimate king

Planetary Clusters Marking the Founding of Xia, Shang, and Zhou

	Sun	Mercury	Venus	Mars	Jupiter	Saturn	Lodge
Feb. 26, 1953	321°	295°	295°	295°	292°	296°	*Yingshi* (Peg/Aqr–Psc)
Dec. 20, 1576	255	234	279	236	234	238	*Wei-Ji* (Sco/Sgr)
May 28, 1059	56	79	82	75	77	82	*Yugui* (Cancer)

The latest, best-documented, and perhaps most illustrative of the three events is the massing of planets in 1059 BCE which occurred just west of the vast constellation known as the *Vermilion Bird*, which extends from lunar mansion *Willow* (δ Hya) to *Chariot Platform* (β Crv) (Figure 6.1). The account of the event in the *Bamboo Annals* chronicle recorded under the reign of the last king of Shang, Di Xin, a few years before the Zhou overthrow of the Shang Dynasty, currently reads, "the Five Planets gathered in *Chamber*; a great red crow alighted on the Zhou altar to the soil."[7] In addition, an independent record of a second astronomical event – a total lunar eclipse – in the "Xiao kai" chapter of the *Lost Books of Zhou* (*Yi Zhou shu*)[8] precisely dates that eclipse to the thirty-fifth year of King Wen of Zhou, first month, day *bingzi* (day thirteen in the cycle of sixty). This dating has proven correct to the day (night of March 12–13, 1065 BCE).[9] In his statement prompted by the ill omen King Wen enjoins officials to begin planning for the succession. King Wen then quotes

---

of the First Babylonian Dynasty, Ammizaduga. Hunger and Pingree (1999, 32–9); Reiner and Pingree (1975); see also Conman (2009, 15).

[7] The parallel account in *Mozi*, "Against Aggressive Warfare" (*Fei gong*), expands on this: "a scarlet crow clasping a jade scepter in its beak descended on the Zhou altar to the soil at Mt. Qi saying, 'Heaven commands King Wen of Zhou to attack Yin [Shang] and take possession of the State.'" The location *Chamber* (Sco) assigned to the planetary massing in the *Bamboo Annals* is a demonstrably late interpolation into the text, after the damaged bamboo slips on which the chronicle was written were recovered from a looted tomb and painstakingly reconstructed by scholars at the court of Emperor Wu of Jin in 281 CE. The planetary massing of 1059 BCE actually occurred in Cancer just ahead of the beak of the *Vermilion Bird* asterism. The reasons for the interpolation of the erroneous location of "Scorpius" were discussed in Pankenier (1981–82, 7–8), and in further detail in Pankenier (1992b, 279 ff.).

[8] For the history of the *Yi Zhoushu*, see Loewe (1993, 229–33).

[9] Zhu Youzeng (1940, 3.31). A recent study of this passage by Li Xueqin (2000) reconfirmed that this is indeed a record of a lunar eclipse in King Wen's thirty-fifth year. The dating of this lunar eclipse was simultaneously put forward by scholars working independently – Li Changhao (1981, 21) Pankenier (1981–82, 7). Another significant contemporary astronomical event may have been an apparition of Halley's Comet. A study by Zhang Yuzhe (1978) initially dated the comet apparition to 1057 BCE. Subsequent more rigorous analysis of Halley's Comet's orbit (Yeomans and Kiang, 1981), correcting for gravitational perturbations using all Chinese historical observations, showed Zhang Yuzhe's result to be in error by two years. Halley's Comet should have appeared in late 1059, some seven to eight months after the planetary massing the previous May. Pankenier (1983, 185). Accounts of the Zhou conquest and watershed Battle of Chengpu in *Huainanzi*, *Lunheng*, and *Yue jue shu* (*Jice kao*) indicate that, contrary to expectation, having the tail of the comet point in one's direction is advantageous (ibid., 194, n. 44).

an apposite adage, "the Brilliant Lights [sc. celestial bodies] are not constant, virtue alone is luminous [*ming ming fei chang, wei de wei ming*]." The accuracy of this eclipse record strongly corroborates the dating of the planetary massing associated with the "Red Bird" in 1059 to King Wen's forty-first year, exactly seven years later.[10] That cluster of five planets was clearly visible for many days after full darkness had fallen as the "great Red Bird" set in the northwest, "clasping in its beak a scepter," which we now recognize as a description of the planetary formation.

From the historical accounts in the *Book of Odes, Bamboo Annals, Discourses of the States, The Grand Scribe's Records*, and elsewhere we know that King Wen promptly undertook political and military actions that clearly revealed his intention to challenge his Shang overlord. But King Wen did not live to see the Shang Dynasty overthrown, instead dying in 1050, the ninth year of the new calendar he inaugurated in 1058, the so-called "First Year of the Receipt of [Heaven's] Mandate" (*shou ming yuan nian*). Not until Jupiter returned again to the heart of the *Vermilion Bird* constellation in early 1046 BCE did King Wen's son and heir King Wu actually launch the successful campaign, which culminated in the conquest of Shang on 20 March 1046 BCE (second month of the calendar for that year, on the first day of the sexagenary cycle, *jiazi*).[11] During the march from Zhou and the decisive battle, Jupiter remained stationary within three to four degreees of the *Bird Star* (α Hya at the heart of the *Vermilion Bird*), as stipulated by the "Discourses of Zhou" chapter of the *Discourses of the States*: "When King Wu attacked King Zhòu [i.e. Di Xin of Shang], Jupiter was in *Quail Fire* . . . our astrological space, the Zhou people." The pre-Qin text *The Pheasant Cap Master* (*Heguanzi*), "Du wan," states explicitly that this bird constellation, variously described as a scarlet crow, quail, pheasant, or vermilion sparrow (*zhu que*), is in fact the Phoenix: "the *feng huang* [phoenix] is the fowl of *Quail Fire*, the quintessence of *yang*."[12]

---

[10] The *Lost Books of Zhou* and the *Bamboo Annals* both consistently indicate that King Wen died in his fiftieth year of rule, nine years after receiving the Mandate and nine years after the planetary portent. Pankenier (1992b, 498–510). The recently discovered Qinghua Warring States bamboo slip text "Bao xun" also mentions King Wen's demise after fifty years of rule. Liu Guozhong (2011, 79).

[11] January 20, 1046 BCE, exactly one sexagenary cycle earlier, is also a possibility, as I initially proposed in 1981 (Pankenier, 1981–82). The fact that March 20 coincided with Jupiter's opposition and stationary episode prompts me to favor the latter date. John Justeson (1989, 104) discusses how stationary episodes of Jupiter and Saturn were used by the Maya to schedule human affairs "ostensibly as sacred mandates for elite decision-making." See also Milbrath (1999, 233–4, 241–2, 247). If the historical record placing Jupiter in Quail Fire (α Hydrae) is correct, and a preponderance of the evidence indicates that it is, then the date of 1045 proposed by some for the Zhou conquest is plainly out of the question. Cf. Shaughnessy (1999), Li (in press).

[12] Chen Jiujin, Lu Yang, and Liu Yaohan (1984, 73). As Chen, Lu, and Liu show, *Heguanzi* reflects astronomical traditions of the Yi minority of Shu in the southwest (a people historically

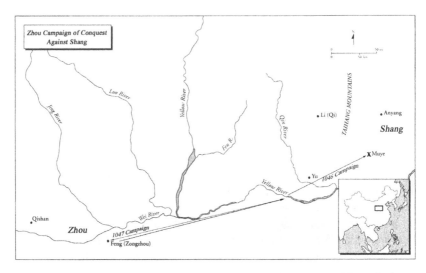

Map 6.1 The Zhou campaign against Shang from Feng to Muye. Redrawn from Loewe and Shaughnessy (1999, 308, Map 5.2), © Cambridge University Press.

Close scrutiny of Jupiter's behavior during the previous campaign season in 1048, nominally the twelfth in the sequence of Jupiter stations, shows that after steadily advancing eastward through the summer, in late 1048 Jupiter suddenly ceased its forward motion toward the *Bird Star*, paused, reversed direction, and then began to back off (i.e. retrograde). This unexpected development could explain why the first Zhou campaign was called off at the last minute after the armies had already reached the Yellow River ford at Mengjin (lit. "Meng Ford") in Henan. According to the "Basic Annals of Zhou" account in *The*

linked to the Qiang, neighbors of the Zhou on the west and southwest). *Heguanzi* is the only pre-Qin text to equate *Quail Fire* with the phoenix. When Sima Qian cites "*Beak,*" "*Gullet,*" and "*Crop*" in the "Treatise on the Celestial Offices" (*Shiji,* 27.1303, Appendix 1) as still current variant astronyms for lodges "*Willow*" (24), "*Seven Stars*" (25), and "*Spread*" (26), he is probably following those Yi traditions, which may have originated in Qiang–Zhou astral nomenclature. Yi astronomy also pays particular attention to lunar phases, dividing the month into two periods of twelve days before and after full moon, which could explain the conspicuous emergence of lunar phases in dating formulas in the early Western Zhou bronze inscriptions when the Shang previously evinced no such interest. Chen Jiujin, Lu Yang, and Liu Yaohan (1984, 121, 254, 263), Kalinowski (1986). The division of the month denoted by the lunar phase terms in Western Zhou bronze inscriptional dates is strikingly similar to the Yi–Qiang scheme. Xu Fengxian (2010–11), Pankenier (1992c). For further discussion of the passage from the "Discourses of Zhou," see Cullen (2001, 47–51). For the archaeological evidence bearing on interethnic relations with the Qiang in Shang and Zhou, see von Falkenhausen (2006, 201–2, 210, 237).

Grand Scribe's Records, it was at this point, after the Zhou allies had declared the time to be right – "all said, [Shang] Zhòu can be attacked" – that King Wu announced to his anxious comrades-in-arms, "you do not know Heaven's Mandate; it may not yet be done," and retreated from Mengjin to the Zhou homeland in the Wei valley.[13] Related to this aborted campaign is a passage from the (now lost) Warring States period (403–221 BCE) military text *Six Quivers* (*Liu Tao*), quoted in Wang Yi's (fl. *c*.100 CE) commentary on the fourth-century BCE text "Heaven Questioned" (*Tian wen*): "On the morning of the first day we took our oath. How did we all arrive in time? When the geese came flocking together, who was it made them gather?"[14] The answer to the first question, clearly, appears to be "they followed the signs in the sky!" This is confirmed by another passage in the *Six Quivers*. King Wen asks his chief counsellor, Tai Gong (Lü Shang), how to respond to Shang king Di Xin's tyranny and cruelty, to which Tai Gong responds,

Let the King cultivate virtue, respect the worthy, show benevolence towards the people, and observe Heaven's Way. If Heaven's Way [displays] no baleful sign, one may not pre-empt it; if Man's Way [suffers] no calamity, one may not go ahead and plot strategy.[15]

The *Six Quivers* then sheds more light on the circumstances of the aborted campaign:

When King Wu [first] went east to attack [Shang] and reached the banks of the Yellow River, it rained heavily and thundered intensely. Dan, Duke of Zhou, came before King Wu and said, "Heaven does not assist Zhou. The import is that my Lord's virtue and comportment are not without flaw, the people are afflicted and complaining. Therefore Heaven sends down calamities on us. I request to withdraw the army." Tai Gong said, "You may not." King Wu and the Duke of Zhou looked in the distance at the ranks of Zhòu [Di Xin's] troops, drew up the army and halted [the advance]. Tai Gong asked, "Sire, why do you not have them charge?" The Duke of Zhou said, "The season of Heaven is not with us. Divination by turtle and firebrand give no sign. The prognostication by milfoil is inauspicious, it is perverse and unfavorable; what is more, the changes in the stars are baleful. Therefore, [I] Dan halted them. How could they [be allowed to] advance?"[16]

Further study of the events surrounding the Zhou conquest of Shang and recon-struction of the chronology of the period has confirmed the ancient astrological traditions and revealed new facts about the Zhou founders' claim to have acted at the instigation of Heaven and to have received Heaven's Mandate.[17] Here it

---

[13] Pankenier (1981–82, 15–16).   [14] Hong Xingzu (1971, 3.19a), trans. Hawkes (1959).
[15] Following the wording of the Yinqueshan bamboo slip MS version. Sheng Dongling (1992, 47). Baleful signs in "Heaven's Way" refers to anomalous celestial phenomena or inauspicious astral omens for the Shang.
[16] Hong Xingzu (1971, 3.19a).
[17] In a recent study, Kai Vogelsang (2002, 193) avers that "the all-important idea of the 'Mandate of Heaven' . . . appears fully developed only in the 'Zhou shu.' While it is invoked eight times in

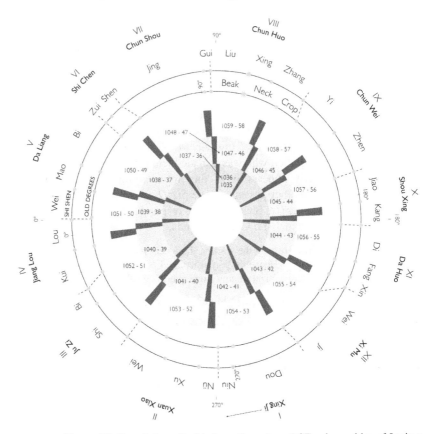

Figure 6.2 Correlation of mid-eleventh-century BCE ephemerides of Jupiter
and the locations of lodges *Willow–Seven Stars–Spread* (aka *Beak–Neck–
Crop*), and the *Bird Star*, α Hya (marked by the small star), in *Quail Fire*,
the heart of the *Vermilion Bird* constellation. The sequence of twelve Jupiter
stations is labeled in the outer ring, with the vernal equinox zero degrees
at the left. The names of the twenty-eight lodges and the positions of their
determinative stars according to Shi Shen's scheme (*c.*450 BCE) are the next
ring, followed by the Old Degree system of determinatives (*c.* sixth century
BCE). Summer solstice and equinoxes are indicated for the mid eleventh
century BCE. The next three rings show Jupiter's periods of visibility during
three synodic periods in the mid eleventh century. Dark sections show when
the planet was unobservable. For example, in 1059 Jupiter was last visible
at dusk in the west on or about June 11. It subsequently reappeared at dawn
in the east in mid-July, just within *Willow*, the *Beak* of the *Vermilion Bird*.
Moving toward the center of the diagram, the following two wedges define
the two subsequent *Quail Fire* years. Note that a sequential count from *Quail
Fire* in 1059–1058 through *Quail Fire* of 1047–1046 makes thirteen stations,
not the twelve that one might expect given Jupiter's nominal twelve-year
sidereal period. This may be part of the explanation for King Wu's launching
a campaign prematurely in 1048–1047 (graphic Sophia Pankenier).

is worth pondering repeated allusions to that "Great Command" preserved in renderings of early pronouncements in core chapters of the *Book of Documents*, such as in the "Great Announcement" (*Da gao*): "Heaven brightly manifests its awesome majesty and supports our very great foundation"; in the "Many Regions" (*Duo fang*) with regard to the dissoluteness of the Shang: "Heaven then searched in your numerous regions and greatly shook you by its severity; it would spare those who had regard for Heaven, but in your numerous regions there were none who were able to have regard for it"; in the "Many Officers" (*Duo shi*): "Their successor in our time [i.e. Shang king Di Xin] greatly lacked a clear manifestation in Heaven . . . He had no consideration for Heaven's manifestations, nor for what the people revere." The epoch-making events surrounding this singularly impressive planetary portent would seem to confirm the early Chinese preoccupation with revealed wisdom in the form of "bright manifestations" (*tian xian* or *tian wei*) displayed by Heaven:[18] the appearance of an auspicious phoenix (*feng huang, luan huang*) in conjunction with the rise of Zhou; the association of the *Quail Fire* space occupied by

the 'gao' chapters . . . only one early Western Zhou inscription, that of the Da Yu *ding*, contains an expression that may be associated with the idea . . . It is no accident that our knowledge of this element of Zhou ideology is based entirely on the *Documents*: it simply cannot be derived from inscriptions." It is asking a lot to demand that the Mandate of Heaven idea be "fully developed" in the extremely terse and formulaic Western Zhou inscriptions. More to the point, one cannot ignore the mid-eleventh-century *He zun* inscription from the reign of King Cheng, in which the King declares, "In the past your fathers were able to aid King Wen, whereupon King Wen received this [Great Command]. When King Wu conquered the great state of Shang, he then made reverent declaration to Heaven, saying: 'Let me dwell in this, the center of the land, and from here govern the people.' Hark! While you are still minors lacking in understanding, look to your fathers' scrupulous respect for Heaven. Comprehend my commands and respectfully follow orders! [Your] sovereign's reverential virtue finds favor with Heaven, which guides me in my slow-wittedness." Presumably, Vogelsang disqualified the *He zun* because the two graphs "Great Command" (*da ling/ming*) are illegible in the published rubbings. However, Ma Chengyuan and Tang Lan, both eminent paleographic experts, personally examined the vessel and partially discerned the graphs and transcribe the inscription accordingly. Ma Chengyuan (1976; 1992, 362), and especially Tang Lan (1976; 1986, 73). If one considers the speech as a whole it is difficult to imagine any other two words that could fit the context, especially since the first three clauses later became stock phrases. King Wu's proclamation is about as fully developed a statement of the concept of Heaven's Mandate as one could ask of an early Western Zhou commemorative inscription. For a translation and discussion of the *He zun* inscription, see Chapter 7 below.

18 That the early texts should refer only obliquely to the phenomena is not in itself remarkable. A similar reticence has been observed by John Justeson (1989, 76) in classic Maya texts: "[As] Maya astronomy concerned the behavior of the sky gods, 'deities whose activities vitally influenced human affairs' . . . astronomical correlates of historical events have begun to be recognized in the essentially historical narratives. Classic texts almost never mention these correlates; seldom do they make any explicit astrological statements, referring instead to associated human events. As in the interpretation of structure alignments, these unstated correlates must be inferred from distinctive patterns and demonstrated by statistical argumentation." He concludes (ibid., 115), "So the features that may correlate with specific astronomical events generally are not reflected by direct textual statements; if their presence is marked at all, it is in accompanying scenes."

the *Vermilion Bird* constellation with the fate of Zhou in the "field allocation" (*fen ye*) astrological scheme; the practice in Zhou times of divining the fate of rulers of states, outcomes of battles, etc. based on Jupiter's location;[19] and, last but not least, the obligatory association of five-planet massings with dynastic transitions in the astral omenology of the early imperial period.

A recently discovered Warring States text – *Qi ye*

The close attention in early Zhou to the behavior of heavenly bodies for the purpose of astral prognostication has now been confirmed by the earliest text concerning events in the conquest period thus far discovered. The recently unearthed Warring States text *Qi ye* among the Qinghua bamboo slips gives an account of the elaborate ceremony (*yin zhi*) in Zhou celebrating the victory over the state of Qi (Li).[a] During the celebration poems were composed and recited by the participants, including "King" Wu and the Duke of Zhou. (As is often the case in retrospective accounts, King Wu is referred to by his posthumous temple name.) Although incomplete, a previously unknown ode recited by the Duke of Zhou and dedicated to King Wu, "Luminous, Luminous, Supernal Lord," *Ming, ming Shangdi*, is of particular importance:

"Luminous, Luminous the Supernal Lord, brightly gleaming descends below, [his] illustrious [presence] drawing near, delighting in the fragrance of the offerings; in [. . . lacuna] the Moon has its waxing and waning, and Jupiter has its pauses and motions; [I] compose this song of felicitation, ten thousand years without end." A second ode, "The Cricket", *Xishuai*, also uses the Sun, Moon, and Jupiter figuratively in a similar fashion and mentions retrograde motion.

Even though crucial lines are lost, enough remains of both odes to confirm several important facts: (i) the Supernal Lord and celestial bodies are being invoked in the context of a victory celebration immediately following the return of the army; (ii) the Supernal Lord is being thanked directly with sacrificial offerings for the victory; (iii) crucial aspects of the changes in the Moon and Jupiter's appearance and movement are singled out, not merely for poetic effect as examples of Heaven's luminosity, but as signs of timeliness and changing circumstances; (iv) observation of Jupiter's movements had already led to a recognition of the planet's stationary episodes and retrograde motion, pointing to intimate knowledge.

---

[19] In Chapter 9 we will see that, mimicking the Zhou conquest precedent, prognostication based on the passage of precisely one Jupiter cycle plays a crucial role in the account of Duke Wen of Jin's restoration in the "Discourses of Jin" chapter of the *Discourses of the States*, 10.11a–12a.

The *Qi ye* begins the account of the victory celebration with "King Wu in the eighth year punishingly attacked Qi," *Wu Wang ba nian zheng fa Qi*, which has heretofore been interpreted to mean that the defeat of Qi (or Li) occurred in King Wu's eighth year. This fits no reasonable chronological scenario, and dating the defeat of Qi *after* King Wen's death would contradict every other known source. The phrase *Wu Wang ba nian*, "King Wu in the eighth year," does not begin with the prepositioned copula *wei*, which was the standard practice when quoting an annalistic record of a dated event – e.g. "it was in the eighth year of King Wu," *wei Wu Wang ba nian*. Therefore this is a narrative on a par with received accounts and it is debatable whether the event is actually being assigned to King Wu's reign. Rather, I would argue that it should be "King Wu in the eighth year [of the Mandate] attacked Qi," in which case the event took place while King Wu's father, King Wen, still had three years to live. This interpretation agrees with other historical sources, and is consistent with the usual pattern wherein the events of the last decade of King Wen's reign are dated in relative terms: "Receipt of the Mandate (*shou ming*) such and such a year; the next year (or second year)...; the next year (or third year)...", with *shou ming* typically elided. All those historical accounts, including the *Bamboo Annals*, date the victory over Qi to the sixth year of the Mandate (1053 in my reconstruction).[b] Thus, in my view, the *Qi ye* copyist likely miswrote the original "six" as "eight," a common enough scribal error.

A challenging issue remains, however, namely that the *Qi ye* prologue goes on to mention that the celebration was held in a temple called *Wen Taishi*, "Great Hall of Culture." Some have taken the "*wen*" to be a posthumous reference to King Wu's father, in which case he was already deceased by the time of the celebration. This, they feel, is borne out by the fact that the Duke of Zhou, the Duke of Shao, and King Wu's siblings all play prominent roles in the proceedings, while King Wen is not mentioned. But this omission could simply be a matter of authorial discretion. It would seem that the *Qi ye* passage refers either to the defeat of Qi in the sixth year of the Mandate and the location is the Zhou Ancestral Hall (i.e. *wenzu*), or else as a historical retrospective the passage conflates two different events, the victory celebration following the defeat of Qi and another similar event sometime after King Wen's demise. In any event, the passage presents several internal contradictions.

[a] Liu Guozhong (2011, 131, 209). This event is also the subject of the chapter in the *Book of Documents* entitled "The Earl of the West Suppressed Li" (*Xi Bo kan Li*). King Wu was never titled Earl of the West. The Qinghua slips also contain graphic forms and terminology that point to a Western Zhou date. For *Sui* as Jupiter, see Li Feng (in press).

[b] Pankenier (1981–82, 33; 1983, 325; 1992a, 278; 1992b). This solution would resolve the contradiction regarding the identity of the reigning king at the time of the capture of Qi

> (Li) between the *Qi ye* and the transmitted texts *Lost Books of Zhou, Bamboo Annals,* and *Shangshu dazhuan* (*Great Commentary on the Book of Documents*). Fang and Wang (1981, 231); Zhu (1974, 84); *Shangshu dazhuan,* 2.16b, 4:5a. See also Du Yu's (222–85 CE) comments regarding King Wen's term of imprisonment in *Zuozhuan* (Duke Xiang, thirty-first year).

The sequence of celestial events accompanying the Zhou conquest remained precedent-setting for well over a millennium thereafter. Consider this passage from a memorial to Cao Pi (187–226), first Emperor of Wei (in waiting), in conjunction with the abdication of the last emperor of the Han Dynasty, the "Proferring Emperor" Xiandi (190–220 CE):

On day *xinyou* [December 1, 220 CE] Consulting Erudites Su Lin and Dong Ba submitted a memorial saying, "The twelve stations of Heaven are astrological fields; each kingdom and principality belongs to such an allotment. Zhou is in *Quail Fire* and Wei is in *Great Span*. Jupiter progresses in succession through each territory's station, the Son of Heaven receives the Mandate and the lords of the states are enfeoffed accordingly. When King Wen of Zhou first received the Mandate, Jupiter was in *Quail Fire*. Down to King Wu's attack on [Shang] Zhòu it was thirteen years and Jupiter was again in *Quail Fire*, therefore the *Spring and Autumn Commentary* [i.e. *Discourses of the States*] says, "When King Wu attacked [Shang] Zhou, Jupiter was in *Quail Fire*; where Jupiter was located is the astrological field allotted to us, the Zhou." Formerly, in the seventh year of [the reign period] Brilliant Harmony [184 CE] Jupiter was in *Great Span* and the Martial King [of Wei, Cao Cao (155–220)] first received the Mandate when he led the suppression of the Yellow Turban [Rebellion]. That year was changed to become the first year of the Pacified Middle reign period. [Subsequently,] in the first year of Established Tranquility reign period [196 CE], Jupiter was again in *Great Span* and [Cao Cao] first rose to become general-in-chief. Thirteen years later [Jupiter] was once more in *Great Span* and [Cao Cao] was first elevated to the post of Chancellor. Now, twenty-five more years [have elapsed], Jupiter is again in *Great Span*, and your Majesty [Cao Pi] has received the Mandate. Thus Wei's obtaining Jupiter's [sanction] corresponds to King Wen of Zhou's receipt of the Mandate.[20]

From one to three – from history to "prehistory"

Using the 1059 BCE date of the Zhou Mandate conjunction as a benchmark, it became possible to interpret similar accounts in pre-Qin sources such as *Mozi*, which takes up the precedent-setting portents associated with all three dynastic foundings in the second millennium BCE.[21] With the help of the relative date of the *Bamboo Annals* placing the Shang founding 517 years before the Zhou event, the curious behavior of the planets recorded at that juncture, "the Five Planets traveled cross-wise" (*wuxing cuo xing*), became comprehensible as a

---

[20] *History of the Three Kingdoms* (*Sanguozhi*), 2.75. For the later history of planetary portents and the Mandate, see Part Four below.
[21] Pankenier (1983–85).

description of the planets' motions in the fall of 1576, when in a matter of weeks the five variously reversed horizons and times of visibility from dusk to dawn and dawn to dusk. Moreover, the description in *Mozi* of a still earlier bestowal of a jade scepter of authority on the founder of the Xia Dynasty in a "Dark Palace" (*xuan gong*) matched that of the Zhou portent.[22] This led to the confirmation of early Chinese observation of the densest massing of planets in over 5,000 years discussed above, that of 1953 BCE. Once again the "jade scepter" bestowed by a supra-sensible agency referred to an extraordinary planetary event.[23] With that the three planetary clusters associated with the founding of the Three Dynasties emerged from the obscurity of millennia.

Derk Bodde took note of the fact that Joseph Needham "points out that conjunctions of Jupiter, Saturn, and Mars recur every 516.33 years, which, he thinks, could be the basis for Mencius' belief" in the appearance of sages every 500 years.[24] Actually, Needham was merely following Herbert Chatley's earlier conjecture,[25] but it is worth noting that, although neither Bodde nor Needham makes this plain, the *only way* the Zhou Chinese could have inferred the existence of a 500-year period would be if both the 1576 and 1059 BCE planetary phenomena had been observed and recorded in annalistic form, from which their separation by 517 years could later be deduced.[26]

Critical objections

In criticizing the use of records of planetary clusters in chronological studies Huang Yi-long treated the planetary events in isolation, divorced from the historical context. He ignored the plethora of historical and chronological evidence adduced in verifying the recorded observations. Instead, Huang offered his own view that later textual accounts, if not entirely correct, must be entirely spurious. Hence, even though the densest massing of the planets in nearly a thousand years in either direction demonstrably occurred in 1059 BCE, Huang concludes, because the actual planetary event occurred in Cancer rather than in Scorpius as some later textual sources say, that the whole report of the observation must be dismissed as a late invention. Even though he admits the recorded date in

---

[22] The "Dark Palace" was identified by the late Warring States period astrologer Shi Shen as lodge *Align-the-Hall* (Pegasus); see *History of the Jin Dynasty, Jin shu* ("Monograph on Astrology," *Tianwen zhi*, 11.301), where Shi Shen's pre-Han astrological nomenclature is preserved.

[23] Pankenier (1983–85), also Weitzel (1945, 159–61).

[24] Bodde (1991, 123). Needham and Wang (1959, 408).   [25] Pankenier (1981–82, 24).

[26] There is no record of the next impressive clustering of all five planets in the same 517-year series, which occurred when Confucius was still a youth, in November–December of 543 BCE. The following cluster in the same series, in April of 26 BCE, was observed and recorded in the *History of the Former Han Dynasty, Han shu*, but as a conjunction of Saturn, Jupiter, and Mars only, since, unlike the three previous occasions, Mercury and Venus remained at a distance. *Han shu*, 3.1310.

Han shu of Liu Bang's much less impressive "Mandate" planetary alignment in 205 BCE was manipulated for political reasons, Huang finds it impossible to imagine that an authentic account of the 1059 event could have been similarly adapted to conform to Han period Five Elemental Phases speculations. Critics like Huang ignore the fact that the true location of the 1059 event in the *Beak* of the *Vermilion Bird* asterism is implicated in the authentically pre-Qin account in *Mozi* and forget that the chronology of the *Bamboo Annals* for the period can be shown to yield the true date, if one simply recognizes the location "lodge *Chamber*" for what it is – a late interpolation arising from Han portentological revisionism (see *infra*). To reduce just one corroborating chronological datum to its simplest terms: Scorpius is eight years later than Cancer (true location of the 1059 cluster of planets) in the Jupiter cycle. If one merely adds eight years to the 509-year span separating the two planetary events recorded in the *Bamboo Annals* to account for Jupiter's true location in Cancer, one arrives at the figure 517 years, which is precisely the period separating the *actual* astronomical events to which I called attention (i.e. 1576–1059 = 517). Given the scholarly consensus that the Zhou conquest occurred in the mid eleventh century BCE and confirming evidence from the Xia Shang Zhou Chronology Project,[27] there is no possibility here of picking and choosing among a variety of much less impressive planetary phenomena as Huang Yi-long implies in his misrepresentation of the situation.[28] Astronomer Zhang Peiyu recently put the matter succinctly:

[27] Xia Shang Zhou duandai gongcheng zhuanjiazu (2000). Ignoring the evidence for a mid-eleventh-century date for the Zhou conquest, Sarah Allan (2009, 6, n. 16) dismisses reliance even on scientifically verifiable records in the *Bamboo Annals*. In lieu of the "Mandate" planetary massing of 1059 BCE, which demonstrably occurred, and the observation of Halley's Comet, Sarah Allan bases her argument (ibid., 6, 38) on the supposed occurrence of some other unrecorded, transient astral phenomenon for which there is no evidence.

[28] Jiang Xiaoyuan (2004, 115, 242) followed Huang Yi-long (1990) in rejecting the authenticity of the early records of planetary massings out of hand without examining the historical evidence. Both Jiang and Huang ignore the record of the *bingzi* lunar eclipse in King Wen's thirty-fifth year recorded in the *Lost Books of Zhou*, "Xiao kai" chapter, mentioned earlier. Though innocent of any expertise in ancient Chinese history or textual criticism, Douglas J. Keenan (2002, 61–8) too raises doubts about the historicity of the Chinese observations and dating methodology. Keenan is cited approvingly by John M. Steele, despite Steele's own admonition (2004, 346) that "in order to use early astronomical records in modern science it is necessary to understand the cultural background to those records." If one disregards the obfuscations and appeals to authority, the fundamental historical question (which Huang, Keenan, and Steele avoid confronting) is: how could the ancient Chinese, more than a millennium after the fact and lacking the ability to retrospectively compute multiple planetary phenomena, have precise knowledge of an eleventh-century BCE planetary grouping (not to mention earlier phenomena), *unless knowledge of those events had been preserved and handed down?* Unbeknownst to Huang, Keenan, and Steele, it is common knowledge among historians of early China that Sima Qian's *The Grand Scribe's Records* (*c*.100 BCE) reproduces a substantially accurate Shang dynasty king list from the last half of the second millennium BCE. Obviously, records survived and a mode of transmission existed. The quotation below from Huan Tan's *New Discourses* proves it.

it is particularly important to point out that starting from the circulation of the *San tong li* [Liu Xin's "Triple Concordance"] calendar, compiled in the first century AD, ancient scholars began to show great interest in the retro-calculation of the exact dates and cycle of planetary conjunctions. However, the computation of planetary trajectories is a complex exercise, and so those early computations contain many inaccuracies: calculating the exact locations of conjunctions of over 1,000 years in the past would have been unthinkable for those early astronomers. Because of this difficulty, I would argue that it would not have been possible for scholars of the Warring States or Han period (when the received classical texts containing reference to those astronomical events were first recorded) to have been able to accurately retro-calculate the exact time and location of the planetary conjunction that correlates to the Conquest of Shang. Since this event can be shown by modern calculations to have actually occurred, and because it was recorded in the historical traditions, we can thus eliminate the possibility of a falsification of records of this conjunction by later hands.[29]

Therefore, these discoveries confirm the historicity, indeed the great antiquity, of chronological and astronomical data in the *Bamboo Annals*, *Mozi*, and other Zhou and Han works, and they explain why such planetary phenomena later came to be viewed as definitive omens of a change of Heaven's Mandate to rule. They also suggest that a concept ancestral to the Zhou "Mandate of Heaven" existed prior to the founding of Shang in the mid second millennium BCE, as implied in the *Book of Documents*: "You know that the earlier men of Yin [Shang] had documents and records of how Yin superseded the Mandate of Xia."[30] Furthermore, the 517-year period of the planetary phenomena of 1576 and 1059 confirms the accuracy of Mencius' (fourth-century BCE) assertion that "just over 500 years" separated Shang Dynasty founder Cheng Tang from the Zhou founder King Wen, and King Wen from Confucius (551–479 BCE).[31]

Small wonder, then, that, from Confucius' time on, Heaven was expected to intervene at any moment to raise up a new sage ruler and dynastic founder to bring an end to the political chaos and internecine warfare of the Warring States period. When this failed to occur, the doctrinal solution among Confucius' disciples was to anoint him "uncrowned king" and produce esoteric exegeses of canonical texts like the *Springs and Autumns Annals* chronicle to buttress this contention. Remarkably, the mid-Han apocryphal or unorthodox "weft" commentaries (*chen wei*) on the Classics contain recognizable accounts of the

---

[29] Zhang Peiyu (2002, 350).

[30] Trans. Karlgren (1950a, 56). In this connection it is worth noting David N. Keightley's (1982, 272, 296) conclusion: "I believe that Akatsuka is correct in discerning a tension, finally resolved in Western Chou in favor of Ti (or T'ien), between the worship of an impartial Ti and the worship of the partial ancestors . . . and in suggesting that Shang Ti did not restrict his assistance to only the Yin court . . . so that one may indeed conceive of a 'Mandate of Ti,' precursor to the Chou 'Mandate of Heaven,' which could inflict such disasters as crop failures and enemy attacks on the Shang."

[31] *Mencius*, IV/B26.

eleventh-century BCE planetary phenomenon, though they also exhibit traces of alteration in response to contemporary doctrinal debates and portentological speculation. However, the kernel of astronomical fact is still clearly discernible. Consider once again this particularly striking account of King Wen's receipt of the Mandate from Huan Tan's (d. 28 CE) "New Discourses":

Afterward there was a Phoenix in the suburbs that grasped a "Writing" in its beak. King Wen said, "The Yin [Shang] Lord does not act according to the Way, [he] tyrannizes and disorders all under Heaven. The August Mandate has already shifted, [he] will not persist for long." Thereupon King Wen composed the "Song of the Phoenix," which goes,

> The Phoenix soars [down] on spreading wings;
> Clasping a "Writing" it comes gamboling, thereby to command Chang [King Wen].
> I gazed up at Heaven and examined the Diagram;
> Yin [Shang] is about to expire [it portended].
> Great Heaven is azure, azure;
> First there [sc. in the heavens] was a presage [lit. "sprouting"].
> The linked essences of the Five Spirits [the planets in Han usage] met in lodge *Chamber* [Sco] to deliberate.[32]

Although this was composed after the planetary massing had already been reassigned to lunar mansion *Chamber*, there is no mistaking the identities of the Five Spirits. The elemental phase associated with Zhou was first changed from Fire to Wood at the end of the Western Han Dynasty, hence it is only from the apocrypha of mid-Han date on that the location of the Zhou planetary omen began to be reported as *Chamber*.[33]

Given the ideological significance of celestial portents in Huan Tan's time it is surprising that, although the Han founder Gao Zu's planetary portent of 205 BCE is mentioned several times, nowhere in *The Grand Scribe's Records* (*c*.100 BCE) nor in the *History of the Former Han Dynasty* (*c*.100 CE) is there specific mention of the precedent-setting planetary clusters explicitly associated elsewhere with the transfers of Heaven's Mandate during the Three Dynasties period.[34] It is evident, therefore, that the apocrypha and other late Zhou and

---

[32] *Taiping yulan*, 84.5b.
[33] Aihe Wang (2000, 137–55) discusses the doctrinal and ideological background of this process in detail. For a passage from Liu Xin's *Canon of the Generations* (*Shi jing*) illustrating Liu's role in this revision of history, see Cullen (2001, 47). See also *Wen xuan* (sixth century CE): "the three humane [officials] deserted the [Shang] state and the five luminosities entered *Chamber*; *Chunqiu yuan ming bao* says: 'In the time of Zhou of Yin [Shang] the Five Planets gathered in *Chamber*. *Chamber* is the essence of the Azure [i.e., Wood] Spirit; based on it the Zhou arose,'" quoted in *Wenxuan*, 59.28. For mention of the Five Essences in connection with Heaven's Mandate in the *Forest of Changes* (*Yilin*), see Gu Jiegang (n.d., Vol. 3, 27, 34).
[34] The planetary massing of 205 BCE in Gemini–Cancer that was taken as a sign of the transfer of the Mandate to Han is recorded in *The Grand Scribe's Records* (*Shiji*, 27.1348 and 89.2581) and also *Han shu* (26.1301 and 36.1964); see Chapter 9 below.

Han texts preserve remnants of ancient traditions which are not represented in the standard historical sources. The survival of such accounts is probably attributable to their transmission as components of popular constellations of texts and beliefs concerning the interventionist role of the archaic sky god (whether Heaven or the Supernal Lord),[35] who bestowed esoteric revelations by means of "iconic images" (*xiang*) or "heavenly patterns" (*tian wen*) in the sky.

In the past, despite the seminal work by Hu Houxuan,[36] Xu Fuguan,[37] Gu Jiegang[38] and others, the relative scarcity of concrete evidence for this kind of cosmological conception in the second millennium BCE posed an obstacle. Hence my claim in 1982 that a record of the planetary cluster of 1953 BCE could have been transmitted for centuries before the appearance in the archaeological record of Chinese writing at first blush appeared to require a leap of faith. But as I have demonstrated, the lack of contemporaneous written records from Xia and early Shang does not pose insurmountable problems for the historian. As we saw in Part Two, Chinese writing probably pre-dates the earliest oracle bone inscriptions by a considerable time. We also know from archaeological as well as epigraphic evidence – the graphic shape of the oracle bone characters for "record" (*dian*), "document" (*ce*), and "writing brush," as well as brush-written characters on pots and bones going back at least to Taosi in 2100 BCE – that the Shang Chinese were undoubtedly writing with brush and ink on perishable materials such as bamboo or wood which have not survived.[39] Then, too, as we saw above, there is the testimony in the *Book of Documents*, *Zuozhuan*, and elsewhere that certain early archives could still be consulted in late Shang and early Zhou. This is borne out by the fact that the portions of the "Canon of Yao" dealing with the gods of the winds and of the four quarters, even though they were unintelligible to the late Zhou compilers of the "Canon," were faithfully copied and transmitted nonetheless. Today, with the benefit of references to them in the Shang oracle bones, we can interpret what the Eastern Zhou and Warring States compilers of the *Documents* could not.[40]

Early antecedents of "Five Elemental Phases" correlations

In what follows, I survey other avenues by which to approach the thought of the Three Dynasties period, in order to establish something of the cosmo-political context in which the astral omens were observed. Recent evidence bearing on cosmological conceptions of the late Neolithic and early Bronze Age, together

---

[35] The spate of bamboo and silk manuscripts discovered in China in recent decades shows that many more texts than previously suspected were in circulation in Warring States times.
[36] Hu Houxuan (1983, 1–29).   [37] Xu Fuguan (1961).   [38] Gu Jiegang (n.d., Vol. 5, 425).
[39] Bagley (2004, 215, 219).   [40] Liu Qiyu (2004, 68).

with comparative and theoretical approaches to the study of the history of religions, may help bridge the historiographical gaps.

Although the idea of correlations between the natural and the human worlds, particularly as they relate to the cycle of seasons and the alternation of light and dark, no doubt emerged spontaneously in antiquity,[41] the second-order linkage between the rise and fall of political entities and a sequence of elemental forces is by no means equally self-evident. Hence it has always seemed that the "mutual production" order of elemental phases (Wood, Fire, Earth, Metal, Water) expounded in many late Zhou and Han texts stood on firmer ground as an interpretive scheme, since it corresponds to the seasonal order in which the elemental forces exhibit their influence, starting with Spring (Earth being later identified with the center).[42] The speculative theories of Zou Yan (fl. mid third century BCE) and the Warring States *yin-yang* specialists, on the other hand, according to which the recurring sequence of five cosmic forces directly influenced dynastic fortunes, have seemed just that – speculations with no sound basis in fact, historical or otherwise. But the discovery of a consistent correlation of a sequence of *three* remarkable planetary observations in paradigmatic succession – winter, autumn, and summer "palaces" of the heavens – with the conferral of Heaven's Mandate on successive dynastic founders, as well as their association in the sources with relevant colorful omens and "elemental forces" (black/Water, white/Metal, red/Fire), offers compelling evidence that cosmological speculation and dynastic politics were already linked at this time.[43] This sequence of basic correlations, present in the accounts cited in *Mozi* and elsewhere, and achievable only through contemporaneous observation of the

---

[41] Above we saw examples of an acute awareness of directionality in the oriention of burials and houses. According to Xu Fuguan (1961, 12), "Everywhere in ancient times conspicuous terrestrial phenomena were employed to speak of heavenly phenomena. The functions of water and fire are different, but they are two things whose uses are also mutually implicated. Appropriation of the two in astronomy to speak about analogous heavenly phenomena is probably very ancient." See also Needham (1969, 261); Kalinowski (2004, 90).

[42] Zhu Kezhen (1979, 12) demonstrated long ago that the disparate sizes of the celestial palaces matches the duration of the corresponding seasons, so that the division of the heavens into palaces would have closely followed the identification of the seasons. On the Fire Calendar (*huoli*) and use of the *Fire Star* (*da huo*, α Sco) as a fundamental seasonal benchmark in Shang and earlier times, see Pang (1978), Ecsedy et al. (1989), Feng Shi (1990b, 19–42, esp. 28 ff.), also Feng Shi (1990c, 109 ff.; 1990a, 55).

[43] The term "elemental forces" reflects the cosmological thinking of a much later period, of course; here it is simply used as a convenient term of reference for a nascent scheme of correspondences. Earlier, Li Ling (1991, 22) reached a similar conclusion: "the Five Elemental-Phases are one category of divinatory techniques whose connection with astronomy is most intimate." Here I am obliged to disagree with Nathan Sivin (1995b, 2), who dismissed the view of Benjamin Schwartz that the correlative mode of thought reflected in Five Elemental Phases concepts "may have existed even in neolithic 'primitive' China." In contrast, Sivin avers, "that the Chou was already focused on the cosmic foundations of the state, and well on its way to the classic formulations . . . can no longer be taken seriously."

actual locations of the planetary events, is precisely that given by the "regu-lation colors" traditionally held to have been adopted by the Three Dynasties: Xia, Shang, and Zhou. These regulation colors are found in virtually all sources dating from before about 100 BCE, including the *Book of Documents*, the *Confucian Analects*, the "Duke Tan" chapter of the *Book of Rites* (*Liji*),[44] the *Luxuriant Dew of the Springs and Autumns Chronicle* (*Chunqiu fanlu*),[45] and the "Basic Annals" in *The Grand Scribe's Records*, as well as in several Han apocrypha. For the "Cinnabar Writing," bestowed on King Wen, for example, there is the *Springs and Autumns of Master Lü* (240 BCE):

In the time of King Wen, Heaven first manifested Fire. A red bird clasping a Cinnabar Writing alighted on the Zhou altar to the soil. King Wen said, "The *qi* of Fire is in the ascendant." Therefore, for his color he exalted red and in his affairs he emulated Fire.[46]

What is remarkable about this association of the three colors with the three dynasties is not merely the fact of the cosmological accuracy of the correlations based on the locations of the planetary events, which can hardly be coincidental, but also what this suggests about the nature of such thinking in the second millennium BCE and the continuity with correlative schema many centuries later. We are well informed about the state of astronomical knowledge in late Zhou and Han when many traditions concerning the Mandate portents first made their appearance in writing, and as we saw above such complex planetary phenomena could not have been retrospectively calculated. The conclusion, therefore, must be that basic correlations and terminology contained in the Zhou accounts must reflect the state of cosmological correlations at the time the original observations were made.[47] This argument in favor of a continuity

---

[44] The regulation colors are also mentioned individually in the *Book of Rites* chapters "Jiao te sheng" and "Ming tang wei"; however, a complete statement is found in the earlier "Tan gong shang": "The Xiahoushi esteemed black; for obsequies they preferred dusk, in war they drove black steeds, and for sacrifices they used black [offerings]; the Shang esteemed white, for obsequies they preferred noonday, in war they drove white steeds, and for sacrifices they used white [offerings]; the Zhou esteemed red, for obsequies they preferred daybreak, in war they drove bay steeds, and for sacrifices they used red [offerings]." Ruan Yuan (1970, 6.12).

[45] Su Yu (1974, 7.10b). In yet another strong confirmation of Zhou traditions concerning the Shang, Qiu Xigui (1989, 70–2) has demonstrated the accuracy of the traditional attribution to the Shang of a preference for the color white in ritual contexts; see also Liu Zhao (2009). For possible color–direction co-ordination in ritual placement of jade discs (*bi*) at Anyang, see Thorp (2006, 163). For the Zhou preference for unblemished red animal offerings, see the *Book of Documents*, "Announcement at Luo," which records the sacrifice of ruddy-colored bulls to Kings Wen and Wu.

[46] Lü Buwei (1966, 13.4a).

[47] The possibility of their astronomical origin was alluded to by Xu Fuguan (1961, 12), who nevertheless still held that there was no direct connection between archaic astrological notions and the later Five Elemental Phases theory. And as Joseph Needham (Needham and Wang 1959, 242) observed, "Zhu Kezhen plausibly infers that the scheme of dividing the heavens

of fundamental conception between Three Dynasties-period cosmology and later Five Elemental Phases theorizing is beginning to appear a good deal less startling since the 1987 discovery of the 5,000-year-old Yangshao burial at Puyang and other equally spectacular finds. In the now famous Puyang burials of shamans or other figures of high social status, already mentioned in Chapter 2, the iconic figures of a tiger and a dragon were carefully laid out in a mosaic of mussel shells and placed, with cosmological accuracy, to the west and east of the principal tomb occupant, the latter being oriented along the north–south axis.[48]

Clear evidence from the early Bronze Age of the association of iconic creatures with the four directions so familiar from the Warring States and Han periods is alleged to have been unearthed at the site of King Wen's predynastic Western Zhou capital of Fengjing. A circular eave tile bearing the graph *feng* depicts a dragon on the left, a fish on top, a bird on the bottom, and a tiger on the right (Chapter 2, Figure 2.15a). The positions of the dragon, tiger, and bird agree with the traditional cosmological correlations and the fish with the dark, "watery" north.[49] Still earlier, the *identical four iconic creatures* are depicted on the Anyang-period Shang dragon basin shown in Figure 2.10. Moreover, there is an oracle bone inscription (*Heji* 14360) interpreted as a record of sacrifices to the *Bird* and *Tiger*, presumably a reference to those astral configurations in the sky and their symbolic correlations.

Table 6.1 illustrates the cosmological correlations of the Three Dynasties of the Chinese Bronze Age so far recovered exclusively from the oracle bone and bronze inscriptions, archaeological investigations, and verification of astronomical phenomena already mentioned. Inconsistencies with the later Warring

along the equatorial circle into four main palaces was growing up already in Wu Ding's time (–1339 to –1281)." Shang cosmology certainly stressed the Four Quarters and the center which they occupied as "middle Shang," *zhong Shang*; see Huang Tianshu (2006b). The sky above and the earth below being continuous, the scheme extended equally to the heavens, the center being the pivot at the celestial Pole.

48 For an assessment of the find, see Zhang Guangzhi (1988). For a more in-depth study of the find's purported astronomical significance, see Feng Shi (1990a, 52–60, 69; and especially 1990c, 108–18). For a detailed discussion tying this graphic representation to subsequent cosmological conceptions and artifacts, see Li Xueqin (1992–3). Feng Shi (1993, 9–17) has also studied the stone altars of the Hongshan culture dating from around 3000 BCE, concluding that they were used for sacrifices to a sky god and/or heavenly bodies such as the Sun and Moon; see also Tian (1988, 21–68).

49 Feng Yunpeng and Feng Yunyuan (1893, Vols. 22–4, available at http://catalog.hathitrust.org/ Record/002252003); Yi Ding, Yu Lu, and Hong Yong (1996, 13). Provenance of the artifact is unknown and it may not be an eave tile since Western Zhou exemplars are typically semicircular, but the character *feng* in the center is unmistakable. For idiosyncratic combinations of cardinal emblems on a Spring and Autumn-period bronze mirror and in the recently unearthed Warring States Chu MS *Rong Cheng shi*, see Tseng (2011, 251, Figure 416) and Pines (2010, 515) respectively. See also the iconic emblems (including the fabled "unicorn" *qilin*) on the silk brocade from the oasis of Niya along the Silk Road in Chapter 10, Figure 10.3, below.

Table 6.1 *Documented Cosmological correlations in the 2ⁿᵈ Millennium BCE.*

Dynasty	Color	Direction	Icon ~ asterism	Palace[a]	Quadrant ~ wind[b] Shang oracle bone	Phase ~ season	Yao dian "peoples"
XIA	black	N ~ up	Fish – Turtle Pegasus	*Xuan* 玄 Dark	宛 *yuan* ~ 役 *yi*	Water ~ winter	陳 *ao*
SHANG	white	E ~ left	Dragon Scorpius	*Biao* 鑣 Bright	析 *xi* ~ 協 *xie*	Metal ~ autumn	析 *xi* *
ZHOU	red	S ~ down	Bird Hydra		夾 *jia* ~ 微 *wei*	Fire ~ summer	因 *yin*
		W ~ right	Tiger Orion		夷 *yi* ~ 彝 *yi* [c]		夷 *yi*

[a] The meridian transits of the four cardinal asterisms of the Bronze Age in the "Canon of Yao," characterized in terms of the length of day and night on the solstices and equinoxes, implicitly distinguish the four seasons as well. One might also add a column for the seasonal sacrifices at different temples, whose regular rotation David N. Keightley (1998, 801, n. 100) describes as reminiscent of later Luminous Hall (*mingtang*) ritual.

[b] Chang Zhengguang (1989a; 1989b); Liu Qiyu (2004, 66). Oracle bones of the Shang king Wu Ding period (thirteenth–twelfth centuries BCE) bearing the relevant inscriptions are *Heji* 14294–5. Voluminous research exists on the four winds and the sacrifices to the deities of the Four Quarters, too extensive to cite here; e.g. Liu Zongdi (2002), http://hi.baidu.com/fdme/blog/item/00dda88fcb3796fb503d92d8.html.

[c] It is of considerable interest that this word is also the name of the Yi minority, which is thought by many to descend from the Western Qiang people who were perennial adversaries of the Shang on the west, and who had also close historical ties with the Zhou. The calendrical astronomy in the *Heguanzi*, the "Lesser Annuary of Xia," the "You guan" chapter of *Guanzi*, and the "Canon of Yao" all have strong affinities with each other and with the cosmography of the *Classic of the Mountains and the Seas* (*Shan hai jing*). Chen et al. (1984, *passim*). Moreover, a close connection can be demonstrated between these texts and the ethnographically documented astro-calendrical traditions and practices of the Yi minority.

States versions (e.g. Shang ≈ Sco/Dragon ≈ Metal; fish as icon of the north in Shang/Western Zhou) and the reverse temporal sequence (Winter > Autumn > Summer) attest to the fact that we are here dealing with an archaic scheme of correlations. In any case, the above table is sufficiently complete to make it clear that correlative cosmological thinking along these lines was already well established by the end of the second millennium BCE.[50]

The dynastic successions from Xia to Shang, and then Shang to Zhou, are now regarded as marking successive shifts of hegemony or ascendancy among more or less coexisting polities. In the process, the cultural and political center of gravity in central Henan was staked out by successive relocations of Xia, Shang, and Zhou ritual and civil administration to that area. By the time of the

[50] *Pace* Schwartz (1985, 351–2).

Zhou overthrow of Shang in 1046 BCE, the powerful attraction of the ancient Xia heartland, the "center" near the confluence of the Wei, Yellow, and Luo Rivers, was irresistible, for cosmological as well as practical reasons. Following the example of the Shang, who had earlier relocated from their homelands nearer the east coast, from the Wei River valley in the far west of the territory nominally under Shang tutelage the Zhou moved to establish a new dynastic capital near present-day Luoyang. In doing so immediately after the conquest, and by reporting this fact to Heaven (as recorded on the *He zun*), the first Zhou rulers affirmed their desire to govern the civilized world from the center located in the former Shang domain. In this way the Zhou set about legitimizing and consolidating their rule by governing from the ancient homeland of the Xia, with whom they claimed affiliation by descent and culture. Early texts have the Zhou ancestor, Gu Gong Tan Fu, leading the Zhou people in a migration to the Plains of Zhou by the Wei River from the Fen River valley in the former Xia territory.[51]

In seeking Heaven's blessing on the new dynasty the Zhou king Wu conducted the most sacred of inaugural state sacrifices at a location called the "Hall of Heaven" (*tian shi*), a possible reference to Mt. Sung, the "Central Peak" (*zhong yue*) or *axis mundi* which rises impressively from the yellow loess plain just southeast of Luoyang.[52] This location was associated with the Pole of the heavens where the celestial deity dwelt and about which all his heavenly minions revolved.[53] When the notion of a "central region" (*zhong yu*) is first made explicit in early Western Zhou inscriptions, we recognize this as a continuation of the Shang concept that the heart of their domain was the center of the cosmos, as well as the physical center of the world.[54] Thus, in the earliest Zhou inscriptional record of state worship of Heaven, reference is made to surveying the four cardinal directions from the vantage point of the *axis mundi*, indicating that one of the first official acts of the Zhou king was to

---

[51] *Book of Odes*, "Mian" (Mao 273); Shaughnessy (1999, 300 ff.).

[52] Lin Yun argues that the inscription on the *Tian wang gui* bronze tureen from the reign of King Wu refers to a grand *feng* sacrifice conducted on Mt. Song by King Wu on his way back from the victory over Shang at Muye. Similar reference to Mt. Song is also found in *Zuozhuan*, Duke Zhao, fourth year. The *Tian wang gui* inscription corroborates accounts of the event found in both the *Lost Books of Zhou* and the "Basic Annals of Zhou" in *The Grand Scribe's Records* (4.129). Zhu Youzeng (1940, 5.70–2); Lin Yun (1998, 167–73).

[53] Sima Qian asserts that the Supernal Lord's all-powerful influence emanates from the Pole by calling the *Dipper* "Di's Carriage" and by portraying the *Dipper's* movements as the efficient cause of transformations of *yin* and *yang*, the Five Elemental Phases, the seasons, and all natural periodicities; see "Treatise on the Celestial Offices," *Shiji*, 27.1291; also Major (1993, 107). For a complete annotated translation and study of the "Treatise," see Appendix, this volume.

[54] Hsu and Linduff (1988, 98); for Shang cosmography, see Allan (1991). For the principle of axial centrality in Han, see Major (1993, 37). For an anthropological perspective, compare Clifford Geertz's discussion of the "exemplary center" model of political organization in traditional Indonesia in Geertz (1973, 222–3), and for "Galactic Polities," see Tambiah (1985).

establish ceremonially the legitimacy of Zhou authority over the Four Quarters. Here we encounter the archaic conception of the Chinese world, already familiar from the Shang oracle bone inscriptions, as a self-contained cosmological whole over which the royal charisma ideally extended centrifugally from the center to the Four Quarters (*si fang*), just as did that of the ruler in the sky.[55]

Although color co-ordination in a ritual or cosmological context is supposed to have made its earliest appearance in 770 BCE, when Duke Xiang of Qin initiated sacrifices to the "White Emperor" on achieving the status of protector of Zhou,[56] and despite Joseph Needham's skepticism a half-century ago about the division of the heavens into five celestial palaces as early as the second millennium BCE,[57] the picture that has now emerged points to the existence of just such symbolic correlations during the Three Dynasties period, as suggested in the much later "Great Plan" (*Hong fan*) and "Gao Yao mo" chapters of the *Book of Documents*. Here, perhaps, we have a less speculative basis for the later theory of a cyclical correlation of elemental "virtues" (*de*) with political destinies – the linkage between cosmic virtue, supramundane agency, and dynastic virtue was established not by subsequent events, but "astrologically" at the start of each dynasty.

While A.C. Graham expressed doubts that the five materials and the five processes (Five Elemental Phases, *wu xing*) were already prime correlates of the colors as early as the second millennium BCE, he underscored the importance of identifying the underlying *empirical relations* between phenomena which Chinese correlative thinking strove to account for:

From a modern viewpoint Chinese proto-science can be discovering significant connexions between phenomena only when there are indeed parallel causal relations between things contrasted as Yin and Yang, *or there are causal relations with the seasons or the directions*, the two strong correlates of the Five Powers.[58]

What the account here shows is that just such relations as Graham prescribes as benchmarks are seen to exist between the planetary phenomena, the cardinal directions, the seasons of observation, their associated colors, and emblematic creatures. Huang Tianshu has already found evidence in the Shang oracle bones

---

[55] Aihe Wang (2000).   [56] *Shiji*, 28.1358; Kaltenmark (1961, 20 ff.).
[57] Needham (1969, 246). Curiously, Needham and Wang (1959, 248) had previously stated that "as the Shang people certainly had two of the quadrantal points, they must surely have been aware of the other two also. The *Shu Ching* text mentions all four." In any event, all the elements needed for a cosmological conception of four quadrantal palaces plus the polar center were already in evidence by late Shang.
[58] Graham (1989a, 346, emphasis mine). Cf. Puett (2002, 146–52). John S. Major (1978, 13), for his part, draws the link between the Five Planets and the Five Elemental-Phases: "it does seem reasonable to believe that something like the Five Phases existed as a cosmological principle in the pre-philosophical state of Chinese thought, expressed in myths about the characteristics of the gods of the five planets. So again we see that a key concept of Chinese science probably can be traced back to Chinese versions of widespread cosmological myths."

of the contrasting use of *yin* and *yang* in their basic topographical senses of "north-facing slope ~ south bank of a river" and "south-facing slope ~ north bank of a river" respectively.[59] Take, as a further example, the fact that the total lunar eclipse of 1065 and the planetary cluster of 1059 BCE, both of which occurred in the *Vermilion Bird* constellation, were also both taken to have direct application to the fortunes of Zhou. In the former case the omen was apparently thought to portend King Wen's demise, so that on witnessing the "untimely" eclipse King Wen ordered his subordinates to begin "deliberating about the succession." This implies that the *Vermilion Bird* was already astrologically linked with the Zhou by mid eleventh century BCE as stated in *Discourses of the States*, just as the *Fire Star* was already linked with the fortunes of Shang, and called *Shang Star*.[60]

If such a precursor to the later "field-allocation" astrological schema (for which see Chapter 9) were already in place by this time, it would help to explain the traditional accounts which tell of desertions by advisers and officials close to Shang ruler Di Xin prior to the conquest, and by elements of the Shang army at the Battle of Muye in early 1046, since the locations of the planetary portent in May 1059 and the possible baleful apparition of Halley's Comet near the *Fire Star* (Antares, the *Shang Star*) eight months later would naturally have been observable throughout the area under Shang tutelage. Zhou king Wen's overtly treasonable activity from the first year of the Mandate (1058) on, directly challenging the Shang king's authority, indicates that he strove fully to exploit his advantage, both militarily and psychologically. When, therefore, the Duke of Zhou is quoted in the "Great Announcement," *Da gao*, in the *Book of Documents* as saying,

Heaven gave its grace to the serene king [i.e. deceased King Wen], and raised our small state Zhou. The serene king followed only the oracle, and was able tranquilly to receive this Mandate . . . Oh, Heaven brightly manifests its majesty, and supports our very great foundation

such language should no longer be discounted as mere hyperbole. Rather, it is an example of the tendency of the ancients "to understand what happened to them as caused by supra-sensible agencies, gods, and demons . . . [A]s the ancients experienced and recorded things, the gods were the very nodes of the causal network that gave events coherence and meaning."[61] For his part, Stanley Tambiah cites Lévy-Bruhl's notion of a "mystical" mentality, which extends this mode of experiencing to all the supra-sensible forces.[62] A crucial facet of Thorkild Jacobsen's "theocratic" mode of experiencing was, as we

---

[59] Huang Tianshu (2006a).
[60] Pankenier (1981–82, 21); Li Xueqin (2005b, 7–11); Liu Qiyu (2004, 58).
[61] Jacobsen (1994, 46); Rochberg (2007, 166).   [62] Tambiah (1990, 85).

have seen, a reliance on the efficacy of divination, whether provoked (initiated to elicit a response) or unprovoked (*ex post facto* interpretation).[63]

Karl Löwith thought that "what separates us most deeply from the ancients is that they believed in the possibility of foreknowing the future, either by rational inference or by the popular means of questioning oracles and of practicing divination, while we do not."[64] But the gulf may not be quite so deep as Löwith believed, or as wide as we prefer to think. Stanley Tambiah observed that

Frazer maintained that the fundamental conception of magic is "identical with modern science," namely "the uniformity of nature." The magician believes that the same causes will always provide the same results, and as long as he performs the ceremony in accordance with the rules laid down, the desired results will inevitably follow. Thus the similarity between magical and scientific conceptions of the world is close: "In both of them the succession of events is perfectly regular and certain, being determined by immutable laws, the operation of which can be foreseen and calculated precisely, the element of chance and of accident are banished from the course of nature."[65]

Given the likely historical basis of cosmo-political thinking in the Three Dynasties period as it applies to the early dynastic succession, and given the remarkable vitality of popular astrological traditions attesting to its validity in late Zhou and Han, it is not surprising that the Warring States thinker Zou Yan's proto-"scientific" theorizing about a cyclical succession of elemental phases won him a sympathetic hearing in the courts of rulers. A major reason for Zou's success was that in the minds of his elite audience his synthesis resonated with and conferred intellectual respectability and order on venerable cosmo-magical beliefs, which had long been received wisdom among scribe-astrologers and their patrons. Once the traditional pieces were fitted together in his dynamic theory the rational implications of the cycle became self-evident. As Xu Fuguan put it, "Zou Yan's theory of the succession of Five Virtues represents a revival in another form of primitive religion. The Virtues of the Five Elements, in their sequential alternation, are a concretization of the 'mandate' (*ming*) of Heaven's Mandate (*tianming*)."[66]

Dynastic changes of Elemental Phase

Sima Qian succinctly formulated the linkage between Five Elemental Phases cosmology and the astrological origins of the mature theory nearly a millennium after the Zhou founders first articulated the doctrine of Heaven's Mandate:

---

[63] Rochberg (2004).   [64] Löwith (1949, 10).
[65] Tambiah (1990, 52).   [66] Xu Fuguan (1961, 52).

When the Five Planets gather, *this is a change of Phase*: the possessor of [fitting] virtue is celebrated, a new Great Man is set up to possess the four quarters, and his descendants flourish and multiply. But the one lacking in virtue suffers calamities or extinction.

The significance of this statement has previously not been correctly understood; indeed, it is usually mistranslated. By Sima Qian's time the direct correlation between the heavenly phenomena and temporal events was self-evident, and so he says:

Ever since the people have existed, when have successive rulers not calendared the movements of the Sun, Moon, stars and asterisms?[67] Coming to the Five Houses[68] and the Three Dynasties [Xia, Shang, Zhou], they continued by making this [knowledge] clear, they differentiated wearers of cap and sash from the barbarian peoples as inner is to outer, and they divided the Middle Kingdoms into twelve regions. Looking up they observed the figures in Heaven, looking down they modeled themselves on the categories of Earth.[69] Therefore, in Heaven there are Sun and Moon, on Earth there are *yin* and *yang*; in Heaven there are the Five Planets, on Earth there are the Five Elemental Phases; in Heaven are arrayed the lodges, and on Earth there are the terrestrial regions.[70]

Echoing Xu Fuguan, Li Ling recently concluded that *yin-yang* and Five Elemental Phases theories

reached their greatest efflorescence in late Warring States, Qin, and Han. Although one encounters new turns of thought and they contain many superfluous embellishments as a result of efforts to regularize and systematize, still they certainly cannot be subsumed under the strange talk of Zou Yan and his ilk. Rather, they are the legacy of the many hemerologists and those who devised theories based on the past, deriving their material from high antiquity. With primitive thought as their backdrop, these ideas flowed straight from exceptionally archaic sources, their impetus in no way being attributable to the mainstream thinking comprising the theories of the various philosophical schools [of the late Warring States period].[71]

---

[67] Sima Qian is paraphrasing the "Canon of Yao" in the *Book of Documents*. In the "Basic Annals of the Five Emperors," Sima Qian glosses "calendared the images" (*li xiang*) as "calculated algorithms" (*shu fa*) (as pointed out to me by Christopher Cullen), indicating that by *li* he meant the deriving of numerical models from their movements. Calendrical science was thoroughly mathematized by Sima Qian's time, so he is reading his presentist interpretation into a text that refers to observational practices in the Bronze Age. According to Liu Qiyu, the *Yaodian* originally wrote not 曆象 but *Li xiang* 歷象, which Liu glosses simply as the "phenomenon of the revolving of the Sun, Moon, and stars in the heavens." "Calculated algorithms," Liu (2004 45) says, is simply Sima Qian's verbal gloss of the original substantive "succession of celestial images" (*li xiang*).

[68] The five predynastic rulers: Huang Di, Gao Yang (Zhuan Xu), Gao Xin (Yao), Tang Yu (Shun).

[69] The "Treatise" is quoting the "Appended Commentary" to the *Book of Changes*.

[70] *Shiji*, 27.1321, 1342. Phenomena involving the Five Planets above resonate with the corresponding operations of the Five Elemental Phases here below. Therefore the expression *yi xing* in Sima Qian's statement above means "change of elemental phase," in a direct reference to Five Elemental Phases theory. "Lacking in virtue" in this context no doubt refers to the consequences of being out of step with the prevailing *virtus*, "power," or *de* governing the era.

[71] Li Ling (1991, 75).

Compare with these views John S. Major's discussion of the cosmological chapters of *Huainanzi* in the light of the dominant idea in the Western Han period that knowledge of the natural world translates into political power:

the credibility of the Huang-Lao School in the early Han may have rested in part on the degree to which it was grounded in widely-shared assumptions that went back to the foundations of Chinese civilization . . . Cosmogony, cosmography, astronomy, calendrical astrology, and other features of cosmology form a seamless web, the principles of which a ruler would ignore only at his peril.[72]

The evidence shows that the eventual absorption of Five Elemental Phases mode of correlative thinking into the Han ideology and the recrudescence of popular and subversive millenarian movements in the Han Dynasty are both ultimately attributable in large measure to the benchmark observations of second-millennium BCE scribe-astrologers. Contrary to the earthbound perspective (based solely on Warring States and Han texts) which holds that correlative cosmological thinking is an innovation of the immediate pre-imperial period, we have seen that, in fact, it has a history extending well back into the Bronze Age.

[72] Major (1993, 43).

7   The rhetoric of the supernal

Heaven's Great Command to King Wen was not [conveyed] sonorously, by means of sound...[1]

The Zhou dynastic founders forcefully reasserted the centrality of their concept of a "Mandate of Heaven" as the explicit justification for their usurpation of Shang power, and implicitly as a theory of history. Zhou pronouncements about the supreme deity, whether their "Heaven," *Tian*, or the Shang Supernal Lord, Shangdi, constitute our most authentic record of the religious and political motives being played out in the historical events of the period. In the past there has largely prevailed a consensus among scholars that the Zhou conception, though perhaps prefigured to some degree in Shang divination texts, was largely an original formulation devised by Zhou ideologues in response to political exigency – the conquest of Shang. Our discussion of verifiable astronomical phenomena dating from long before the conquest period strongly suggests the existence of a belief in a species of interventionist sky god centuries before the Zhou overthrow of Shang, a conclusion that is also supported by traditional accounts of the period and by studies in comparative religion. For example, Mircea Eliade found that "the notion of universal sovereignty ... owes its development and its definition of outline largely to the notion of the sky's transcendence":

Even before any religious values have been set upon the sky it reveals its transcendence. The sky "symbolizes" transcendence, power and changelessness simply by being there. It exists because it is high, infinite, immovable, powerful ... the whole nature of the sky is an inexhaustible hierophany. Consequently anything that happens among the stars or in the upper areas of the atmosphere – the rhythmic revolution of the stars, chasing clouds, storms, thunderbolts, meteors, rainbows – is a moment in that hierophany. When this hierophany became personified, when the divinities of the sky showed themselves, or took the place of the holiness of the sky as such, is difficult to say precisely. What is quite certain is that the sky divinities have always been supreme divinities ... that their hierophanies, dramatized in various ways by myth, have remained for that reason sky

---

[1] *Shangshu dazhuan*, 4.5b.

hierophanies; and that what one may call the history of sky divinities is largely a history of notions of "force," of "creation," of "laws" and of "sovereignty."[2]

This evidence of a connection between celestial events and the founding of the earliest dynasties makes a closer examination of the theological motives expressed by those dynastic founders all the more relevant.

In the view of philosopher Kenneth Burke, there is present in all remarks about the deity a rhetorical element that provides us with clues to how the use of language predisposes religious thinkers to think thoughts of ultimate purpose in particular ways.[3] The Zhou belief in heavenly intervention, which they represented as the motive for their successful overthrow of Shang, can be understood as a working out in temporal terms of the implications of a logical relation among the "coeternal realms of Heaven and the natural order," on the one hand, and the human sociopolitical order, on the other. The Zhou founders did not express themselves this way, of course. At most they exhibited self-consciousness about proposing legitimations fundamentally at variance with the ideology of the Shang theocratic state, which had enjoyed a longevity of centuries.[4] But the analogical use of language borrowed from the sociopolitical realm by which the Zhou characterized their relationship with the high god, like that of the earliest Shang records, speaks volumes about their theology and about how their conceptualization of problems of meaning was self-motivating in important respects.

The following discussion, which is much indebted to Kenneth Burke, will sketch out some of the implications of the logical relations between the temporal and the eternal realms discernible in Shang and Zhou thinking, in an attempt to adumbrate how certain features of that theology are implicated in the evolution of the concept of Heaven's Mandate. In order to do so it will first be necessary to consider briefly how it happens that the realm of language can itself be a source of theological motive. In essence, this "rhetoric of religion," as Kenneth Burke has characterized the analogical use of language in theology, concerns the implications arising from the use, for example in myth, of "quasi-narrative terms for the expressing of relationships that are not intrinsically narrative, but 'circular' or 'tautological.'"[5] A most important

---

[2] Eliade (1958, 39, 40).

[3] Burke (1970, esp. 1–42). For a classic sociological study of religion which, as well as being more ecumenical, reinforces that of Burke in many respects, see Berger (1990, esp. 1–51).

[4] Not incidentally, the difference in their names for the high god shows clearly that Shang and Zhou were culturally, if not ethnically, distinct, although the Zhou were obviously profoundly influenced by the high culture of the Shang, including their written language and sacrificial ritual.

[5] Burke (1970, 258). The mental process Burke here refers to is the uniquely recursive property of human thought and memory. As Corballis (2011, 180) has it, "The prior significance of recursion may therefore lie, not in language itself, but rather in the nature of human thought that guides language and supplies much of its content."

consequence of this terminological ambiguity, for our purposes, is the blurring of the distinction between the temporal and eternal realms and the resultant transference of motives from the one to the other.[6] According to Burke, "the supernatural being by definition the realm of the ineffable," all words for "God" and for human relationships to the divine must be used analogically. That is to say, all language about the supernatural is borrowed from among our words for the kinds of things we can talk about literally. As a result, the stuff from which our language for the supernatural can be analogically constructed is limited to three empirical sources: (i) natural objects and processes (the "sweep and power of the natural," in Kenneth Burke's apt phrase), including structural consistencies and symmetries; (ii) the sociopolitical order; that is, "the dignities and solemnities of office – and the intimacies of the familial"; (iii) the verbal, including meta-linguistic terminology, or words about words, and the symbolic in general:[7]

Since "God" by definition transcends all symbol-systems, we must begin, like theology, by noting that language is intrinsically unfitted to discuss the "supernatural" literally. For language is empirically confined to terms referring to physical nature, terms referring to socio-political relationships and terms describing language itself. Hence, all words for "God" must be used analogically – as were we to speak of God's "powerful arm" (a physical analogy), or of God as "lord" or "father" (a socio-political analogy) or of God as the "Word" (a linguistic analogy). The idea of God as a "person" would be derived by analogy from the sheerly physical insofar as persons have bodies, from the socio-political insofar as persons have status and from the linguistic insofar as the idea of personality implies such kinds of "reason" as flower in man's symbol-using prowess (linguistic, artistic, philosophic, scientific, moralistic, pragmatic).[8]

Although Kenneth Burke chiefly concerns himself with the Western religious tradition, empirical categories are represented in much the same way in the theological vocabulary of ancient China. One need only call to mind Shang usage like "Supernal Lord" Shangdi, who "commands," *ling*; "approves," *nuo*; "sends down," *jiang*; "bestows," *shou*; and supervises "ministers," *chen*, or Zhou usage like "Heaven," *Tian*, which "decrees," *ming/ling*; "punishes," *fa*; "brightly manifests," *xian xian*; "overawes," *wei*; "inspects," *jian*; "ascends and descends," *zhi jiang*; "protects and directs," *bao yi*; "hears," *wen*; has an "eldest son," *yuan zi*; and so on. Thus in Shang and Zhou China as well we

---

[6] Kenneth Burke (1969, 43) makes a similar point with respect to the belief in the possibility of influencing the supra-sensible realm by means of symbolic behavior, rituals such as magic or divination: "Magic therefore is 'primitive rhetoric,' it is rooted in an essential function of language itself, a function that is wholly realistic, and is actually born anew; the use of language as a symbolic means of inducing cooperation in beings that by nature respond to symbols."

[7] Burke (1970, 37).

[8] Ibid., 15.

can identify a variety of analogies typically drawn from the sociopolitical and linguistic realms.[9] The distinction between the temporal and the eternal realms, when treated in linguistically analogous form, may be compared to "the distinction between the unfolding of a sentence through the materiality of its parts and the unitary, non-material essence or meaning of the sentence."[10] Thus, just as the meaning arising from a narrative structure of language can present itself as transcending, or as in some sense logically prior to, the temporal arrangement of the words themselves, in their use of language to characterize the Supernal Lord in a particular way the ancient Chinese were unconsciously "discovering" the logical priority of the numinous realm. Kenneth Burke provides another classic example:

We can see this more clearly when thinking of the relation between the practical use of a language and a book on the theory of its grammar and syntax. Under natural conditions, people learn languages long before the rules of grammar and syntax are explicitly formulated. These rules are "discovered" relatively late in the development of linguistic sophistication, and sometimes not at all. Yet there is a sense in which they "have been there" from the start, implicit in the given symbol-system. In this sense, to "discover" them is but to formulate what one somehow knew before one ever began to ask about such "forms."[11]

Similarly, according to Peter Berger:

Whenever the socially established nomos attains the quality of being taken for granted, there occurs a merging of its meanings with what are considered to be the fundamental meanings inherent in the universe. Nomos and cosmos appear to be co-extensive. In archaic societies, nomos appears as a microcosmic reflection, the world of men as expressing meanings inherent in the universe as such... Whatever the historical

---

[9] Speaking about the Presocratic Greeks, Eric Havelock (1987, 43–4) makes much the same point: "Cumulatively, in these brief cosmic visions, a world of mobile and dynamically shifting phenomena is reduced to a political order under a dominant authority... By an act of cosmic projection, they translated the human mind into the cosmos, as it were by a Hegelian effort. It was left to Parmenides clearly to grasp the truth that the dimensions of this mind lie in the human thought process." See also Corballis (2011, 137).

[10] Burke (1970, 3). Recall now the observation of Stanley J. Tambiah (1990, 85) (quoting Lévy-Bruhl) that "the 'savage'... made no demarcation between a domain of nature as opposed to the supernatural, it was better to describe his view of certain beings, forces or powers as 'supra-sensible' rather than as beliefs in 'supernatural beings.'"

[11] Tambiah (1990, 238). One could think about the discovery of writing in much the same way. Writing was a cosmic secret waiting to be discovered. This would account for the sacred or magical qualities attributed to the written characters, which Léon Vandermeersch dubbed *chiffres magiques*. See Burke's reference to the "genius of the verbal" below.

variations, the tendency is for the meanings of the humanly constructed order to be projected into the universe as such.[12]

In the analogical use of language lies a paradox, says Burke, which arises from the fact that once a theological terminology has been derived from everyday experience to express ideas about the supra-sensible, the order can be reversed and the same terminology can be borrowed back again or "resecularized." But in the process these terms inevitably bear with them the added freight arising from their supra-sensible connotations: "They are thus 'technically prior' in a way that would be quite analogous to the Platonic view of 'archetypes' already existing in the 'memory' that vaguely 'recalls' them in their ideal 'perfection.'"[13]

We saw above a number of examples of Shang and Zhou expressions for divine activity as it impinged on human consciousness, much of which language implicitly ascribes human-like motives to Heaven or Shangdi. These could easily be multiplied by examples of attributes like "anger," "benevolence," "mildness," "pleasure," etc. The result of such usage is further illustrated by Kenneth Burke as it relates to the understanding of personality:

personality as an empirical concept is composed of ingredients distributed among the three empirical orders (words about nature, words about the socio-political, words about words). And personality as a term for deity is extended by analogy from these empirical usages . . . at this stage a theological dialectic strategically reverses its direction. That is, it conceives of personality here and now as infused by the genius of the analogical extension . . . empirical personality can be looked upon as sharing in the spirit of the supernatural personality . . . [the supernatural] is thus treated as in essence "prior" to the other three, and as their "ground" . . . The terms for the supernatural, themselves derived by analogy from the empirical realm, can now be borrowed back, and reapplied – in analogy atop analogy – to the empirical realm, as when human personality here and now is conceived in terms of "derivation" from a transcendent super-personality.[14]

In China this process proceeded in stages, culminating in the late Warring States Confucian conception of human nature or personality as essentially "endowed" by Heaven, by which formulation Confucians implicitly affirmed the derivation of human personality from that of the supernal power. For example, there is the famous passage in "The Doctrine of the Mean": "What is mandated by Heaven is called [human] nature" (*tian ming zhi wei xing*). In this same vein, Peter Berger points to the transformation in classical China of the pre-existing macrocosm/microcosm scheme legitimating the social order, typical of archaic societies, beyond a strictly mythological worldview: "In China, for instance, even the very rational, virtually secularizing, demythologization

---

[12] Berger (1990, 24). See also the discussion of the origin of the supra-sensible in Swanson (1964, 27).

[13] Burke (1970, 238).   [14] Ibid., 36.

of the concept of Dao (the 'right order' or 'right way' of things) permitted the continuing conception of the institutional structure as reflective of cosmic order."[15]

One of the most consistent activities of the deity in the early period of particular relevance is that of "commanding" (*ling*) or "decreeing/mandating" (*ming*), the earliest and most significant usage quite obviously deriving from the sociopolitical realm.[16] Because the two graphemically indistinguishable words come from the exercise of sovereign power in the realm of human sociopolitical experience, they carry with them connotations of "ruler," "hierarchy," "authority," and "obedience," and perhaps "sanction," "covenant," and so on, from their very earliest attested use in the Shang oracle bone inscriptions. Thus, in the divinations, besides being used to denote verbal charges issued by the king to his subordinates, *ling* is also used to denote the analogous interaction between the Supernal Lord and natural phenomena, and between him and his "Minister Regulators" (*chen zheng*). Considerably less otiose is Heaven in the Zhou conception, which by all accounts issues non-verbal commands, *ming/ling*, not just to natural entities but also to humanity – that is, to the dynastic leadership – as did Heaven to Xia and Shang dynasts as well in the Zhou recounting of history. The distinctive medium of such non-verbal "commands" remains, however, as with the Shang, through the particular phenomena of nature, whether unpredictable or regular.[17] Here we might recall the epigram from the head of this chapter: "Heaven's Great Command to King Wen was not [conveyed] sonorously, by means of sound ..."[18]

In both Shang and Zhou, therefore, we have the projection into the suprasensible realm of a sociopolitical model of sovereignty and hierarchical order. And, as Burke points out, by a strategic ambiguity the term "order" comes to apply both to the realm of nature in general and to the special realm of human sociopolitical organization. This process whereby ideas of the natural order become infused by characteristics of the sociopolitical order carries with it the implication that phenomena in nature are in some sense *actualizations of the will of the divine personality*. Things do not simply occur of themselves. As

---

[15] Berger (1990, 35). Michael C. Corballis (2011, 137), citing Robin Dunbar, argues that "it is through theory of mind [i.e. recursive mentation allowing for multiple orders of intentionality] that people may have come to know God, as it were. The notion of a God who is kind, who watches over us, who punishes, who admits us to Heaven if we are suitably virtuous, depends on the understanding that other beings – in this case a supposedly supernatural one – can have human-like thoughts and emotions."

[16] Schaberg (2005, 23–48).

[17] Much later, in the fourth-century BCE "Chu Silk Manuscript," the Supernal Lord actually speaks directly in admonishing the people, the only instance of its kind known to me. Li Ling (1985, 31 ff.).

[18] *Great Commentary on the Book of Documents*; *Shangshu dazhuan*, 4.5b. Sarah Allan (2009), seemingly confusing agent with agency, identifies the Zhou high god *Tian* or "Heaven" as a specific physical phenomenon witnessed in the sky, which she does not identify.

Kenneth Burke so aptly put it: "Although the concept of sheer 'motion' is non-ethical, 'action' implies the ethical – the human personality . . . 'Things' can but move or be moved. 'Persons' by definition can 'act.'"[19] It is this view of the phenomena of nature as in some sense the manifestation of divine order that also imbues the phenomena with an ethical quality. As Marcel Granet perceptively observed about the use of metaphor and allegory by Chinese writers,

> Imagery is not employed merely to simplify the idea or to make it more attractive: *in itself it has a moral value.* This is evident in the case of certain themes. For example, the picture of birds flying in couples is, in itself, an exhortation to fidelity. If, then, metaphors borrowed from Nature are used to give expression to the emotions, it is due not so much to a consciousness of the beauty of Nature as to the fact that it is moral to conform to Nature.[20]

A.C. Graham too characterized the synthesis of fact and value in Chinese correlative thinking this way: "a cosmos of the old kind has also an advantage to which post-Galilean science makes no claim; those who live in it know not only what is but *what should be.*"[21]

Heaven's Mandate

Once the high god was conceived as "commanding" by analogy with the empirical theocratic hierarchy, the stage was set for the theological dialectic to reverse direction, whereupon the conclusion was drawn that the sociopolitical realm known to human experience, in which such "commanding" prominently figures, was in fact divinely enjoined. In other words, the very model of "commanding/decreeing," with all its sociopolitical implications, was thought to be as much an actualization of the divine will in the empirical realm as are manifestations in the natural realm such as rain or thunder, or, for that matter, planetary phenomena. And so, Burke says, "the sheerly natural order contains a verbal element or principle that, from the purely empirical point of view, could belong only in the socio-political order. Empirically, the natural order of sheerly astro-physical motion depends upon no verbal principle for its existence. But theologically, it does."[22] Verbal "commands" as were issued by the temporal Shang king thus acquired an implicit "spiritual" quality or sanction, their legitimacy deriving from their "prior" supra-sensible source. Conversely, by conceiving of the Supernal Lord as issuing "commands" the ancient Chinese "hit upon a vision of a natural order now infused with the genius of the verbal and socio-political orders."[23]

---

[19] Burke (1970, 41, 187).   [20] Granet (1932, 50, italics mine).
[21] Graham (1989a, 350, italics mine).   [22] Burke (1970, 185).   [23] Ibid.

This process has important consequences for understanding the development of the concept of Heaven's Mandate. For if it is true that the Supernal Lord is represented as rather aloof in the earliest Shang documents, it is nevertheless also true that the dialectical transformation hinging on the ambiguous relation between the temporal and supernal realms had by then already occurred in Shang theology.[24] Just as in early Judeo-Christian tradition, Shang

> standard usage bridges [the] distinction between the realms of verbal action and non-verbal motion when it speaks of sheerly natural objects or processes as "actualities." Here ... we can discern a trace of the theological view that sees nature as the sign of God's action – and thus by another route we see the theological way of merging the principle of the natural order with the principle of verbal contract or covenant intrinsic to legal enactment in the socio-political order ... If, by "Order," we have in mind the idea of a command, then obviously the corresponding word for the proper response would be "Obey."[25]

What better way to express this merging of disparate orders than the symbolic identification of exceptional planetary configurations with the jade scepter of office that signaled the delegation of legitimate authority to rule? It is no accident, therefore, that the historical record of this sort of communication, when translated into the linguistic realm of "writings" or "diagrams" such as revealed to dynastic founders, begins with celestial observations, like the "bright manifestations" alluded to in the *Book of Documents*.[26]

Early Zhou examples of ritual rhetoric: *He zun*, *Tian Wang gui*, and the Hall of Heaven

The He zun[a]

The *He zun* was cast during the reign of King Cheng, second ruler of Western Zhou (1042–1021), making it one of rare inscribed bronzes from

---

[24] For the developmental continuum of legitimations ("socially objectivated knowledge that serves to explain and justify the social order") according to historical circumstances, from pre-theoretical through theoretically self-conscious, see Berger (1990, 29, 31–2).

[25] Burke (1970, 186).

[26] As we saw above, the "Treatises on Astrology" in the early dynastic histories explicitly identify the "River Diagram," *Hetu*, as a text in which was recorded revealed wisdom concerning the heavenly bodies. For an account of the transmission of esoteric knowledge in the form of "River Diagrams," *Hetu*, or "Luo Writings," *Luoshu*, see *Han shu*, 27.1315. Above we cited a specific example from Huan Tan's *New Discourses*. For a discussion of the military application of such revealed "texts" in the Warring States period, "in imitation of divine patterns that inform the cosmos," see Lewis (1990, esp. 98 ff. and 137–63). Mark E. Lewis argues for a Warring States date for the "textualization" of the cosmos. In view of the evidence presented thus far, Lewis's contention that the notion of a cosmic kingship (wherein the legitimate authority of the king is derived from his ability to "read" the hieroglyphics of the cosmos) is a Warring States invention is scarcely tenable.

the very beginning of the dynasty. This spectacular *zun* beaker was first unearthed near Baoji, Shaanxi, in 1965, but its important inscription was not discovered until the vessel was cleaned and corrosion removed in 1975. In short order, China's most eminent historians and paleographers had published studies analyzing the 122-character-long inscription and discussing its historical significance. The *He zun* commemorates a ritual feast following a solemn *Feng* sacrifice to the king's ancestors. The inscription records King Cheng's hortatory declaration to He, the maker of the vessel and a member of the royal lineage(s) present at the feast, calling on him to emulate his father in steadfastly supporting the king. The maker, He, then records the king's bestowal on him on this occasion of thirty strings of cowries with which he financed the casting of this prized ritual vessel.

Apart from the baroque magnificence of the vessel itself, the *He zun*'s great historical importance lies in the inscription's corroboration of accounts in early texts like the "Announcement of the Duke of Shao" (*Shao gao*) chapter in the *Book of Documents*, the "Laying out the City" (*Duyi jie*) chapter in the *Lost Books of Zhou*, and elsewhere, concerning the founding of the capital of Luo on the north bank of the Luo River, soon after the overthrow of the Shang Dynasty in 1046. First, the inscription contains one of the earliest references to the receipt of Heaven's Mandate by the dynastic founder, King Wen (1099–1050), and the ritual focus on the "Chamber of Heaven" (*tian shi*) temple complex. Second, the *He zun* confirms that the building of an administrative center in the heart of the former Shang domain was the express intention of King Cheng's father, King Wu (1049–1044), who died only two years after conquering Shang. Third, it confirms that it was during the fifth year of King Cheng's reign (1038) that construction of the new capital was begun, here for the first time named "Achieved Zhou" (*Chengzhou*).

Although obscured due to corrosion, the words "Great Command" (*da ling*) in the second line of the inscription immediately following "received this" can be convincingly restored on contextual and rhetorical grounds, as well as physical inspection. The founding of Luo in the king's fifth year is now confirmed by the *He zun*, so that the *Great Commentary on the Book of Documents, Shangshu dazhuan*, account of King Cheng's "creating rituals and music" in the sixth year, and ruling in his own right in the seventh year, is entirely plausible. Thus the ordinal sequence refers to the early years of King Cheng's reign, which coincide with the Duke of Zhou's regency. The king's exhortation to He and other junior members of the lineage to support him, shortly after the suppression of a rebellion and just as King Cheng reached majority and assumed the full power of the kingship, fits the historical context. One could hardly ask for a better illustration of the rhetoric of command – obedience, first from Heaven to the Zhou kings and thence to their subordinates:

It was when the King began laying out his royal seat at Chengzhou. [The King] returned from extolling King Wu in the Feng sacrifice, with sacrificial meat from the [Hall of] Heaven. In the fourth month, on day bingxu, *the king exhorted the scions of the royal clan in the ancestral temple, saying: "In the past, your fathers were able to aid King Wen, whereupon King Wen received this [Great Command]. When King Wu conquered the great city Shang, he then made reverent declaration to Heaven, saying: 'Let me dwell in this, the central region, and from here govern the people.' Hark! While you are still minors lacking in understanding, look to your fathers' scrupulous respect for Heaven. Comprehend my commands and respectfully follow orders! [Your] sovereign's reverential virtue finds favor with Heaven, which guides me in my slow-wittedness." The King's exhortation having finished, [vessel maker] He was presented with the thirty strings of cowries used to make this treasured sacrificial vessel for [his father] Sire [X]. It was the King's fifth year.*[b]

Tian wang gui[c]

The *Tian Wang gui* (aka *Da Feng gui*) was discovered at Qishan, the Zhou ancestral homeland in Shaanxi, during the Daoguang reign period (1821–55). The scholarly consensus is that the vessel dates from the very end of King Wu's reign (1049–1044) just after the Zhou conquest (1046). The *Tian Wang gui* and its remarkable seventy-seven-character inscription have been thoroughly studied, and it now serves as a "standard type-vessel" in early Western Zhou bronze typology. In addition to providing an important account of a three-day series of the highest state sacrifices, the *Tian Wang gui* inscription is also the earliest to make systematic use of rhyme, displaying careful attention to composition and prosody. The translation below is arranged to highlight the prosodic composition, rather than reproducing the eight lines of the original (end rhymes are shown in bold). The grammar and syntax are slightly opaque in parts, and after more than a century some characters still defy decipherment. Fortunately, the most important content is accessible. The inscription concerns a series of large-scale state rites conducted by the king, beginning with the Great *Feng* sacrifice in the Chamber of Heaven (*tian shi*) on the first day (also encountered in the *He zun* and *Zuoce Mai fang zun*)[d] followed the next day by an elaborate *yi*-sacrifice) to King Wu's deceased father, King Wen, in company with the Supernal Lord. This is followed on the third day by a grand feast hosted by the king featuring the sacrificial meats and liquors. Comparison with the more detailed account of the *Feng* sacrifice in the *Mai fang zun* indicates that the Hall of Heaven formed part of the Mingtang or Luminous Hall ritual complex, with its circular moat (*bi yong*) and Spirit Terrace (*ling tai*), where the most important sacrifices to the Supernal Lord (or Heaven) and the ancestors were conducted. The maker of this vessel, Tian Wang, plays a leading role in the sacrifices and is richly rewarded, so that it is clear he must have been very prominent and close to the king. It may be that this is none other than Tai Gong Wang (aka Lü Shang) himself, the high-ranking vassal of the Shang who went over to the Zhou cause, contributed vital strategic and political counsel to King Wu during the conquest campaign, and was rewarded with his own appanage, the domain of Qi in Shandong. This identification is supported by the fact that an *yi*-sacrificial rite like the one mentioned here also figured in the Shang royal cult, as did, of course, the Supernal Lord.

"On day [yi-]*hai* [12], *the King* [Wu] *performed the Great Feng rites, offering to the three* [four?] **directions**. *The King sacrificed in the Hall of Heaven, then* **came down**.

Tian Wang assisted the King in performing the yi-*panoply of sacrifices to his illustrious father King Wen, offering millet ale to Shangdi and King Wen, attending on* **high**.

'[Our] *illustrious King* [Wen] *raised up virtue;* [our] *great succeeding King* [Wu] **carried on**, *bringing an end to the Yin kings' sacrificial rounds* [i.e. *"years"*].*'*

On day dingchou [14], *the King served a feast of the sacrificial viands; the King* **descended**

to [award] *Wang . . .* [list of gifts, titles].

'Receiving these rewards, I hasten to extoll the King's grace on this precious tureen.'"[e]

[a] *Jicheng* 4.6014; Tang Lan (1976); Zhang Zhenglang (1976); Ma Chengyuan (1976); Itō (1978); Shirakawa (1964–84, Vol. 48, 171–84); Yang (1983); Ma Chengyuan (1989, 20–2); Hsu and Linduff (1988, 96–9); Chen Gongrou (2005); Zhu Fenghan (2006); Gassmann and Behr (2011, 148).
[b] Trans. DWP and Zhou Ying.
[c] *Jicheng* 3.4261: 77-0-195; Chen Mengjia (1955, 137–75); Sun Zuoyun (1958); Tang Lan (1958, 69); Huang Shengzhang (1960); Yu Xingwu (1960); Yin Difei (1960); Ma Chengyuan (1989, 23); Hsu and Linduff (1988, 99–100); Lin Yun (1998); Guo Moruo (1999, Vol. 3, 1a–2b); Shirakawa (1962–84, Vol. 1, 1–38); Sun Zhichu (1980); Tang Lan (1986); Shirakawa (2000, 1–26); Zhou Xifu (2002); Gassmann and Behr (2011, 146).
[d] *Jicheng* 4.6015: 164-196-67.
[e] Trans. DWP and Zhao Lu.

* * *

The Supernal Lord as depicted in the earliest oracle bone inscriptions still stands at the head of a supernal hierarchy with the power "to make or break the dynasty," in David N. Keightley's phrase. The Supernal Lord's was by then an ancient power, still awesome and apparently approachable only through the intercession of the ancestral spirits, whose role was soon to be enhanced, along with the dynastic fortunes, at the expense of the Supernal Lord and lesser nature spirits. But there can be little doubt of the priority originally attributed to the power and influence of the Supernal Lord as the ultimate cosmic source both of the dynastic lineage in the mythical genealogy and of its sociopolitical organization.[27] My portrayal of the Supernal Lord and Heaven as essentially sky divinities contradicts assertions that the Shang did not possess a high god and that *Di* in the oracle bone inscriptions refers to a corporate body of royal "fathers."[28] In my view, the corpus of late Shang oracle bone inscriptions, given their predominantly cultic focus, should not be assumed to inform us adequately about the whole spectrum of Shang religious thinking. Robert Eno alludes in passing to the evidence of "asymmetry with regard to the powers of ancestors and nature deities," which "may reflect a historical process whereby Shang kings extended the legitimacy of the royal Tzu clan by broadening the powers of their ancestors, tending toward a coincidence between the pantheons of the clan and the state." But he nevertheless concludes that because Shangdi and the ancestors are portrayed in the inscriptions as having powers over both human and natural realms, therefore "our conclusion must be that whatever '*Ti*' was, it was something nearer to an ancestral figure than a natural one."[29] An alternative interpretation would be to affirm in this apparent overlap of functions, exhibited within the context of the late Shang ancestral cult, evidence of the process of gradual encroachment by the royal ancestors on the former prerogatives of the high god.[30]

From the above discussion we can draw several conclusions with regard to the early history of astrology and its association with political developments. As Eliade has shown, what happened among the stars was understood as a "signal moment in the inexhaustible hierophany of the sky divinity," and dramatizations in myth of such phenomena typically involve notions of "sovereignty," "power,"

---

[27] Curiously, there is no clear evidence of sacrifices explicitly dedicated to Shangdi in any Shang sources. The *2nd Year Ge Qi You* bronze vessel from the second year of the last Shang king, Di Xin, is cited (e.g. Allan 2009, 15) as a unique instance of a libation offered to Shangdi, but this is probably a misreading. The vessel records a *yong* rite dedicated to "Consort Bing," wife of the dynastic founder Da (or Tai) Yi. No date is given besides "first month." However, the term *Di*, "Lord," was already in common use by this time to refer to royal ancestors. The *jia*-day (first of the sixty-day cycle) immediately preceded *yi* and *bing* days when sacrifices to Founder Tai Yi and Ancestress Bing were performed, so the greater likelihood is that "Shangdi" here is instead a reference to Shang Jia (or to Da Yi in the hierarchy), to whom the libation would have been offered on the immediately preceding *jia* or *yi* day.

[28] Eno (1990, 1–26).   [29] Ibid., 4, n. 7.

[30] For discussion of the historical processes involved in the inflation of the Shang ancestral cult, see Keightley (1982, 294 ff.).

"law," and the like. Kenneth Burke's analysis of the rhetoric of religion shows too that the realm of the supernal was conceptualized by analogy with human sociopolitical experience, and that once so conceived it was understood to be the ultimate source of the verities of that experience. Given the existence of a Supernal Lord in early Shang, a divinity up above who from his dwelling place in the center of the sky commanded processes of nature, meteorological phenomena, and a hierarchy of subordinated supra-sensible entities (thereby controlling human destiny), it follows that the god would also have been thought responsible for the movements of heavenly bodies and phenomena associated with them.

My general argument is consistent with Peter Berger's observation that

probably the most ancient form of [religious] legitimation is the conception of the institutional order as directly reflecting or manifesting the divine structure of the cosmos, that is, the conception of the relationship between society and cosmos as one between microcosm and macrocosm. Everything "here below" has its analog "up above."[31]

The subservient relationship is clearer in the case of the Zhou, who expressly ascribe supreme authority and priority to the *ming*, "commands/mandate," of Heaven. Moreover, their more extensive ideological statements refer explicitly to Xia historical antecedents nearly a millennium earlier. With the Zhou the scope of Heaven's activity, like the universal kingship itself, was broadened at the expense of the ancestral cult in a manner consistent with the pattern followed by similar religious restorations elsewhere, which "brought back to life ancient supreme gods of heaven who had been turned into *dei otiosi*,"[32] but the conception of the personality of the Supernal Lord apparently remained basically unchanged. The Zhou approach to problems of meaning is more generally stated, more comprehensive than the opportunistic, piecemeal approach of late Shang theology because of the emphasis on the universality of Heaven's authority in contrast to the Shang preoccupation with their cult of royal ancestors. The processes and phenomena of the natural order, from the great rhythms of the seasons to the occasional and unpredictable anomalies, were all seen essentially as actualizations of the will of Heaven. By attributing human-like personality to Heaven, and by vigorously reviving the conception of phenomenal nature as an index of Heaven's engagement, the Zhou, culturally distinct as they probably were, were able to reimbue nature with an ethical quality. This feeling for the ethical dimension comes most strongly to the fore in the early Zhou texts and inscriptions, but ultimately it pre-dates the Zhou.

---

[31] Berger (1990, 34). Needham (Needham and Wang 1959, 171) called it "that sense of the unity and even 'ethical solidarity' of the universe."

[32] Eliade (1958, 75).

In view of the discussion thus far, Sarah Allan's assertion that "when the Zhou conquered the Shang, they had no religious ideology that clearly separated them from the Shang," appears unlikely.[33] Her claim overlooks the predynastic Zhou foundation myths, the interpretive context of the astral omens, the ideological import of very early Zhou dynastic bronze inscriptions like the *He zun* and *Tian Wang gui*, and the theoretical perspectives of comparative religion.[34] Even King Wen's receipt of "Heaven's Great Command" in the guise of the planetary phenomenon of 1059 pre-dates the Zhou conquest by thirteen years.

If, moreover, one lends credence, as Sarah Allan does for the Shang, to the historical role of a mythic "ten suns" master narrative (originally put forward by Tung Tso-pin), one is hardly justified in denying similar consideration to the Zhou foundational myths. One cannot credibly claim, as Allan does, that "an historical reconstruction should not be based upon mythological materials" and that "much of the material in the traditional historical texts is mythical in character and thus unreliable as the basis of historical reconstruction," while simultaneously interpreting Shang religious belief based on those same sources.[35] According to Sarah Allan,

> The essence of my theory is that the ten cyclical characters, later known as *t'ien-kan* 天干 (celestial stems), were originally the names of ten suns, who rose from the Mulberry Tree in the East, one on each day of the week, to fly across the sky as birds. These names are also used in Shang oracle bone inscriptions to designate the ten days of the *hsün*-week (旬), and the Shang ancestors were classified according to the ten sun categories.[36]

In fact, the oracle bone inscriptions never identify the celestial stems as suns, nor do they mention myths about mulberry trees or sun-birds. Allan's theory about Shang "totemism" is inferred from passages in the literature of a millennium

---

[33] Allan (2007, 488). In her recent account of the formation of Chinese civilization Sarah Allan ignores the non-Shang lineage-founder myths of competing polities of the Bronze Age, as well as the late Neolithic Longshan sites of Xinzhai and Taosi, the latter marginally smaller than Erlitou but the largest urban center in the immediate "predynastic" period. As Hwang Ming-chorng (1996, 172) shows, the Shang "ten-sun cosmogony is not a universal concept across the Yellow River Plain, rather it is a 'tribal' cosmogony" among others, which also have religious and ideological significance. As Rousseau famously said, "No state has ever been founded without Religion serving as its base." Quoted in Lewis-Williams and Pearce (2005, 38, 202).

[34] Of particular relevance with regard to the emergence of high gods are Guy Swanson's findings (1964, 20, 55–81) that there is a strong correlation between deities and the constitutional structure of sovereign groups. High gods are present, almost invariably, in societies where there are three or more types of such sovereign groups ranked in hierarchical order (e.g. household, clan-village, chiefdom). The model of a powerful, centralized government may be a sufficient condition for the appearance of a high god, but Swanson's data show it is not a necessary one, as some have argued, e.g. Kwang-chih Chang (1976, 190). Sarah Allan's doubts (1981; 1984) about the evidence for Three Dynasties religious ideology and mythology hark back to earlier work, but lately (Allan 2010) she too has begun to acknowledge the existence of a Xia Dynasty. For the myths relating to Yu the Great and the founding of the Xia, an insightful study is Porter (1993). For the transition to early statehood, see Liu (2007).

[35] Allan (2009).   [36] Allan (2010, 5–6).

later where the myths are not even attributed to the historical Shang Dynasty but seemingly to a much earlier mythological epoch. The "ten suns" Shang mythos cannot be recovered from the oracle bone inscriptions, but derives instead from much later material. The only specific allusion to a connection between Shang kings and the Sun is a legend about the penultimate Shang king, Di Yi, using a blood-filled leather bag for archery practice while sarcastically claiming to be shooting down the Sun, hubris hardly suggestive of a reverential attitude. The Shang did indeed employ a ten-day week whose days were enumerated using the heavenly stems. But, as we saw in Part One, those cyclical signs may well have pre-dated the Shang Dynasty, and the rationale for their use as posthumous designations for their dead kings remains obscure.

The Supernal Lord's planetary minions

It now seems clear that by the twentieth century BCE the ancient Chinese had already distinguished the somewhat erratic movements of the Five Planets from the regular motion of the fixed stars, as well as from transitory celestial phenomena like comets, meteors, auroras, and the more common meteorological phenomena. Once discovered, these five bright objects moving independently of the background of fixed stars would certainly have commanded attention as they wandered among the seasonal constellations along the ecliptic. Occasionally they met briefly in groups of two or three, more rarely four, and, rarest of all, five planets. Given a conceptualization of the supernal realm by analogy with human sociopolitical experience, it is hardly to be wondered that this relative freedom of action could be likened to that of the king's own deputies, who were dispatched to distant locations on the king's business and who gathered on occasion to deliberate policy. Hence, since the powerful arbiters of time and light – Sun and Moon – are identifiable in the oracle bone inscriptions as deserving of special treatment by the Supernal Lord, while the more subordinate wind and rain were subject to his direct command, it is likely that the Lord's "Five Minister Regulators" (*wu chen zheng*) refers specifically to the Five Planets whose behavior and function qualified them for high rank in the supernal hierarchy.

Chen Mengjia, for his part, was convinced that the Five Minister Regulators belonging to the Supernal Lord in the oracle bone inscriptions correspond to the later *wu gong chen* in *Zuozhuan* (Duke Zhao, seventeenth year), where they figure as officials in charge of the seasons of Heaven (*zhang tian shi zhe*). Elsewhere (*Zuozhuan*, Duke Zhao, twenty-ninth year), they become cosmic functionaries in charge of five kinds of useful materials, precursors of the Five Elemental Phases.[37] In the "Treatise on the Celestial Offices," Sima Qian is

---

[37] Chen Mengjia (1956, 572).

explicit about the celestial identity of Heaven's five ministers: "These Five Planets are Heaven's Five Assistants."[38] By the Han, illustrations of the Five Planets with associated gods are found in which each is depicted holding a construction tool of one kind or another. From this John S. Major concludes, "there is perhaps also a hint that the planetary gods are the architects of the sub-celestial world as it comes into being in its multiplicity of forms."[39] In view of the historical responses to exceptional clusters of the Five Planets, it seems that "deliberations" of the five planetary spirits were taken to signal momentous shifts in interventionist policy at the highest level of the supernal realm. The occurrence of such deliberations in the very regions of the sky with pre-existing astrological links to specific terrestrial powers (i.e. *Great Fire* with Shang and the *Vermilion Bird* with Zhou) must have lent extraordinary force to the directives the Supernal Lord was understood to be handing down. In the case of the Zhou conquest in particular, the accuracy of this interpretation is reinforced by analysis of specific military and political actions taken by the Zhou leaders in the short span of thirteen years preceding the decisive Battle of Muye in early 1046 BCE. It is further underscored by the intense preoccupation of the early Zhou rulers with "proper attention to Heaven's awesomeness," as recorded in the *Book of Documents*.

It is unclear whether all Five Planets were already associated at this early date with all Five Elemental Phases, colors, and other correlates of the mature system of late Zhou. But in the remarkably precise identification of certain colors and elements with the sequence of three planetary clusters in the second millennium BCE, it is possible to discern the beginnings of what also came to be thought to be a preordained pattern. As yet, the agent of causation is still the high god rather than a theoretical cosmic imperative as in the cyclical scheme elaborated by Zou Yan in the third century BCE. But like the principles of sociopolitical order and the concept of Heaven's Mandate itself, the essential basis of Zou Yan's theory of historical process appears to have been disclosed empirically, and later Five Elemental Phases theory derives much of its persuasiveness from a resonance with popular beliefs and religious legitimations of the social order harking back to the early Bronze Age.

Shang and Zhou contrasts

Quite apart from the new linguistic coinage pertaining to Heaven that makes its appearance in early Zhou (*tian, tianming, tian wei, tian xian*, etc.), attitudes concerning the relationship of Heaven to mankind begin to be explicitly formulated in terms of a coherent and conclusive political ideology. History in early Zhou meant the study of precedents, an examination of the complex amalgam

---

[38] *Shiji*, 27.1350.   [39] Major (1993, 27; 1978, 12).

of genealogical, cosmo-magical, and factual knowledge about the past from which, it was hoped, lessons might be learned: "The mirror is not far," it was said; "it is in the generations of the lords of Xia and the lords of Yin [Shang]."[40] A broadening of archaeological and archaeoastronomical horizons obliges us to examine in a new light the "mirror" of Three Dynasties history, especially the history of ideas put forward by those early Zhou ideologues. From the Zhou founders' appeals to precedent going beyond the founding of Shang all the way back to early Xia, and from the confirmation this version of history has received from an unexpected direction – actual events in the skies – it is now apparent that the associated impression that historical change correlates with manifestations of the divine will in nature must have a history at least as long, even if theoretical formulations based on such insights first emerged much later.

Belief in the validity of heavenly intervention would have been powerfully reinforced with each repetition of the remarkable planetary phenomena to which I have drawn attention. Certainly, from the perspective of the Zhou, the *third* historical appearance of a planetary omen of the Mandate's conferral and the successful overthrow of Shang would have provided more than sufficient grounds for the kind of conclusive formulation of the doctrine of Heaven's Mandate found in their earliest pronouncements.[41] Philosophy and psychology provide support for this interpretation. For example, as Klaus E. Müller has written,

If several events are linked in a sequence, a chain of dependence is formed, each link of which increasingly strengthens the others; the impression emerges of a serial, law-governed succession, and "apparent continuity." The longer the sequence, the more reliable the connection, and the more established the position of the governing link – uninterrupted continuities have a legitimizing function. This function increases with the significance of the first link, which founded the sequence and lent it the "causal thrust" upon which legitimization is established. Divine ancestors, hegemonic genealogies, heroes as creators of important institutions, legendary city founders, founders of religions and "fore-runners" all stand for this.[42]

That the Zhou would have been most anxious to develop complex legitimations to account for historical developments is to be expected, since by overturning the Shang hegemony they also undermined the foundations of the

---

[40] From an exhortation by King Wen of Zhou to his followers. Tu (2000, 168).
[41] By the Han Dynasty eight centuries later the appearance of a fourth planetary alignment as a sign of Heaven's conferral of the Mandate had become a necessity. See Chapter 9 below.
[42] Müller (2002, 46). David Hume identified this as a "universal principle of conceptual combination." Sir James George Frazer was the first "to show that what are involved here are the two fundamental principles of magic: the – as he put it – 'law of similarity' and the 'law of contact or contagion.' Psychologists who have specifically looked into the principle of contiguity, and who have thereby been able to provide the clearest proof of its effectiveness in experimental terms, describe it as 'a significant and deeply rooted characteristic of thought,' which gives rise to an 'apparent dependence'" (ibid., 45).

long-standing cosmo-political order to which they had earlier professed allegiance. Indeed, they even continued certain Shang rituals. If the Warring States and Han portrayals of the Zhou dynastic founders like the Duke of Zhou are anachronistic, it is in representing them as self-consciously concerned with categories of meaning. The enunciation of the concept of Heaven's Mandate as a form of politico-religious legitimation, on the other hand, fits the historical context. As Peter Berger has shown, the rootedness of religion in the practical concerns of everyday life means that

the religious legitimations, or at least most of them, make little sense if one conceives of them as productions of theoreticians that are then applied *ex post facto* to particular complexes of activity. The need for legitimation arises in the course of activity. Typically, this is in the consciousness of the actors before that of the theoreticians... To put it simply, most men in history have felt the need for religious legitimation – only very few have been interested in the development of religious "ideas."[43]

* * *

The contrast between the particularistic and opportunistic approaches to fundamental spiritual issues taken by Shang divinations as against the incipiently ideological disposition epitomized by the doctrine of Heaven's Mandate is a striking one. In spite of a number of cosmo-political continuities, study of early Zhou sources suggests that at some level a significant conceptual watershed has been crossed. With the advent of religious reform and the universal sovereignty of Zhou, clear progress has been made toward the time, still some centuries ahead, when

the ubiquity, the wisdom and the passivity of the sky god were seen afresh in a metaphysical sense, and the god became the epiphany of the order of nature and the moral law... the divine "person" gave place to the "idea"; religious experience... gave place to theoretic understanding, or philosophy.[44]

My portrayal of the emergent contrast between late Shang and early Zhou religious dispositions echoes Clifford Geertz's elaboration (following Max Weber) of the disparity between "traditional" and "rationalized" religions, the latter coming to fuller expression in the "Axial Age":

Traditional religions consist of a multitude of very concretely defined and only loosely ordered sacred entities, an untidy collection of fussy ritual acts and vivid animistic images which are able to involve themselves in an independent, segmental, and immediate manner with almost any sort of actual event. Such systems... meet the perennial concerns of religion, what Weber called the "problems of meaning" – evil, suffering, frustration, bafflement, and so on – piecemeal. They attack them opportunistically as they arise in each particular instance – each death, each crop failure, each untoward

---

[43] Berger (1990, 41).   [44] Eliade (1958, 110).

natural or social occurrence – employing one or another weapon chosen, on grounds of symbolic appropriateness, from their cluttered arsenal of myth and magic... As the approach to fundamental spiritual issues which traditional religions take is discrete and irregular, so also is their characteristic form. Rationalized religions, on the other hand, are more abstract, more logically coherent, and more generally phrased. The problems of meaning, which in traditional systems are expressed only implicitly and fragmentarily, here get inclusive formulations and evoke comprehensive attitudes. They become conceptualized as universal and inherent qualities of human existence as such, rather than being seen as inseparable aspects of this or that specific event... *The narrower, concrete questions, of course, remain; but they are subsumed under the broader ones, whose more radically disquieting suggestions they therefore bring forward. And with this raising of the broader ones in a stark and general form arises also the need to answer them in an equally sweeping, universal, and conclusive manner.*[45]

A Han Dynasty retrospective interpretation

In the Western Han, Dong Zhongshu (*c.*179–*c.*104 BCE), after Zou Yan, characterized the Shang–Zhou transition as a shift in a binary cycle away from a preoccupation with "substance" (*zhi*) toward an emphasis on "pattern" (*wen*). Sima Qian, following Dong Zhongshu, tellingly discerned a symptomatic spiritual decline during the Shang, a falling away from a traditional attitude of "reverence" toward Heaven and the natural realm in favor of superstitious preoccupation with the spirits of the ancestors. With regard to the divine sanction underpinning the universal kingship the key shift is marked by a de-emphasis of legitimacy based on the principle of contiguity – that is, membership in the royal lineage – toward a focus on legitimacy premised on emulating Heaven as the paradigm of order and harmony, an ethos inspired by an archaic, fundamentally metaphorical idea about the congruence obtaining between the supernal and temporal realms. In structuralist terms one might characterize this transition as an ideological shift away from the "metonymic-connexion axis" back to the "metaphoric-similarity axis."[46] In this respect the choice of *wen*, "pattern, form, style," to denote the cosmological paradigm represented by the concept of the Mandate of Heaven, and the "vehement reassertion of the transcendental power of the high god" which it entailed, appear particularly apt. By implication, of course, this same paradigm applies equally to the zeitgeist of Xia, with whom the Zhou explicitly identified themselves.[47]

---

[45] Geertz (1973, 172, italics mine). While I find this characterization of the distinction particularly apt in the case of the contrast between Shang and Zhou religion, this should not be taken as unreserved endorsement of Weber's interpretation of Chinese religion, not least in view of its latent Eurocentrism. Hobson (2008, 14–19).

[46] For insightful application of structuralist analysis to ancient Chinese correlative thinking, see Graham (1989a, 315). Graham's account is in the tradition of Frazer, Jacobson, and Lévi-Strauss.

[47] Schwartz (1985, 38). Note especially Benjamin Schwartz's choice of the word "reassertion" above. See also his remarks (ibid., 48 ff.) concerning the re-emphasis on correct performance

In contrast, the window on the world of the Shang provided by the oracle bone inscriptions clearly seems skewed by the particular preoccupations of late Shang divinatory theology. Remarkably, cosmology and astrology figure only marginally, the natural powers finally not at all, in a magico-religious practice largely devoted during the final decades of the dynasty to the routine observances of the ancestral cult. We have scant information about the two centuries following the founding of the dynasty by Cheng Tang, before "reverence" deteriorated into "superstition," as Sima Qian said, but discernible in the institutionalization and routinization of divination long after the heady years of the founding is a flagging attention to entire categories of supra-sensible agency which seemed to grow less and less relevant. David N. Keightley characterizes the taming of Shang belief by convention this way: "These later inscriptions record, I would suggest, the whisperings of charms and wishes, a constant bureaucratic murmur, forming a routine background of invocation to the daily life of the last two Shang kings, who were now talking, perhaps, more to themselves than to the ultra-human powers."[48] Certainly the worship of the Supernal Lord (not to mention the seasonal calendar) seems to have been eclipsed by the cult of the royal ancestors during the last century and more of the Shang Dynasty.[49]

The view presented here of a fundamental continuity of cosmo-political conceptions throughout the ancient period carries with it the implication that, whatever other common threads with later tradition there are – reverence for the ancestors and sacrifices to the high god – the late Shang probably represented a significant departure in important respects, and not merely because of the cultic focus of the oracle bone inscriptions, but evidently also because of "imperial overreach." The conception I have in mind as reasserting itself in early Zhou is of a balanced management of the spiritual regime that did not privilege the ancestors above the high god in ritual or theology. Arthur Wright succinctly expressed the worldview this way:

Underlying all this was a kind of primitive organicism: a belief that the worlds of the gods and of men were interconnected, that it behooved men to respect the natural forces and natural features over which the gods presided . . . and that the ancestors, particularly those of great lineages, continued to play an important role in the affairs of their descendants as surrogates of the God on High. Thus gods, men, and nature, the living and the dead – all were seen as interacting in a seamless web.[50]

---

of ritual with the advent of the Zhou, especially "the striking observation of the *Book of Ceremonies* . . . [that] the Shang people had put the spirits in the first place and the rites second, while the Chou put the rites first and the spirits in second place."

[48] Keightley (1988, 382).

[49] See Keightley (1998, 811), who remarks on the Shang ritual cycle becoming so structured as to supplant the calendar in secular records.

[50] Wright (1977, 41).

The Zhou people (among others) must have been culturally distinct from Shang in important respects, shown not least by their worship of a different high god "Heaven/Sky" (*Tian*) in contrast to the *Di* of Shang, but also by the separate myth of their progenitor Hou Ji's miraculous birth, and even by their different order of enumeration of the four directions.[51] It is hardly surprising that they did not share the Shang king's optimism and confidence in the power of his royal ancestors. The Zhou rulers identified with their "virtuous" predecessors in Xia, whatever structural form that polity may have taken (as well as nominally with the "virtuous" Shang founder, Cheng Tang). Their tireless speechmaking on behalf of Heaven was no mere pious posturing, nor was their exculpatory message a dissembling pose. They saw themselves as the agents of the restoration of the prestige of Heaven, hence the anxious preoccupation with preserving the Mandate for posterity that suffuses the rhetoric of early Zhou bronze inscriptions and received texts: "Heaven's Mandate is not easy to keep; it is not to be counted on," it was said.[52] One possible reason for their anxiety, as Karl Löwith incisively remarked, is that

the conjunctionist thesis leads imperceptibly and unconsciously to the idea of variability and pluralism in religious and political regimes. If changes depend on the movements and conjunctions of the upper planets with certain signs of the zodiac, then major historical events can only be considered "providential" in the metaphorical sense.[53]

David N. Keightley has argued that by the end of Shang "the balance of caution against confidence, of negative doubt against the need for positive action, of religious reflection against secular activity, of Neolithic pessimism against Bronze Age optimism . . . shifted in favor of optimism, confidence, and human control."[54] If this is so, the religious and ideological correctives applied by the Zhou founders in response to the social, psychological, and cultural strain of assuming dynastic hegemony may be said to manifest in their inspiration a return to Neolithic pessimism. Yet their comprehensive, conclusive cosmo-political formulation marks a significant departure from the opportunistic, piecemeal approach to problems of meaning taken by Shang divinatory theology. In striving to emulate Heaven, which encompassed the world and provided the rhythmic backbeat to the phenomena, the Zhou reasserted the primacy of universality and inclusiveness, in marked contrast to their former overlords' indulgence in exclusivist hegemony, itself a logical outgrowth of the Shang preoccupation with the cult of their royal ancestors.

That is not to say that cosmic religion supplanted ancestor worship – it surely did not, as the Zhou were hardly less conscious of their lineage origins or less reverent toward their forebears. It was the imbalance between the twin

---

[51] Hwang (1996, 172, 478).   [52] Chan (1963, 7).
[53] Löwith (1949, 20).   [54] Keightley (1988, 388).

pillars of Bronze Age religiosity, reverence for the supra-sensible powers and for the ancestors, caused by the Shang rulers' elevating their dead ancestors (and even living kings) to parity with *Di*, that required redress. The epoch-making Shang–Zhou dynastic transition consequently surfaced a fundamental tension at the heart of ancient Chinese political thinking about the Mandate and the nature of legitimate succession that was henceforth to manifest itself repeatedly in the political life of early China. More important, in view of the evidence of a fundamental consistency between late Zhou cosmological conceptions and their second-millennium BCE antecedents, the Zhou claim to have re-established the continuity of an ancient tradition that took its cues from Heaven and the natural order appears well founded.[55]

[55] "The Western Zhou lasted for a period of some four hundred years. During this period, the concept of *tian ming* developed from a specific astronomical event during the reign of King Wen, which supported the Zhou claim of legitimacy, to the idea of a changing mandate of heaven, closely associated with a theory of dynastic cycle. Since Shang Di ruled the sky, it was he who gave the command, demonstrating his intention in the sky. Since the mandate was displayed in the sky, it could also be described as a 'celestial' or 'heavenly' mandate (*tian ming*). As time passed, this gradually became more abstracted into a cosmological theory in which the 'sky's command' or, as it is usually translated, the 'mandate of heaven,' became a more abstract idea. Moreover, it was interpreted as part of a repeating cycle. This was the origin of the idea of a changing mandate of heaven." So writes Sarah Allan (2007, 43), succinctly restating my own conclusions in Pankenier (1995).

8   Cosmology and the calendar

I numbered the days, measured the months, and established the years so as to match the phases of Sun and Moon.[1]

Il est un fait établi que le calendrier a occupé en Chine une place prépondérante.[2]

In a seminal study, Marc Kalinowski says that ancient accounts of the creation and ordering of the world

follow a similar narrative pattern . . . all of them assign a determining function to the calendar and the seasonal cycle in the process of world formation. This function appears to be a distinctive feature of the oldest Chinese cosmogonic accounts and provides an interesting contrast to what can be found in the cosmological literature of later times.[3]

The goal implicit in this mode of thought was to achieve a universal order, a perfect congruence between the natural and human rhythms, for which the ideal paradigm was found in the heavens. The constancy and regularity of celestial motions and their phenomenal manifestation in the seasonal rhythms initially provided patterns of permanence and timeliness to contrast with the constant flux of more mundane events which insistently evoked the irregularity and capriciousness of nature.[4] As I suggested at the outset, the ability to comprehend the celestial motions and to sustain a reciprocal conformity between their regular variations and human activity – that is, the discernment necessary to "mimic heaven" (*xiang tian*) – was a fundamental qualification of kingship. Similarly, according to Peter Berger,

Where the microcosm/macrocosm understanding of the relationship between society and the cosmos prevails, the parallelism between the two spheres typically extends to specific roles. These are then understood as mimetic reiterations of the cosmic realities for which they are supposed to stand. All social roles are representations of larger complexes of objectivated meanings.[5]

---

[1] Civilizing claim attributed to the Yellow Emperor, Huang Di, in the Mawangdui MS *Shiliu jing*; *Mawangdui Hanmu boshu* zhengli xiaozu (1980, 61), 1.78b–79a.
[2] Kalinowski (2004, 87).   [3] Ibid., 122.   [4] Needham (1969, 336); Urton (1978, 157).
[5] Berger (1990, 38). We saw how this process is mediated linguistically in the preceding chapter.

242

By way of comparison, David Carrasco characterized the ancient "cosmo-magical" mode of thinking this way:

This locative view, which has been discerned in the traditional societies of Mesopotamia and Egypt, in which everything has value and even sacrality when it is in its place, is an imperial view of the world designed to ensure social and symbolic control on the part of the king and the capital. It is informed by a cosmological conviction consisting of five facets which dominated human society for over 2,000 years in the Near Eastern world, including (1) there is a cosmic order that permeates every level of reality; (2) this cosmic order is the divine society of the gods; (3) the structure and dynamics of this society can be discerned in the movement and patterned juxtaposition of the heavenly bodies; (4) human society should be a microcosm of the divine society; and (5) the chief responsibility of priests and kings is to attune human order to the divine order.[6]

This theme is reiterated often in the early literature. In the "Canon of Yao" in the *Book of Documents*, for example, there is the famous passage in which the legendary Yao delegates responsibility for certain of these crucial heavenly observations to the pair of officials Xi and He: "[Yao] commanded Xi and He, in reverent accordance with August Heaven, to track and delineate the Sun, Moon and stars, and seasonal asterisms, and so to respectfully bestow the seasons on the people."[7]

Following this there is the much-studied catalog of seasonal culminating stars that were to be observed in order to establish the approach of the solstices and equinoxes, widely observed junctures of great practical and religious significance. Again, in the "Canon of Shun," Yao's successor on his accession is said to have used the stars *Jade Pivot* and *Jade Transverse* – that is, Ursa Major – to bring the seasons, months, and days into conformity with the movements of the seven Regulators, i.e. the Sun, Moon, and Five Planets.[8] Here we also recall the predynastic-era myth of Gao Xin's (Di Ku's) appointment of Yan Bo and Shi Chen as regulators of two of the most important seasonal asterisms at opposite ends of the sky, Scorpius and Orion. The central premise of the parable – the incompatibility of the two feuding siblings and their consequent exile to the ends of the Earth – reveals the etiological function of the myth.

---

[6] Carrasco (1989, 49). Carrasco's authority for this locative view "that guarantees meaning and value through structures of conjunction and conformity," is Smith (1978). Paul Wheatley (1971, 414–16) expressed a similar view about ancient China. For a sociological analysis of the microcosm/macrocosm mythos as the most ancient form of religious legitimation, see Berger (1990, 24 and 266).

[7] Trans. Karlgren (1950a, 3). On the mythology and astral lore encapsulated in this passage in the "Canon of Yao," see Liu Qiyu (2004); Kalinowski (2004).

[8] Compare this with Sima Qian's version in the "Treatise on the Celestial Offices": "The *Dipper* is the [*Supernal*] *Lord's Carriage*. Revolving in the center it oversees and controls the four directions, separates *yin* from *yang*, establishes the four seasons, equalizes the Five Elemental Phases, shifts the seasonal nodes and angular measures, and fixes the various cycles – all are tied to the *Dipper*." *Shiji*, 27.1293.

It accounts (à la Burke), by means of a familial analogy, for the diametrical opposition between the brothers' astral correlates, which cannot appear in the sky simultaneously.[9] Reverence for the normative patterns of Heaven is also attributed to the "former wise kings" of both Xia and Shang. Yan Bo and Shi Chen, together with paragons like Xi Zhong, A Heng, Fu Yue, Shiwei, Zao Fu, Wang Liang (both royal charioteers), Juzi (aka Chang Yi or Chang Xi, who fled to the Moon with the elixir of immortality, the legendary beautiful consort of Di Ku), as well as the venerable Emperor Zhuan Xu, Yu the Great, and his father Gun, all were immortalized through identification with particular asterisms or chronograms.[10] In the "Announcement Concerning Wine" (*Jiu gao*) in the *Book of Documents*, a pointed contrast is drawn between the pious rectitude of early Shang rulers and their irresponsible heirs: "I have heard it said that anciently Yin's former wise kings in their conduct stood in awe of Heaven's brightness [*tian xian*] and of the small people. They practiced virtue and held on to wisdom."[11] These were clearly individuals who knew the value of watching closely what transpired in the heavens. The rationale for this close observation of the skies is expressed most succinctly in the "Great Plan" (*Hong fan*) chapter of the *Book of Documents* of Warring States date:

What the king scrutinizes is the year, the dignitaries and noblemen the months, the many lower officials the days. When in years, months, and days the seasonableness has no changes, the many cereals ripen, the administration is enlightened, talented men of the people are distinguished, the house is peaceful and at ease.[12]

The level of observational and calendrical expertise reflected in texts such as these harks back to the stage characterized as "observing the patterns and

---

[9] Barber and Barber (2004). As Eric Havelock (1983, 14) showed in his classic study, "all cultures preserve their identity in their language, not only as it is casually spoken, but particularly as it is preserved, providing a storehouse of cultural information which can be reused. It is easy to understand how this works in a literate culture … [b]ut how is such information preserved in an oral culture? It can consist only in the individual memories of persons, and to achieve this the language employed – what I may call the storage language – must meet two basic requirements, both of which are mnemonic. It must be rhythmic, to allow the cadence of the words to assist the task of memorization; and it must tell stories rather than relate facts: it must prefer mythos to logos."

[10] Porter (1993). As an example, Fu Yue's star near the tail of the *Dragon* figures prominently in the *Chuci* poem "*Aisui*" as a way station in the author's celestial peregrination. For *Xi Zhong, Fu Yue, Zao Fu, Wang Liang* and other notable astral paragons, see Jiang Xiaoyuan (1992, 43–8).

[11] Trans. Karlgren (1950a, 45/9, trans. modified). "Heaven's bright manifestations," *tian xian*, is more or less synonymous with "Heaven's awesome displays," *tian wei*, i.e. more unpredictable demonstrations of heaven's awesomeness. In the "Metal-Bound Coffer" (*Jin teng*) chapter of the *Book of Documents* (compiled in Warring States time), "Heaven's awesome displays" refers explicitly to disastrous meteorological phenomena, while the "Metal-Bound Coffer" dwells on how Kings Wen and Wu came to know Heaven's awesomeness and to wield it in decimating their enemies.

[12] Karlgren (1950a, 33/29, trans. modified).

presenting the seasons" (*guan xiang shou shi*) by historians of Chinese astronomy (paraphrasing the "Canon of Yao").[13] The first stellar configurations about whose function we have historically had detailed information, the four cardinal asterisms mentioned in the "Canon of Yao" – *Topknot* (Pleiades), *Bird* (α Hya), *Fire* (α Sco), and *Ruins* (β Aqr) – could have served the functions assigned to them during much of the second millennium BCE, though they may go back earlier.[14] Their use, and the recognition of the seasonal orientation of the *Dipper*'s handle and the *Dragon*, give evidence of a transition in progress by the late Neolithic to an astronomically based calendar from an annuary based primarily on seasonal "indications" (*wu hou*) derived from the plant and animal kingdoms, meteorology, etc. Traces of the earlier annuary type of calendar still survive in sources such as the *Lesser Annuary of Xia* (*Xia xiao zheng*), in the "Explicating the Seasons" (*Shi xun*) chapter in the *Lost Books of Zhou*, and in passages such as the following from the *Discourses of the States*:

When the constellation *Horns* [lodge 1, α Vir] appears, the rain stops. When the constellation *Root of Heaven* [lodge 2, λ Vir] appears, the rivers dry up. When the constellation *Base* [lodge 3, α Lib] appears, the plants shed their leaves. When the constellation *Quadriga* [lodge 4, ρ Sco] appears, frost falls. When the constellation *Fire* [lodge 5, α Sco] appears, the clear winds warn of cold. Thus the teachings of the former kings said, "When the rains stop, clear the roads. When the rivers dry up, complete the bridges. When the plants shed their leaves, store the harvest. When the frost falls, get ready the fur garments. When the clear wind comes, repair the inner and outer defense walls and the palaces and buildings."[15]

Even though the modern phrase "the stars move and the *Dipper* revolves" (*xing yi dou zhuan*) is now merely a cliché for the passage of time, it evokes an epoch when the circumpolar stars were everyman's timepiece.[16]

---

[13] "Bestowing the seasons," *shou shi*, is still the name of China's National Time Service Institute, the equivalent of the US Naval Observatory in Washington, DC and the Bureau des longitudes in Paris.

[14] For analyses of efforts to date the astronomical portions of the "Canon of Yao," see Li Changhao (1981, 8–12); Liu Qiyu (2004).

[15] *Guoyu* 2.9a; Hart (1984, 49). For examples from *Lesser Annuary of Xia*, see Zheng Wenguang (1979, 44). Note that in the *Discourses of the States* passage quoted here only the first five lodges are adduced in sequence to illustrate the autumn indications. Similar seasonal and occupational correlations certainly existed for the remaining twenty-three lodges as well.

[16] *Heguanzi* says, "When the handle of the Dipper points to the east (at nightfall), it is spring to all the world. When the handle of the *Dipper* points to the south, it is summer to all the world. When the handle of the *Dipper* points to the west, it is autumn to all the world. When the handle of the *Dipper* points to the north, it is winter to all the world. As the handle of the *Dipper* revolves above, so affairs are set below." Graham (1989b, 517). Chen Jiujin, Lu Yang, and Liu Yaohan (1984, 71) show that the ancient practice of using the handle of the *Dipper* as a seasonal indicator originated with the Di–Qiang people, long-time adversaries of both Shang and Zhou and putative ancestors of the Yi minority now found in southwest China. There is otherwise no trace of this practice in Shang and Zhou. The authors also discuss the

From earliest times the patterns and rhythms displayed by the sky were as normative and meaningful as the rest of one's natural surroundings and social existence. There was no positive distinction between them. The process of cosmicization, through which the institutional order came to be seen as directly reflecting the divine structure of the cosmos, also entailed the monopolization of cosmological knowledge by the ruling elites, the "former kings" credited above in the quote from *Discourses of the States*. As the fourth-century BCE Chu Silk Manuscript states in no uncertain terms, "if men do not know the yearly cycle, they have no way to render homage to the gods."[17]

Ultimately this led to a narrowing of the acceptable channels through which everyman's relationship with the sacred could legitimately be mediated, a process which came to be encoded in myth and other information systems discussed in Part One.

As Merlin Donald has said,

The construction of early calendars always implied some theoretic development, in the sense that implicit models were being built of astronomical events. Many of the elements of modern science were already present in primitive astronomy: systematic and selective observation, and the collection, coding, and eventually the visual storage of data; the analysis of stored data for regularities and cohesive structure; and the formulation of predictions on the basis of those regularities. All early agricultural societies, out of necessity, had calendars based on astronomical science. Therefore, procedurally, the groundwork for scientific observation and prediction had already been constructed 5,000 to 10,000 years ago, not necessarily in written symbols but in a different kind of visuographic invention, which represented an analog *system* of knowledge.[18]

* * *

The separation of Heaven and Earth

One of the most frequently cited traditions bearing on the relation between ancient Chinese religion and the expanding power of the centralized state is found in "Discourses of Chu" in the *Discourses of the States*.[19] The passage in question is of singular importance both because it contains the earliest use of the terms "shamaness" and "shaman" and because it is the source of the myth concerning the separation of Heaven and Earth accomplished by Zhuan Xu, the legendary emperor who appointed Chong and Li to have charge of the affairs of Heaven and Earth:

considerable astro-calendrical content of *Heguanzi* and demonstrate its close affiliation with the *Lesser Annuary of Xia*.

[17] Cf. Kalinowski (2004, 115). For study of the derivative relationship between the Chu Silk Manuscript with its seasonal spirits and the more bureaucratic "bestowing the seasons" passage in the "Canon of Yao," see Xing Wen (1998).

[18] Donald (1991, 339).   [19] *Guoyu*, 18.1a ff.

Anciently, men and spirits did not intermingle. At that time there were certain persons who were so perspicacious, single-minded, and reverential that their understanding enabled them to make meaningful collation of what lies above and below, and their insight to illumine what is distant and profound. Therefore the spirits would descend into them. The possessors of such power were, if men, called *hsi* [shamans], and if women, *wu* [shamanesses]. It is they who supervised the positions of the spirits at the ceremonies, sacrificed to them, and otherwise handled religious matters. As a consequence, the sphere of the divine and the profane were kept distinct. The spirits sent down blessings on the people, and accepted from them their offerings. There were no natural calamities. In the degenerate time of Shao-hao ... however, the Nine Li [a troublesome tribe] threw virtue into disorder. Men and spirits became intermingled, with each household indiscriminately performing for itself the religious observances which had hitherto been conducted by the shamans. As a consequence, men lost their reverence for the spirits, the spirits violated the rules of men, and natural calamities arose. Hence the successor of Shao-hao, Zhuan Xu, charged Chong, Governor of the South, to handle the affairs of Heaven in order to determine the proper places of the spirits, and Li, Governor of Fire, to handle the affairs of Earth in order to determine the proper places of men. And such is what is meant by "cutting the communication between Heaven and Earth." Subsequently, the Three Miao revived the [disorderly] virtue of the Nine Li. Yao again raised up the descendants of Chong and Li. He caused those who had not forgotten the old ways to again have charge of the offices of Heaven and Earth. From then on, down to the Xia and Shang, accordingly, for generations the Chong and Li were the ones who ordered Heaven and Earth and kept their jurisdictions separate.[20]

Of this account K.C. Chang has written:

This myth is the most important textual reference to shamanism in ancient China, and it provides the crucial clue to understanding the central role of shamanism in ancient Chinese politics. Heaven is where all the wisdom of human affairs lies ... Access to that wisdom was, of course, requisite for political authority. In the past, everybody had had that access through the shamans. Since heaven had been severed from earth, only those who controlled that access had the wisdom, hence the authority – to rule. Shamans, therefore, were a crucial part of every state court; in fact, scholars of ancient China agree that the king himself was actually head shaman.[21]

K.C. Chang's assessment reflects his view of the central role of "state" shamanism and of the king as head shaman, a view that may have led him to pay too little attention to exactly what sort of wisdom was sought in Heaven, and whether in fact there is any necessary connection between such closely held knowledge and that customarily associated with shamanism.[22] In Chapter 2 we

---

[20] Bodde (1961, 390–1); cited by Kwang-chih Chang (1976, 162).
[21] Kwang-chih Chang (1983, 44). See the remarks in reference to this passage in Von Falkenhausen (2006, 47).
[22] More recently, in the most penetrating study to date of shamanism in Bronze Age China, David N. Keightley (1998, 763–831) argues convincingly against the notion that Shang spirituality was shamanistic in any meaningful sense.

saw how according to Cai Mo the "care and feeding" of *Dragons* was a most crucial element of such "wisdom."

I suggest that the "Discourses of Chu" passage, rather than providing a clue to the central role of shamanism in ancient Chinese politics, actually accounts for the co-optation and subsequent decline of such practices among the ruling elite as a consequence of the development of a new kind of esoteric and highly specialized knowledge of heaven more commensurate with centralized state formation – the kind of knowledge associated with calendrical astronomy, of which archetypal incumbents of the hereditary office of court astrologer-cum-calendar specialist like Chong, Li, Wu Xian, Yan Bo, Shi Chen, and Cai Mo were the custodians. As Jean Levi concluded regarding this passage, "far from creating an opposition between officials and sorcerors [*wu*]," the text "is bent on nothing but placing the latter under the control of the bureaucrats."[23]

No further mention of shamans is made in the passage after the description of the revolution brought about by Zhuan Xu and his appointment of Chong and Li, the officers who were charged with the same duties as the astronomers Xi and He in the "Canon of Yao." This new knowledge, for which Chong and Li were responsible, apparently so crucial to the exercise of political authority, is clearly differentiated here from the sort of direct communication with the supra-sensible previously accessible to "everyman." Instead, such knowledge becomes the preserve of the king and specialist hereditary offices charged with its collection and transmission.[24] Given the cosmic bailiwick of figures like Chong and Li, perhaps it is also possible to see in this account a rationalization of the consequences for the sky divinity of what Eliade denoted the "progressive descent of the sacred into the concrete," a regular historical process whereby "the supreme divinities of the sky are constantly pushed to the periphery of religious life where they are almost ignored; other sacred forces, nearer to man, more accessible to his daily experience, more useful to him, fill the leading role."[25] Often the leading role comes to be occupied by the cult of the dead

---

[23] Levi (1989, 223); quoted in Keightley (1998, 823). This is echoed by V.N. Basilov (1989, 33): "The appearance of the state, which took on itself the management of society, was the decisive stage in the process of suppression of shamanism," quoted in Keightley (1998, 826, n. 165). For the relationship between astronomical observation, astrology, and government in ancient China, see Jiang Xiaoyuan (2004, 53–8, and especially 83 ff.). The "Who's Who" of heaven-observing functionaries from the dawn of history in *The Grand Scribe's Records* lists all of the above astrologers and more in chronological order. Among these, Zhuan Xu, Chong, and Li figure prominently in the legendary genealogy of the Chu royal house. "Treatise on the Celestial Offices," *Shiji*, 27.1343 (Appendix 1 to this volume); and the "Hereditary House of Chu," *Shiji*, 39.1689. See also and Liu Qiyu (2004).

[24] This was also the conclusion of Yang Xiangkui: "The kings severed the communication between heaven and the people, and they took as their monopoly the great right of communicating with God." Quoted in Kwang-chih Chang (1983, 164), Yang (1962, Vol. 1, 164), also Jiang Xiaoyuan (2004, 91).

[25] Eliade (1958, 43).

ancestors, with the supreme divinity of the sky being relegated to a more specialized role such as that of guarantor of the harvest, seasonable weather, and the like. In Eliade's words, "men only remember the sky and the supreme deity when they are directly threatened by a danger from the sky; at other times, their piety is called upon by the needs of every day, and their practices and devotion are directed towards the forces that control those needs."[26]

Consider now the judgment of Sima Qian, who discussed the same political and religious developments in tracing the ancient origins of calendrical astronomy:

It was Huang Di who first examined into and determined the calendrical constellations, established the Five Elemental Phases, began the [alternation of] repose and activity, and regulated the intercalary epact. Thereafter there were the offices in charge of Heaven and Earth, spirits and people, and the categories of sacrificial objects; these were called the Five Offices. Each attended to ordering its own affairs without interfering with the other. In consequence, the people were able to have their beliefs, and the spirits were able to have their efficacious virtue. The people and the spirits had their separate tasks, each respecting the other without encroachment, therefore the spirits sent down excellent harvests, and the people offered up things [wu, i.e. ritually appropriate offerings and foodstuffs] in sacrifice; disasters and calamities did not arise, and what each sought was not lacking. During the degenerate time of Shao-hao, the Nine Li brought virtue into disorder, the people and spirits intermingled and interfered with each other and could not count on the sacrificial objects; disasters and calamities arrived in concert, and none lived out his allotted span of years. When Zhuan Xu succeeded him [Shao-hao] he mandated that Chong, Regulator of the South, should manage the affairs of Heaven in order to take charge of the spirits, and also that Li, Regulator of Fire, should manage the affairs of Earth in order to take charge of the people, and to restore to those affairs the constant order of the past when there was no mutual encroachment. Subsequently, the Three Miao [again] emulated the [disorderly] virtue of the Nine Li, whereupon the two offices [of Chong and Li] completely neglected their duties. The intercalations were misplaced, the first month of spring was misidentified, the *Assistant Conductors* lost their periodic function,[27] and the calendrical calculations became disordered. Yao again contacted those successors to [the offices] of Chong and Li who had not forgotten the old [methods] and had them once again take charge, and [for this purpose] he set up the offices of Xi and He to make clear the correct measure of the seasons, whereupon the yin and yang became moderated, the wind and rain timely, an abundance of *materia vitalis* [qi] arrived, and there were no unnatural afflictions among the people. In his dotage [Yao] abdicated in favor of Shun, and in the Temple of

---

[26] Ibid., 50.

[27] This is an indication of empirical awareness of the effect of precession, even though the change is here attributed to negligence. *Han shu*, 21A.973, also contains a similar statement. Liu Xin identifies this and other calendrical dislocations as portents of dynastic change in *Hanshu*, 36.1964 (my thanks to Juri Kroll for this latter reference). On the significance of this indication of the obsolescence of Arcturus as a seasonal marker, see Needham and Wang (1959, 251–2). For an account in *Hesiod* (700 BCE) of Arcturus's use as a seasonal signpost in second millennium BCE Greece and its obsolescence there, see Worthen (1991, 210).

Cultured Ancestors warned and admonished [Shun] saying, "The calendrical calculations of Heaven reside in your person." Shun [in his turn] also charged Yu [with the same
words]. From this may be observed what rulers have judged of the greatest import.[28]

Even if Sima Qian's discussion is framed in terms of anachronistic Western
Han conceptions of the unified nature of the political order in the second millennium BCE, his general description of early difficulties in achieving mastery of
the calendar and its promulgation throughout the region under Xia and Shang
tutelage has the ring of truth. Inevitably, there would have been setbacks of
the kind described, because of the recalcitrance of subjugated peoples reluctant to submit to authority; in part due to political enmity, decline, or disorder
and incompetence at the center; and in part as a result of the poorly understood obsolescence of time-honored seasonal asterisms due to the effects of
precession.[29]

Sima Qian's main point is clear – the revolution alluded to in the above
passage from *Discourses of the States* has less to do with the role of shamanism
than with the emergence of bureaucratic control of time and the knowledge
of Heaven as pre-eminent concerns of the state.[30] Even allowing for the distinctively Han retrospective flavor of Sima Qian's portrayal of Bronze Age
developments, the ascendancy among the ruling elite of a fundamentally cosmological worldview and of the doctrine of Heaven's Mandate, together with
a positive lack of evidence of shamanistic practice in the strict sense, indicates
that astrological and shamanistic practices probably did not, in fact, coexist
at that level.[31] What is being described in this passage is perhaps the Bronze
Age transition when those charged with the responsibility of interpreting the
heavens began to institute regular schemes of human devising like the calendar,
itself a concrete manifestation of the impulse to domesticate the phenomena. As
has been noted by Guy Swanson, "we begin to see in the connection between

---

[28] *Shiji*, "Treatise on the Calendar," *Li shu*, 26.1257. Note the striking similarity to Cai Mo's
account from *Zuozhuan* in Chapter 2.

[29] For another version of the cosmic conflict and subsequent disaster seen as marking the end of
the urzeit of cosmogony, see Major (1993, 26, 44 ff.). The version given here parallels that
found in the *Book of Documents* where the breaking of the communication between Heaven
and Earth was an act of volition on the part of Huang Di, the legendary Yellow Emperor: "The
charge was given to Chong and Li to break the communication between Earth and Heaven
so that there was no descending or ascending." A chief difference between the two versions,
apart from the etiology of the rupture, lies in the different methodologies admitted by each for
resolving the resulting dilemma, the one quasi-mystical or religious and the other, presented
here, fundamentally cosmo-political or bureaucratic. For analysis of the cosmic disaster and its
cultural legacy, see Porter (1996, 27–56).

[30] Jiang Xiaoyuan (2004, 98) reached a similar conclusion to that presented above: "the early
astrologers evolved out of those ancient shamans and shamanesses who communicated with
Heaven," see also Hwang (1996, 510). In this connection it is useful to recall that Cai Mo (i.e.
Zuo Qiuming) and Sima Qian were members of the same profession.

[31] O'Keefe (1983); I am grateful to Lionel Jensen for calling my attention to this work.

sovereign groups and spirits a direct empirical link connecting the independent decision-making structures in a society – the structures by which goals are chosen and responsibilities allocated – and the supernatural."[32]

The introduction of the calendar, along with other technologies such as writing and the keeping of lists and records, greatly accelerated the process of nibbling away at the frontiers of the unknown, by definition the domain of the chaotic and unmanageable. For the elite who manipulated such knowledge, though the potential for untoward irruption of irreality into the sphere of the mundane was an ever-present threat, by increasingly taking matters into their own hands, so to speak, a fundamentally optimistic, human-centered disposition began to evolve, self-reflective and burdened though it was by the heavy responsibility to maintain ritual regularity: "Time had been extracted from nature and molded to fit religious and bureaucratic uses."[33] As Merlin Donald has shown, this is a product of the breaking free of the mythic mode of thought and the emergence of the theoretic mode – "myth must be destroyed," as he says. The dawning came of the realization that products of human ingenuity, like the calendar, could supplant the earlier immersion in the world of mythic understanding:

As technology and social organization depended more and more on some form of record keeping (that is, external memory devices), visual symbolic devices emerged in increasing numbers . . . the human mind began to reflect upon the contents of its own representations, to modify and refine them. The shift was away from immediate, pragmatic problem solving and reasoning, toward the application of these skills to the permanent symbolic representations contained in external memory sources.[34]

This is what Cai Mo was talking about in explaining why *Dragons* could no longer be taken alive. He cast his explanation in the form of a sociological theory, decline from a prior ideal state, but in reality the change was the result of growing human self-confidence leading to the emergence of theoretic self-consciousness. Such cognitive awareness, once acquired, like literacy is virtually impossible to divest – there's no going back. (You cannot *not* read the same billboard every time you pass by, it registers in spite of yourself.) Similarly, you cannot *not* attempt to fit new information into existing theoretical frames of reference. One might even say that the search for oneness with the Dao is in part a striving to reverse the consequences of self-awareness, a quest to return to mythic consciousness.

---

[32] Swanson (1964, 190). Thomas Worthen (1991, 16) expands on the motives of this development: "Culture seeks to bring nature into its own sphere, which is regular and always prescriptive. Rituals are performed not to imitate the regularities in nature but to induce nature to imitate a culturally effected regularity."
[33] Aveni (2002, 85).   [34] Donald (1991, 333).

Mismanagement of the calendar

As I noted at the outset, there is every reason to believe that by the early second millennium BCE there was firmly established in China a mindset characterized by a self-conscious dependence on regularly scrutinizing the patterns of Heaven for guidance. Of the various royal roles perhaps this was the most crucial, for on its competent performance depended all the others. A clear indication of this is given in the remainder of the "Great Plan" passage quoted above: "When in days, months, and years the seasonableness has changed, the many cereals do not ripen, the administration is dark and unenlightened, talented men of the people are in petty positions, the house is not at peace." In other words, when there is a failure in this bellwether function of the theocrat, not only are the direct economic consequences severe, but this very breakdown is symptomatic of a more generalized decline in leadership and competence at all levels: "the truth is that people are unwise, not that dragons are truly wise," as Cai Mo said. This order of things is virtually paradigmatic in traditional accounts of the dynastic cycle.

Similarly, in response to a question put to him by his father, King Ling of Zhou, in what purports to be a conversation dating from 549 BCE, Prince Jin, speaking of failed regimes of the past, in a wide-ranging monologue says,

Those who were destroyed – how was it that they were without blessings, for they were all descendants of the Yellow Emperor or Yen Ti? ... *It is just because they did not follow the measures of Heaven and Earth, did not accord with the order of the Four Seasons, did not take as models the principles of living creatures*: it was for these things that they were destroyed, and had no posterity, so that today their sacrifices are not performed.[35]

Though this explanation is redolent of Warring States rationalization, there is more to the particulars of the indictment than a superficial reading might suggest. Significantly, the famous "Chu Silk Manuscript" (*c.*300 BCE) deals at considerable length with calendrical and astrological themes, exemplifying the perennial preoccupation with the ominous consequences of neglecting the calendar, thereby offending the gods and incurring disaster. One major section of that illustrated document neatly encapsulates this aspect of the astromantic thinking of the time. Part A of the text dwells on the theme of "reverence for Heaven and bestowal of the seasons." There the Supernal Lord is uniquely portrayed as directly and explicitly admonishing the people that if their sacrifices are not reverential and constant, if they miss the seasons of Heaven, then Heaven above will send down punishment, causing the four seasons to lose their proper sequence, the heavenly bodies to move confusedly, the grass and

[35] Hart (1984, 40, italics mine).

trees to grow irregularly, provoking meteorological abnormalities, landslides, floods, downpours tainted with soil, military conflict, etc. The Supernal Lord of Heaven is portrayed as a deity whose basic nature is to act as an arbiter of fate who dispenses benevolence (*de*) or chastisement (*te*) in response to human behavior.[36] The crucial indicator of good temporal order is correct maintenance of the calendar.

Early Zhou precursors of this theme are also in evidence. In the "Many Officers" chapter of the *Book of Documents* the Duke of Zhou is portrayed as citing precedent in berating the now subject former officers of Shang for recalcitrance:

I have heard it said that the Supernal Lord would guide the idly sportive, but the lord of Xia did not moderate his idle sports. Then the [Supernal] Lord descended and ascended and approached that Xia [king], but he did not care about the Lord. He was greatly licentious and dissolute and had notoriety. Then Heaven did not care about or listen to him. And when he neglected its great Mandate, it sent down and applied punishment. And so it charged your ancestor Cheng Tang to depose the Xia, and through talented men of the people regulate the Four Quarters. From Cheng Tang to [King] Di Yi there were none who did not make bright their virtue and carefully attend to the sacrifices. Heaven also grandly established them, and protected and directed the lords of Yin. Of the Yin kings also none dared neglect the Lord. There were none who were not counterparts to Heaven in benefiting [the people]. Their successor in our time [i.e. Shang king Zhòu Xin] greatly lacked a bright manifestation in Heaven [*wang xian yu tian*]. Still less was he willing to think of how the earlier kings toiled for the house. He was greatly licentious in his dissoluteness. He had no consideration for what Heaven manifests [*tian xian*], nor for what the people respect. Then the Supernal Lord did not protect him, and sent down destruction as great as this.[37]

What is particularly noteworthy in this passage is that the final rulers in the dynastic line, Xia Jie and Di Xin (Shang Zhòu), are indicted specifically for their inattention to the sacrifices directed toward Heaven (or the Supernal Lord), and for a lack of consideration for "Heaven's (bright) manifestations." Further, they failed to benefit the people or to show proper consideration for "what the people respected." As we have seen, sacrifices to Heaven were seasonally related rituals intended to reinforce the cosmo-magical role of the king in maintaining correspondence between the natural and human realms, the regularizing congruence referred to above. The chief instrument for regulating such rituals is a calendar that maps out the seasons of the tropical year. Neglect of the sacrifices to Heaven therefore seems to imply an indifferent, or at least ineffective, approach toward keeping track of this facet of sacred time. Of course, in the early period it was only by regularly scrutinizing the bright features of Heaven like the *Dragon* that such timekeeping was possible at all.

---

[36] Li Ling (1985, 31). See also Li Xueqin (1982, 68–72); Harper (1999, 847).
[37] Trans. Karlgren (1950a, 55/5–10, trans. modified).

Consider another example. Here is the more specific charge of negligence Sima Qian levels at the royal astronomers of the Xia king: "In the time of Emperor Zhong Kang, Xi and He were drunken and dissolute; [they] disregarded the seasons and confused the days. Yin proceeded to punish them and composed the 'Rectifying Yin' Yin zheng [chapter of the *Book of Documents*]."[38] So not only was the calendar allowed to lose its synchronization with the seasons, even the cyclical count of the days – the most basic reckoning of all – was catastrophically disrupted.

As I have already suggested, in the early period the count of the days, months, and years, and time-honored calendar customs, those "cultural offshoots of time," in Mary Barnard's apt phrase, would have provided the pre-eminent experience of control over the temporal dimension and its phenomenal manifestations in the natural rhythms. The calendar, and its proper calibration by the Sun, Moon, and stars, would have constituted the secure lifeline connecting the present with the past, just as it supplied the cultural template with which an ambiguous future could be made to fit a pattern of human devising.[39] To have it otherwise, to contemplate time unmeasured and undifferentiated save by the often inscrutable supra-sensible forces governing the natural world, would have been profoundly unsettling, even unthinkable.[40] Nature is first brought into the sphere of culture by observing patterns and regularities and the causal relations among phenomena, which is the beginning of science. But, as Thomas Worthen reminds us, "Ritual and science regularize things, but . . . [o]ur science and our rituals and religions do not really satisfy all of our doubts about the mysteries of life and the beyond. There is still the tension of the unknown, and chaos lurks behind every formula."[41] Thus it is not surprising that a mismanagement of time caused by a failure to conform to the images displayed in the heavens should figure prominently among the shortcomings of deposed dynasties, though not as mere portents, as traditional interpretations of accounts like those in *Mozi* would suggest, but as precipitating causes.

---

[38] *The Grand Scribe's Records*, "Basic Annals of Xia," *Shiji*, 2.85. In *Zuozhuan* (Duke Zhao, seventeenth year) the specific failing of the astronomers that led to their punishment is quoted from the *Books of Xia*: "the *Dipper* [handle] did not alight on the [proper] quarter [stellar location]," (reading *fang*, "chamber," as *fang*, "quarter, direction").

[39] This is perhaps the most concrete meaning of the *Book of Documents* passage: "You [the sovereign] should not set an example of laziness or desires to the possessors of states; it is fearsome, it is awe-inspiring, in one day, in two days, there are ten thousand first signs of happenings [which you should be prepared for]. Do not empty the various offices. The works of Heaven, it is man who carries them out on its behalf," trans. Karlgren (1950a, 9/5, trans. modified). It is significant, as A.C. Graham (1989a, 18) points out, that Confucius did not use the term "Way of Heaven," *tian dao*, probably because by then the diviners and skywatchers had already appropriated the term to refer to the movements of the celestial bodies.

[40] For an analysis of the myth of the flood as a symbolic response to the disruption of established cosmological verities and the threatening irruption of chaos, see Porter (1993, 82 ff., 92); Worthen (1991, passim); Berger (1990, 23, 26, 27).

[41] Worthen (1991, 74).

As Mischa Titiev has remarked, "calendrical observances are invariably communal or broadly social." He goes on to say that

> analysis of their intent reveals that they are designed primarily to strengthen the bonds of cohesion that hold together all of a society's members or else to aid the individuals who form a social unit to adjust to one another and to their external environment... calendrical rites ordinarily disappear when a society diminishes in power or loses its identity.[42]

As Marc Kalinowski observed of the "Canon of Yao," "quand on envisage le texte dans son ensemble, le récit d'aménagement de l'espace et du temps placé au début de la Règle de Yao institue le calendrier comme régulateur des équilibres naturels et fondement d'une liturgie saisonnière à vocation politique."[43] This explains why dynastic transitions in China inevitably entailed reform of the calendar. Understandably, this theme is a dominant one in portrayals throughout most of the first millennium BCE of the shortcomings of failed regimes.

Evolution of Shang theology

The evidence is clear in the oracle bone inscriptions of a major shift in the magico-religious theology of the Shang, one important aspect of which was a drastic reduction in temporal flexibility. According to David N. Keightley,

> Concern with timeliness was characteristic of the divinations of Wu Ting, who reigned when the sacrificial schedule was still being formulated, when the date of each sacrifice might still be submitted for spiritual approval. By period V temporal flexibility had been lost. The ritual schedule was now rigidly formulated; the day on which a particular ancestor would receive sacrifice was already established; the divination was no longer concerned with determining the auspicious time, but only with announcing that the sacrifice would take place as scheduled.[44]

In the final period, by which time the regular cycle of rituals approximated 360 to 370 days in length and the sexagenary cycle became tied to the lunar months, the sacrificial schedule provided continuity and stability by substituting a temporal dimension of its own for a timeliness contingent on external factors – "uncertainty had been replaced by pattern and order."[45] At the same time, it is no doubt also indicative of a major theological change that the Shang kings by this same period had ceased completely divining about sacrifices to the various nature powers and had arrogated to themselves the title *Di*, previously reserved for the Supernal Lord, governor of the phenomena, including, of course, the celestial arbiters of time, Sun, Moon, stars, and planets.

---

[42] Titiev (1960, 294–5, 297).   [43] Kalinowski (2004, 114).
[44] Keightley (1984, 18).   [45] Ibid., 14.

In Sima Qian's view, the decline of Shang was ultimately traceable to changes detrimental to the dynastic fortunes that arose during the reign of King Zu Jia (in Tung Tso-pin's period II), the very period during which the regular ritual schedule of the New School began to take increasing effect.[46] Significantly, Sima Qian also faulted the Shang for allowing their early attention to "pious reverence" to degenerate "until mean men had made it a superstitious concern for the spirits of the dead ancestors."[47] Taken together, these various developments appear to suggest a less than reverential attitude toward the high god and the spirits of the natural order in late Shang, and the possibility that, at the very least, an equally perfunctory approach was taken toward the calendar and its associated ritual observances as that which produced the routinization of divination. This is the thrust of the Duke of Zhou's accusation that the Shang king "despised Heaven's command, and did not perpetually and reverently think of the sacrifices."[48] He was obviously not talking about the sacrifices to the Shang royal ancestors. In the *Book of Documents* chapter "Wei zi," the Prince of Wei is even more explicit in his indictment: "now the Yin people steal the auspicious and faultless sacrificial animals of the Spirits of Heaven and Earth, and use them to make fine their repasts, without (fear of) disaster."[49] One might well ask, as Anthony Aveni does of the Maya, "What kind of people dare to write their history alongside that of the gods – to use time as a vehicle to legalize and canonize the authority of their rulership of the state?"[50]

Thus it is not surprising that the signs given by *Mozi* of Heaven's displeasure with Xia – "the Sun and Moon were untimely, and cold and heat came irregularly... the Five Grains were malformed and people were greatly agitated" – and with Shang – "the sacrifices were not according to the seasons"[51] – suggest that periodic obsolescence of the calendar, or perhaps in the case of Shang its reckless disregard, had important political implications. Since the record of contemporaneous planetary observation has proven correct, perhaps Mozi's statements also reflect the situation with regard to the sacrifices that were properly pegged to the calendar.[52] Therefore it is conceivable that the last Shang king Di Xin's alleged impiety in connection with the sacrifices consisted of a willful negligence tantamount to sacrilege. His religious

---

[46] *Shiji*, 2.104; *Guoyu*, 3.21a.
[47] *Shiji*, "Basic Annals of Emperor Gaozu," *Gaozu benji*, 8.324. Compare this conception with that cited above from the *Book of Documents*, "Announcement Concerning Wine."
[48] Trans. Karlgren (1950a, 63/3).   [49] Ibid., 27/6, trans. modified.   [50] Aveni (2002, 187).
[51] *Mozi*, Chapter 19, "Against Aggressive Warfare," *Fei gong xia*.
[52] Recall David N. Keightley's (2000, 44) observation of a strong likelihood that "the start of the year could have involved more than one kind of year. The Shang diviners might have pegged the first moon of their luni-solar calendar to the first lunation after the winter solstice, while the peasants might have tied their agricultural calendar to the observation of stars and constellations... the first, liturgical system, not the second, agricultural system... gave rise to the numbered moons recorded in the divination inscriptions."

posture appears to have been characterized by punctilious adherence to the late Shang cycle of the ancestral cult, but by gross negligence when it came to conforming to the patterns of Heaven, and to respect for the natural spirits and for the time-honored seasonal customs and festivals of the general populace ("what the people respected"). This seems to be the implication of the contrast that is repeatedly drawn between the early Shang "wise kings" and the depravity of the final ruler. It is also worth recalling the traditional accounts of the late Shang king Wu Yi's deliberate sacrilege in shooting full of arrows a blood-filled leather sack said to represent "Heaven," and Di Xin's purportedly contemptuous dismissal of the risk of being dethroned in light of his divine appointment to rule.[53] Whatever the justice of this indictment, it is a fact that rectification of the calendar, redefinition of the "correct" beginning of the year, and "getting right with Heaven" by restoring the seasonal sacrifices were all accorded great symbolic importance in connection with the Zhou founding and every subsequent dynastic transition or new beginning.

Nevertheless, such reform seems not to have gone much beyond the symbolic. In the *Lost Books of Zhou* there is the statement:

when it came to our Zhou kings, [they] were brought to attack the Shang, and to change the First Month of the year and the royal regalia, to make manifest the Three Fundaments. But as for "respectfully bestowing the seasons on the people" [*jing shou min shi*], royal progresses and sacrifices, [they] still followed the Xia.[54]

Thus there appears to be a time-honored distinction between the ritual–bureaucratic calendar, with all its political symbolism, and a more popular calendar of seasonal festivals, market days, and the like based on something like the *Lesser Annuary of Xia*, which the common folk continued to follow irrespective of the changes at the top. This distinction is also apparent in the same "Great Plan" passage quoted above, where the observational habits of the people are clearly distinguished from those of the elite: "What the common people [scrutinize] are the stars. There are stars which favor wind, there are stars which favor rain. [Owing to] the course of Sun and Moon there is winter and summer. According as the moons follow the [various] stars, there is wind and rain."[55] Even as late as the sixth century BCE Confucius famously deplored mismanagement of the calendar as an indirect way of expressing his disapproval of the powers that be in a classic "set piece" in which astro-calendrical officials serve as proxies for the rulers:

Ji Kangzi asked Confucius, "This is the Zhou twelfth month, the Xia tenth month, and still there are katydids, why is that?" Confucius answered, "I have heard 'after the *Fire Star* has set, the hibernators have all gone.' But now the *Fire Star* is still declining in

---

[53] *Shiji*, "Basic Annals of Yin," *Yin benji*, 3.104.   [54] *Yi Zhou shu*, 6.87.
[55] Trans. Karlgren (1950a, 33/32); Major (1993, 91).

the west. The officials in charge of the calendar are wrong." Ji Kangzi said, "By how many months are they off?" Confucius said, "By the tenth month of the Xia calendar the *Fire Star* has already set. But now the *Fire Star* is still visible; [they have] missed intercalating twice."[56]

How the calendrical misreckoning described in *Mozi* manifested itself we do not know, though Xia Dynasty king Zhong Kang's difficulties mentioned above suggest a perennial problem, which was evidently common even as late as the sixth century BCE.[57] In the Bronze Age, such mismanagement could certainly have been claimed by rival contenders for power to reveal the incompetence of the royal or hereditary "priestly" lineage charged with the sacred task of correctly scheduling ritual observances and promulgating the calendar. Of course there would have been other proximate political causes of a decline in dynastic prestige, but, in the case of the Zhou conquest of Shang in particular, a scenario such as described above could account for the demonstrative piety of the successful usurpers in these matters. Lapses in managing time, whether due to negligence or to the cumulative effects of precession, may well have figured prominently among the actual failures of rulers during the final reigns of both Xia and Shang.

---

[56] *Kongzi Jiayu*, "*Bianwu.*" Compare *Zuozhuan*, Duke Xiang, twenty-seventh year. The implication, of course, is that the calendar is two months "fast."
[57] Or the Han dynasty for that matter, see Cullen (2001, 33).

Part Four

Warring States and Han astral portentology

9   Astral prognostication and the Battle of Chengpu

For drama and enduring historical significance few episodes in ancient Chinese history can compare with the struggle for supremacy between the great states of Chu and Jin in the early Spring and Autumn period (722–481 BCE). Besides constituting an epochal challenge to the dominant northern Hua–Xia cultural heritage, the military conflict in 633–632 BCE between Jin and Chu was also the final act of the intense competition for ascendancy that accompanied the decline of the eastern state of Qi, after the demise of the illustrious Hegemon Duke Huan (r. 685–643): "The deer was loose," as commentators of a later epoch would put it when the imperial dignity was the game. In the mid seventh century the contenders all knew that Duke Huan's erstwhile status as Hegemon (*ba*), ostensibly protector of the ruling Zhou Dynasty (1046–256 BCE), was up for grabs.

As if this scenario did not provide drama enough, Jin's ascendancy is inextricably linked with the fascinating personal history and celebrated adventures of Prince Chong Er, soon to be Duke Wen of Jin (r. 636–628). Intrigued against at the Jin court, Chong Er fled the state with his maternal relatives, spent nineteen years in exile – twelve among his mother's people, the so-called "barbarians" (*rong di*) – all the while seeking patronage as an itinerant royal guest at successive state courts, accompanied by a handful of loyal retainers. Recognition of his personal qualities and potential usefulness led to magnanimous treatment by the rulers of some of Jin's major rivals, including Qin and Chu. Ultimately, at age sixty-one, Chong Er was restored to power as Duke Wen of Jin in 636 BCE, with military backing from Qin. Not long after, Jin was drawn into military confrontation with Duke Wen's nemesis, Chancellor Zi Yu of Chu, when the armies of Jin and Chu clashed.

The decisive battle took place in early 632 BCE at Chengpu near the Yellow River in western Shandong (Map 9.1). Chu had the advantage of better tactical position, but superior generalship by Jin's commanders (supported by troops from Qi and Qin) resulted in a total rout of the Chu expeditionary forces, which had been marauding in western Shandong.[1] The Chu army was below strength,

---

[1] Ronald Egan (1977, 332 ff., 343) analyzes the *Zuozhuan*'s narrative portrayal of the contrast between Duke Wen and Zi Yu, the exemplary leader versus the rash and self-serving figure. For

Map 9.1 China in the Spring and Autumn period. After Loewe and Shaughnessy (1999, 548, Map 8.1), © Cambridge University Press, reproduced by permission.

although supported by the small states of Cai and Chen, which were sandwiched between their much larger and more aggressive neighbors. Chu king Cheng (r. 671–626 BCE) had dispatched inadequate reinforcements in response to his commander Zi Yu's requests. By some accounts King Cheng was displeased with Zi Yu's belligerence in pursuing a grudge against the former Prince Chong Er, and also, apparently, because the king saw no compelling reason to seek a decisive confrontation with Jin now in the ascendancy and enjoying the support of powerful Qin. The precise timing of the Battle of Chengpu and the

detailed description of the circumstances of the Battle of Chengpu – tactics, personalities, and diplomatic maneuvering – see Kierman (1974, 47–56). Mass burials of over 600 young males aged twenty to twenty-five discovered near the site of the battlefield at Xishuipo, Puyang, are presumed to be casualties of the conflict. Von Falkenhausen (2006, 413, 414, Figure 98).

willingness of Jin to confront Chu at this juncture will shortly be the focus of interest. In the event, however, all of the Chu army but the center under Chancellor Zi Yu's command went down to defeat, and Zi Yu himself, one of Chu's ablest military commanders, was shamed into committing suicide during the retreat to Chu.

Here is how one modern historian, Tong Shuye, characterizes this epic confrontation:

The Battle of Chengpu was the first great battle of the early part of the Spring and Autumn period, one which fully concerned the whole situation in the Central Plain. At the time Chu was projecting its power throughout the Central Plain and had invaded the great states of the lower reaches of the Yellow River like Qi and Song, while Lu, Wei, Zheng, Chen, and Cai had already capitulated to Chu. On another front, barbarian *Di* forces had also attacked the royal lands, forcing the Zhou king into flight. By this time Duke Huan of Qi's career as Lord Protector was over. The age was truly one in which "aggression by southern and northern barbarians coincided, and survival of the central states hung by a thread." Had Duke Wen of Jin not risen to prominence in the north and taken the situation in hand, the royal Zhou house and the lords of the central states would have been swept away long before the Warring States period. After the total defeat of the Chu armies the influence of the southern Yi barbarians receded from the Central Plain and the incursions of the northern *Di* also gradually declined. Thus the survival of the Hua–Xia states and their culture was secured, and Duke Wen of Jin is to be credited with this great achievement![2]

Historical accounts of Duke Wen of Jin's ascendancy

Despite the early date, the events of Chong Er's life and this epic confrontation are well documented in the historical record. Although some intertextual dependency is apparent in late accounts, especially in Sima Qian's *The Grand Scribe's Records*, there is a complementarity between the Warring States narratives in the *Zuozhuan* and the *Discourses of the States*, the latter account being unique in a number of important respects. Having said that, there is no doubt that the *Discourses* version of events is later than the *Zuozhuan*, parallel passages from the latter having clearly been redacted and in some cases debased.[3] Nevertheless, detailed comparison of the two sources, and of the contents of the *Discourses of the States* in particular, leads to the conclusion that the author of the *Discourses* was well acquainted with Jin history, partial to Jin, and very likely from one of Jin's successor states (i.e. after the partition of Jin in 376 BCE).[4] The versions of events in the *Springs and Autumns Annals* and the *Zuozhuan* commentary are straightforward accounts of the

---

[2] Tong Shuye (1975, 180); for a different perspective, see Kierman (1974, 56).
[3] Shen Changyun (1987, 139).   [4] Ibid., 137, 138.

high points of Chong Er's career, culminating in a series of precisely dated military and political activities during the months before and immediately after his restoration and the Battle of Chengpu. These are deemed most reliable.[5] Most of the accounts, differing only in minor detail, are reproduced in the "Hereditary House of Jin" (*Jin shi jia*) chapter in *The Grand Scribe's Records*. The historical events are corroborated in crucial respects by the recently discovered royal commendation inscribed on a set of bells cast by Chong Er's maternal uncle and close confidant, Zi Fan. Besides serving him faithfully during his long exile, Zi Fan was also the brilliant strategist who helped engineer Chong Er's restoration and the subsequent defeat of Chu.[6]

For our purposes the precise dates of significant events in the first year of Duke Wen's reign, 636 BCE, and in the spring of 632 will prove particularly informative because of what they reveal about considerations of timing. Of special interest will be the version in the *Discourses of the States*, "Discourses of Jin," which, in addition to retelling the story of Chong Er's exploits, recasts the history of the period in astrological terms, as the playing out of a sequence of events preordained by Heaven. The only similar passage in the *Discourses of the States* is Ling Zhoujiu's famous "musicological" analysis in the "Discourses of Zhou" of the sequence of events leading up to the Zhou conquest of Shang in 1046 BCE, four hundred years earlier. There the text reports the locations of principal heavenly bodies during the Zhou campaign of conquest and their significance in terms of the "field-allocation" system of astrological correlations. Earlier, we saw how certain of these positional observations, most significantly that of Jupiter in the Zhou astrological space called *Quail Fire* (the immediate vicinity of the star Alphard, α Hya), are accurate and help to explain much about the timing of major events, as well as the import of the Zhou leaders' pronouncements concerning the conferral of Heaven's Mandate on the Zhou Dynasty.

Later it will become apparent why these two watershed events, the Zhou conquest and the Battle of Chengpu, are singled out for special attention and astrological interpretation in *Discourses of the States*. For the moment, suffice it to say that celestial events accompanying the Zhou conquest established important astrological precedents against which subsequent claims of Heavenly endorsement must be measured. What I intend to show here is that sufficiently informative traditions concerning those astrological precedents were known to the author of *Discourses of the States* in the late Warring States period, as were the astrological circumstances of Chong Er's ascendancy and the Battle of Chengpu. Before discussing those considerations in detail, however, it will be helpful briefly to review what is known about applied field-allocation astrology as it was practiced in the mid to late Zhou.

---

[5] Egan (1977, 323).   [6] Feng Shi (1997); Zhang Guangyuan (1995).

Theory of field-allocation astrology

As we have seen in the preceding chapters, preoccupation with the correlation of celestial and terrestrial phenomena preceded the elaboration of systematized cosmo-political theories, establishing the conceptual parameters within which such theories would develop. In addition to conceptions about a generalized celestial influence on terrestrial events, individual celestial bodies like the Sun and Moon were directly linked to specific meteorological, physical, and political phenomena, as were eclipses, comets, meteor showers, planetary events, and the like, most of which appear in the records from the Shang period (1554–1046 BCE). Not long after the emergence of the concept of a binary *yin-yang* complementarity and Five Elemental Phases correlative cosmology in the Warring States period (403–221 BCE), one also finds explicit references to specific solar and lunar effects on the natural world.

If the causal action of celestial bodies could be mediated through the medium of *materia vitalis* and make itself felt on all inanimate and animate things, this explains how such reflexive action could be thought operative between the supra-sensible forces and human society. Moreover, if celestial bodies were considered to exhibit different qualities, exert different influences, and be correlated with (and influenced by) different institutions, terrestrial regions, peoples, etc., all essential elements of late Zhou astrology, it follows that by virtue of the intrinsic periodicities of certain celestial bodies, knowledge of past events and their associated astrological conditions would render the omen interpreter capable not only of predicting future celestial events, but also of forecasting in detail the terrestrial events likely to be produced by them.[7] From conceptions such as these, therefore, flow a number of concrete conclusions about the implications of phenomena, and ultimately about the possibility of foreknowledge of events, both prerequisites for the formulation of a chronosophy[8] such as attributed to Zou Yan (fl. third century BCE).

According to David Pingree, the earliest form assumed by chronosophy in the ancient world was generally a natural theology of history, which provided history "with intelligibility and/or meaning by looking for sufficient reason outside history itself and outside the world human beings are living in." As Pingree

---

[7] A famous remark by Mencius (4B/26) shows how near at hand such thoughts were at the time: "Consider the heavens so high and the stars so distant. If we seek out former instances [*gu*] we may, while sitting still, have command of a thousand years' worth of solstices." That Xunzi (third century BCE) found it necessary to argue against the notion that celestial events directly influence terrestrial affairs clearly shows how pervasive such thinking was at the time. See especially Xunzi's "Discussion of Heaven" in Watson (1967, 81 ff.). For Zou Yan, see Harper (1999, 824*);* Sivin (1995a).

[8] A chronosophy integrates past, present, and future into a coherent scheme by means of some procedure supposed to afford the possibility of predicting the future; *Encyclopaedia Britannica*, Vol. 2, s.v. "Astrology," 219 ff.

points out, as a natural theology of history, astrology is logically incompatible with a theocentric view. Thus, given the presumed immutability of the sequential alternation of the "Five Virtues" (*wu de*; with "virtue" understood in the sense of *vertus*), Zou Yan's mutual-conquest theory addresses itself only to the intelligibility of the historical process. So far as we can determine, he made no claim about historically transcendent meaning. The tension between intelligibility and the expression of divine intentionality in history does, however, emerge in the perennial problem latent in Chinese dynastic ideology; that is, the tension arising from ancient Chinese cosmo-political views about legitimacy attributable to manifest heavenly sanction (i.e. through direct revelation of Heaven's Mandate) as opposed to that conferred by heredity. Once it was understood in late Zhou that eclipses of the Sun and Moon were in fact periodic phenomena, the same astrological motives naturally lent increased impetus to efforts to improve the accuracy of prediction of those ominous celestial events, efforts which continued under imperial auspices throughout Chinese history.

Such archaic predilections certainly contributed to the formation of the influential *yin-yang* and Huang-Lao thought of the late Warring States and Han periods (206 BCE–220 CE), a dominant principle of which was that knowledge of the natural world translates directly into political power. As we saw in a previous chapter, in his discussion of the cosmological chapters of the *Huainanzi* John S. Major shows how this assumption underlies the worldview that "cosmology, cosmography, astronomy, calendrical astrology, and other forms of cosmology form a seamless web, the principles of which a ruler would ignore only at his peril."[9] In a similar vein, Mark Edward Lewis has described the popular Warring States period *mythos* in which a nonheroic figure with no combat skills gains victory over a great warrior through the possession of a divinely or magically revealed military treatise, showing that "military theorists thought of their doctrines as an esoteric wisdom that expressed divine patterns inherent in the cosmos."[10] At its simplest, the most basic principle of what Lewis denotes a "calendrical model of warfare" held that killing was consonant with the cosmos only when carried out according to the proper seasons of the year. Other

---

[9] Major (1993, 14) also argues that "the credibility of the Huang-Lao School in the early Han may have rested in part on the degree to which it was grounded in widely shared assumptions that went back to the foundations of Chinese civilization." Cf. e.g. *Huainanzi*: "The furred and feathered are the kinds which fly and run, and therefore belong to the *yang*; the shelled and scaly are the kinds which hibernate and hide, and therefore belong to the *yin*. The sun is ruler of the *yang*, and for this reason in spring and summer the herd animals shed hair, and at the solstice the deer shed their horns; the moon is ancestor of the *yin*, which is why when the moon wanes the brains of fishes diminish, and when the moon dies the swollen oyster shrinks." Trans. Graham (1989a, 333). See also the first twelve chapters ("Twelve Intervals") of the *Springs and Autumns of Master Lü*, where the seasonal "ordinances" are described in detail. In no way can these be recent insights.

[10] Lewis (1990, 98).

accounts that describe revealed texts, such as the so-called "River Diagram" and the "Luo Writing" discussed in Chapter 5, also point to the connection of these texts with reading astral omens and political ascendancy.

More to the point, however, the *Huainanzi* ranks astrological factors first among those to be taken into tactical consideration, in this way pointing directly to the agency by means of which patterns of cosmic order having military application were revealed. The *Huainanzi* says, "Clearly understanding the motions of the planets, stars, Sun, and Moon; the knack of expedient [application] of punishment and moral suasion; the advantage of facing to the front or rear, or going left or right; these are helpful in battle."[11] Such ideas were ridiculed by theorists like Hanfei and Xunzi. Hanfei was particularly harsh:

Initially, for several years Wei turned eastward to attack and finish off Wey and Tao. For several years later it then turned westward [to attack Qin] and lost territory. This does not show that the *Five Thunder Spirits*, *Supreme One*, the six *Sheti* spirits, and *Five Chariots*, the *Sky River*, *Spear of Yin*, and Jupiter [all auspicious] were in the west for several years. Nor does it indicate that *Heavenly Gap*, *Hu'ni*, *Punishing Star*, Mars, and *Stride Terrace* [all inauspicious] were in the east during subsequent years. Therefore, I say that turtle and milfoil, ghosts and spirits are not able to assure victory, and that [positioning oneself] to the left, to the right, in front, or behind [them] does not suffice to determine [the outcome of] a battle. There is no greater stupidity than to put one's faith in this.[12]

Nevertheless, such conceptions are reinforced by still earlier textual accounts in *Zuozhuan* and *Discourses of the States* concerning the correlation of political and military actions with celestial events, most notably Jupiter's motion. In this regard, it is important to note that when one reads in the *Discourses of the States* that "the space *Shi Chen* [Orion], this is the dwelling place of the Jin people,"[13] such statements should not be taken as metaphorical, but rather as indicative of the *functional identity* at a conceptual level between the celestial and terrestrial realms. By the same token, as indicated above, a basic axiom of astrological omenology was that celestial bodies partook of the particular emanations (qi) of terrestrial regions as they traversed their corresponding celestial spaces, and so it is perfectly natural for Sima Qian to say, "[the state] wherein Jupiter is located may not be attacked, but may attack others."[14] By the early imperial period in the third and second centuries BCE, clusters of the Five Planets such

---

[11] *Huainanzi*, "Treatise on Military Affairs" (*Bing lue xun*), 15.5a. See Lau and Ames (1996, 153, 155), where it is asserted that astrological calculations can assure victory in six battles out of ten, and that he who has mastered the way of warfare "understands the course of the heavens above and the topography of earth below."

[12] *Hanfeizi*, "*Shixie*" chapter, *Xinbian zhuzi jicheng*, Vol. 5, 88–9. For similarly critical opinions in later periods, see Yates (2005, 21).

[13] *Guoyu*, 10.1a.

[14] *Shiji*, 27.1312. Sima Qian attributes this axiom to the Warring States astrologer Shi Shen (fl. fourth century BCE).

as those that occurred during Xia, Shang, and Zhou had come to constitute the definitive sign of the bestowal of Heaven's Mandate on a new dynasty.

Principles of field-allocation astrology

Generally speaking, in all its early manifestations "field-allocation" as a concept refers to the correlation between terrestrial regions and their celestial counterparts based on certain patterns of correspondences, for the purpose of divining the good or ill fortune of the terrestrial locations.[15] The regular field-allocation scheme whose application is described in the *Rites of Zhou* (see Table 9.1) dates from late Warring States times. In fact, however, as we saw in Table 6.1, there is considerable evidence to suggest that the nucleus of the system – a nascent correlation of the four celestial quadrants with the cardinal directions, together with the sequence of seasons and their associated colors – was already in place by the late second millennium BCE.[16]

The general principles and motives underpinning this astrological scheme are succinctly stated in the job description in the *Rites of Zhou* of the Astrologer Royal (*Bao zhang shi*):

[The Astrologer Royal] concerns himself with the stars in the heavens, keeping a record of the changes and movements of the stars and planets, Sun and Moon, in order to examine the movements of the terrestrial world, with the object of distinguishing [prognosticating] good and bad fortune. He divides the territories of the Nine Provinces of the sub-celestial realm in accordance with their dependence on particular celestial bodies. All the fiefs and principalities are connected with distinct stars, and from this their prosperity or misfortune can be ascertained. He makes prognostications according to the twelve years [of the Jupiter cycle], of good and evil in the terrestrial world.[17]

The ambiguity of the above passage with regard to how the astral and terrestrial correspondences were established has led to varying interpretations, all based on fragmentary and inconclusive evidence from other late Zhou and Han texts. The most frequently cited operative principles include: (i) relating the Nine Provinces to the seven stars of the *Dipper*, (ii) tying the Five Planets and their associated terrestrial regions to stellar locations, (iii) the regular system of allocating the twenty-eight lodges among the terrestrial polities, (iv) defining the astral correlate of an ancient feudatory according to the celestial location

---

[15] A comprehensive survey of all historical materials relating to field-allocation astrology is Li Yong (1992, 22–31).

[16] In his history of Chinese astronomy, Chen Zungui (1955, 89) also held that the basic concept of astral–terrestrial correspondences goes back to the "primitive" period, and that the correlation of specific stars with certain terrestrial locations cannot be later than the Spring and Autumn period and may date to the Western Zhou (1046–771 BCE).

[17] *Shisanjing zhushu*, 26.181; trans. Needham and Wang (1959, 190, trans. modified); Biot (1851, Vol. 2, 113 ff.).

Table 9.1 *The field-allocation system of astra-terrestrial correlations.*[a]

Jupiter station	Chronogram	State	Province	Lodge	Direction
Xingji[b]	*chou*	Wu–Yue	*Yang*	*Dipper, Ox-leader* (Sgr-Cap)	NNE
Xuanxiao	*zi*	Qi	*Qing*	*Girl, Ruins* (Aqr)	N
Juzi	*hai*	Wey	*Bing*	*Roof, Hall, Wall* (Aqr–Peg)	NNW
Jianglou	*xu*	Lu	*Xu*	*Stride, Pasture* (And–Ari)	WNW
Daliang	*you*	Zhao	*Ji*	*Stomach, Topknot* (Ari–Tau)	W
Shi chen	*shen*	Wei (Jin)	*Yi*	*Net, Owl, Triaster* (Tau–Ori)	WSW
Chunshou	*wei*	Qin	*Yong*	*Well, Ghost* (Gem–Can)	SSW
Chunhuo	*wu*	Zhou	*San He*[c]	*Willow, Stars, Spread* (Hya)	S
Chunwei	*si*	Chu	*Jing*	*Wings, Chariot* (Cra–Cor)	SSE
Shouxing	*chen*	Zheng	*Yan*	*Horns, Neck* (Vir)	ESE
Dahuo	*mao*	Song	*Yu*	*Base, Chamber, Heart* (Lib–Sco)	E
Ximu	*yin*	Yan	*You*	*Tail, Basket* (Sco–Sgr)	ENE

[a] The scheme presented in Table 9.1 is from Zheng Xuan's (127–200 CE) commentary on the *Rites of Zhou* and is essentially identical to that attributed to Shi Shen's (fl. fourth century BCE) *Canon of Stars*, which is reproduced in the astrological treatise in the *History of the Jin Dynasty* (*Jin shu*, 11.307 ff.), and also adduced in a commentary to *Shiji*, 27.1346. According to Zheng Xuan (127–200), an earlier lost work had elaborated the astral correlates of the Nine Provinces but all that could be reconstructed in his time were the terrestrial correlates of the twelve Jupiter stations (*ci*; lit. "stage of an army's march") in the more familiar field-allocation system. The correlations given in the *Han shu* "Treatise on Astrology" differ in minor respects as a consequence of the addition of an anomalous thirteenth division corresponding to the "Yangtze and Lake [District]," identified with lodge *Southern Dipper*. In his commentary, *Zhouli shu* (*Shisanjing zhushu*, 26.181), Jia Gongyan (fl. 650 CE) takes "allocated asterisms," *fen xing*, to refer to the stars of the *Northern Dipper* and reproduces the scheme of correlations shown below in Table 9.2 from the Han Dynasty apocryphal work *Chunqiuwei wen yao gou*. Little else is known about the principles of this system except that its prognostications presumably relied on perceived changes in the brilliance, color, or visibility of the stars of Ursa Major; see, for examples, the commentaries in *Shiji*, 27.1293 ff. See also Major (1993, 135).

[b] In Warring States times the astral field allocated to Yue was apparently *Ximu*, not *Xingji* as in Han and later schemes. Chen Jiujin (1978, 62).

[c] The addition of "Three Rivers," *San He* (the Yellow River, the Xizhi, and the Hungzhong), reflects the apparent recognition in the scheme of the political reality of Qin and Han westward expansion and the resulting troublesome relations with border peoples (above all the Qiang) in Qinghai.

of Jupiter at the time of investiture, (v) identifying the celestial correlate of a locality as the asterism to which ancient inhabitants of that place principally offered sacrifice. Of these, Zheng Qiao (1104–62) accepts only the last as having a sound historical basis.[18]

[18] Zheng Qiao's criticism of the various theories is quoted in *Gu jin tushu jicheng*, 57.1341. For a recent study of the problem, see Li Yong (1992, 26–7).

Table 9.2 *The Nine Fields of Heaven and their astral correlations.*[a]

Direction	Field	Province – Lodges	States	
Center	Revolving	*Yan*	*Horns, Neck, Base*	Han, Zheng
East	Azure	*Yu*	*Chamber, Heart, Tail*	Song, Yan →
Northeast	Changing	*Yang*	*Basket, Dipper, Ox-Leader*	Yan, Wu, Yue →
North	Dark	*Qing*	*Girl, Ruins, Roof, Hall*	Yue, Qi, Wey →
Northwest	Somber	*Xu*	*Wall, Stride, Pasture*	Wey, Lu →
West	Brilliant	*Ji*	*Stomach, Topknot, Net*	Lu, Zhao
Southwest	Vermilion	*Liang*	*Owl, Triaster, Well*	Jin, Qin →
South	Fiery	*Yong*	*Ghost, Willow, Stars*	Qin, Zhou →
Southeast	Luminous	*Jing*	*Spread, Wings, Chariot*	Zhou, Chu

[a] The *Springs and Autumns of Master Lü* (*Lüshi chunqiu*, 13.lb.) reproduces the Nine Fields of Heaven correlations. The arrows in the far right column indicate that in several cases this scheme's allocation of astral fields results in some lodges straddling different Provinces. Although the passage does not supply the names of the Nine Provinces, its tabulation of the Nine Fields of Heaven giving the cardinal and inter-cardinal directions of each, together with their associated lodges, makes the correspondences unambiguous. The terrestrial correlations are those assigned by Gao You (168–212) in his commentary. They conform to the parallel passage in *Huainanzi*, 3.2b. Major (1993, 69); Li Yong (1992, 22, 25). Several idiosyncratic variants for half of these standard province names appear in a recently discovered fourth- to third-century BCE Chu manuscript which recounts the career of Yu the Great; see Pines (2010, 512); Dorofeeva-Lichtmann (2010; 2009, 629 ff.).

Table 9.1 reproduces the conventional scheme of correspondences among the twelve Jupiter stations, equatorial solar chronograms (roughly equivalent to signs of the zodiac), twenty-eight lodges, and terrestrial regions. A similar tabulation found in Chapter 3 of the *Huainanzi* differs in minor detail, on the one hand by omitting the state of Jin and reversing (possibly erroneously) the lunar mansion correlates of the states of Wei and Zhao, and on the other by separating the states of Wu and Yue so that they correspond to *Serving Girl* (10, Aqr) and to *Southern Dipper/Ox-Leader* (8–9, Sgr/Cap) respectively. This curious Han Dynasty variation produces an idiosyncratic list of thirteen divisions, rather than the otherwise conventional twelve.[19]

The pre-eminence of Jupiter's role in astrological prognostication based on this system is evident in the lines immediately following in the *Huainanzi* passage on which Major comments:[20] "as shown by the prognostication rules given . . . . all of the attention given here to the description and apportionment of the lodges is for the purposes of the '*Yin-yang* Militarist' School, to indicate good or bad military fortune for the several states." Despite this, and unlike

[19] John Major (1993, 128) speculates that this greater discrimination may be a reflection of the *Huainanzi*'s "southern orientation." For recent studies of the Nine Provinces scheme of terrestrial apportionment and the relevant sources, see Dorofeeva-Lichtmann (2010; 2009).

[20] *Huainanzi*, 3.15a–b.

the standard listing shown here which begins with the first Jupiter station "Star Period" *Xingji* in Sagittarius, the *Huainanzi* begins the list with the first lodge *Horns* (α Vir), presumably because that is the sequence the author follows in the passage immediately preceding which records the angular dimensions of the lodges. As we saw above, the military value of calendrical astromancy is made even more explicit in the *Huainanzi* treatise on military affairs.

The scheme of apportionments shown in Table 9.1 is obviously the product of a process of systematization, which necessarily dates from sometime after the de jure partitioning of the state of Jin by Zhao, Han, and Wei in 403 BCE. The older scheme (Table 9.2) based on the Nine Provinces (*jiu zhou*) has now been expanded to take account of the political and military realities of the early Warring States period. Three additional "provinces" have been carved out of the older scheme of nine (Bing, San He, You) and the polities each has supplanted (Wey, Zhou, Yan) have been reassigned accordingly.

The *Dipper* in astral divination

Although astrological prognostication involving the stars of the *Dipper* does not play a role in the discussion to follow and is poorly documented prior to the Han Dynasty, it is worthwhile to point out here the correlations previously alluded to between the *Dipper*'s stars and the Nine Provinces. Sima Qian implies that some aspect of this system of astrological correlations was historically used in tandem with the more familiar field-allocation system: "The twenty-eight lodges govern the twelve provinces and the *Dipper*'s handle seconds them; the source [of this scheme] is very ancient."[a] In referring specifically to the handle of the *Dipper* he had in mind the latter's use as a sort of celestial clock and alignments of its stars with certain lodges which divide the sky into quadrants (Chapter 12). There is evidence to suggest that the *Dipper*'s use as a seasonal indicator goes back to the Neolithic, as suggested by the arrangement of the cosmo-priest's tomb discovered at Xishuipo, Puyang, Henan.[b]

Province	Name	*Dipper* Star	
Yong	Kui shu	*Bowl Pivot*	Dubhe
Ji	Xuan	*Gyrator*	Merak
Qing/Yan	Ji	*Jade Device*	Phecda
Yang/Xu	Quan	*Balance Weight*	Megrez
Jing	Heng	*Balance Arm*	Alioth
Liang	Kaiyang	*Yang Initiator*	Mizar
Yu	Yaoguang	*Twinkling Brilliance*	Benetnash

[a] *Shiji*, 27.1346.
[b] Yi Shitong (1996, 22–31).

Here each of the Nine Fields corresponds to three lodges, except *Dark Heaven, Xuantian*, which has four. As may be seen from the punctuation and arrows in the fourth and fifth columns, this apportionment results in a straddling of boundaries by some lodges, which ordinarily referred to a single terrestrial location. For example, the state of Yan in column five ordinarily corresponds to province *You* – lodges *Tail* (Sco) and *Basket* (Sgr) – but here those two lodges are allotted separately to provinces *Yu* and *Yang* (column 3) respectively, and province *You* does not yet appear. This is a clear indication that the pattern of augmentation of the divisions, first from Four Quarters to Nine Provinces,[21] and ultimately to twelve states, came about over time, the second *before*, and the last *after*, the distribution of the twenty-eight lodges among the twelve chronograms (Jupiter stations) was conventionalized. In transforming the scheme of nine divisions into twelve (Table 9.1), one new province, Bing, was interposed between Qing and Xu and assigned to the state of Wey; another, San He, was placed between Yong and Jing and assigned to Zhou; and a third, You, was placed between Yu and Yang and given to province Yan.

In addition, the name of the west-southwest province was changed from Liang to Yi and made to correspond to the territory of the state of Wei, one of the three successor states of Jin. An implicit correlation between province Liang and the state of Jin in the Nine Fields system suggests a different identity for Liang than that given in later texts. In the "Tribute of Yu" (*Yu gong*) chapter of the *Book of Documents*, probably of Warring States date, Liang is identified with southernmost Shaanxi province, down through Gansu and Sichuan, its only boundaries given as Mt. Hua and an unidentified tributary of the Han River known as the Black River.[22] But the state of Jin and its immediate predecessor Tang were unquestionably located east of the Fen River in western Shanxi, above the Yellow River's abrupt eastward bend near its confluence with the Wei River. Just west of the confluence of the Fen River with the Yellow River is Mt. Liang, one of the peaks demarcating the watershed in the border region between provinces *Ji* and *Yong* regulated by the legendary Xia Dynasty founder, Yu the Great.[23] One might conjecture that Liang was originally located in this area in late Shang and early Zhou and that the name was only later reassigned to the southwest after Liang's extermination by Qin during the Spring and

[21] For the early history of astrological correlations between specific asterisms and terrestrial polities in Shang and Zhou and evidence of nascent field-allocation astrology in the late second millennium BCE, see the conclusions of Guo Moruo and Zheng Wenguang cited in Li Yong (1992, 27).

[22] Nienhauser et al. (1994, 27).

[23] Ibid., 22. *Literary Expositor* (*Erya*, "Explicating Earth" chapter) has Provinces You and Jing, but lacks Qing and Liang. The *Rites of Zhou* (*Zhouli*) includes Provinces You and Bing but lacks Xu and Liang. Obviously, in late Warring States times the identity of a Province Liang was very problematical, notwithstanding the authority of the "Tribute of Yu" chapter of the *Book of Documents*.

Autumn period and the inclusion of the area from southern Shaanxi through Sichuan into the Hua–Xia cultural orbit. This was done in the interest of filling a lacuna in the southwest, thereby filling out the three-by-three grid of the so-called "well-field" (*jing tian*) system, a schematic topography central to late Warring States cosmographic speculation.[24]

On the whole, the two systems' apportionments into nine and twelve astral–terrestrial correlations are in basic agreement. It is clear that the expansion by three terms did not alter pre-existing astrological relationships. Instead, the motive was no doubt to accommodate ancient tradition to new political realities by making terminological adjustments "on the ground," as it were. Chief among the historical changes demanding accommodation, besides the breakup of Jin in the fifth century BCE, was the eastward removal and subsequent decline of Royal Zhou after 771 BCE, coupled with the rise of Qin in the former Zhou homelands in the Wei River valley in Shaanxi. Taken together, these adaptations point at the latest to a Western Zhou or early Spring and Autumn period date (tenth to eighth centuries BCE) for the Nine Provinces precursor to the field-allocation scheme.[25]

Derivation of relevant astrological correspondences

As we have seen, the antecedents of field-allocation astrology are to be found in certain conceptions already influential in the second millennium BCE. Apart from the rare planetary conjunctions already described, several passages in pre-Qin works preserve the remnants of etiological myths and traditions that establish the existence of definite connections between certain celestial locations and terrestrial polities or peoples in the second millennium BCE. One of the most famous is the legend of Yan Bo and Shi Chen, which is preserved in the *Zuozhuan* (Duke Zhao, first year):

Formerly, Gao Xin [Di Ku] had two sons, the eldest was named Yan Bo and the younger Shi Chen. They lived in Kuanglin but could not get along, daily taking up shield and lance against one another. In the end, Gao Xin could no longer condone it and removed Yan Bo to Shangqiu to have charge of [the asterism] *Chen* [*Great Fire*, α Sco]; the ancestors of the Shang people followed him, therefore *Great Fire* is the Shang asterism. [Gao Xin] removed Shi Chen to Da Xia to have charge of *Shen* [Orion's belt], so the people of Tang followed him, and there served the houses of Xia and Shang. The last

---

[24] Tan Qixiang (1982, Vol. 1, 17–18). Despite some uncertainty about the precise identities and boundaries of provinces along the eastern seaboard named in the *Rong Cheng shi* manuscript, Province Liang in the west still does not figure in that text. Pines (2010, 512); Dorofeeva-Lichtmann (2009, 625).

[25] The 2002 discovery of the unprovenanced *Bin Gong xu* bronze with inscription (*c.* ninth century BCE?) seems to confirm that Yu the Great's cosmogonic exploits in hydraulic engineering were celebrated in early Zhou. See Qiu Xigui (2004); Jao Tsung-yi (2003); Liu Yu (2003); Li Xueqin (2005a); Zhou Fengwu (2003); Jiang Linchang (2003); Xing Wen (2003).

of their line was Tang Shuyu. When [Zhou] King Wu's wife, Yi Jiang, was pregnant with Tai Shu [i.e. Tang Shuyu], she dreamed that the Supernal Lord told her, "I have named your son Yu and will give Tang to him, make Tang belong to *Shen*, and cause his descendants to flourish." When the child was born he had the character *yu* on his hand, and so his name was called Yu. When [Zhou] King Cheng extinguished the old house of Tang he enfeoffed Tai Shu here, hence *Shen* is the star of Jin. From this we can see that Shi Chen is the spirit of *Shen* [Orion].[26]

and again (Duke Xiang, ninth year):

The ancient Regulator of Fire was remunerated either [at the season of] asterism *Heart* [principal star Antares] or asterism *Beak* [α Hya] in order [that he should] take out and bring in the fire.[27] For this reason *Beak* is [called] *Quail Fire* and *Heart* is [called] *Great Fire*. Tao Tang's [i.e. legendary Emperor Yao's] Regulator of Fire, Yan Bo, dwelt at Shangqiu and sacrificed to *Great Fire* [α Sco, Antares], using Fire to mark the seasons there. Xiang Tu [the grandson of Xie and father of the Shang people] continued in like manner, and so the Shang mainly attended to *Great Fire*. They observed that the incipient signs of their calamities and defeats invariably began with Fire.[28]

The same legend is alluded to in an equally famous passage in the "Discourses of Jin" in *Discourses of the States*: "I have heard that when Jin was first enfeoffed, Jupiter was in *Great Fire*, which is the star of Yan Bo; in truth it marked the periods of the Shang."[29]

This famous nexus of Jin astral lore about the feuding brothers Yan Bo and Shi Chen, which has abundant echoes in later literary tradition, weaves together elements of cultural significance in telltale fashion. At bottom, it is a classic example of the kind of etiological myth dating from the preliterate period that served both to explain and to transmit vital astro-calendrical knowledge.[30] In this pithy story we can discern a euhemerized tale about the human origins of the deities associated with the principal constellations of spring and autumn, Scorpius and Orion, which are diametrically opposed or "at odds" in the heavens and hence cannot appear in the sky simultaneously. As if unable to abide each other's presence, Yan Bo invariably ducks beneath the western horizon just before Shi Chen rises in the east. These personified asterisms are then linked to the cardinal directions (East and West); to seasonal activity for which they

---

[26] Hung (1966, Vol. 1, 344); cf. Legge (1972, 580).
[27] This would have been in spring and autumn. Use of fire in winter, the "watery" season, was strictly regulated so as not to upset the natural order of things.
[28] Hung (1966, Vol. 1, 266); cf. Legge (1972, 439). A Chu bamboo slip version of this passage dating from the Warring States period was recently unearthed. It is slightly defective but essentially the same. Cao Jinyan (2011, 5–6).
[29] *Guoyu*, 10.3a.
[30] Joseph Needham (Needham and Wang 1959, 248, 282) and others have noted that Scorpius (lodge 5, Antares) and Orion (*Shen* 21) marked the equinoctial points in the early Bronze Age, giving an indication of the antiquity of the myth of the feuding brothers.

anciently served as harbingers (the carrying out and in of the hearth fires, marking the beginning and end of the agricultural season);[31] and to the dominant political entities of the archaic period (Shang and Xia), their successor states, and the hereditary lines of astro-calendrical specialists who served them. In this and other examples of such astral lore, notably texts linking the above-mentioned *Vermilion Bird* component asterism *Beak* to the fortunes of the Zhou people, it is possible to discern the archaic astrological correlations and operative principles that underpinned the prognostications preserved in pre-Qin texts. In them we can discern the historical nucleus of the system of astrological correlations, which over time was to become amplified and standardized in the configuration shown above in Tables 9.1 and 9.2.

The prominence and appropriateness of Orion and Scorpius as seasonal harbingers has frequently been remarked upon because of their ancient links to the equinoxes. More tellingly, however, Scorpius's celestial location and ancient calendrical function may link this constellation with the flood mythos and the cosmogonic exploits of Yu the Great and his father Gun.[32] What principally concerns us here, however, is the fact that both Scorpius and Orion are similarly located adjacent to the *Sky River* (Milky Way), though, as we have already noted, at opposite ends of the sky. If we consider the relationship between the asterisms and their terrestrial correlates in the Heavenly Fields scheme in Table 9.2, it will be noted that *Great Fire* corresponds to east and province Yu, identified with Song, the successor state of Shang in eastern central Henan, while *Shen* or Orion corresponds to the southwest, province Liang, and the state of Jin, consistent with the legend of the feud between Shi Chen and Yan Bo.

Although the connection between Shang and the east is well established, how are we to account for the placement of Jin in the southwest? Or, for that matter, for such glaring geographical anomalies as the association of ancient states like Zhou in the west and Qi in the Shandong peninsula on the east coast with cardinal south and north respectively? To answer this question we must plot the provinces and their celestial counterparts on a map of the north China plain. In this way the incongruities begin to assume a paradigmatic character arising from a pre-eminent concern with celestial rather than terrestrial topography. Now the scheme may be seen to preserve at its core distinct traces of the original cosmographic organizing principle that over time became obscured.

---

[31] Appropriately, the supergiants Alphard (α Hya) and Antares (α Sco) are noticeably orange to reddish in hue.

[32] Porter (1996).

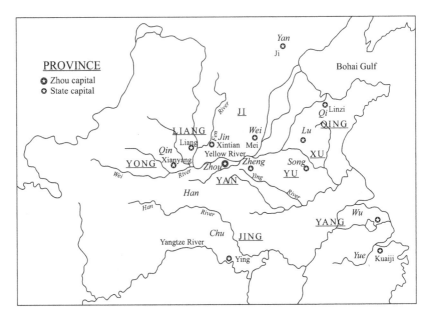

Map 9.2 The Nine Provinces in relation to the Yellow River. The standard directions correlated with the provinces (e.g. *Qing* ~ "north" and *Yong* ~ "south") are those that follow from the normative positions of their astral fields in relation to the Milky Way.

The *Sky River* as cosmographic divide

The Milky Way arches across the dome of the heavens from Sagittarius in the northeast (juncture of the northern and eastern quadrants of the sky) to Gemini in the southwest (juncture of the southern and western quadrants), with Scorpius and Orion strategically located at opposite extremes. If we imagine that ancient Chinese astrologers, practicing their art in the Yellow River drainage envisioned the Milky Way as the celestial analog of the great river, an explanation for the apparently incongruous astrological correlations immediately presents itself. When the generally southwest to northeast course of the Yellow River (Map 9.2) is compared to the southwest to northeast orientation of the *Sky River* (Map 9.3), the correlations of the Nine Provinces in the Spring and Autumn period make sense. The "fields" of Qi (province *Qing*) in the "north" and Qin/Zhou (province *Yong*) in the "south" determine the generally "north to south" orientation of the major cardinal axis, while the area above the Yellow River now becomes "west" and that below becomes "east." Jin's location (province *Liang*) in the astrological "southwest" can now be seen as a logical outgrowth of this paradigmatic celestial frame of reference, as does Chu's

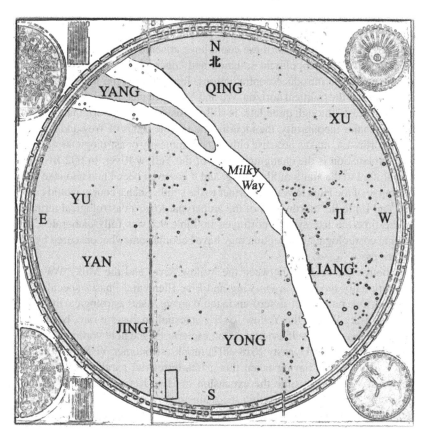

Map 9.3 Song Dynasty planisphere with the astral fields labeled with the names of the Nine Provinces, showing their distribution in relation to the Milky Way. Adapted from *Zhongguo shehui kexueyuan kaogu janjiusuo* (1980, 101, Figure 97), reproduced by permission.

(province *Jing*) location in the "southeast." Further examination of the figures confirms that the scheme's essential fidelity to celestial topography rather than to terrestrial directions arises from the archaic conception in which the Yellow River and the Milky Way were functionally identical in astrological terms.[33]

---

[33] An analogous conception may have arisen in ancient Egypt: "In considering the Milky Way as a river, the sky must have taken on a different directional orientation for the ancient Egyptians. We know that in contrast with their own Nile, the Egyptians considered the Euphrates river to flow north by flowing south. Thus, it is likely that they considered the river in the sky to flow north by flowing east." Kozloff (1994, 174).

The *Sky River*, thus, is primary. The mature Heavenly Fields scheme presents a few minor inconsistencies with this portrayal of the system's origins, but it is not difficult to imagine how these could have arisen over time as the growing complexity of the geopolitical situation and concomitant astrological ramifications tended to introduce contradictions. Interestingly, the identification of latecomers on the political horizon, Wu and Yue (province *Yang*), with astrological "northeast," though quite late, is still consistent with the original concept. The one major incongruity, the identification of the states of Wey and Lu with north-northwest, makes no sense either geographically or astrologically. A possible explanation is the changing course of the Yellow River. In 602 BCE the river shifted to the south of Shandong, and it has even been known to take over the course of the Huai River still farther to the south. Such a course would cause Wey and Lu to lie "northwest" of the archetypal *River* in astrological terms.[34] By the time the late system portrayed in Table 9.1 was fully elaborated, the original cosmological conception may have become somewhat obscured by the accretions.

Although the analogy between the Yellow River and the Milky Way was conventionally repeated in *yin-yang* and Five Elemental Phases speculations from this time on, explicit assertions in the Warring States astrological literature that the Milky Way is the Yellow River's celestial analog are rare, because at the time it was either self-evident or too esoteric.[35] One noteworthy exception is the brilliant Tang Dynasty (618–907) monk-astronomer Yi Xing (fl. 725), who is recorded as having taken this astral–terrestrial parallelism as fundamental. By Yi Xing's time the expansion of Hua–Xia civilization had long since brought south China within the mainstream of Chinese culture, so that *both* of China's major river systems now had to be taken into account, obscuring the archaic one-to-one correspondence between the Yellow River and the Milky Way. Nevertheless, the fundamental conception is still discernible in his explanation. According to the Song Dynasty encyclopedist Zheng Qiao (1104–62),

How excellent was Yi Xing's discussion of the twelve stations in the Tang Dynasty! From beginning to end he relied on the *River of Clouds* [Milky Way] to talk about them. The *River of Clouds* is the *materia vitalis* of the Yellow and Yangtze rivers. He discerned the Two Divisions [of China] by the mountain ranges and river systems and recognized the rising and setting of the *River of Clouds* at the Four Nodes [NE, SE, SW, NW]. On comparing below to the ancient Han Dynasty commanderies and kingdoms,

---

[34] Liu (2004, 21, Figure 2.1).

[35] Like most of his contemporaries, the famous Han Dynasty astronomer and polymath Zhang Heng (78–139 CE) subscribed to the conventional view that the "essence of [elemental] Water becomes the Sky Han (River)." Zhang Heng, *Ling Xian*, *Taiping yulan*, 1:8.10a. In a similar vein, the Han apocryphal text *Hetu kuodi xiang* makes the astral–terrestrial correspondence plain: "the essence of the Yellow River rises to become the Sky Han." *Taiping yulan*, 1:8.11a.

differentiating and locating the field allocations was as easy as pointing to one's palm. On the whole, the stars are like *materia vitalis*, nothing more; the *River of Clouds*, the *Northern Dipper*, the Five Planets, there is nothing which is not *materia vitalis*. How profound was Yi Xing's learning![36]

Jupiter in applied field-allocation astrology: *Zuozhuan* examples

In order to clarify the main features of applied field-allocation astrology of particular interest in Warring States and Han times, it will be useful to examine typical passages in *Zuozhuan* involving Jupiter before considering the examples in *Discourses of the States*. Essentially the same principles apply in the case of other heavenly bodies and celestial manifestations; however, because Jupiter plays the dominant prognosticating role in military and political contexts, the discussion will focus on that planet only. Reproduced here are three of the eight instances in *Zuozhuan* where Jupiter's location is given in relation to the astrological fields associated with various states.[37] The texts display two principal features of field-allocation prognostication: (i) forecasting of events is based on the passage of one or more integral twelve-year cycles of Jupiter; (ii) Jupiter's presence in a state's astrological space confers military advantage.

1. In the eighth year of Duke Zhao (533 BCE) the *Springs and Autumns Annals* records the destruction of Chen by Chu. *Zuozhuan* elaborates:

The Marquis of Jin asked the historiographer Zhao, "Will Chen cease to exist after this?" and was told, "not yet." "Why is that?" asked the Duke. [The historiographer] replied: "[The house of Chen] is descended from Zhuan Xu. Jupiter was in *Quail Fire* and [the dynasty of Zhuan Xu] was extinguished; it will be the same with the extinction of Chen. Now [Jupiter] is in the [Milky Way] *Ford at Split Wood*, [Chen] will be restored again. Moreover, the branch of the House of Chen which is in Qi will obtain the government of that state and only after that will Chen perish.[38]

The restoration of Chen by Chu occurred in 529, its annihilation by Chu in 479. A venerable historical–genealogical tradition is adduced as the basis for the

---

[36] Zheng Qiao's approving comments conclude his discussion "Distinguishing field allocations" in *Gujin tushu jicheng*, Chapter 57, "Stars and Asterisms: General Discussion," 1341. For the geographical details of Yi Xing's analytical scheme, see the "Monograph on Astrology," *Tianwen zhi*, in the *New History of the Tang Dynasty*, *Xin Tang shu*, 21.817. By the Ming–Qing period this basic heuristic was apparently either lost sight of or ignored, so that scholars proposed significant revisions to the obsolescent field-allocation system. Fang Yizhi (1611–71) even observed that although "the southerly Yangzhou region comprised about half of China's land area, its heavenly 'field' included only three of the twenty-eight lodges." Henderson (1984, 223). For the riverine divisions of China, see ibid., 221, Figure 8.15.

[37] The eight instances are found in Duke Xiang, third year; Duke Xiang, twenty-eighth year (three examples); Duke Zhao, eighth year; Duke Zhao, ninth year; Duke Zhao, tenth year; Duke Zhao, eleventh year; Duke Zhao, thirty-second year.

[38] Hung (1966, Vol. 1, 623).

prediction that the time of Chen's demise had not yet come. The prognostication is explained in the *Zuozhuan* commentary for the next year, as follows:

2. In the ninth year of Duke Zhao (532 BCE) the *Springs and Autumns Annals* records a fire in the capital of Chen. *Zuozhuan* adds,

In the fourth month there was a fire in Chen. Pi Zao of Zheng said: "In five years the state of Chen will be restored, and after fifty-two years of restoration it will finally perish." Zi Chan asked the reason and [Pi Zao] replied, "Chen belongs to [Zhuan Xu's element of] Water. Fire is antagonistic to Water, and the state of Chu [descended from Regulator of Fire Zhu Rong] emulates Fire. Now the *Fire* [*Star*] has appeared and set fire to Chen [indicating] the expulsion of Chu and the establishment of Chen. Antagonistic [relations] reach fulfillment in fives, therefore I said "in five years." Jupiter will reach *Quail Fire* five times and after that Chen will finally perish. That Chu will then be able to possess it is the Way of Heaven. Therefore, I said "fifty-two years."[39]

Here, the correlative scheme of the Five Elemental Phases is invoked to explain the antagonism between Chen and Chu, based on their archaic astrological linkage with Watery and Fiery asterisms and corresponding quadrants of the sky. The spring appearance of the *Fire Star*, Antares, is said to be the cause of the conflagration in Chen.

3. In the thirty-second year of Duke Zhao (510 BCE) the *Springs and Autumns* laconically reports, "in summer Wu attacked Yue." *Zuozhuan* amplifies:

This was the first instance of a [regular] expedition by Wu against Yue. Scribe-astrologer Mi said, "In less than forty years Yue is likely to possess Wu! Yue has obtained [the advantage of] Jupiter [being located in its astrological space], yet Wu attacks it – [Wu] is certain to suffer misfortune from this."[40]

The prognostication is of the defeat of Wu by Yue in 478 BCE and Wu's annihilation the following year. Here Wu acts rashly, in defiance of the dictum that the state within whose astrological space Jupiter is stationed cannot be attacked with impunity.

The *Discourses of the States* account of Duke Wen's restoration

The *Discourses of the States* account in the "Discourses of Jin" contains two passages that relate the circumstances of Chong Er's exile and subsequent restoration as rightful ruler of Jin. The first recounts a famous episode that purports to date from 644 BCE, when Chong Er found himself in dire straits while traveling through the territory of Wei:

[39] Ibid., 370.   [40] Ibid.

On passing through Wulu [Chong Er] begged a rustic for food. The man picked up a clod of earth and offered it to him. The Prince became incensed and was about to scourge the man, but Zi Fan said, "This is a gift from Heaven; if the people offer you the land, what more could you ask of them? The workings of Heaven are invariably prefigured [in portents]. [This one means that] in twelve years you must capture this ground. You officers, mark my words. Jupiter is currently in *Longevity Star*; when it reaches *Quail's Tail*, won't [the Prince] possess this land? Heaven has decreed it in this way. [When Jupiter] is once again in *Longevity Star*, [Prince Chong Er] must gain the Lords of the States. It is the Way of Heaven."[41]

The second, longer narrative concerns the sequence of events surrounding Chong Er's restoration eight years later in 636:

[Scribe-astrologer] Dong Yin [of Jin] came out to meet the Duke [Wen, i.e. Chong Er] at the Yellow River. The Duke asked, "May I cross?" [Dong Yin] replied, "Jupiter is in *Great Span* and about to complete its heavenly travel. Thus, in your first year you will start by gaining the star of *Shi Chen*. As for the astral space *Shi Chen*, [its terrestrial correlate] is the abode of the people of Jin and it is that whereby Jin arose. Now that my Lord's return coincides with *Shi Chen* [it means] you have but to cross! When my Lord departed, Jupiter was in *Great Fire*, which is the star of Yan Bo. [*Great Fire*] is called the *Great Seasonal Indicator*. The *Great Seasonal Indicator* is for bringing goodness to fruition; Hou Ji [legendary progenitor of the Zhou people] emulated it, and Tang Shu was enfoeffed during it. The historical records of the blind [historiographer] say: 'Successors will continue the ancestral line like the increase of grain; there must needs be a Jin state.' Your servant divined by milfoil and obtained all eights for [hexagram] *Tai* [Peace], [whose judgment] says, 'Heaven and Earth unite in receiving sacrifice; the smaller departs and the greater approaches. [Good Fortune. Success].' The present [circumstances] correspond to this, so how could you not cross? What's more, you left under *Chen* [*Great Fire* = Sco] and you return under *Triaster* [*Shi Chen* = Orion]. These are both auspicious signs for Jin; they are Heaven's great seasonal markers. If you cross and complete your grasp [of power in Jin], you are sure to dominate the Lords of the States as Hegemon. Your descendants are depending on it; have no fear, my lord." The Prince crossed the River and summoned [local headmen]. [The localities of] Linghu, Jiushuai, and Sangquan all capitulated. The people of Jin [i.e. in the capital] were frightened, and Duke Huai [of Jin] fled to Gaoliang, leaving Lu Sheng and Ji Rui in command of the army. On day *jiawu* [31] [the Jin army] encamped at Luliu. The Earl of Qin sent Gongzi Zhi into the army [to inform its command]; then the army withdrew and took up station at Xun. On day *xinchou* [38], Hu Yan [i.e. Zi Fan], together with grandees from Qin and Jin, entered into a covenant at Xun. On day *renyin* [39], Duke [Wen] went among the Jin army. On day *jiachen* [41], the Earl of Qin returned [home]. On day *bingwu* [43], [Duke Wen] entered Quwo [former Jin capital]. On day *dingwei* [44], he entered [the present Jin capital] Jiang, and ascended the throne in the temple of Duke Wu. On day *wushen* [45], [Duke Wen had the deposed] Duke Huai assassinated at Gaoliang.[42]

---

[41] *Guoyu*, 10.1b; *Zuozhuan*, Duke Xi, twenty-third year, Hung (1966, Vol. 1, 121).
[42] *Guoyu*, 10.11a–12a; cf. *Zuozhuan*, Duke Xi, twenty-fourth year, Hung (1966, Vol. 1, 122).

Table 9.3 *Astrological sequence of Zi Fan (Hu Yan) and Dong Yin.*

Date	Event – Jupiter's location
655 BCE	Chong Er flees Jin; Jupiter in *Great Fire* (Sco)
644	Begs food at Wulu; Jupiter in *Longevity Star* (Vir)
637	Dong Yin meets at Yellow River; Jupiter in *Great Span* (Pleiades)
636	Duke Wen of Jin first Year; Jupiter in *Shi Chen* (Orion)
635	Jupiter in *Quail's Head* (Gem–Can)
634	Jupiter in *Quail Fire* (Leo [Hya])
633	*Chun Wei* (Cor)
632	Battle of Chengpu; Jupiter in *Longevity Star* (Vir)
631	Jupiter in *Ford at Split Wood* (Sgr)

There is much of historical and literary interest in these two passages but our focus here will be on the prognostications and their implications. First, it will be noted that both passages assume the main operative principles of field-allocation astrology identified above; that is, the strategic advantage conferred by Jupiter's presence in a state's astrological space and prediction based on integral twelve-year cycles. From the point of view of Jin, Jupiter's twelve-year cycle begins and ends in *Shi Chen*, the astrological space allotted to Jin. In explaining his prediction, the historiographer Dong Yin invokes the venerable traditions about the significance of *Shi Chen* and *Great Fire* in Jin history, described above in our account of the origins of field-allocation correlations. Reduced to their essentials, the astrological prognostications (in fact, fourth-century BCE *ex post facto* reconstructions) may be tabulated as in Table 9.3.

According to the first passage containing the prediction of Hu Yan (Zi Fan), because the omen at Wulu occurred while Jupiter was located in the Jupiter station *Longevity Star*, Heaven has decreed that when Jupiter next returns to the same location twelve years hence Chong Er will capture Wulu from Wei, then win the allegiance of the Lords of the States and be confirmed as Hegemon. The second, more elaborate astrological passage, again in the form of wise counsel proffered by an astrologically expert historiographer, is consistent with the first and invokes the legendary roles played by *Great Fire* and *Shi Chen* in Jin history. Jupiter's location in Jin's astrological space *Shi Chen* at the very moment Chong Er is to make his bid for power is underscored as especially auspicious. Historical and genealogical analogies, together with an apposite judgment text from the *Book of Changes* alluding to the flight of Duke Huai of Jin at Chong Er's approach, are adduced to persuade Chong Er not only that the recommended action is consistent with astrological precedent, but that success is virtually assured because of the propitiousness of the moment. Indeed, not to act as advised is portrayed as a dereliction of Chong Er's filial duty to his ancestors as well as to his descendants.

After the great victory over Chu, Zhou king Xiang richly rewarded Zi Fan for his role in restoring Duke Wen and reasserting royal authority over the Central States. The King's commendation of Zi Fan as recorded in the recently discovered *Zi Fan he zhong* inscription reads:

The King's fifth month, first auspicious [heavenly stem] day *dingwei* (44). "Zi Fan assisted the Duke of Jin as adviser and brought about the restoration of his state. [When] the several Chu [adherents] did not come to hear [the Royal] commands in the King's presence, Zi Fan, together with the Duke of Jin, led the Six Armies of the West [i.e. Jin] in extensively punishing Chu and achieved great merit. Chu is bereft of its armies, its commander(s) annihilated. Zi Fan assisted the Duke of Jin as adviser in mediating among the several Lords, causing them to appear before the King, and so was able to settle the King's throne. [Now, therefore,] the King bestows on Zi Fan . . . "

Because day *dingwei* was the same lucky day chosen for Chong Er's investiture as Duke Wen of Jin in 636, four years earlier, and because the narration begins with this event, Feng Shi argues that the date refers to that occasion.[a] Li Xueqin, on the other hand, presents compelling evidence that the recorded date refers to the date of casting and dedication of the bell, some time after the convocation and possibly as much as three years later. Either way there is little doubt that the portion of the inscription quoted here records the actual wording of the Zhou king's commendation.[b]

[a] Feng Shi (1997, 63).
[b] Li Xueqin (1999c). Stem day *ding* was deemed highly auspicious from at least the Shang Dynasty.

Apart from the prescience of Zi Fan and Dong Yin in accurately forecasting the future, the fictional nature of the prognostications and their reconstruction after the fact is shown by a four-year error in identifying Jupiter's location. Table 9.4 is based on analysis of the planet's true longitudes in the years indicated and shows where Jupiter ought to have been placed by contemporary observers. It is apparent that instead of being located as claimed, in Jin's astrological space in Duke Wen's first year, Jupiter's location was most likely identified as *Shi Chen* in 644 and again twelve years later in 632, at the time of the Battle of Chengpu.

Table 9.4 *Jupiter's actual locations between 655 and 631 BCE.*

BCE	Event	Jupiter station
655	Chong Er flees Jin	*Chunshou, Quail's Head* (Gem–Can)
644	Begs food at Wulu	*Shi Chen* (Orion)
637	Dong Yin at river	*Xing Ji* (Sgr–Cap)
636	Duke Wen, first year	*Xuan Xiao* (Aqr)
635		*Juzi* (Peg)
634		*Jiang Lou* (And)
633		*Da Liang, Great Span* (Pleiades)
632	Battle of Chengpu	*Shi Chen* (Orion)
631		*Chunshou, Quail's Head* (Gem–Can)

In some cases there may be ambiguity in determining what contemporary observers would have denoted a given year based on Jupiter's location because of the phenomenon of "station drift," arising from the incommensurability of Jupiter's synodic period with a solar year (398.88 versus 365.24 days). Jupiter's continued motion during its annual seven-week period of invisibility behind the Sun causes the planet to reappear a few degrees farther along in a given station on each successive reappearance. After about seven twelve-year cycles the accumulated drift will approximate one year's travel, and if the planet's location is routinely assigned on the basis of a simple sequential year count rather than regular observation, at some point actual observation of Jupiter's reappearance will reveal the planet to have unexpectedly skipped a station, a phenomenon known as "missing a station" (*shi ci*) (the phenomenon is illustrated graphically in Figure 6.2 in Chapter 6).[43]

Shi Chen years, in particular, pose an identity problem because that station's range in longitude is just under fifteen degrees, or only about half of Jupiter's annual motion, so that in any given *Shi Chen* year the planet will actually spend half the time outside *Shi Chen*'s nominal boundaries. That being the case, determination of what constitutes a *Shi Chen* year can depend vitally on what the immediately preceding years have been understood to be. In the present instance, the difficulty is compounded by the fact that much of Jupiter's travel through *Shi Chen* during the cycle in question took place while the planet was obscured behind the Sun during May and June of 633 BCE. Reappearing on or about June 21, 633, Jupiter was already past the nominal eastern boundary of

---

[43] This is precisely the situation referred to in *Zuozhuan* (Duke Xiang, twenty-eighth year), where Jupiter's reappearance in station *Xing Ji* was anticipated, but the planet "transgressed" and appeared in *Xuan Xiao* instead.

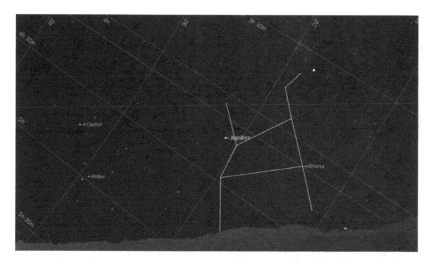

Figure 9.1 Jupiter's location in station *Shi Chen* in 633 BCE a few minutes east of the star Alhena (at RA 4h 31m) marking the nominal eastern edge of station *Shi Chen* (Starry Night Pro 6.4.3).

Shi Chen, according to the "old degree" (*gu du*) system of determinative stars which marked the boundaries of the lodges.[44]

In order to conclude with confidence that Jupiter's twelve-month period of visibility (which can begin in any month) is correctly described as this or that Jupiter year, we need benchmarks against which to compare the planet's behavior on a given occasion with positional records showing the range of acceptable variability. Fortunately, in the present case such benchmarks exist. It is instructive to compare Jupiter's true location in 632 BCE with the circumstances in the years 242 and 206 BCE (Figures 9.2 and 9.3), which agree with the positional indications recorded in the excavated Mawangdui manuscript "Prognostications of the Five Planets" (second century BCE).[45] The figures graphically illustrate the variability of Jupiter's position at its dawn reappearance in *Shi Chen* before and after three Jupiter cycles had elapsed between 242 and 206 BCE. Note that in the latter case, in 206, Jupiter's position is identical with that in 633 BCE shown in Figure 9.1. In the absence of detailed knowledge about how Jupiter's location was identified in the seventh century BCE, this record

---

[44] For the "old degree" system of reference stars in use prior to the second century BCE, see Wang and Liu (1989, 59–68).
[45] Xi Zezong (1989b, 55).

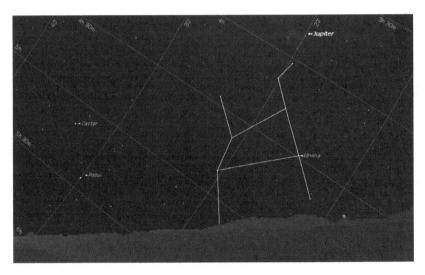

Figure 9.2 Jupiter's July 242 BCE reappearance in station *Shi Chen* between Orion and Gemini. The star Alhena (RA 4h 31m) marks the nominal eastern edge of station *Shi Chen* (Starry Night Pro 6.4.3).

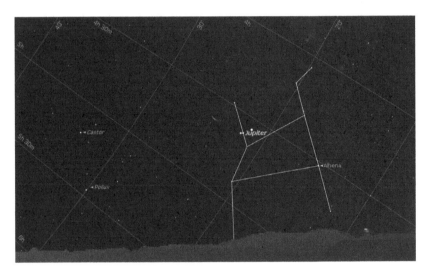

Figure 9.3 Jupiter's July 206 BCE reappearance between Orion and Gemini just east of Alhena (RA 4h 31m) marking the eastern edge of station *Shi Chen* (Starry Night Pro 6.4.3).

from the late third century provides confirmation that identifying Jupiter's location as *Shi Chen* during its period of visibility in 633–632 is at least consistent with later practice.[46]

An astronomical dating of the *Discourses of the States*

If we assume this portrayal of the circumstances to be accurate, the argument has interesting implications for the date of composition of the *Discourses of the States*. The dates of Duke Wen of Jin were known in Warring States times. Any error in the *Discourses of the States* identification of Jupiter's location would have arisen from retrospective calculation at the time of that text's compilation. An author working in late Warring States times would necessarily have assumed a twelve-year period for Jupiter, the same figure given in the "Prognostications of the Five Planets" in the early second century BCE. Since Jupiter's actual synodic period is 11.86 years, a retrospective extrapolation compounding the 0.14 year (∽51-day) error per twelve-year cycle would have to span some twenty-eight to twenty-nine cycles (336 to 348 years) to accumulate a four-year error. This would place the approximate date of the text's compilation at the very end of the fourth century BCE or the beginning of the third (i.e. $636 - 336 = 300$; $636 - 348 = 288$). This result is consistent with scholarly opinion regarding the date of compilation of *Discourses of the States*, based on considerations of grammar, usage, conceptual framework, and historical allusions.[a]

[a] E. Bruce Brooks (1994, 50) has proposed a date for the text of 305–307 BCE. William Hung (1966, Vol. 1, lxxxiv), basing his calculation on evidence from *Zuozhuan*, dated the beginning of retrospective chronological calculation using the Jupiter cycle to about 364 BCE.

As we saw in Chapter 6, the account in "Discourses of Zhou" locating Jupiter in the Zhou astrological space *Quail Fire* (*Chun Huo*) in the year Shang was conquered is correct. Jupiter's astrological role at the time is also implicated in the newly discovered Qinghua ms. Qi Ye. Observation of Jupiter on the morning of the Battle of Muye in early 1046 is explicitly mentioned in the famous *Li gui* inscription, contemporaneous with the Zhou conquest. However, the absolute date of the Zhou conquest was no longer precisely known in late Warring States times, and even if it had been no retrospective calculation using a twelve-year period for Jupiter could possibly have correctly placed

[46] Recent discoveries of contemporaneous bronze inscriptions dated according to the Jupiter cycle leave no doubt that the system was certainly in conventional use by the sixth century BCE. Wang and Hao (2009, 69–75).

Jupiter in *Quail Fire*, so that bit of astral lore had to have been based on received tradition. Therefore, it is not unreasonable to suppose that the author of *Discourses of the States* also had access to astrological portents associated with Duke Wen's restoration and the Battle of Chengpu, events more than four centuries closer to his own time than the Zhou conquest of Shang. For example, tradition might have held that "when Jin became Hegemon, Jupiter was in *Shi Chen*," or something similarly ambiguous, which the author of the astrological excursuses of Dong Yin and Zi Fan would have interpreted and embellished on the basis of his other sources and retrospective calculation. At a minimum, it is apparent that the author of the *Discourses* was not fashioning his astrological account from whole cloth.

The coincidence of *Shi Chen* with Duke Wen's first year as emphasized in the text, and in all probability with the Battle of Chengpu, suggests that field-allocation astrology played a more prominent role in military strategy in the early Spring and Autumn period than has previously been recognized. The remainder of this discussion will take a closer look at the timing of datable events and their coincidence with significant celestial phenomena in an effort to learn what, if any, astrological considerations may have influenced decision making.

Correlation of political and military activity with celestial phenomena

The appendix to this chapter provides a detailed timeline of events from the time of Chong Er's restoration to the victorious Jin army's return home after the decisive victory over Chu at Chengpu. All dates given in the various sources have been reconciled and their astronomical accuracy checked using astronomical software as well as reconstructions of pre-Qin calendars, specifically the tables of new Moons and intercalations for the Spring and Autumn period (722–481 BCE).[47] Another suggestive indication of the attention being paid to celestial phenomena in scheduling events is the correlation between full Moons and ritually significant investiture ceremonies, first at the time of Chong Er's return to the old Jin capital of Quwo and assumption of power as Duke Wen of Jin in the fifth month of 636 BCE (May 10), and subsequently at the time of the elaborate investiture and covenant ceremony confirming Duke Wen as Hegemon in the presence of the Zhou king and the assembled lords in the fifth month of 632.[48] In addition, the fact that a number of other dated events coincide with either

---

[47] Zhang Peiyu (1987).
[48] Following up on the reference in *Zuozhuan* to the precedent-setting investiture of Duke Huan of Qi in 679 BCE, the famous "Sunflower Hill," *Kuiqiu*, covenant in the eighth month of that year indicates that it too may have been scheduled to coincide with a full Moon; see *infra*.

new or full Moons suggests that certain kinds of military and ritual activity required such timing, for reasons we now can only conjecture. For example, on its victorious return march to Jin in the sixth month of 632, the Jin army forded the Yellow River during a full Moon, perhaps to take advantage of the evening moonlight, while the subsequent triumphal entry into the Jin capital, instead, coincided with a new Moon (first day of the month). The Battle of Chengpu itself occurred the day following a New Moon, with the new Moon possibly still invisible. Such correlations might seem purely coincidental were it not for the fact that fully one-third of the dated events coincide with either new or full Moons and others fall within a day of the precise time of syzygy (Sun–Moon conjunction).[49]

As interesting as such lunar correlations may be from a tactical or ritual standpoint, however, they are prosaic by comparison with the planetary phenomena coincident with the Battle of Chengpu and the subsequent ceremonies. We saw above how 633–632 BCE was the first occasion after Duke Wen assumed power that Jupiter took up station in *Shi Chen* (Orion–Gemini), the astrological space allotted to Jin. In astrological terms this means that this would have been the ideal opportunity for Jin to undertake military adventures, since Jupiter's presence in a state's astrological space confers powerful advantage. Moreover, this opportunity would recur only after another twelve years had elapsed, and only after the astrological advantage conferred by Jupiter had passed, first to Qin, then to Chu. Therefore, to the extent that astrological considerations figured in strategic planning, as the *Discourses of the States* strongly suggests they did, it would have been most difficult, given the political and military circumstances, for Duke Wen and his advisers to let such an opportunity pass.

But the spring and early summer of 632 BCE witnessed an even more remarkable coincidence of celestial and terrestrial events, for as the months progressed it would have become obvious to astute observers that a planetary gathering of some significance was in the offing. Figure 9.4 shows Jupiter prominent in Gemini in the evening sky just at the time of the Battle of Chengpu in March. During the period March through May three more planets approached Jupiter's position, culminating in an impressive alignment of Venus, Mercury,

---

[49] It is noteworthy in this connection that the remarks of Sun Bin quoted above concerning the use of astrological calculations in warfare come from the chapter of *The Art of War* entitled "The Moon and Warfare." History records that similar considerations figured importantly in ancient Greece. When the Athenians sent to Sparta for help in repelling the Persian army at Marathon (490 BCE), "the Spartans said they were of a mind to assist the Athenians, but they were unable to do so immediately, because they did not wish to violate their religious prohibitions. For it was the ninth day of the first decade of the lunar month, and they said the army could not set forth on an expedition on the ninth day while the disc of the Moon was not yet not full. And so they waited for the full Moon." Herodotus, *The Histories*, Book 6, Chapter 106. For similar considerations connected with the Battle of Plataia (479 BCE), see Paul Stephenson (forthcoming).

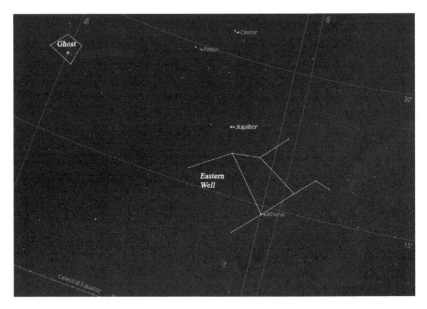

Figure 9.4 Jupiter's location just east of the star Alhena during the Battle of Chengpu, March 12, 632 BCE (Starry Night Pro 6.4.3).

Jupiter, and Mars in Cancer in late May (Figure 9.5), precisely at the time of the investiture ceremony and covenant in the Zhou king's presence that formally confirmed Duke Wen's status as Hegemon.

This cluster of planets was neither as complete nor as compact as the one in 1059 BCE that signaled the conferral of Heaven's Mandate on King Wen, but, as these things go it would certainly have commanded attention, and what is more, one might say that it was "ritually appropriate," in that as Hegemon Duke Wen would not have merited a clustering of all five planets. Remarkably, and this would have been all the more striking to observers in 632 who had knowledge of the Zhou Mandate precedent, this grouping of the four planets *took place in the identical location as the Zhou Mandate conjunction* some four centuries earlier (Figure 9.6). Hardly less striking is the fact that the famous alignment of all five planets in May 205 BCE, recorded by Han historians as signaling Heavenly approval of Liu Bang's founding of the Han Dynasty, also was *centered on this very same location* in the sky (compare Figures 9.5–9.7).[50]

---

[50] For the earliest record of this planetary event, see *Shiji*, 27.1348. See also comments by Liu Yunyou (Xi Zezong) (1974, 35).

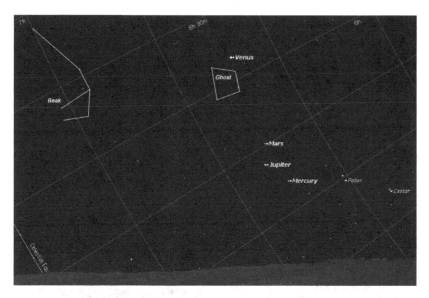

Figure 9.5 Alignment of four planets ahead of the "*Beak*" of the *Vermilion Bird* in late May 632 BCE at the time of Duke Wen's investiture as Hegemon (*ba*) (Starry Night Pro 6.4.3).

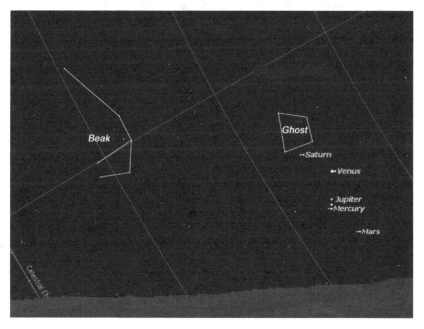

Figure 9.6 The Zhou Mandate cluster of the Five Planets ahead of the "*Beak*" of the *Vermilion Bird* in late May 1059 BCE (Starry Night Pro 6.4.3).

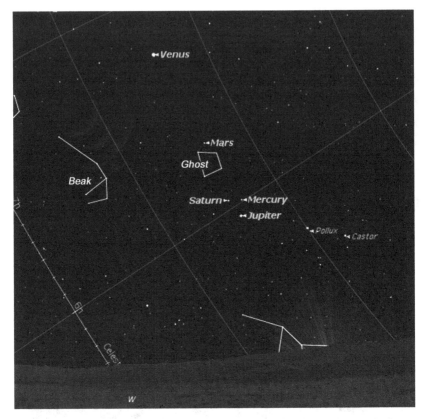

Figure 9.7 The Han Dynasty founder's alignment of planets in late May 205 BCE, also ahead of the *Beak* of the *Vermilion Bird* (Starry Night Pro 6.4.3).

Astrological parallelism in two accounts in *Discourses of the States*

Recalling now that a regularly applied operative principle in field-allocation astrology is the twelve-year Jupiter cycle or "Great Year," it will be helpful to compare the Zhou Mandate precedent with the prognostications concerning Duke Wen of Jin. In the latter case, we saw how fulfillment of the prediction was based on Jupiter's return to a given location after one twelve-year cycle had elapsed. In the passage from "Discourses of Jin," the author reconstructed this cycle as beginning and ending in *Longevity Star* (*Shouxing*), with Duke Wen's first year coinciding with *Shi Chen*. In actual fact, contemporary observers would probably have seen Jupiter's presence in *Shi Chen* as coinciding with the portent at Wulu (644) and the Battle of Chengpu (632). On this basis we concluded that the *Discourses* author had at his disposal accurate, though perhaps ambiguous, astrological accounts of the period which hinted at the auspicious role of Jupiter and *Shi Chen*, leaving it to the narrator's imagination to reconstruct the precise details.

Turning to the earlier Zhou Mandate precedent in the mid eleventh century BCE, although rarely alluded to in the standard sources, the astrological circumstances of the conquest period recounted in detail in "Discourses of Zhou" must have been widely known from the Warring States period on. In the Eastern Han period, Ban Gu's (32–92 CE) "You tong zhi fu" has the line "[when his] 'Neighbor to the East' [i.e. Shang Zhòu] tortured and killed the humane, King [Wu] conformed his positions to the Three and the Five." Explication by Ying Shao (140–206) has survived in Yan Shigu's (581–645) commentary on the rhapsody, which again quotes the *Discourses of Zhou* astrological passage in which the formula "the Three and the Five" refers to the positions of the planets. Here, then, is my reconstruction of the sequence of events leading up to the Zhou conquest:[51]

1059, May 28   Dense massing of the Five Planets in *Quail's Head* (Can–Gem), near the *"Beak"* of the *Vermilion Bird.*

1058, March   Jupiter on station at the *Bird Star* in *Quail Fire*, King Wen declares "First Year of the Mandate."

1048, Autumn   Jupiter advances toward *Quail Fire*, then suddenly stops and retrogrades; the Zhou army's advance is aborted at Mengjin Ford on the Yellow River, despite the Zhou allies' entreaties. King Wu admonishes, "You do not know Heaven's Mandate."

1046, March   For some weeks Jupiter is stationary within three degrees of the *Bird Star* in *Quail Fire*; Battle of Muye on March 20, 1046 (Zhou "Mandate thirteenth year," second month, day *jiazi*).

From 1058 *Quail Fire* to 1046 *Quail Fire* is thus one complete Jupiter cycle or "Great Year."

From this chronology it is clear that the astrological circumstances of the spring of 632 BCE recapitulate in detail the salient features of the Zhou Mandate precedent in the mid eleventh century:

(1) Jupiter is in the station astrologically associated with the state assuming military advantage;

(2) the timing of events (or prognostication) correlates with the passage of one Jupiter cycle;

(3) Jupiter is stationary at East Station and/or about to resume direct motion at the time of the decisive military engagement;

---

[51] Pankenier (1981–82; 1995).

(4) an impressive massing of planets occurs near the beak of the *Vermilion Bird* constellation, signaling endorsement of what has just occurred (or will shortly occur, in the case of King Wen).

Conclusion

Records of the planetary massing in 1059 BCE and of Jupiter's location in *Quail Fire* at the time of the Zhou Conquest survived in Han and Zhou texts. Persistent traditions regarding the appearance of a "Red Bird" (Phoenix) in connection with the Mandate's conferral in 1046 BCE indicate that knowledge of the actual location of the Zhou portent in the *Vermilion Bird* was known in Warring States and earlier times. As we saw above in Chapter 6, this paradigmatic astrological sequence was still well understood in the third century CE at the time of the last Han emperor's abdication. Only now has reconstruction of the actual astronomical circumstances of the Mandate's conferral and the Zhou Conquest confirmed the precise correlation of Jupiter's twelve-year period from *Quail Fire* to *Quail Fire* (i.e. the astrological space allotted to Zhou in field-allocation astrology) with the watershed events of the Zhou conquest period.

The recurrence of such portentous planetary phenomena in precisely the same location in connection with Duke Wen's victory over Chu and his elevation to the status of Hegemon must have deeply impressed those in the know who witnessed the events of 632 BCE. Certainly, to the extent that the participants embraced these astrological precepts, field-allocation astrology would have figured importantly in strategic and tactical planning. Indeed, Jupiter's presence in the astrological space allotted to Jin in 632 could help to explain the eagerness of Jin to engage Chu and, conversely, might also account for King Cheng of Chu's evident lack of enthusiasm for confrontation with Jin. In any event, it seems likely that Duke Wen's investiture as Hegemon, and probably the Battle of Chengpu itself, were scheduled to coincide with what was happening in the skies.[52] Given the extraordinarily prestigious, precedent-setting events at the founding of Zhou, and given that these are the only two instances in *Discourses of the States* where the astrological circumstances of epoch-making events are elaborated to this degree, it seems fair to conclude that the astrological parallels were present in the mind of the author of the pertinent passages.

[52] The parallels between the Battle of Muye in 1046 and the Battle of Chengpu in 632 BCE are actually even more precise, since at the time of both engagements Jupiter was either paused at its east station or just resuming direct motion after this annual stationary episode. There are highly suggestive parallels in Maya planetary astrology; see, for example, Justeson (1989, 104), who discusses how stationary episodes of Jupiter and Saturn were used to schedule human affairs "ostensibly as sacred mandates for elite decision-making." For further examples, see Milbrath (2003, 301–30).

We are led to the conclusion, therefore, that field-allocation astrology figured in military planning during the Spring and Autumn period, and that precedent-setting events at the time of the conquest in early Zhou established the principal parameters for astrological prognostication in such contexts. Echoes of these astrological practices and principles, especially regarding the role of Jupiter and its twelve-year cycle, survive in Warring States and later historical narratives such as *Zuozhuan*, *Discourses of the States*, and *History of the Three Kingdoms*, and demonstrate the importance of astral–terrestrial correspondences in the conceptual world of the Spring and Autumn and Warring States periods (late eighth through third centuries BCE). Finally, the timely recurrence of impressive planetary conjunctions in the same highly symbolic celestial location on three momentous occasions – 1059, 632, and 205 BCE – can hardly have failed to reinforce in the minds of witnesses to the events the already strong conviction that Heaven manifested a special interest in the fate of Hua–Xia civilization.

Appendix: timeline of Duke Wen of Jin's restoration and events surrounding the Battle of Chengpu (March 12, 632 BCE)

637–636 BCE (Duke Xi of Lu's twenty-third to twenty-fourth year in *Springs and Autumns*)

Discourses of the States contains astrological prognostications based on field-allocation astrology and the twelve-year Jupiter cycle; *The Grand Scribe's Records*[53] (following *Zuozhuan*) has:

* **Duke Wen of Jin, first year (636),** spring, Chong Er "reaches the Yellow River," then:

– **Second month,** *jiawu* **(31)** "Jin army encamps at Luliu" [dated only in *Zuozhuan*]

  xinchou **(38)** "Grandees of Qin and Jin covenant at Xun."

  renyin **(39)** "Chong Er joins the Jin army."

– **third month,** *jichou* **(26)** *hui* "last day of the month; attempt on Duke Wen's life." [Only *Zuozhuan* gives lunar phase.][54]

– **[fourth month],** *bingwu* **(43)** "Chong Er enters Quwo."[55]

  dingwei **(44)** "Chong Er installed as Lord of Jin in Wu Temple."

  wushen **(45)** "Duke Wen orders Duke Huai killed."

---

[53] *Shiji*, 39.1661.

[54] There is a one-month discrepancy between the *Zuozhuan* (Lu) dates given here and the calendar reconstructed for the period by Zhang Peiyu, probably due to the insertion of an intercalary first month in Lu at the beginning of Duke Xi's twenty-fourth year. See Zhang Peiyu (1987, 137). For an alternative explanation of the calendrical discrepancy based on the date recorded in the *Zi Fan he zhong* inscription, see Feng Shi (1997, 63–4).

[55] The Moon was 15.9 days old the preceding evening, May 9, at which time it was ninety-nine percent illuminated.

634–632 BCE The *Springs and Autumns* and *Zuozhuan* have:

* **Duke Xi 26; 634 BCE** – Lu covenants with Wei and Ju; Qi displeased, invades Lu; Lu appeals to Chu for help in attacking Qi.
- **Xi 26** Song turns on Chu and sides with Jin, so Chu troops first attack Song and besiege Minyi.
- **Xi 26** Lu leads some Chu troops to attack Qi and captures Guyi (upstream from Pu) and garrisons the town with Chu troops.
* **Xi 27** winter; Chu king Cheng personally leads troops of Zheng, Chen, Cai, and Xu in besieging Song; Lu participates in covenant on Song territory.
- Song appeals to Jin for help; Cao and Wei recently collaborated with Chu (Wei and Chu recently established marriage ties).
- Jin strategy is to goad Chu into confrontation: first attack Cao and Wei, this way Chu will be forced to lift the siege of Song, halting Chu incursion into Qi.
* **Xi 28, 632 BCE** – Jin first establishes the Three Armies; raises troops to invade Cao and attack Wei, captures Wulu from Wei (near Puyang; cf. *Discourses of the States* prophecy of twelve years earlier).
* **Xi 28, second month** – Jin and Qi covenant at Lianyu (Wei territory near Puyang).
- Duke Cheng of Wei wants to take part in covenant but Jin refuses him for duplicity; the people of Wei drive him out to Xiangniu in the south to curry favor with Jin.
- Lu sends troops to defend Wei; Chu troops are unsuccessful in relieving Wei; Lu fears Jin, kills Chu grandee Gongzi Jia to placate Jin, dissembles in telling Chu it was because he was not resolute in relief of Wei.
* **Xi 28, third month, *bingwu* (43)** (February 17, 632) – Jin army occupies Cao; Chu siege of Song becomes acute; Song again appeals to Jin for relief.
- Duke Wen of Jin hesitates to break openly with Chu because Qi and Qin are still unwilling to co-operate fully.
- Duke Wen's adviser Xian Zhen persuades with a stratagem: have Song bribe Qi and Qin to approach Chu about making peace; Jin will capture Cao lord and give Cao and Wei land to Song. Chu protects Cao and Wei, so will not agree to peace with Song despite Qi and Qin request; the latter will become incensed and Jin thus can gain Qi and Qin co-operation against Chu.
- Jin proceeds as planned: captures Cao Bo and delivers him to Song (Cao is a former vassal of Song).
- King Cheng of Chu withdraws to Shen; orders Chu troops to leave Qi town Guyi and calls on Premier Zi Yu to lift the siege of Song; King Cheng, unwilling to oppose Jin, retires from the field.
- Zi Yu, refusing to do as ordered, sends a deputy to plead with Chu king Cheng to declare war on Jin; Zi Yu's self-serving arrogance offends the Chu king, who sends only token reserves and lets Zi Yu act on his own.

– Zi Yu sends an emissary to Jin demanding that Wei ruler be allowed to return and Cao be re-established, after which he promises to lift the siege of Song.

– Adviser Xian Zhen persuades Duke Wen of Jin with stratagem: privately allow lords of Cao and Wei to return home, causing Cao and Wei to desert Chu, then arrest the Chu emissary to provoke Chu.

– Strategem works: Cao and Wei desert Chu; Zi Yu, predictably outraged, attacks Jin army.

– Duke Wen honors an old promise, made while a guest of King Cheng of Chu, to retreat three stages (ninety *li*) to avoid confronting the Chu army; the majority of Chu forces are reluctant to pursue. Zi Yu, willful and vainglorious, insists on attacking Jin.

* **Xi 28:**

– **fourth month, *wuchen* (5)** (March 11) – Jin, Song, Qi, and Qin troops occupy Chengpu; Chu army takes advantageous tactical position and encamps. [*Zuozhuan* date = fourth month began on *dingmao* (4), March 10.]

– **fourth month, *jisi* (6)** (March 12, Moon age 1.6d] – Jin and Chu declare intention to fight. **Battle of Chengpu.** Jin's superior tactics and deception carry the day; only Zi Yu's army of the middle withdraws undefeated; Jin army feasts on captured Chu provisions for three days. [*Springs and Autumns, Zuozhuan, The Grand Scribe's Records* all provide the date.]

– *jiawu* (31) (April 6; Moon age 26.8d in Psc) – Jin army withdraws to Hengyong (days 10–30); Duke Wen builds "King's Palace" at Jiantu and invites Zhou king; Zheng, which had been colluding with Chu, in fear now sues for peace with Jin. [Only *Zuozhuan* gives date.]

– **fifth month, *bingwu* (43)** (April 18; month began on *dingyou* (34)) – Jin and Zheng covenant at Hengyong. [Only *Zuozhuan* gives the date.]

– *dingwei* (44) (April 19) – Zhou king arrives; Jin presents Chu captives; Zheng ruler comes to assist Zhou king in officiating in the ceremonies; Zhou king makes awards and declaration. [Only *Zuozhuan* gives the date.]

– **fifth month, *guichou* (50)** (April 25; Moon age 15.9d at 20:00, 98% full in Sco) – Jin grandee Wangzi Hu convenes the feudal lords; first big covenant since Duke Huan of Qi at Sunflower Hill;[56] Jin Qi, Lu, Song, Wei, Zheng, Cai, Ju all covenant; Chen Hou comes to observe; after declining the required three times, Duke Wen of Jin is confirmed as Hegemon by Zhou king Xiang with elaborate investiture ceremony.[57] [*Springs and Autumns* gives

---

[56] The calendar for the covenant at Sunflower Hill in Duke Xi's ninth year, ninth month, *wuchen* (5), as recorded in *Zuozhuan*, shows that in 651 BCE the ninth month new Moon was *bingchen* (53) or August 6. Therefore *wuchen* (5) was August 18. On that evening the waxing Moon was ninety-eight percent full by LT 20:00 at age 13.6 d. Precise full Moon was on *gengwu* (7), August 20, at LT 6:21, thirty-four hours later.

[57] Sima Qian slips in quoting a passage from the *Book of Documents*, "Charge to Wen Hou" (*Wen Hou zhi ming*), containing a charge from Zhou king Ping (770–720) to his contemporary Wen

guichou (50); *Zuozhuan* and *Shiji* both have *guihai* (60); precise full Moon was at LT18:19 on *renzi* (49) the preceding day; Jupiter now brilliantly prominent in Cancer, to be joined by Mecury, Venus, Mars within days.]

– **[sixth month],** *renwu* **(19)** (May 24; Moon age 15.8d at 20:00, 100% full in Sgr) – Jin army fords Yellow River on homeward march.

– **autumn, seventh month,** *bingshen* **(33)** (June 7; new Moon) – troops enter the capital of Jin in triumphal array. As of May 23, the four planets are all tightly grouped in Cancer and spectacularly visible in the evening sky. [Only *Zuozhuan* gives the date.]

Hou of Jin (780–746). Sima Qian took the *Book of Documents* passage to refer to the ceremony investing Chong Er over a century later in 632 BCE. *Shiji*, 39.1666.

10   A new astrological paradigm

In previous chapters we explored the origins and development of the system of astral–terrestrial correspondences that took shape during the first millennium BCE. One of the noteworthy features of the system is its unabashedly sinocentric conception and application. The Chinese world constituted the known universe and no accommodation was made in the scheme for non-Chinese. By the Western Han Dynasty (206 BCE–8CE), however, some concession had to be made to the new political reality. Leaving no room for prognostication concerning non-Chinese peoples was an anachronistic bias that astrology could no longer afford if it was to have a claim to relevance in the imperial period. Here we consider the innovative revision to the field-allocation scheme that first appeared in Sima Qian's "Treatise on the Celestial Offices" in *The Grand Scribe's Records* (*c.*100 BCE). By creative application of then-prevailing *yin-yang* cosmology to the older astral–terrestrial paradigm, this latest stage in the development of astrology attempted to adapt the former preoccupation with a multivalent Chinese world to the circumstances of the early empire, with its bipolar "us-versus-them" perspective on contemporary power relations. In addition to describing the effort to decant the ancient scheme into a new conceptual bottle, we will look at a concrete illustration of its application to military strategy by Emperor Xuan (r. 74–49 BCE).

Although field-allocation (*fenye*) astrology is hardly mentioned in the Western Han Dynasty philosophical works, numerous allusions in fourth- to second-century BCE narratives like *Zuozhuan*, *Discourses of the States*, the *Springs and Autumns of Master Lü*, and the *Huainanzi*; and an abundance of archaeologically excavated Han Dynasty artifacts ranging from diviners' mantic astrolabes (*shi*) to the Mawangdui silk manuscripts, all make it clear that both a theory of astral–terrestrial correspondences and the idea of mutual resonance (*gan ying*) between the two realms were fundamental to the cosmological thought of the early Han period.[1] Unlike Hellenistic astrology, the Chinese did not stress the unidirectional influence expressed by Ptolemy's famous axiom "as above, so

---

[1] For a general discussion of these and other aspects of pre-imperial and Han Dynasty astrology and cosmology, see Harper (1999, 831 ff.).

here below," so much as the reciprocal "as here below, so above" reflected in the belief that celestial anomalies and other ominous manifestations of Heaven's displeasure were seen as an index of temporal misrule. For this reason, astral omenology bore directly on the security of the state and hence closely controlled activities from the early imperial period on, and probably earlier as well. This is one explanation for the hereditary, even hermetic, character of the scribe-astrologer's profession.

As we have seen, such observation was not the result of disinterested stargazing. Original records of regular astronomical observation ranging from the mundane (sunrise and sunset, solstices, individual stars and planets) to the exceptional (lunar and solar eclipses, comets, sunspots, supernovae, etc.) appear as early as writing itself in the Shang oracle bone inscriptions, but for many years the conventional view has been that astrology played no significant role in the history of ideas in China before the late Warring States period.[2] As we saw in Chapter 6, however, the process of cosmicization had already occurred by the late Shang Dynasty. The very fact that astronomical phenomena such as eclipses were divined about in Shang Dynasty oracle bones, and that the *Lost Books of Zhou* and the *Springs and Autumns Annals* accurately report numerous eclipses (as well as three comets) and the distress they caused, should be sufficient to give pause, but until now the role of astral–terrestrial correspondences in the very earliest period has not been adequately explored.

In the Western Han Dynasty, the imperially authorized practitioner of astrological prognostication was the Prefect Grand Scribe-Astrologer, *Tai shi gong*, whose duty it was to know the historical precedents, to follow the movements of the heavenly bodies, and to advise the emperor on the implications of developments, especially unanticipated changes or anomalies. Sima Qian's "Treatise on the Celestial Offices" (*c.*100 BCE; Appendix) provides a comprehensive survey of the cosmological and astronomical knowledge in the keeping of his office, as well as its practical application. This included plotting the locations, movements, and changes affecting the stars and planets, together with interpreting the significance of their appearances based on the by then well-established system of astral–terrestrial correspondences (field allocations).

Astrological portents typically had implications for the ruler, high dignitaries, and major affairs of state. As Sima Qian says, "in all cases of celestial anomalies, if the [regular] measures are overstepped, then prognosticate ... The ultimate superior cultivates virtue, the next level practices [good] government, the next level carries out relief efforts, the next level conducts sacrificial rituals, below that nothing [is done]."[3] Because of their extreme rarity, conjunctions of multiple planets, especially dense groupings involving all five naked-eye

---

[2] Sivin (1995b, 5–37).   [3] *Shiji*, 27.1350.

planets (Mercury, Venus, Mars, Jupiter, Saturn), ranked as the most portentous of all celestial phenomena and as such had dynastic implications. As we have seen, this pre-eminence was based on the historical association of planetary alignments with epochal dynastic transitions, culminating in the most recent such alignment in 205 BCE, which occurred in precisely the same location in Gemini–Cancer as the Zhou precedent of 1059 previously discussed. This latest sign in the heavens in 205 BCE was officially recognized and later memorialized in *The Grand Scribe's Records* as the astral omen signaling the rise of the Han Dynasty.[4] From Sima Qian's account of the significance of planetary massings, it is evident that by the beginning of the Han Dynasty heavenly endorsement of the transfer of the Mandate to a new dynasty in the form of a conjunction of the Five Planets had become ideologically de rigueur.

In a conservative "science" like astral prognostication this kind of axiomatic premise could not take shape and win general acceptance overnight; it must have the sanction of long tradition behind it. Clearly, by the founding of the Han Dynasty the connection was self-evident. Sima Qian's concluding summary of the astrological knowledge of his day in his "Treatise on the Celestial Offices" displays both ancient conceptual roots and the Han theoretical reformulation based on the prevailing *yin-yang* and Five Elemental Phases correlative cosmology:

Ever since the people have existed, when have successive rulers not calendared the movements of the Sun, Moon, stars and asterisms? Coming to the Five Houses[5] and the Three Dynasties [Xia, Shang, Zhou], they continued by making this [knowledge] clear, they differentiated wearers of cap and sash from the barbarian peoples as inner is to outer, and they divided the Middle Kingdoms into twelve regions. Looking up they observed the figures in the heavens, looking down they modeled themselves on the categories of the earth.[6] Therefore, in Heaven there are Sun and Moon, on Earth there are *yin* and *yang*; in Heaven there are the Five Planets, on Earth there are the Five Elemental Phases [Wood, Fire, Metal, Water, Earth]; in Heaven are arrayed the lodges, and on Earth there are the terrestrial regions. The Three Luminaries are the essence of

---

[4] "When Han arose, the Five Planets gathered in *Dongjing* [Eastern Well; lodge 22, Gem]"; *Shiji* 27.1348. See also *Han shu*, 26.1301: "First year of Emperor Gaozu of Han, tenth month, the Five Planets gathered in *Dongjing* [LM 22]. Extrapolation based on the calendar [indicates] they followed [the lead of] Jupiter. This was the sign that August Emperor Gao received the Mandate. Therefore, a retainer said to Zhang Er, '*Dongjing* is the territory of Qin. When the King of Han [i.e. Liu Bang, soon to be Emperor Gao] entered Qin, the Five Planets, following Jupiter, gathered together, signifying [he] ought to gain all under Heaven by means of Righteousness.'" The *Han shu* date "tenth month" of 206 BCE for the event is an obvious revision based on the date of the Qin heir Wangzi Ying's surrender to Liu Bang at Xianyang, the Qin capital. The actual planetary alignment occurred the following year, in May of 205 BCE. Sima Qian is more circumspect, saying only "when Han arose." For the theoretical statement that conjunctions of the Five Planets in which Jupiter takes the lead portend the rise of a "righteous" dynastic founder, see *Shiji* 27.1312.

[5] The five predynastic rulers: Huang Di, Gao Yang, Gao Xin (Di Ku), Tang Yu (Yao-Shun).

[6] Sima Qian is quoting the "Appended Commentary" to the *Book of Changes*.

yin and *yang*, their *materia vitalis* originates on Earth, and the Sage comprehensively brought order by them.[7]

Therefore, Sima Qian says,

When the Five Planets gather, *this is a change of Phase*: the possessor of [fitting] virtue is celebrated, a new Great Man is set up to possess the Four Quarters, and his descendants flourish and multiply. But the one lacking in virtue suffers calamities or extinction.[8]

In Map 9.3 above we saw that the earlier field-allocation system of the Warring States period (403–221 BCE) was exclusively sinocentric in conception. The Chinese world was all that mattered, so that the identity between the Milky Way (the *"Sky River"*) and the Yellow River provided the basic paradigm for the entire scheme of correlations between the skyscape and the topography of the terrestrial provinces below. No accommodation was made in the heavens for non-Chinese peoples.[9] Therefore, in view of the increasingly ominous threat to the unified Han empire posed by aggressive non-Chinese peoples on the periphery like the Xiongnu, the Simas assert in the "Treatise" that in macro-astrological terms the warlike nomadic peoples are *yin* with respect to the *yang* of the Chinese world. As such, they correspond to the northern and western quadrants of the heavens, while the Chinese world corresponds to the south and east. By way of theoretical support, Sima Qian adduces the historically powerful Chinese border states of Jin and Qin as cases in point of "hybrid" Chinese polities whose martial proclivities clearly reflected the influence of non-Chinese peoples with whom they had been in intimate contact for generations:

Coming to Qin's swallowing up and annexing the Three Jin [i.e. Wei, Zhao, Han], Yan, and Dai [Shandong], from the [Yellow] River and Mount [Hua] southward is

---

[7] *Shiji*, 27.1342.

[8] *Shiji*, 27.1321. That Sima Qian's was the conventional conception is confirmed by Shen Yue's (441–513) reiteration of this principle in *Song shu* (*Tianwen zhi*, 25.735), where he quotes *Shiji* and amplifies on Sima Qian's last line: "the one lacking in virtue suffers punishment, is separated from his household and kingdom, and devastates his ancestral temple." Shen then says, "Now, in my judgment, based on surviving texts, there have been three conjunctions of the Five Planets. Zhou and Han [each] relied on [such a one] to rule as King, as did Qi as Hegemon. [Duke Huan of] Qi [685–643 BCE] finally ended up as Hegemon, and in the end there was no epochal change. Therefore, there has never occurred such a thing as a conjunction of planets with no [concomitant] change of Phase." Similarly, arguing that dynastic change may be portended by massings of only four planets, Shen Yue later points to record of the transfer of the Mandate from the Han dynasty to Wei in 220 CE: "Emperor Xian of Han, twenty-fifth year of the Jian'an reign period: Emperor Wen of Wei [i.e. Cao Pi; r. 221–37 CE] received his [Emperor Xian's] abdication; this constitutes the change of Phase [portended by] the four planet's three [recent] gatherings." *History of the Song Dynasty*, *Song shu*, 25.736.

[9] Di Cosmo (2002, 305–11).

China [*zhongguo*]. [With respect to] the area within the Four Seas, China therefore occupies the south and east as *yang* – *yang* is the Sun, Jupiter, Mars, and Saturn.[10] Prognostications [about China are based on astral locations] situated south of *Celestial Street* [κ Tau], and *Hunting Net* [lodge 19, ε Tau] governs them. To the north and west are the Hu, Mo, Yuezhi, and other peoples who wear felt and furs and draw the bow as *yin* – *yin* is the Moon, Venus, and Mercury. Prognostications [about them are based on astral locations] situated north of *Celestial Street*, while *Topknot* [Pleiades, lodge 18, 7 Tau] governs them... On the whole, China's mountain ranges and watercourses run north and east, their head[-waters] in [Mount] Long and Shu [Gansu and Sichuan] and tail at the Bo[-hai Gulf] and [Mount] Jie[-shi] [Shanhaiguan]. For this reason, Qin and Jin are fond of using weapons; furthermore, their prognostications [depend on] Venus, governor of China, while the Hu and Mo, who have repeatedly invaded and despoiled, are uniquely prognosticated [based on] Mercury. Mercury's appearances and disappearances are swift and sudden, so as a rule it governs the Yi–Di barbarians. These are the guiding principles. They are modified according to who acts as the guest and who the host. Mars means order,[11] externally [north and west], the army should be mobilized, but internally [south and east], the government should be put in order. Therefore it is said, "though there may be a perspicacious Son of Heaven, one must still look to where Mars is located."[12]

Astral prognostication on this new binary macro level clearly departs in important respects from the earlier astral omen reading. The new scheme reflects the animosity between the Chinese and frontier peoples that intensified greatly during Emperor Wu's reign. Essentially, *yin* celestial fields north and west of the Milky Way as archetypal *River* (i.e. 10–18, provinces Yang through Liang, or Capricorn through Taurus) correspond to the historical fields of activity of the peripheral "barbarian" (non-Chinese) peoples. In contrast, the *yang* celestial fields south and east of the Milky Way (i.e. 19–15, provinces Yong through Yu, or Taurus through Scorpius) correspond to the Hua–Xia heartland (mainstream Chinese culture). This binary scheme surfaces again in the interpretation of important planetary phenomena, for as we read in the "Treatise," "when the Five Planets are disposed in mid-heaven and gather in the east, China benefits, when they gather in the west, foreign kingdoms using

---

[10] In the early Han Mawangdui MS "Prognostications of the Five Planets," *yin* and *yang* are commonly used as directional terms in the sense of "north and west" and "south and east" respectively, but only with respect to relative locations of the kingdoms of the Warring States. For example, in discussing prognostications based on Venus's position in particular astral fields, the "Prognostications" says, "Yue, Qi, Han, Zhao, and Wei are *yang* with respect to Jing (Chu) and Qin. Qi is *yang* with respect to Yan, Zhao, and Wei. Wei is *yang* with respect to Han and Zhao. Han is *yang* with respect to Qin and Zhao. Qin is *yang* with respect to the *Di* barbarians. They are prognosticated on the basis of [Venus's lying] north or south, advancing or retreating." See Liu Lexian (2004, 86). Sima Qian's use of *yin* and *yang* here with respect to the geography of the empire is an innovation.

[11] Following Wang Shumin (1982, 1157) in reading *li* for *bo*.   [12] *Shiji*, 27.1347.

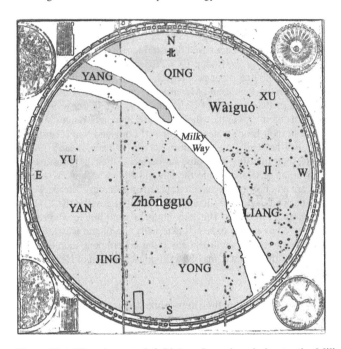

Figure 10.1 The nine astral fields/provinces in relation to the Milky Way.
Shown here are the relative positions of the astral fields/provinces and the
yin-yang binary division of the sky in the new macro-astrology. Adapted
from Zhongguo shehui kexue yuan kaogu janjiusuo (1980, 101, Figure 97),
reproduced by permission.

weapons gain."[13] Remarkably, given astrology's resistance to change, and in
spite of a pious nod in the direction of their esteemed predecessors, the Simas'
account bears witness to a major reformulation of astrological theory and practice, in which the former preoccupation with a multivalent sinocentric world
is adapted to the circumstances of the Han empire with its "us-versus-them"
view of contemporary power relations. By referring again to the map of the sky
(Figure 10.1) it becomes apparent that this broad generalization still invokes the
parallelism between the skyscape and the geopolitical realities of north central
China in the pre-imperial and early Han periods. It is noteworthy that even in
this revised scheme the Simas' geographical focus is still on north China and
the Yellow River watershed. They give surprisingly short shrift to the Yangtze

---

[13] *Shiji*, 27.1328. Precisely because of the macro-astrological context, I do not think it is anachronistic to translate *zhongguo* as "China" in the "Treatise."

region and southward despite witnessing Emperor Wu's aggressive expansion into this area as far south as present-day northern Vietnam.

Astrological prognostication on this new binary macro level clearly departs in important respects from the earlier astral omen watching. To cite just one telling example, linking the powerful southeastern coastal state of Wu with the southern constellation *Vermilion Bird* represents a radical break with the earlier scheme in which Wu and Yue, then seen as the peripheral south, were originally associated with lodges *Southern Dipper* and *Ox-Leader* in the winter or northern quadrant. This move also had the drastic consequence of dispossessing the longest-ruling and culturally most influential of the pre-imperial dynasties, the Zhou (1046–256 BCE), by disconnecting it from the huge *Vermilion Bird* constellation (Can–Cor).

A case study: General Zhao Chongguo and the campaign of 61 BCE

Throughout the reign of Emperor Wu of Han (141–87 BCE), as betokened by that ruler's posthumous epithet of "Martial," imperial policy toward peripheral areas and their nomadic inhabitants was expansionist and aggressive. This was particularly true in the northwest, where long and costly campaigns against the Xiongnu, initially in response to increasingly bold border raids and incursions, gradually became the norm.[14] Much manpower and treasure were expended in fighting the Xiongnu and in alternately coercing and bribing rival groups to collaborate with Han efforts to restrain the Xiongnu (and each other), with decidedly mixed results given the fluid power relations among competing forces along the ill-defined frontier. By the early first century BCE the drain on the imperial treasury had long been a serious problem, tarnishing Emperor Wu's legacy and leading to a partial Chinese withdrawal from northern Korea in 82 BCE, barely three decades after annexation. But the practice of periodically dispatching comparatively small task forces into the far northwest continued until about 65 BCE, when a shift in policy began to make itself felt.

The Han court still faced a serious challenge in pacifying the frontiers, both in terms of creating buffer zones between frontier settlements and potentially troublesome tribal peoples, and in terms of provisioning the Chinese garrison forces whose long-term presence served to pacify the border areas. Provisioning Chinese forces in this area was especially problematical because of the extremely long supply lines and the great quantities of feed grain required by the draft animals and the horses of mounted troops.[15] In an attempt to achieve

---

[14] For a detailed study of military policy and border affairs in the early Han Dynasty, see Loewe (1974).

[15] What this meant quantitatively is spelled out in a memorial by Zhao Chongguo and analyzed by Loewe (1974, 97).

a less costly and more permanent solution, imperial policy shifted from expansion toward a strategy of pacification through gradual colonization of frontier areas. Late in 61 BCE, Zhao Chongguo, a veteran of many years of service in Central Asia and campaigns against the Xiongnu, proposed a new way of consolidating Chinese influence by permanently establishing self-supporting agricultural garrisons, *tuntian*, along the frontiers. An informative sampling of the administrative records of just such a garrison at Cherchen from the years 49–48 BCE has been analyzed, and these materials, together with other finds excavated from the civilian communities of Niya and Loulan, provide a look in unprecedented detail into life on the frontier in the early imperial period.[16]

Of particular interest here, however, is a Chinese campaign into the northwest that lasted throughout 61 BCE. In April–May of that year in response to border unrest and at the venerable age of seventy-six, General of the Rear Zhao Chongguo was dispatched at the head of 60,000 troops to quell an uprising by the Western Qiang people, who frequently allied with the Xiongnu against the Han.[17] Emperor Wu had established the frontier commanderies of Dunhuang, Jiuquan, and Zhangye in 104 BCE precisely with the intention of separating the two peoples and preventing their joining forces. The Qiang were ancient adversaries of the Chinese and may have been the same ethnic group referred to by that name in the Shang Dynasty (1054–1046 BCE) oracle bone inscriptions. They enjoyed the unhappy distinction of being among the captives most frequently sacrificed in bloody rites dedicated to the Shang royal ancestors. Unlike the Xiongnu, however, the Qiang never coalesced into a tribal federation, and their pronounced tendency toward internecine conflict was already pointed out by Zhao Chongguo in a memorial to Emperor Xuan (74–49 BCE) written in 63 BCE: "it is relatively easy to bring the Qiang under control because they are divided into many warlike tribes and always attack each other. It is not in their nature to become unified."[18] Traditionally nomadic pastoralists, in the second century BCE the Western Qiang were active in a broad swath of territory, extending along the Kunlun Mountains from Dunhuang in the east to the Pamir in the west. By the mid first century BCE some had turned from herding to agriculture and significant numbers lived interspersed with Chinese settlers in the frontier areas. According to authoritative contemporary sources, abusive treatment of the Western Qiang by rapacious Chinese frontier officials was a chief cause of their frequent revolts.[19]

---

[16] Atwood (1991, 161–99). Atwood studied the Niya Kharosthi documents dating from 236–321 CE discovered by the explorer Sir Aurel Stein between 1906 and 1931. See also Loewe (1967).

[17] Dubs (1938–55, Vol. 2, 241).   [18] Trans. Yü (1986, Vol. 1, 422); *Han shu*, 69.2912.

[19] Yü (1986, Vol. 1, 424–5).

By the autumn of 61 BCE, concerned about the impending onset of winter and the mounting cost of the campaign, Emperor Xuan became impatient with Zhao Chongguo's apparent lack of progress. The emperor wrote a letter upbraiding his field commander for dilatoriness, urging him in the strongest terms to engage the enemy expeditiously. Zhao, a seasoned veteran of many border campaigns, was not intimidated. In his response he defended his tactics with an eloquent disquisition on military strategy and border affairs, especially the merits of waiting out this particular enemy, citing Sunzi's *The Art of War* as authority: "one who is skilled at warfare makes his opponent come to him, and is not made to come by his opponent."[20] In the end, General Zhao's well-considered strategy was successful. What is especially interesting about this particular exchange, however, is the factor strongly underscored by Emperor Xuan, but ignored by General Zhao in his response. In concluding his letter, Emperor Xuan emphasized that the astral omens favored immediate action against the enemy:

At present the Five Planets appear in the east, [signifying that] China will be greatly benefitted, while the Man and Yi [barbarians] will be utterly defeated. Venus appears on high, and so [the time] is auspicious for the one who employs troops to penetrate deeply and boldly do battle, but unpropitious for the one who dares not engage.[21]

Emperor Xuan clearly had in mind the maxim concerning the Five Planets enunciated a half-century earlier by Sima Qian: "when the Five Planets are disposed in mid-heaven and gather in the east, China benefits, when they gather in the west, foreign kingdoms using weapons gain."[22] On this occasion, in 61 BCE, of the Five Planets, Venus and Mercury were particularly close to each other, rising before dawn on November 21 in lodges *Base* (3) and *Chamber* (4) respectively. Emperor Xuan stresses Venus's location, but he may have had more of the passage from Sima Qian's "Treatise" in mind, since Mercury also figures in the auspicious prognostication: "When [Mercury] and Venus rise together in the east, both red and rayed, foreign kingdoms will be utterly defeated and China will gain the victory."[23]

So here we have a revealing example of the application of astrological theory to tactical decision-making in early imperial history. In point of fact, by November of 61 BCE when Emperor Xuan was observing the omens, the Five Planets were actually strung out across well over half the sky between Taurus and Scorpius (conventionally southwest through east). Rather than being clustered in the eastern quadrant of the heavens, the planets were all located

---

[20] *Han shu*, 69.2981.   [21] Ibid.   [22] *Shiji*, 27.1328.
[23] *Shiji*, 27.1328. On the political and cultural opposition between China and its foreign adversaries at this epoch and its antecedents, see Di Cosmo (2002).

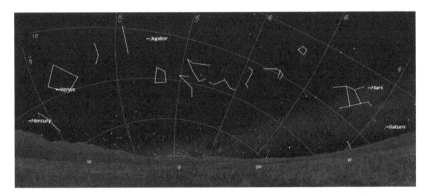

Figure 10.2 "All Five Planets in the *yang* sky (spring–summer) benefit China," November 24, 61 BCE (Starry Night Pro 6.4.3).

in the "southern" or *yang* half of the heavens associated with China in the new theory set forth by Sima Qian. Emperor Xuan's prognostication, in which the planets are "on our side," is clearly based on Sima Qian's binary *yin-yang* reformulation of the ancient scheme of astral–terrestrial correspondences. Given the planets' dispersal over more than 160 degrees in longitude, almost from horizon to horizon, this array, like "strung pearls" (*lian zhu*), according to the Han cliché, could not possibly qualify as a massing in one lodging (i.e. in a space of less than fifteen degrees), so it would not have been considered to have dynastic consequences. Clearly, we are not dealing here with the most spectacular and rarest of planetary massings of the kind that portended major political upheaval. Instead we now have a more prosaic prognostication that provides a "domesticated" role for the planets, one that had emerged in part to render such portentous and unpredictable events less threatening to the ruling dynasty.

It was not until mid-November of 61 BCE that the Five Planets were all west of the Sun, strung out across the predawn sky (Figure 10.2). Venus had previously reached maximum western elongation from the Sun in mid-September, but at thirty-five degrees west elongation would still have shone brilliantly in the southeast as the *Morning Star*. Emperor Xuan was thus quite right to take particular note of Venus "high in the sky," and his interpretation that this boded well for China also derives from Sima Qian's "Treatise," where Venus is denoted "ruler of the Middle Kingdom" and governs prognostications concerning the northwest frontier. Mercury, in contrast, had been east of the Sun through most of the summer and early autumn. Thus Emperor Xuan's statements about the easterly location of all five planets and Venus's simultaneous appearance "on high" could only have been made after about mid-November, by which time Mercury had switched horizons from west to east and begun to appear in the predawn sky. It follows, therefore, that Emperor Xuan's letter was written after mid-November, at least seven months after Zhao Chongguo was dispatched to the far frontier, time enough for the emperor to become more than a little anxious about the cost of Zhao's mission and its outcome.

Given the circumstances, "appear in the east" and "disposed in mid-sky" must refer to the planets' predawn appearance "strung like pearls" across the *yang* half of the sky. Emperor Xuan clearly did not mean to say that the planets were located precisely in the seven lodges (Vir–Sgr) associated with the eastern quadrant of the heavens, but rather, as the new astrological paradigm in the "Treatise on the Celestial Offices" says, the eastern and southern palaces of the heavens belong to China as *yang*, while the western and northern palaces as *yin* belong to "foreign kingdoms that use weapons."[24] Thus the astrological circumstances could hardly have been more propitious for China.

Another indication of the role of astral prognostication is the attitude toward the planetary omens displayed by the principal actors in this exchange. Emperor Xuan, ensconced in his palace far from the events on the ground, is more inclined to give weight to theoretical astrological considerations than is his veteran commander in the field, Zhao Chongguo. One gains the impression from the emperor's letter that he thought the exceptional planetary omen would be sufficient to convince the general to attack promptly so as not to lose his unique advantage. Coming at the very end of the letter, mention of the planetary omen seems intended to clinch Emperor Xuan's admonition to act expeditiously. Zhao Chongguo's response maintains a diplomatic silence in regard to the astral omen. As an expert on border affairs and a veteran of numerous campaigns in Central Asia, from Zhao's tone it is clear that he values patient diplomatic maneuvering and carefully planned military strategy higher than planetary astrology. This should come as no surprise, though perhaps a less capable general than Zhao would have had more difficulty persuading the emperor to allow him a free hand. In any case, we have here clear documentary evidence that, given the occurrence of a significant portent,[25] planetary astrology was given the most serious consideration at the highest level in the early empire, even if in this instance the astral omens were trumped on the ground by superior tactics and generalship.

---

[24] *Shiji*, 27.1328; Appendix 1.
[25] As it happens, the *Han shu* also records the appearance of a comet in the eastern quarter in the sixth month of 61 BCE, though the astrological interpretation of this apparition is not mentioned by Emperor Xuan, presumably because comets were generally seen as inauspicious. Dubs (1938–55, Vol. 2, 241).

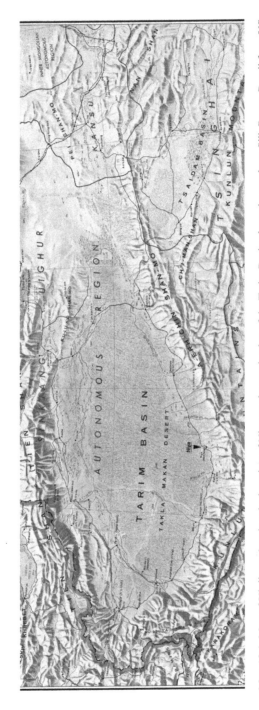

Map 10.1 Map of Xinjiang showing the location of Niya on the south edge of the Tarim Basin along the southern Silk Route. Detail from US Central Intelligence Agency, *Atlas of China*, October 1967.

An archaeological confirmation

In 1995, one of the most surprising archaeological discoveries in recent years drew attention once again to the ancient oasis settlement of Niya in the Takla-makan desert (Map 10.1). Located on the southern branch of the Silk Route in the shadow of the Kunlun Mountains, and buried by shifting sands since about the end of the fourth century, Niya was the westernmost settlement in the small desert kingdom of Loulan on the north shore of Lop Nor. Niya became famous after Sir Aurel Stein publicized the results of his 1901 expedition to Xinjiang (followed by those of 1906, 1913, and 1931).[26] Along with his contemporary, the Swedish explorer Sven Hedin, Stein reported the discovery of a confluence of diverse cultural influences – Buddhist, Chinese, Hellenistic, Iranian – in these remote regions of Central Asia located precisely at the interface between the Xiongnu, Kushan, and Han empires. The reported traces of long-buried Buddhist desert kingdoms caused a sensation. Subsequent expeditions led to the discovery of caches of numerous documents written on wooden strips, mainly in Karosthi and, of course, Stein's and Paul Pelliot's recovery of thou-sands of medieval manuscripts from one of the Caves of a Thousand Buddhas in Dunhuang.

After a long hiatus, further discoveries of the ruins of Han Dynasty agri-cultural garrisons in nearby Cherchen, Kroraina (Loulan), and other sites, including Han administrative records on wooden strips dating from as early as 49–48 BCE, revealed a good deal about the organization of Chinese garrison colonies and the lives of soldiers stationed in the area.[27] Joint Sino-Japanese archaeological expeditions resumed in earnest in the 1980s and continued for several seasons despite the extremely challenging working conditions. These efforts culminated in 1995 with the excavation of a Niya burial ground that once again demonstrated the unique mix of ancient cultural influences in the Tarim Basin and the remarkable degree of preservation of artifacts buried in the desert soil.[28] Lying in the tomb of a beautifully dressed Europoid couple, excavators

---

[26] Stein (1980; 1981; 1990). The Chinese documents found by Stein were later studied and published by Henri Maspero (1953, 169–252) and Édouard Chavannes (1913a, 721–950), Whitfield (2004).

[27] Yü (1986, 420). Chinese materials excavated by Stein and Hedin, though somewhat later, were also directly related to the activities of the Chinese troops stationed in Loulan and present a picture of a self-contained community which hired local residents on occasion as the need arose. Brough (1965, 582–612, esp. 605).

[28] For a survey of the history of exploration in the Niya region, see Wang Binghua (1997) and Wang Yue (1997). For a detailed analysis of the composition and unusual weaving technique exhibited by the silk brocade, consult Yu Zhiyong (1998, 187–8, 194). The consensus of Chinese experts is that this piece and others like it found at Niya probably made their way there as gifts from rulers in north central China. Yang Boda and specialists at the Suzhou Silk Museum identified the silk brocade remnant from which the armguard was fashioned as having originally come from Sichuan.

Figure 10.3 Eastern Han Silk brocade bowman's armguard bearing the legend "when the Five Planets appear in the East it is beneficial for China," found in a grave at Niya in the Taklamakan desert. Note the prominently featured iconic creatures tiger, *qilin* ("unicorn"), and quail in the curious sequence south–north–west; only the dragon is missing. After Pankenier (2000), reproduced by permission.

found an Eastern Han silk brocade artifact whose striking multicolored motifs and rare state of preservation made it one of the ten most important archaeological discoveries of 1995 (Figure 10.3). Not only were the colors still fresh and bright, but woven into the decorative pattern of this unique textile remnant, now recognized as a bowman's armguard, is also the remarkable legend, "when the Five Planets appear in the east it is beneficial for China."[29]

Given the remarkable coincidence between the historical record and the Niya "Five Planets brocade," it is tempting to speculate about a possible relationship between the Niya armguard and General Zhao's campaign in the very region

---

[29] It is noteworthy, however, that the legend on the bowman's armguard is actually truncated. The fortuitous discovery of a second, intact piece of the same brocade shows that the epigram originally read, "When the Five Planets appear in the east it is beneficial for China *to suppress the Southern Qiang*." See www.xjww.com.cn/article/article.php?articleid=1868 – top.

where the Han Chinese contended for *Lebensraum* with border peoples like the Western Qiang and Xiongnu. The area at risk in 61 BCE included Dunhuang, and the Han authorities recruited several thousand mercenaries for the campaign from among the non-belligerent ethnic groups living in the general area, including the Lesser Yuezhi and other Qiang tribes like the Ruo (var. Chuo) Qiang.[30] We know nothing about the ethnicity of the tall fair-haired bowman from Niya buried with the Five Planets brocade, bow, and arrows, except that he was clearly Europoid. Linguistic evidence provided by the names of a thousand residents of nearby Loulan suggests they were Tocharians (i.e. Indo-Europeans). Whether the bowman actually understood the significance of the epigram on his armguard is, perhaps, doubtful, although whoever abbreviated the epigram may well have. The various brocade artifacts found in Niya tombs were definitely not made to order but were fashioned locally from textile remnants or trade goods. The current consensus among archaeologists is that the colorful examples found at Niya are typical late Eastern Han products, in which case they would have been produced long after Zhao Chongguo's campaign. All that we may fairly conclude, in the absence of more precise dating of the tomb, is that the only connection between the Niya brocade armguard and General Zhao's campaign is the astrological epigram and the target of the suppression.[31]

Conclusion

We have traced the development of thinking about planetary influences with the help of theoretical discussions in authoritative sources and practical application in specific historical circumstances. This latest stage in the development of planetary astrology in the Han represents a reformulation of the earlier Warring States astrological theory, adapting the former preoccupation with a multivalent world to the circumstances of the early empire with its binary "us-versus-them" view of contemporary power relations. At the same time, by comparison with the situation in early Zhou and Warring States times, thinking in astrological terms appears also to have moved "downscale." Judging from the appearance of the epigram in the Niya brocade, what began as an esoteric, even hermetic,

---

[30] The Chinese considered the people of the Shanshan kingdom, whom they called the Lesser Yuezhi, to be related to the inhabitants of the Kushan Empire to the west, whom they called the Greater Yuezhi. On Ruo Qiang collaboration with the Han in punitive campaigns against other Qiang tribes, see Twitchett and Loewe (1986, 425). At the time the Ruo Qiang inhabited the region southeast of the Cherchen River and present-day Ruoqiang, about midway between Dunhuang and Niya. Tan Qixiang (1982, Vol. 2, 37–8). On the numerous mummified bodies from as early as 2000 BCE unearthed in the region, see Mallory and Mair (2000).

[31] Burrow (1935). Although these documents date from some two centuries later, given the identical burial customs and grave goods present in the tombs, it is likely that the Han period population of the area was much the same.

science of astral prognostication enjoyed such widespread popularity by the second to third centuries CE that one of its central maxims could even become an element in the decorative repertoire of makers of luxury goods. One could hardly ask for more eloquent testimony to the ubiquity of astrological conceptions in the early empire.

Part Five

One with the sky

11 Cosmic capitals

In Part Two we examined the role of astronomical alignment in shaping the built environment, showing that centuries before establishment of the empire in 221 BCE the Chinese had already developed practical applications of astronomical knowledge capable of establishing cardinal orientation. The general principles outlined so far had broad application in the theory and practice of geomancy or "siting" (*fengshui*). As Steven Bennett has shown,

the construction of reality according to a celestial terrestrial axis not only had implications for astrology and astronomy, but served as a paradigm for siting. The middle chapter of the "Vivid bag classic," *Ch'ing nang jing*, contains a passage that is clearly an adaptation of the above. The revised form has served as the standard explanation of the cosmic nature of land shapes: "In the sky there are Five Planets, on earth there are Five Phases. The sky is divided into asterisms and lunar lodges, the earth is arranged into hills and streams. *Ch'i* flows on the earth. The shapes [of the constellations] are affixed to the sky. We depend on the shapes to determine the *ch'i*, so as to establish the human order." If the siting expert can find a set of land forms which corresponds to the shapes and proportions dictated by the heavenly bodies, then the state truly will be an expression of the ideal celestial order. Since similar things resonate with each other, what could bring greater well-being than a site harmonized with the very structure of heaven? To fulfill his task the siting specialist must know the proper celestial forms. It is not surprising, therefore, that much of the literature is devoted to this topic.[1]

The cosmological identification of the imperial center with the celestial Pole and an intense focus on the circumpolar skyscape are manifested in the symbolic orientation of early imperial capitals. Conscious imitation of the celestial patterns is emblematic of the skyward orientation of rulership in China from earliest times, ultimately gaining physical expression in the imperial capital. Here we look more closely at the replication of the skyscape in the first imperial capitals.

The Qin Dynasty (221–206 BCE) cosmic capital – Xianyang

In 221 BCE the First Emperor of Qin succeeded in annihilating the last of the warring states and establishing an empire of "all under Heaven." In addition to

---

[1] Bennett (1978, 16).

arrogating to himself the quasi-divine title of *Di*, "Lord," Qin Shihuang went to great lengths also to solidify his symbolic role as August Lord by reinstituting ancient traditions and rituals, including royal progresses, reminiscent of the exploits of the legendary Yu the Great and the Zhou Dynasty founders.[2] Among the most extraordinary of his undertakings was an ambitious building program, which in addition to creating an extensive network of roads and defensive walls and hundreds of imperial palaces, also included the reconstruction of the palaces of conquered states in the Qin capital and the relocation there of their wealthy elites.

Of particular interest to us here is the general plan of the Qin capital, Xianyang. The "Basic Annals of the First Emperor of Qin" in Sima Qian's *The Grand Scribe's Records* (*c*.100 BCE) describes the layout of the imperial center:

> In his [the First Emperor of Qin's] twenty-seventh year (220 B.C.)... He built the Temporary Palace to the south of the Wei River. Shortly afterward, he renamed the Temporary Palace the [Celestial] Culmen Temple to symbolize the *Celestial Culmen* [~ *Pole*]. From the [Celestial Culmen] Temple a road led to Mount Li, where he built the Forehall of the Sweet Springs [Palace]. He connected it to Xianyang [Palace].[3]

Subsequently, however,

> In his thirty-fifth year (212 B.C.)... the emperor considered Xianyang too overcrowded and the palaces of the former kings too small. [He said:] "I have heard that King Wen of Zhou [d. 1050 BCE] located his capital at Feng, King Wu [r. 1049–1045 BCE] had his at Hao. The area between Feng and Hao is [suitable to be] the capital of an emperor. Thereupon he laid out and started to build audience halls on the south bank of the Wei River in the Shanglin Park. He began with the *E-pang* Forehall of the Palace, which was five hundred paces [*bu*] from east to west and fifty spans [*zhang*] from north to south. Ten thousand people could be seated on its terraces, and below, flagpoles fifty spans high were erected. Running around the Forehall, elevated passageways reached straight to South Mountain. The summit of South Mountain was designated the [capital] gate-tower. He had a covered causeway built from the *E-pang* [Forehall] across the Wei River to connect it to Xianyang [Palace] in imitation of the *Screened Causeway* [*Gedao*, asterism in Cassiopeia], which extends from the *Celestial Pole* across the [*Sky*] *River* to connect with [lodge] *Align-the-Hall* (*Yingshi*) (Figure 11.1a–b).[4]

---

[2] The First Emperor's titling himself *Di* was not, in fact, unprecedented. His great-grandfather, King Zhao Xiang (r. 306–251 BCE), had previously proclaimed himself "Emperor of the West" (*Xi Di*) in 288 BCE. *Shiji*, 5.212.

[3] Nienhauser et al. (1994, 138, trans. modified); Schafer (1977, 260). The Temporary Palace was initially where the First Emperor held court and conducted state ceremonies. After it became the Northern Culmen Temple it was used for sacrifices, in particular to Heaven. Shi Xingbang (1993, 111); Xu Weimin (2000, 137).

[4] Xu Weimin (2000, 148, trans. modified). Shanglin Park refers to a vast parkland enclosing fourteen square kilometers, containing some nineteen Qin palaces. It was later expanded during Sima Qian's own lifetime and filled with exotica by Emperor Wu of Han (r. 141–87 BCE). Shi Xingbang (1993, 111) confirms the existence of the causeways.

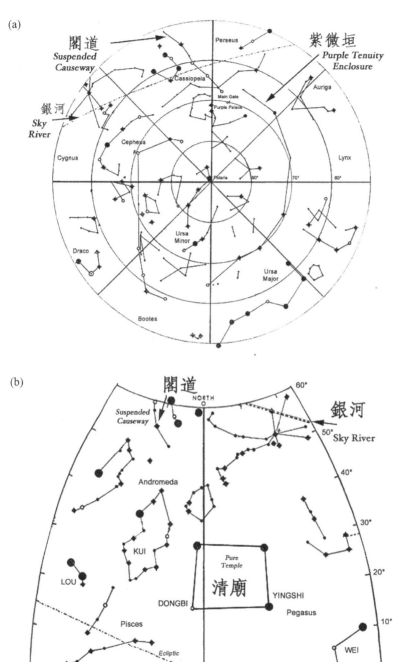

Figure 11.1 Star charts showing the *Screened Causeway* leading (a) from the *Purple Tenuity Enclosure* surrounding the Pole across the *Sky River*, (b) down to the *Celestial (Pure) Temple (Ding* in Pegasus). Redrawn from Ho (1966).

The Screened Causeway (*Gedao* in the sky) spanned the river and permitted the emperor to move from the Xianyang Palace to the *E-pang* Palace in secret. The *E-pang* was the largest palatial complex ever built. Begun in 212 BCE, only the *E-pang* ("raised"?) Forehall of the enormous complex was actually completed before the First Emperor's death. The Han period dimensions given by Sima Qian correspond to 693 spans long by 116.5 paces wide.[5] Modern studies of the actual ruins of the foundation show that its rammed-earth platform actually measures 1,320 meters east to west, 420 meters north to south, and 8 meters in height, very close to Sima Qian's figures.[6] In the above accounts we see how in relation to the layout of the imperial capital the Wei River is transformed into the terrestrial analog of the *Sky River* (Milky Way), just as earlier the Yellow River was the terrestrial analog of the Milky Way in the Warring States period (403–221 BCE) "field-allocation" scheme of astral–terrestrial correspondences which prevailed in the Central Plain states.[7] With remarkable hubris, the First Emperor showed no compunction in identifying his now cosmicized imperial capital, Xianyang, with the celestial Pole, both terminologically and in mimicking the communication between the Pole and the *Pure Temple* via the *Screened Causeway*. The vast *E-pang* Palace, truly monumental in conception, was identified with the prominent asterism *Align-the-Hall* (i.e. *Ding*, lodges 13–14) comprising the Great Square of Pegasus.[8] This cosmological parallelism, evoking the celestial correlate of the imperial charisma at the Pole, became fundamental in the dynastic ideology of the Chinese empire.

The *Yellow Plans of the Three Capital Commanderies* (*Sanfu huangtu*, third century, with some later interpolations), a widely circulated text compiled from Han sources and frequently quoted down through the Song Dynasty (960–1279), confirms that this astral–terrestrial correspondence was commonly understood at the time. For example, Zhang Shoujie (fl. *c*.725) quotes from the *Sanfu huangtu*:

The *Yellow Plans of the Three Capital Commanderies* says, "When Qin first unified the sub-celestial realm [the Emperor] made Xianyang his capital. Because he laid out his palace and audience hall on North Hill, the [circumpolar] *Palace of Purple Tenuity* resembled the Emperor's Palace. The Wei River ran through the capital, simulating the

---

[5] In the Western Han one pace (*bu*) measured about 1.38 meters and one span (*zhang*) 2.3 meters.
[6] Xu Weimin (2000, 134); Steinhardt (1999, 50–3). For the psychology of such monumentality in the Qin dynasty, see Wu Hung (1997, 108 ff.). The volume of fill it would take to construct the foundation, discounting compaction, amounts to about three times the volume of the Great Pyramid of Giza.
[7] For the conceptual basis of "field-allocation" astral prognostication, see below.
[8] *Align-the-Hall* commonly referred to the entire Square of Pegasus (*Ding*), and not merely the western side of the square defined by β and α Pegasi (Scheat and Markab), which later became the convention.

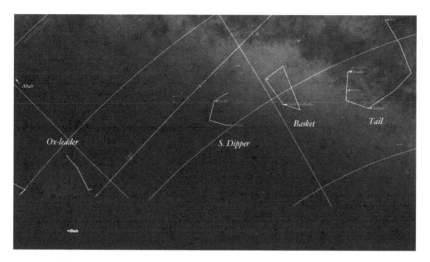

Figure 11.2 Location of lodge *Ox-Leader* marking the intersection of the ecliptic, equator, and *Sky River* where Sun, Moon, and Five Planets cross into the winter sky (Starry Night Pro 6.4.3).

Sky River, and the Transverse Bridge crossed [the Wei River] to the south, on the model of [lodge] *Ox-Leader* [β Cap]."[9]

The *Yuanhe junxian tuzhi* describes the Transverse Bridge in more detail:

The Mid-Wei Bridge was twenty-two *li* southeast of [Xianyang] county seat. Originally called Transverse Bridge, it was built across the Wei River. When the First Emperor made Xianyang his capital the Wei River ran through it like the *Sky River*. The Transverse Bridge crossed to the south, on the model of [lodge] *Ox-Leader*. South of the Wei River was the Palace of Enduring Happiness, and north of the river was the Xianyang Palace. Desiring to [facilitate] communication between the two, [the Emperor] built this bridge.[10]

The stars of *Ox-Leader* in Capricorn are on the opposite side of the Milky Way from the Pole and mark where the Sun, Moon and Five Planets ford the *Sky River* (Figure 11.2). The suspended gallery known as the Screened Causeway, analog of the eponymous asterism *Gedao* in Cassiopeia, was thus not the only such causeway across the Wei River. From the perspective of cosmography it is significant that both *Ox-Leader* and *Align-the-Hall* were represented. To reach

---

[9] *Sanfu huangtu* (quoted in Zhang Shoujie's *Shiji zhengyi, Shiji,* 86.2535). Note the revealing phrasing – "the [circumpolar] *Palace of Purple Tenuity* resembled the Emperor's Palace."
[10] Quoted in Qu Yingjie (1991, 200).

either of the two, nearly at opposite ends of the sky, one must cross the *Sky River* on different bridges.

Taiping yulan quotes the *Yellow Plans* as saying that gates were built on all four sides of the cardinally oriented palaces in imitation of the circumpolar *Purple Palace Enclosure (Palace of Purple Tenuity)*.[11] Therefore, given the foregoing we have it on the best authority that the identification of imperial Xianyang with the celestial Pole was in the minds of the First Emperor and his architects.[12] Likewise, given the First Emperor's fixation with the precedent-setting nature of his universal empire and the monumentality of its physical expression, together with his assumption of the title *Di* previously reserved for the Supernal Lord, the cosmographic parallels he took pains to reify were clearly more than merely figurative. From North Hill on the north bank of the Wei River to South Mountain on the south, from the Transverse Bridge in the east, past the Mid-Wei Bridge to the Screened Causeway on the west, he created a simulacrum of the heavens, just as he is reputed to have done in the colossal tomb he was building for himself. As Stanley Tambiah has remarked on the "galactic polity" model of state organization, "one of the principal implications of the cosmological model is that the center ideologically represents the totality and embodies the utility of the whole."[13]

Between Western Zhou (1046–771 BCE) and the founding of the Qin empire the picture remains somewhat confused, and confusing. A vast amount of new archaeological information has emerged since Paul Wheatley's pioneering 1971 study of ancient Chinese urbanism, but the data on cardinal alignment and especially its regional diversity have yet to be systematically compiled and analyzed.[14] Palaces and capitals from the earliest period have only sporadically been identified, but there do exist notable examples of precise cardinal orientation, such as the Eastern Zhou (eighth–seventh centuries BCE) royal city of Wangcheng.[15] It is worth reiterating that, as we saw in Chapter 3, significant changes in cardinal alignment can be strongly indicative of major political or cultural transitions. This was clearly so when the Western Zhou Dynasty devolved power on the rising state of Qin as Protector of the West in the mid eighth century BCE. As Lothar von Falkenhausen has observed,

---

[11] *Taiping yulan*, 73.3b; Sun and Kistemaker (1997, 70). Cf. Shi Xingbang (1993, 111), whose identifications of *Purple Palace* and *Screened Causeway (Gedao)* are not entirely correct. The *Screened Causeway (Gedao)* comprises four stars, not one, and leads across the *Sky River* from the main gate of the *Purple Palace* (Figure 11.1a). Xu Weimin (2000) follows Shi. Wang Xueli's (1999, 150) discussion of the astral–terrestrial correspondences, in contrast, is correct.

[12] For a description of Emperor Wu's (r. 141–87 BCE) similar ambitions in the Western Han dynasty, see Lewis (2006, 178; 2007, 94), who describes how the emperor relocated the cosmic center from the capital itself to the Shanglin Park.

[13] Tambiah (1985, 266).   [14] Wheatley (1971).   [15] Von Falkenhausen (2006, 172).

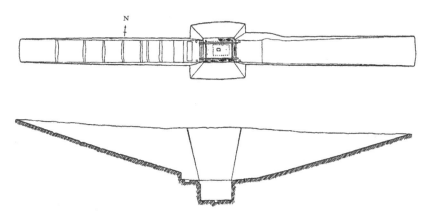

Figure 11.3 The large early Qin ducal tomb M2 (eighth–seventh century BCE) aligned east to west, recently excavated at Dabuzishan (the arrow indicates magnetic north). Redrawn from Dai Chunyang (2000, 75, Figure 2).

Qin tombs differ in two respects from Eastern Zhou-period tombs elsewhere in the Zhou culture sphere: they are overwhelmingly oriented east–west rather than north–south, and they feature flexed rather than extended burial. These idiosyncracies have been taken as markers of an alien ethnic identity of the Qin people. And indeed it is impressive to observe how the predominant tomb orientation at central Shaanxi cemeteries suddenly shifted by 90 degrees at the transition from Western to Eastern Zhou, when the Qin took over the area from the royal Zhou.[16]

This tradition must be very old, because the same custom is observable as early as the Qin Dabuzishan burials from early Western Zhou. So far as I am aware, no mention has been made of the significance of the precise orientation of the very imposing ducal tombs found there, but a published illustration (Figure 11.3) clearly displays the east–west axial alignment.[17] The old Qin capitals, on the other hand, exhibit a marked preference for alignment towards the northwest. Yongcheng served as the capital of the state of Qin for 294 years, from 677–383 BCE, followed by Yueyang from 383 to 350. Excavation of the surviving walls of Yongcheng at Fengxiang xian in Shaanxi indicate that its main gate faced west and its main axis was aligned west of north. The same orientation is observable at Yueyang. Both are aligned some thirteen degrees west of north.[18]

---

[16] Ibid., 215.
[17] Ibid., 75, Figure 2. Dai Chunyang (2000, 79). See also http://longnan.cncn.com/jingdian/dabaozishan/profile.html.
[18] Qu Yingjie (1991, 194).

Celestial resonances of the name Xianyang – a conjecture

The origin of the name of the Qin capital, Xianyang, is obscure. A traditional gloss asserts that the name derives from the fact that the city lay south of Mt. Jiujun and on the north bank of the Wei River (*xian* can mean "all").[19] In fact, however, Jiujunshan (site of Tang Dynasty tombs) lies at least fifty kilometers northwest of Xianyang, seemingly too far for Xianyang to qualify as lying on its south-facing *yang* slope. As with so many traditional etymologies, this could be a late rationalization, perhaps from the Tang when the mountain seems to have played a larger role in geomancy, as evidenced by the placement of the Zhao ling imperial tomb and others. Toponyms of the type "X-*yang*" typically do refer to a location on the south-facing or *yang* bank of river "X," of course, as for instance in the case of the Eastern Zhou capital, Luoyang, and innumerable other places. Initially, Xianyang was certainly located on the north bank of the Wei River, so one might reasonably expect it to be have retained the former name of the place, Weiyang.

Above we saw that "from the [Northern] Culmen Temple a road led to Mount Li, where [the First Emperor] built the Sweet Springs Forehall." This road or causeway must have been some forty kilometers long. Mount Li is the location of the famous Huaqing Hot Springs (*Huaqing chi*), whose mineral-rich waters have soothed tired rulers' bodies since Zhou king You (d. 771 BCE) reputedly built the Li Palace there 2,800 years ago. This was also where Emperor Xuanzong (685–762) of the Tang Dynasty first caught sight of the beautiful Yang Guifei and gave the spring its present name. Besides meaning "all, complete," of course, *xian* also means "salty," so that, for example, a *xian hu* is a salt lake. *Gan*, on the other hand, means "sweet" not "salty," but the seeming contradiction is easily resolved. Traditionally, savvy peasants would taste the soil or water to gauge its pH; if there was a sweet taste or smell it was alkaline. Therefore *gan* in this context actually refers to alkalinity. The "Sweet Springs Palace" (*Ganquan gong*) of Qin time is probably an allusion to the quality of the abundant spring water emanating from various locations on Li Shan, and a possible origin of the name of the capital, Xianyang.[20] Can we further substantiate this in any way?

We have seen evidence of the First Emperor's predilection for mimicking the heavens and ancient Qin self-identification with the west in Gansu, where it had its beginnings. Perhaps there are also cosmographic resonances for the name Xianyang. In the "Treatise on the Celestial Offices," Sima Qian says, "the *Purple Palace, Chamber* (lodge 4) and *Heart* (5), *Weight and Balance Beam*

---

[19] Ibid., 195

[20] For the Sweet Springs Palace's location directly opposite the Xianyang Palace, see Xu Weimin (2000, 127 ff.).

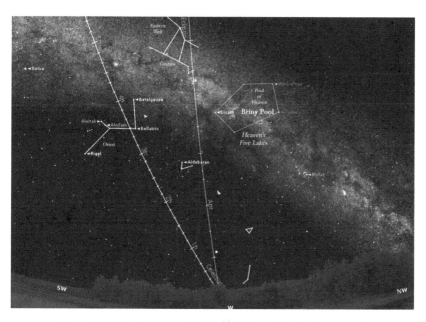

Figure 11.4 The location of *Xianchi* or *"Mineral Spring"* (aka *Pool of Heaven*) source of the *Sky River* (Starry Night Pro 6.4.3).

(25–6), the *Mineral Spring (Xianchi)*, *Ruins* (11), and *Roof* (12), are *Sectional Asterisms* within the array of lodges; these are the seats of Heaven's Five Offices."[21] In other words, these five locations on the celestial equator mark the cardinal palaces of the heavens, the *Palace of Purple Tenuity* representing the circumpolar center, while the *Mineral Spring* represented cardinal direction West. The "Treatise" further states, "the *Mineral Spring* of the *Western Palace* is said to be [within] *Heaven's Five Lakes* [*tian wu huang*]."[22] In our terms, the *Mineral Spring* in Auriga is located in the Milky Way between Orion and Gemini, where the ecliptic intersects the *Sky River* (Figure 11.4). In the *Huainanzi*, "Heavenly Patterns" (*Tianwen*) chapter, and in the much older Chu poem "Encountering Sorrow" (*Lisao*, late fourth century BCE), the *Mineral Spring* in the west is where the Sun god bathed after completing his journey

---

[21] *Shiji*, 27.1350.

[22] *Shiji* 27.1304. The *Pool* is marked by λ, μ, ρ Aur. *Heaven's Five Lakes* is the ring of five stars in Auriga that surrounds the *Mineral Spring*, sometimes identified as the Emperor's five carriages. Sun and Kistemaker (1997, 179) and later sources identify the *Spring* with three stars in Auriga. See the verses from the poem "Dragon Pool," *Long chi pian*, by the early Tang poet Shen Quanqi (*c*.650–729), quoted in Chapter 13 below (see also Figure 13.7).

across the sky.[23] (Less reliably, other sources say *Mineral Spring* is the name of a musical piece attributed to the Yellow Emperor or Emperor Yao.)

In Chapter 9 we saw how the Yellow River was the analog of the Milky Way in that old scheme of astral–terrestrial correspondences (Maps 9.2 and 9.3).[24] In redesigning Xianyang as his "cosmic capital," the First Emperor obviously repurposed that pre-existing Warring States design by substituting the Wei River for the Yellow River. He was, of course, emulating Zhou king Cheng's establishment of his new capital, Chengzhou, at Luo in early Western Zhou (*c*.1036 BCE) "from which central region to govern the people," to paraphrase the contemporaneous *He zun* inscription. Remarkably, Qin Shihuang not only substituted the Wei River for the Yellow River as archetypal correlate of the *Sky River*, he also redefined the very idea of "center" or *axis mundi* by relocating it from "East of the Hangu Pass" (*Guandong*) to "Within the Pass" (*Guanzhong*)!

The Qin kingdom, of course, was originally located on the western periphery of the Hua–Xia world, well to the west of the bend in the Yellow River and west of the Hangu Pass. Now, the *xian* of the celestial "*Briny*" (*Mineral*) *Pool Xianchi* is the same *xian* as in Xianyang and can simply mean "mineral-rich."[25] It is quite likely that the *Xianchi* in the sky is not "briny" after all, but alkaline, and thus more appropriately read *Jianchi* and translated *Mineral Spring*. The Huaqing Springs at Mount Li where kings bathed for centuries could thus be the terrestrial correlate of the celestial *Mineral Spring* in the west where the mythological solar deity eased his aching body at the end of the day. The name Xianyang, therefore, could have originally been read Jianyang and referred to the waters of Mount Li while simultaneously evoking the *Mineral Spring*. In fact, that *Jianchi* in the heavens is the *Sky River*'s chief source in the west. Since the First Emperor (as well as his predecessors, no doubt) was certainly aware of the Warring States astral–terrestrial correspondences, it would not be surprising if Qin earlier strongly self-identified with the West. Map 9.2 above showed Province Yong (Guanzhong) nominally located due "south" in terms of field-allocation astrology, but it must be remembered that Qin relocated to this area in the eighth century BCE from Gansu, Province Liang, nominally west-southwest. In constructing his capital the First Emperor emulated Zhou king Wen, who had been appointed Protector of the West by the Shang.

---

[23] Major (1993, 81, 94, 102).   [24] Pankenier (1998a; 2005).

[25] *Xián* *ɦɣɛm 咸, "salty" (variant 鹹), is here *jiǎn* *kɣɛm, "alkaline" 碱 (variant 鹼). The soils in the Wei Valley around Xianyang were notoriously alkaline, prompting the construction of the 125-kilometer Zheng Guo Canal (*c*.245 BCE), which made the plain north of Xianyang richly productive, contributing importantly to Qin's rapid growth. *Shiji*, "He qu shu." Alkali in various forms (e.g. borax) has widely been used for laundering and in soap making. When soap was first brought to China it was colloquially referred to as "foreign alkali" *yangjian*. This, of course, explains why Yang Guifei enjoyed bathing, and especially washing her hair, in the Huaqing Springs.

In the minds of knowledgeable counsellors like Shang Yang (395–338), the astral–terrestrial resonance of the name Xianyang surely ought to have evoked its namesake, the celestial *Mineral Spring*.[26] The conception could date from the time of Duke Xiao of Qin (381–338 BCE), who moved the capital to Xianyang in 350 BCE at the instigation of Shang Yang (aka Lord Shang of Wei, the powerful successor state of Jin). Xianyang was founded well over a century before unification, so that if this inference is correct it points to a much earlier preoccupation with astrology and cosmography at the Qin court, predating the founding of Xianyang and clearly disclosed by the layout of early Qin capitals and tombs.[27] The identification of imperial Xianyang as cosmic *Center* commenced with the establishment of the Qin Empire in 221 BCE.

Reconstruction of the actual layout of Qin-time Xianyang is complicated by the northward migration of the Wei River by hundreds of meters during the intervening twenty-two centuries.[28] We saw that the longitudinal axes of the earlier Qin capitals, Yongcheng and Yueyang, were both aligned slightly west of north. Qu Yingjie speculates that Xianyang also shared this alignment, and initially perhaps it did. But given the above evidence of polar mimicry (also displayed by Chang'an), a precise north–south alignment of the imperial capital by the First Emperor seems more likely and is apparently suggested by archaeological surveys (Figure 11.5).[29] In present-day Xi'an, Meridian Avenue (*zi wu da dao*), on the north–south line with the old Xiangyang Palace, still runs due south out of the city toward the Qinling Mountains.

The First Emperor's reshaping of the political map of the Warring States period and his hubristic cosmo-political reorientation of the imperial center necessarily entailed the abandonment of the pre-existing Yellow River = *Sky River* astrological paradigm in favor of the Wei River = *Sky River* analogy. This new orientation is further underscored by the axial north–south alignment of post-unification Qin tomb mounds, including that of the First Emperor himself.[30] Indeed, it is likely that Qin Shihuang's reconstruction at Xianyang of replicas of his defeated enemies' capitals also reflects this intention to create a simulacrum of the whole empire at the cosmic center. Perhaps the only surprising feature of the new imperial capital's symbolism is the retention of the original name, Xianyang, whose likely astrological allusion to a western peripheral location beside the *Mineral Spring* would seem to be somewhat out of place in the new cosmic order. But in this, perhaps, the First Emperor thought

---

[26] The name "Mineral ~ Alkaline" for the area of Xianyang is ancient, having been found inscribed on potsherds dating from before the founding of the capital.

[27] Shi Xingbang (1993, 110–11) explains that Qin capitals had already begun to expand to the south bank of the Wei River in late Warring States times. For Warring States cosmography and its conceptual basis, see especially Dorofeeva-Lichtmann (1996).

[28] See Xu Weimin (2000, 111, Figure 14).

[29] Qu Yingjie (1991, 201); Xu Weimin (2000, 113, 135).   [30] Tian Yaqi (2003, 300).

(a)

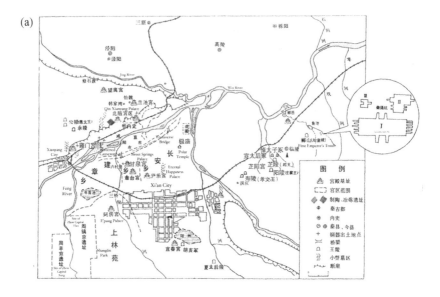

(b)

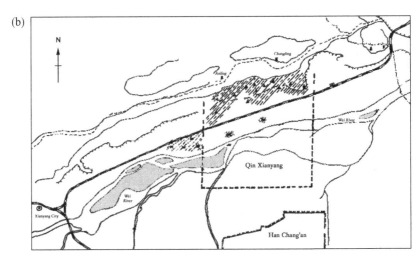

Figure 11.5 (a) Principal sites in and around Qin Xianyang. In Qin and Western Han times the bed of the Wei River lay more than two kilometers south of the river's present course. Redrawn from Wang Xueli (1999, 101). (b) Location of Chang'an in Western Han. Redrawn from Steinhardt (1999, 53, Figure 47).

to remain true to the legacy of his remarkable great-grandfather Zhaoxiang (r. 306–251 BCE), at one point called Emperor of the West (*Xi Di*). Not surprisingly, the array of imperial carriages buried beside King Zhaoxiang's massive tumulus also face west.

Cosmic congruence of time and space

But there is more, for in the First Emperor of Qin's time, in late November, the brilliant silvery ribbon of the *Sky River* arched across the sky from southwest to northeast just like its terrestrial analog, the Wei River. Imagine yourself standing on a parapet of the Xianyang Palace looking south across the Wei toward South Mountain in the distance, on the evening of the first day of the New Year. Below you the Wei River flows languidly southwest to northeast. Looking northeast and southwest, the *Sky River* and the Wei appear to meet at the horizon, creating the impression of a *single continuous stream* rising up among the stars and arching across the sky. Lowering your gaze, you see the Transverse Bridge, Mid-Wei Bridge, and Screened Causeway leading across the Wei River to the Sweet Springs, Enduring Happiness, and *E'pang* palaces. The locations of the Screened Causeway and the *E'pang* Palace perfectly match their astral counterparts, the *Screened Causeway* (Cassiopiea) leading from the Pole behind you to the *Pure Temple* in the sky (Square of Pegasus) due south. That *Pure Temple* (*Ding*), just culminating on the meridian and perpendicular to the horizon, would only at this time be capable of fulfilling its ritual function of correctly aligning structures in the sub-celestial realm with the celestial Pole, as documented in numerous pre-Qin sources (Chapter 4).[31]

Here, I suggest, we have the explanation for the Qin Dynasty's choice of precisely this month to begin the New Year. This was the unique, highly symbolic moment during the year when Heaven above and the sub-celestial realm below were joined together at opposite ends of the sky by the confluence of the terrestrial river and the *Sky River*, affording direct physical communication between Heaven above and the universal empire below (Figure 11.6).[32]

The Han Dynasty (206 BCE–220 CE) cosmic capital

Following the overthrow of the Qin Dynasty after a mere fifteen years, the Han founding emperor Gaozu (*c*.250–195 BCE) and his immediate successor, Emperor Hui (r. 195–188 BCE) were confronted in their turn with the imperative to establish the legitimacy of the Han Dynasty's assumption of the Mandate

---

[31] Ban Dawei (2008); Pankenier (2011).
[32] Yi Ding et al. (1996, 172–5); Schafer (1974, 404–5; 1977, 257–69).

Figure 11.6 The evening skyscape looking south from the capital, Xianyang, at the Qin New Year in late autumn (tenth month). The *Sky River* flowing from southwest to northeast is the analog of the Wei River (Starry Night Pro 6.4.3).

of Heaven. Three centuries later, the Eastern Han scholar Zhang Heng (78–139 CE) immortalized the Han founder's lofty ambitions in his "Rhapsody on the Western Capital" (*Xijing fu*):

When our Exalted Ancestor first entered [the Pass], the *Five Wefts* [*wuwei*, Five Planets] in mutual concord migrated into *Eastern Well* [lodge 22 in Gem] . . . Heaven opened his mind [through this sign], and . . . when it came time for the Emperor to plan [the siting of his capital], in those intentions he indeed gave thought to the spirits and powers, [and deemed] it appropriate to settle [on Chang'an] as the Celestial City.[33]

As Wu Hung points out, the Han founder Gao Zu's symbolic acts were intended to resonate with Zhou Dynasty precedents of eight centuries earlier. Although construction of the new capital was begun by Gaozu in 202 BCE with the building of the Everlasting Palace (*Weiyang gong*), the capital was not actually walled until a decade later, between 194 and 190 BCE, under Emperor Hui. Archaeological studies during the past few decades have revealed how

---

[33] For the dynastic significance of clusters of the *Five Planets*, see Chapter 3 above. *Eastern Well* was the astral correlate of the state of Qin in the field allocation astrology of the time. As Wu Hung (1997, 147) points out, the Han founder's symbolic acts were intended to resonate with Zhou precedents of a thousand years earlier.

Chang'an appeared during its first century as the Han imperial capital.[34] Some of those studies allude in passing to the tradition that the idiosyncratic shape of the city walls was inspired by the stars, but without discussing archaeological or other evidence. Forty years ago Paul Wheatley published an illustration of his interpretation of the tradition, at the same time dismissing it as unsupported by archaeological evidence.[35] Wheatley's interpretation, however tentative, is undermined by his apparent assumption that the Han Chinese perceived the same configurations of circumpolar asterisms as in the West, namely two facing dippers, the smaller emptying into the larger. In fact, only Ursa Major, the *Northern Dipper*, was seen by the Chinese to resemble either a ladle or a carriage, while the stars in Ursa Minor were represented quite differently.[36]

Detailed study by Stephen Hotaling, based on scale drawings of the archaeological excavations, suggested that the contours of the northern wall of the city did indeed reproduce the shape of the *Northern Dipper*, while the southern wall represented lunar lodge *Southern Dipper* (*Nandou*, Π Sgr), where the ecliptic intersects the Milky Way.[37] Subsequent investigations confirmed the accuracy of Hotaling's reconstruction of the configuration of Chang'an's walls.[38] In his analysis, Hotaling quotes an account from the same text cited above, the *Yellow Plans of the Three Capital Commanderies*: "The south of the city wall was in the shape of the *Southern Dipper*, the north was in the shape of the *Northern Dipper*; for this reason until now people still refer to the city wall of the Han capital as the *Dipper* (*dou*) wall."[39]

In contrast, the east wall of the city was accurately aligned on true north, while the imperial palaces inside the city, such as the famed Everlasting Palace, were rectilinear and cardinally oriented.[40] Wang Zhongshu describes the city wall's impressive dimensions:

[34] A full account can be found in Wu (1997, 143–87); also Xiong (2000, 8–11); Lewis (2007, 75–101).

[35] Wheatley (1971, Figure 26).

[36] It appears Wheatley mistook the *Southern Dipper* of the tradition as referring to Ursa Minor rather than lodge 8, *Nandou* (φ Sgr). Wheatley's error is perpetuated every time his illustration is reprinted and his opinion cited without critical discussion. See e.g. Wu (1997, 158); Steinhardt (1999, 66); Kelley and Milone (2011, 324, 326).

[37] Hotaling (1978, 39, Figure 22).

[38] Wu (1997, 150); Wang Zhongshu (1984a; 1984b). See also Liu Qingzhu (2007, 115).

[39] Hotaling (1978, 6); *Sanfu huangtu*, 1.7a–b. Cf. Arthur Wright (1977, 44), who asserts there was no astral symbolism in the "haphazard building of Chang'an." He ventures no explanation for the "*Dipper* wall" tradition in *Sanfu huangtu*. Nancy Steinhardt (1999, 65–6) infers that the walls of Chang'an were unplanned and that their irregular shape was caused by the flow of water around the site, even though the oldest maps show a sizeable artificial canal running straight through the city (ibid., 62–3). The massive water diversion project on the Min River called the Dujiang weir was constructed by Li Bing during the reign of King Zhao of Qin (324–250 BCE), and the east–west Zheng Guo Canal irrigation system just north of Xianyang (built *c*.246 BCE) is 125 kilometers long. Wang Xueli (1999, 65). Obviously, the requisite hydraulic engineering skill was not lacking had diverting the streams around Chang'an been a desideratum – other considerations must have played a role.

[40] Liu Qingzhu (2007, 116).

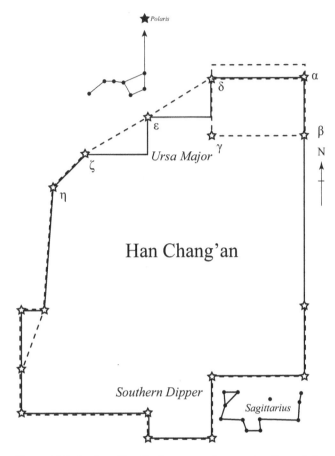

Figure 11.7 Stephen Hotaling's proposed reconstruction of the walls of Chang'an. Redrawn from Hotaling (1978, 39, Figure 22), reproduced by permission.

When archaeologists investigated Chang'an's ruins, they found that most of the wall was still exposed above ground. Many wall sections had collapsed, but underground the foundations still remained. According to the two surveys of 1957 and 1962 . . . the total length of the four walls was 25.7 km . . . The city wall, which was built of rammed yellow earth, was over 12 m high, and the width at its base was 12 to 16 m . . . Outside the city wall was a moat, about 8 m wide and 3 m deep.[41]

At the upper left in Figure 11.7 is Stephen Hotaling's small inset drawing of Ursa Major showing the stars Dubhe and Merak in the *Dipper*'s "bowl"

---

[41] Wang Zhongshu, *Handai kaoguxue gaiyao*, quoted in Wu (1997, 156).

pointing toward Polaris to the north, the present location of the celestial Pole. Hotaling argues that it was this orientation of the *Dipper* that was replicated in the outline of the north wall of Chang'an. But there are problems with this interpretation no less serious than with Wheatley's. One is that Polaris was not the Pole Star until the Ming Dynasty; in fact, the precessing celestial Pole was still several degrees away from Polaris in Han time, though in roughly the same general direction with respect to the *Dipper*. More problematical is that the *Southern Dipper* in Sagittarius (Π, λ, μ, σ, τ, ζ Sgr), whose shape the south wall of Chang'an supposedly replicated, should not lie due south directly *behind* the *Northern Dipper*. Instead it should lie well to the north of the southwesterly direction in which the "handle" portion of Chang'an's north wall points in Hotaling's reconstruction. Most problematical of all, if the design of the north wall of Chang'an had been conceived as Hotaling suggests, then the fictive Pole such a configuration implies would necessarily lie a considerable distance *outside* Chang'an to the north, much as would Kochab (β UMi), the brightest star near the Pole in Han times. But placing the celestial Pole and hence the *axis mundi* outside the walls of the imperial capital is completely at odds with the symbolic intentionality at work here, and clearly contradicts the explicit testimony concerning the precedent-setting analogy between the abode of the Supernal Lord and the First Emperor of Qin's capital, Xianyang.

Hotaling's suggested configuration is one that would result from looking up at the Pole, copying the *Dipper* on a surface, and then placing this drawing face up on the ground in order to transfer the stellar pattern. But proceeding in this fashion actually inverts the orientation of the *Dipper*, which is fine if the purpose is merely to draw a chart of the constellation. But to *exactly replicate* the stellar pattern on the ground one must place the sketch of the *Dipper* face down, as if the circumpolar stars had gently floated down to the ground. Only this procedure exactly mimics the configuration of the circumpolar sky on the ground surface, yielding a precise correspondence between the wall of the imperial capital and the *Northern Dipper* at the Pole. Thus Hotaling's insight is conceptually flawed in a crucial respect. The resulting contradictions can easily be resolved, however, if one imagines the *Dipper* emptying *into* Chang'an rather than *outside* it; that is, configured in a manner identical to its depiction on *shi* mantic astrolabes (Figure 11.8) and stone reliefs of the period such as the famous example from the Wuliang shrine (Figure 3.3). The revision proposed here would simply entail flipping the north–south positions of the pairs of "bowl" stars, Megrez and Phecda, Dubhe and Merak, with the result that the Pole and all the "imperial" stars of UMi like Kochab, "Heaven's Great August Emperor," would then lie *within* the walls of Chang'an (Figure 11.9). Admittedly, the position of the last star in the handle of the Dipper, Alkaid (η UMa), would then appear awkwardly out of place in the northwest

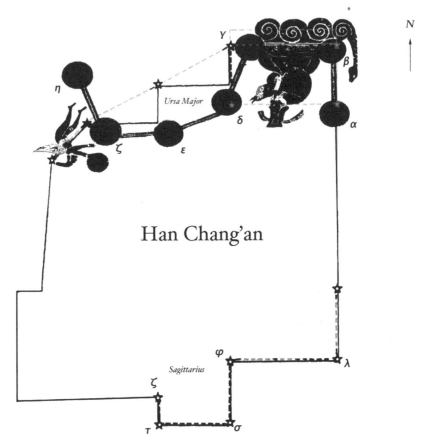

Figure 11.8 The image of the Supernal Lord's *Dipper*-carriage from the Wuliang Shrine (mid second century CE) superimposed on the north wall of the Han capital, Chang'an.

corner, but it was the reconstruction of precisely this section of the wall that posed the greatest problems, leading to Hotaling's characterization of this part as "tentative."

Significantly, this modification of Hotaling's reconstruction also resolves the seemingly problematical identification of the south wall of Chang'an with the *Southern Dipper* (Π Sgr), because now the *Southern Dipper*'s location vis-à-vis the north wall's *Northern Dipper* corresponds to its true position in the sky. (On the Han instrument shown in Figure 11.9, lodge *Southern Dipper* is located between the seven and eight o'clock positions on the round *Dipper*

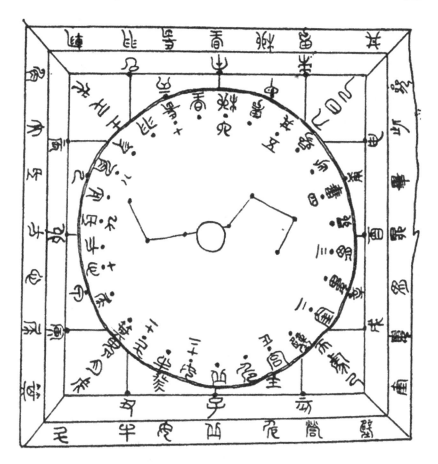

Figure 11.9 Early Han mantic astrolabe from the tomb of the Marquis of Ru Yin (*c*.168 BCE) with the *Dipper* at the center of the round rotating Heaven Plate. After Loewe and Shaughnessy (1999, 840, Figure 12.5), © Cambridge University Press, reproduced by permission.

dial.)[42] Our proposed configuration explains the curious fact, which confounded Hotaling, that the moat along the south wall of Chang'an is said to have pierced the "scoop" of the *Southern Dipper* where it protruded from the wall and

---

[42] For the ancient Chinese practice of keying lodges to the stars of the *Dipper*, in this case linking the determinative star of *Southern Dipper* (φ Sgr) to Alioth (ε UMa), see Needham and Wang (1959, 233); Feng Shi (2007, 276). The opposition between the *Northern* and *Southern Dippers* figures importantly in a Koguryo tomb as well. Kim Il-gwon (2005, 26).

contained the gate through which the main north–south thoroughfare passed. Given the precedent established by the First Emperor of Qin, who exploited the Wei River's course so that it would actually bisect the capital south of his Xianyang palace, and given the fact of the *Southern Dipper*'s actual location in Sagittarius *within* the "silvery river" of the Milky Way, this curious feature of Chang'an's south wall now also fits the pattern, as does the artificial canal to which it would have been linked.

What could be the rationale for representing the *Dipper* as a mirror image of the figure we are accustomed to seeing in the sky? It is questionable whether the diviners who made astrological devices like the mantic astrolabe, or the engineers who built Chang'an's walls, could have imagined themselves looking down on the Pole from a vantage point in the realm of the formless beyond; that is, by imagining themselves standing in the void *outside* the cosmos "looking over the shoulder" of the Supernal Lord.[43] Such a perspective is not found until celestial globes (planispheres) appear much later, and star maps from the Tang (618–907) and Song (960–1279) dynasties do not depict the *Dipper* this way.[44] Some other principle must be at work. The rationale would seem to be that, as the *Appended Commentary to the Book of Changes* says, the ancients, "looking up, took the images from Heaven," and meticulously brought them down unmediated to earth. That is, the architects of Chang'an were not about mapping the circumpolar sky in the modern sense, but about creating a simulacrum of the celestial Pole so that it would magically embody the numinous power of the cosmic center.[45]

---

[43] Earlier I shared the opinion of Harper (1999, 839) that the mantic astrolabe represents the *Dipper* "as seen from above the dome of Heaven." On further reflection I now find that improbable. For an interesting Han account of the use of the mantic astrolabe, see Dubs (1938–55, Vol. 3, 463–4). Interestingly, examples of the *Dipper* in reverse also appear in Koguryo tombs in Korea, for example in the Tokhung ri tomb from 408 CE. Kim Il-gwon (2005, 26).

[44] Sun and Kistemaker (1997, 28, 91). In other contemporaneous contexts the *Dipper* was drawn precisely as it appears in the sky. See, for example, the "*Dipper Cloud*," *beidou yun*, depicted on the Mawangdui MS "Diverse Prognostications on Heavenly Patterns and Formations of *Materia Vitalis*," *Tianwen qixiang za zhan*, from 168 BCE. Csikszentmihalyi (2006, 174); Harper (1999, 844).

[45] Nathan Sivin and Gari Ledyard (1995, 30) make a similar point about the technique employed by Chinese cartographers: "Their essentially numerical approach to astronomical prediction did not oblige them to decide whether the earth was flat, discoidal, or spherical. It is not precisely that mapmakers were convinced the earth was flat: that question did not arise in connection with their work. They simply acted as if they were transferring points from a very large flat surface to a smaller one." John B. Henderson's chapter, "Chinese Cosmographical Thought: The High Intellectual Tradition," does not mention Hotaling's study. Instead, Henderson cites Wright: "cosmological conceptions apparently had little influence on the general shape of Chinese cities until at least the end of the Han era. According to Wright [1977, 42–4], early Chinese cities seem to have been irregular and asymmetrical in form. This was true even of the first great imperial capital in Chinese history, Chang'an in the Former Han era (206 B.C.–A.D. 8)." Henderson (1995, 210). In light of the above, such misapprehensions are in need of revision.

A Neolithic precedent?

In Part One above I mentioned one of the most extraordinary archaeological finds in China in recent memory: the 1987 discovery of a "cosmologically" inspired Yangshao culture burial at Puyang, Xishuipo, in Henan Province (35.7 degrees north, 115 degrees east), dating from the late fourth millennium BCE.[a] The elite individual buried in the center of the unprecedentedly large, cardinally oriented grave M45, thought to be a tribal chief or cosmo-priest, is aligned with his feet to the north and head to the south, and accompanied by three youthful sacrificial victims. What makes the find sensational, however, are the mussel shell mosaics carefully laid out beside the chief occupant of the tomb – a dragon to his east, a tiger to his west, and a third figure comprising a mosaic triangle and two human tibias at his feet in the north (Figure 11.10).[b]

Similar, if less elaborately arranged, mosaics have been found in associated burials several meters due south, including one containing the skeleton of an individual whose lower legs are missing and presumed to be those found in the main burial. Another adds mosaics of a deer and a spider accompanying the dragon and tiger.[c] Still another depicts a human figure sitting astride a dragon. These Yangshao burials have excited great interest because of the cosmological associations evoked by the animal "familiars" accompanying the tomb occupant. This is so because the cardinal directions associated with the dragon and tiger are precisely the same as in the correlative cosmology of some 3,000 years later, with the *Cerulean Dragon* to the east (the constellation comprising Vir–Sco) and *White Tiger* to the west (corresponding to And–Ori). At that time the *Dragon* was associated with spring and the *Tiger* with autumn, although their seasonal appearances are retarded by one season today. The third smaller mosaic in the north has been taken to be a representation of the *Northern Dipper*, which was then nearer the Pole than at any time since. Not surprisingly, the whole composition has been interpreted by some as a map of the heavens. Inevitably, that interpretation has occasioned much debate. Despite the very early date, some see in tomb M45 an actual sky map, others a pictorial representation of a cosmological worldview, some simply a "shamanic" burial.

Here we will simply consider some salient facts about the orientations of items in the tomb and their cosmological associations. There is no denying that the head-to-head and tail-to-tail placements of the iconic figures of dragon and tiger, their creaturely identities, as well as the cardinal arrangement of all three images, are consistent with the well-documented conventional cosmological correlations of the late first millennium BCE. Both

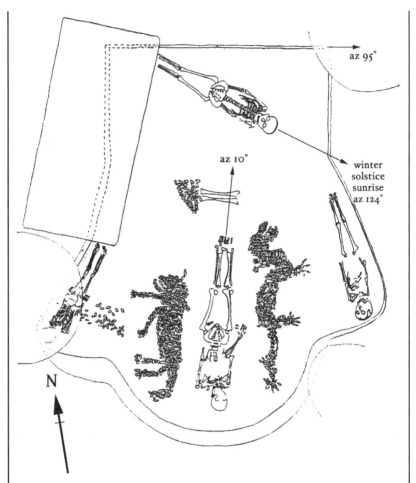

Figure 11.10 The fourth-millennium BCE Puyang "cosmo-priest's" tomb. The north arrow shows magnetic north (redirect slightly eastward to correct for –3.5° declination). Redrawn from Sun and Kistemaker (1997, 116, Figure 6.2), reproduced by permission.

dragon and tiger appear in the Shang oracle bone inscriptions as recipients of sacrifice, both probably identifiable as asterisms;[d] Moreover, the opposition between *Dragon* and *Tiger* (Sco and Ori) constellations is singled out in etiological myth as highly significant. It is precisely these stellar configurations plus the *Dipper* that are stressed in classical texts as the three Great Seasonal Indicators (*da chen*), crucially important seasonal signposts recognized by the late Neolithic and probably earlier.[e]

Many have pointed out that Orion marked the vernal equinox around 7,000 years ago and Antares in Scorpius the autumnal equinox around 5,000 BP, which dates bracket the epoch of the Puyang tomb. It is interesting from the perspective of the preceding discussion of polar mimicry that the *Dipper*-like object placed at the feet of the tomb's chief occupant is depicted in precisely the same orientation as in the graphic representations we saw above from the Han Dynasty; that is, as a mirror image of the *Dipper* in the sky. Not only that, the orientation of the *Dipper* in the tomb is inconsistent with respect to the other two mosaics, *Dragon* and *Tiger* (if assumed to represent the constellations), a clear indication that the representational motives are not the same.

If we consider that a chart of the heavens is not a portrayal of something that can actually be seen at a given moment in time but an imaginative composition of the succession of stellar patterns passing overhead, then given the extraordinarily early date it seems implausible that the Puyang burial could have been intended as a map of the heavens in anything like the modern sense.[f] Instead, what we appear to have in this tomb is a representation of the prime importance of the principal constellations, Sun and Moon in their timekeeping roles, and allusion to their management by uniquely empowered individuals closely connected with rulership. The *Dipper*, in particular, is supremely emblematic of the mysterious power at the center of the sky to which the cosmo-priest paid homage in life and with which he perhaps hoped to be united after death. Its depiction in reverse may, in fact, be specifically evocative of this relationship, as such inversions are common in certain mystical contexts. Just as in the parallel case of Han Chang'an three millennia later, the numinous power of the *Dipper* is being conjured by mimicry in the tomb of this Neolithic cosmo-priest.

It is also important to recognize that although the longitudinal axis of the cosmo-priest's skeleton, taken together with the dragon and tiger mosaics as a group, is about ten degrees east of true north, the straight north edge of the tomb is more accurate, only five degrees south of east. Furthermore, the skeleton placed at an oblique angle near the north edge of the tomb has been carefully arranged with the head pointing southeast to azimuth 124 degrees, the direction of winter solstice sunrise. This is only 4 degrees from the true figure of 120 degrees for the actual azimuth of sunrise at the time and thus surprisingly accurate. Even more impressive, however, is that taking the burial as a self-contained unit, the angle between the north edge of the tomb running east–west and the axis of the skeletal solstitial pointer is, in fact, 29 degrees; i.e., azimuth 119°. This precise alignment can only have been intentional and empirically established by paying close attention to the Sun's rising points along the horizon throughout the year. Since this burial ground may not have been the customary location of solar observation·

and sacrifice, this raises the question whether the interment was scheduled to coincide with the winter solstice, or whether azimuth ~119 degrees was preserved in some fashion so as to be replicable in different contexts. The fact that certain alignments within the tomb are internally self-consistent would seem to suggest the latter. Remarkably, this burial may be more than a millennium older than the solar observation terrace at Taosi discussed in Chapter 1 where some burials (e.g. TG5M23, TG4M24) exhibit the same orientation toward winter solstice sunrise.

[a] Chen Jiujin (1993, 53); Feng Shi (1996, 159–62; 2007, 278, 285). Yi Shitong (1996, 31) relates that he provided the samples for carbon dating by a lab in the Netherlands which yielded the uncalibrated date of 6465 ± 45 years BP, which confirmed the Beijing lab results, but he was informed by the laboratory that in their experience marine shells often prove to be 1000 to 1,500 years younger when calibrated.

[b] The most thorough study of the burial from the perspective of archaeoastronomy is Feng Shi (2007, 374–409), though it must be said that Feng's claims about the sophisticated geometrical relations embodied in the tomb are implausible.

[c] Feng Shi (1996, 161, Figures 38, 39; 2007, 406–7, Figures 6-21, 6-22).

[d] Jao (1998, 33, 36, 39).

[e] Needham and Wang (1959, 248, 282); Yi Shitong (1996, 27–9) dates the beginnings of Chinese astronomy to 6,000 years ago at the latest.

[f] Cf. Yi Shitong (1996, 28).

Compare now the curious configuration of Tang Dynasty Chang'an (Figure 11.11) with that of Han Chang'an (Figure 11.8), and it will be apparent that there is a striking resemblance.[46] The Tang Daming Palace on the north, added in the 660s, was not part of the original plan, according to Arthur Wright, who comments on the cosmological symbolism of the capital: "The emperor, as cosmic pivot, saw to the harmonious operation of these natural forces [i.e. *yin* and *yang*, winds, seasons, etc.], and he did so from the great main palace called T'ai chi Tien [*Taiji Dian*], whose name symbolizes the astral center of the universal order."[47] *Taiji Dian*, of course, means "Polar Hall" and recalls the First Emperor's "[Northern] Culmen Temple." The Daming Palace, whose odd shape and location caused it to be seen as an ad hoc excrescence on the neat rectilinear layout of the wall, can now be more appropriately regarded as fulfilling the same symbolic function as the idiosyncratic north wall of Han Chang'an. Given the precedent of the "*Dipper* wall" of the Han capital, it seems likely that the Daming Palace and its adjacent structures were purpose-built to replicate the

[46] See too the general plan of the Eastern Han capital of Luoyang with its similarly idiosyncratic north wall. Lewis (1999, 99, Map 9).

[47] Wright (1977, 56).

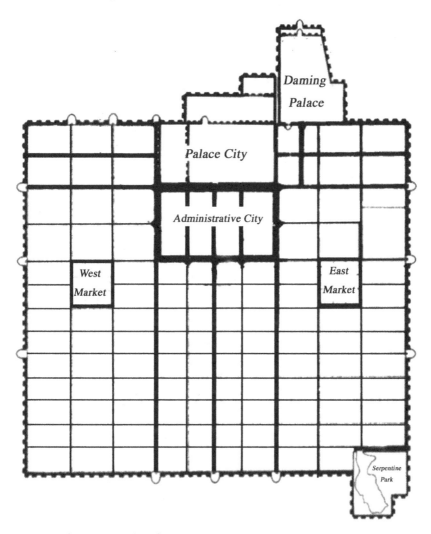

Figure 11.11 Plan of Tang Dynasty Chang'an. Redrawn from Wright (1977, 58), reproduced by permission.

shape of an inward-facing *Dipper*. The parallel seems unmistakable and confirms the analysis above regarding the direction the *Dipper* ought to be facing. Moreover, the seemingly superfluous bulge in the south wall and its enclosed body of water once again presumably evoke the *Southern Dipper* immersed in the *Sky River*. Further evidence of the continuity of this tradition is found in the Sui Dynasty. In the first year of the Great Undertaking (*Da ye*) reign period of the Sui (569–618), Emperor Yang moved his capital to Luoyang. Because the Luo River ran through the city, he had a floating bridge built across it, which he called "Celestial Ford Bridge" (*Tianjin qiao*) in imitation of the *Celestial Ford* across the *Sky River*.[48] Nearly three centuries later, the memory of this famous bridge was still evoked in the Tang Dynasty by Bai Juyi (772–846) in his poem "Spring Feelings with a Friend in mid-Luo" (*He you ren Luo zhong chun gan*). The materialization of the model is less exacting, perhaps, more suggestive than precise, but the overall configuration nevertheless evokes the same cosmological symbolism locating the celestial Pole (Polar Throne Room) within the walls of the capital in its proper position relative to the *Dipper*. After the Tang the feature no longer appears in imperial capitals.

The archetypal celestial temple – Luminous Hall

Throughout the Western Zhou, major royal commissioning and investiture ceremonies were conducted in a temple called "Hall of Heaven" (*Tian shi*) or "Grand Ancestral Temple" (*Tai miao*). We know from copious references in pre-Han sources that this temple, which may have had its double on Mt. Song (*Song shan*), the "Central Peak" (*Zhong yue*) just south of Luoyang, was reserved for the most elaborate sacrifices to the Supernal Lord and other powerful spirits, often in company with the Zhou royal ancestors.[49] The earliest reference to the Hall of Heaven is from a bronze tureen, the *Tian Wang gui* (discussed above in Part Three), which dates from the reign of King Wu (r. *c.*1049–1045/4) who overthrew the Shang. The inscription records the first performance of a royal *Feng* sacrifice immediately after the conquest of Shang, in conjunction with elaborate sacrifices by King Wu to the dynastic founder King Wen (*c.*1099–1050 BCE) and to the Supernal Lord:

On day [*yi*]*hai* [12] the King performed the Great *Feng* [sacrifice]. The King sailed the three quarters. The King sacrificed at the Hall of Heaven, then descended. Tian Wang seconded the King in the Grand Sacrifice to the King's great illustrious deceased father King Wen, and in performing the immolation rite to the Supernal Lord. King Wen watched from on high. The great illustrious King [Wen] was the model; the great

---

[48] The great eastern port city by the same name, Tianjin, was not founded until 1404 during the Ming Dynasty by the Yongle Emperor (r. 1402–24), ostensibly to commemorate his fording of the river at that point, but this name may also invoke astral resonances. *Shiji* 27.1309.

[49] Lin Yun has argued that the Hall of Heaven is none other than Mt. Song, for which he adduces textual support. However, there is also persuasive evidence that Hall of Heaven lay in close proximity to the ancestral temple of the Zhou kings, which would seem more practical than having that pre-eminent sacred space located seventy kilometers from Luoyang on a mountaintop. Since there is a strong tradition suggesting the periodic *Feng* sacrifice must be performed on a sacred peak, it may be that there was a Hall of Heaven shrine on the highly symbolic Central Marchmount (*zhong yue*), which was used for those occasional observances. Lin (1998).

disciplining King [Wu] succeeded him; [he] was grandly able to terminate the Yin king's sacrifices...

Evidence from other inscriptions and early texts suggests that the Hall of Heaven was none other than the Luminous Hall (*ming tang*),[50] associated with or identical to the Grand Ancestral Temple, and surrounded by the moat or circular lake called the *Biyong*.[51] Indeed, the inscription on the *Mai zun* a half-century after the Zhou conquest states explicitly that "at the *Biyong* moat, the King boarded a boat and performed the Great *Feng* [sacrifice]," so the parallel with the *Tian Wang gui* is unmistakable. In his exhaustive study documenting the history, numinous function, and cosmological symbolism of the Luminous Hall, Hwang Ming-chorng analyzed the *Mai zun* inscription and concluded that the account of the Great *Feng* sacrifice, archery contest, boating excursion, etc., all point to the identity of Fengjing or Haojing as the large ceremonial complex located somewhere on the outskirts of Zongzhou, the ceremonial capital built specifically for the worship of Heaven.[52] From the *Mai zun* inscription and other similar records, both inscriptional and literary, it is clear the complex included a circular moat or lake, symbolically laid-out structures labeled "palaces" (*gong*), and a park-like menagerie stocked with wild animals, all distinctive features of the compound referred to in the early texts as a Luminous Hall.

The Luminous Hall

Following the establishment of the new Zhou Dynasty capital at Luoyang in about 1036 BCE described above, the "Announcement of Kang" (*Kang gao*) chapter of the *Book of Documents* records a precedent-setting assembly of all the leaders of the realm convened by the Duke of Zhou. Warring States and Han texts consistently identify the location of this assembly as the Luminous Hall of the Zhou, the same sacred precinct at the capital of Haojing where King Wu

---

[50] Although conventionally translated "Hall of Light," this somewhat pedestrian rendering of *mingtang* fails to capture the numinous quality of *ming*, "bright, luminous, enlightened, numinous." Given the sacred function of the place, *ming* here must be the same as in "spirits" (*shen ming*) and "spirit vessel" (*ming qi*), the term applied to the artifacts and implements buried in tombs. The intimate connection with the brilliantly spangled heavens also suggests that a rendering such as "luminous" (if not "numinous") is more apt.

[51] A detailed, though fragmentary, description of the physical layout of the Luminous Hall is found in the chapter of the same name in the *Record of Rites of Dai the Elder* (*Da Dai liji*). In contrast, the account in the "Art of Rulership" chapter in *Huainanzi* portrays the *Mingtang* as "the site of not only of ancestral rites but of a comprehensive model of history that underpins sage rule." Lewis (2006a, 265). Lewis discusses the spatio-temporal significance of the *Mingtang* in detail (ibid., 260–73); also Wright (1977, 49–51); Wu Hung (1997, 176–87); Tseng (2011). Fu Xi'nian (1981, 38, Figure 5) shows a reconstruction of a large Western Zhou temple-like building at Fufeng in Shaanxi with round turret atop a square structure, matching the description of a *mingtang*.

[52] Hwang (1996, 220, 236–7).

presented war trophies on his victorious return in the spring of 1046 BCE after conquering the Shang in the Wilds of Mu (*Muye*). The Luminous Hall was also the location of the ceremonial events performed by later kings and recounted in the early Zhou bronze inscriptions cited above. This is not the place for a comprehensive survey of the cosmological symbolism of the Luminous Hall in tradition and practice, particularly since the subject has been extensively studied.[53] But it is in order to briefly consider the astral associations of the Luminous Hall as Grand Temple.

According to the "Announcement of Kang," following the establishment of the new Zhou Dynasty capital at Luoyang in the mid eleventh century BCE, a precedent-setting assembly of all the vassals of the new regime was convened.[54] Classical texts consistently identify the location of this assembly as the sacred dynastic precinct called the Luminous Hall. The most authoritative early discussion of the design and function of the *Mingtang* is that of the famous Eastern Han Dynasty scholar, Cai Yong (133–92 CE), found in his *Excursus on the Luminous Hall and the Monthly Ordinances*:

> The *Mingtang* is the Grand Ancestral Temple of the Son of Heaven, wherein the Emperor sacrifices to his ancestors in the company of the Supernal Lord. The lineage of Xia called this place Chamber of the Generations; the Shang people called it Multi-storied Chamber; and the people of Zhou called it *Mingtang*, Luminous Hall. The eastern [Spring chamber] is called Green *yang* Energy; the southern is called *Mingtang*; the western [Autumn chamber] is called Assemblage of Emblems; the northern [Winter chamber] is called Somber Hall, and the central chamber is called Grand Hall [of Heaven]. The *Book of Changes* says, "*Li* is brightness, the hexagram of the south. The Sage faces south and attends (to affairs), all under heaven face the brilliance and are ordered. For the ruler of men there is no more true position than this" ... Therefore, although there are five appellations, principal among them is *Mingtang* [Luminous Hall] ... Compare this to the *Northern Asterism* [*Dipper*] which dwells in its place while all the myriad stars circle it and the ten-thousand things are regulated by it. [It is] the source from which springs governance and instruction, and is the origin of all change and transformation, manifesting unity. Therefore, it is said of the *Mingtang* that its affairs are great and its meaning profound. If one invokes the aspect of purity, it is called Pure Temple; if one invokes its aspect as the hall of governance, then it is called Grand Ancestral Temple; if one invokes the aspect of veneration, then it is called Grand Hall; if one invokes its aspect of facing toward the light, then it is called *Mingtang*; if one invokes the aspect of the schools of the four gates, it is called the Great Learning; if one invokes the aspect of being surrounded on the four sides by [a body of] water, round like a jade *bi*-disc, it is called Circular Moat. They are all different names for the same thing – it is one thing.[55]

---

[53] Ibid., 236–7, 262–3; Wu Hung (1997, 176–87); Henderson (1995, 212–13); Tseng (2011). For Emperor Wu of Han's construction of a Luminous Hall in 109 (or 106) BCE as part of his reinstatement of a system of cosmic sacrifices, see the references cited in Kern (2000, 22)

[54] Karlgren (1950a, 39).

[55] *Mingtang yueling lun*, 3.6a–b. Cai Yong held that the various terms anciently designated a single edifice. For centuries academicians and scholiasts disputed the question of the design

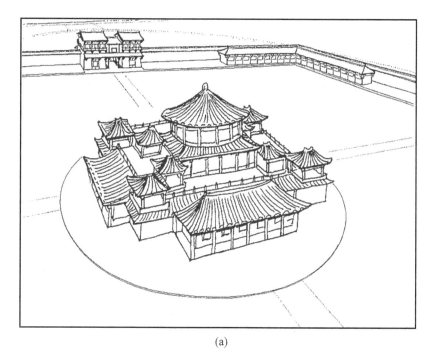

(a)

Figure 11.12 (a) Emperor Wang Mang's Xin Dynasty (9–23 CE) Luminous Hall; reconstruction based on archaeological excavation; (b) plan of Wang Mang's Luminous Hall. Adapted from Wu Hung (1997, 179, 3.16).

Summing up, Mark E. Lewis put it this way:

[the Luminous Hall] is a microcosm in which both cosmos and state are completely realized. It is a ritual complex that combines rites to ancestors and cosmic deities; an administrative center where all officials are gathered and all policies enacted; and an educational institution in which all true teachings are presented. It is also the summation of the ritual structures of earlier dynasties. As a chart of the cosmos, the source of order, and a summation of history, it becomes the perfect image of power.[56]

The Luminous Hall as celestial simulacrum

It will be important to consider in more detail features of the Luminous Hall that have a direct bearing on the notion of a normative celestial temple. The political and religious significance attaching to the Luminous Hall, held to inhere in its

and layout of the *Mingtang* complex, but not its symbolism. The most comprehensive study of the history of the *Mingtang* is Maspero (1948–51).
[56] Lewis (2006a, 271, 303).

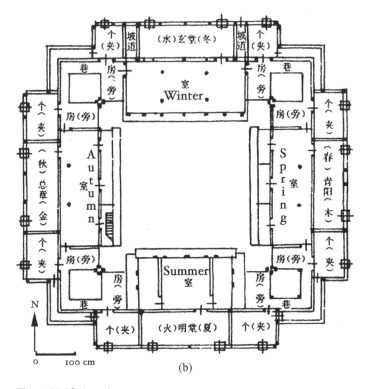

Figure 11.12 (*cont.*)

very design and layout, indicates that in addition to the functions named above, the solar and lunar observations essential to calendrical astronomy would also have been performed within these precincts. Figure 11.12a and b illustrate the design of this architectural icon, the cosmological temple, its round upper story symbolizing Heaven and square ground floor representing Earth, with each of the four sides facing the cardinal directions. The two examples shown are from the Han and the middle of the Spring and Autumn period, but a structure dating from middle Western Zhou (*c.* ninth century BCE), among the largest buildings ever discovered in the Zhou homeland of Zhouyuan, is strikingly similar.[57] It is evident from Figure 11.13 that the combination of symbolic features making up this Luminous Hall complex has a very long history, in all likelihood dating back at least to the epoch of the sacrificial rituals at the beginning of Western Zhou.

---

[57] Fu Xi'nian (1981, 38, Figure 5).

(a)

Figure 11.13 (a) Fifth-century BCE lacquer lid from a Warring States tomb at Linzi, Shandong, displaying a two-dimensional plan of a ya 亞–shaped building with twelve side chambers, identical in design to the Luminous Hall. Adapted from Wang Shixiang (1987, Plate 3). (b) Reconstruction of a mid-Western Zhou (*c.* ninth-century BCE) structure, among the largest structural foundations ever discovered in the Zhou homeland of Zhouyuan. Note the resemblance to the Luminous Hall in Figure 11.12. Adapted from Fu Xi'nian (1981, 38, Figure 5).

Immediately following the passage above, Cai Yong quotes from the "Records of Monthly Ordinances":

The Luminous Hall is that wherein the unification of all things by Heaven and Earth is manifest. The stellar image in Heaven through which the Luminous Hall communicates is called the [*Northern*] *Asterism* [*Dipper*]. Therefore, its twelve palaces here below are the [twelve solar] chronograms. The water surrounds it on the four sides, emblematic

(b)

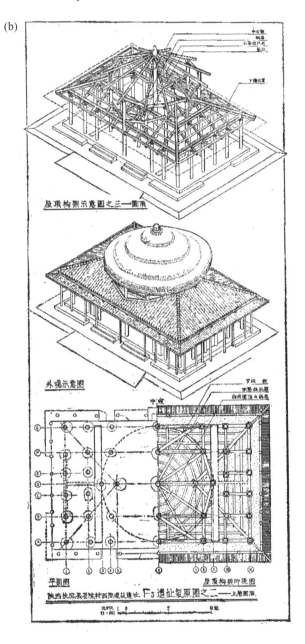

Figure 11.13 (*cont.*)

of the king's acting as the model for all under Heaven, his virtue reaching abroad to the Four Seas, like this water.[58]

Here we have it explicitly stated that the correspondence between the Luminous Hall and Heaven is not merely one of cosmological analogy, but that, in fact, this sacred space is precisely the *axis mundi* through which the terrestrial sovereign communicates with his celestial counterpart at the Pole. Still another Han source, the "*Yin-yang* Record of the Luminous Hall of the *Classic of Rites*," elaborates on the details of this resonance between the temporal and celestial realms:

The *yin* and *yang* of the Luminous Hall are the means by which the kingly ruler responds to Heaven. The scheme of the Luminous Hall is that it is surrounded by water, the water swirling leftward in imitation of [the *Sky River* of] Heaven. In the interior is the Great Hall [of Heaven], in imitation of the *Palace of Purple Tenuity* [circumpolar stars in UMa and Draco]; emerging [from it] to the south there is the Luminous Hall, in imitation of the *Palace of Grand Tenuity* [stars in Leo and Virgo]; emerging [from it] to the west there is the Assemblage of Emblems in imitation of the *Five Lakes* [stars in Auriga]; emerging [from it] to the north there is the Dark Hall in imitation of *Align-the-Hall* [Square of Pegasus]; emerging [from it] to the east there is the *Green Yang*, in imitation of the *Celestial Market* [stars in Ophiucus and Hercules]. [Each of] the Supernal Lord's four seasons governs its own palace, the kingly ruler too in carrying out Heaven's unification of all things attends to the affairs of the kingdom from the [appropriate] quarter.[59]

If this sounds improbably idealized, consider geographer Li Daoyuan's (d. 527) striking description in his *Annotated Water Classic* (*Shui jing zhu*) of the design of the Luminous Hall in the Northern Wei Dynasty capital, Pingcheng (present-day Datong) in the early sixth century:

The Luminous Hall was round above and square below and on the four sides there were twelve doors and nine rooms, without common walls. Outside the rooms, within the columns and beneath the silk atrium awning, were installed mechanical wheels and pale blue-green silk decorated with blue semi-precious stones – looking up it resembled the sky. [On it] were painted the *Polar Asterism* and lunar lodges, so that it resembled the canopy of Heaven. Each month as the [*Northern*] Dipper pointed to [successive] chronograms, it revolved to correspond to the way of Heaven; in this respect [the Luminous Hall] departed from the ancient [model]. On top [of the Luminous Hall] was added a Spirit Terrace (*ling tai*), and below water was led in to form a Circular Moat. Along the water's edge stones were laid to form embankments, in this respect according with the ancient scheme. This is what was laid out and built during the Taizhong reign period [477–99].[60]

---

[58] *Yueling ji*, 3.6a–b.
[59] *Liji mingtang yinyang lu*, quoted in *Sui shu, Niu Hong zhuan*, 49.1304; cf. *Taiping yulan*, 533.2b. The names and cardinal directions assigned here to the "halls" are the same as in the late fourth-century BCE *Guanzi* chapter "Dark Palace," *Xuangong*, which may be an extract of the captions from a diagram of the Luminous Hall. For discussion and comparison with the Chu Silk Manuscript, see Li Ling (1985, 37).
[60] *Shui jing zhu ji shi ding'e*, 13.10b.

Conclusion

It is evident that the ancient Chinese remained keenly interested in the circumpolar region, especially the celestial Pole, from the very beginning of the imperial period. Study of the role of astronomical alignment in shaping the built environment shows that the Chinese had developed practical geometrical applications of astronomical knowledge to achieve highly symbolic polar alignments at the imperial center, monumental embodiments of the identity between the celestial Pole and the locus of temporal power. This persistent intentionality – an age-old preoccupation with the circumpolar skyscape – is widely represented in philosophy as well as applied astronomy.

There are innumerable references in classical Chinese literature to the vital necessity of maintaining conformity with the normative patterns of the cosmos. It has been argued that this focus on the circumpolar sky and the analogy between the ruling center and the celestial Pole followed the establishment of the universal empire in the late third century BCE. But long before the core ideas and practices outlined here became enshrined in the imperial ideology, the archaeological record clearly shows that similar predilections were already present during the formative period beginning in the second millennium BCE. An ancient imperative to conform to Heaven's intentions made it essential to devise practical methods of achieving that objective. The practice of divination is one modality that exemplifies this impulse. Devising a calendar is another. The design and cosmological symbolism of sacred precincts is yet another. So, it is to conceptualization of the cosmos prefigured in ruling metaphors that we now turn.

12   Temporality and the fabric of space–time

Time [flashes by] like an arrow, Sun and Moon [fly back and forth] like the shuttle of the loom, and another year is past.

Years ago Joseph Needham published a celebrated essay, "Time and Knowledge in China and the West," in which he surveyed Chinese ideas about time and temporality. Needham left virtually no realm untouched in discussing time in Chinese philosophy and natural philosophy – time, chronology, and historiography; time measurement; biological change in time; concepts of social evolution and devolution; recognition of technological development over time; science and knowledge as co-operative cumulative enterprises; and so on. The subject is vast, particularly in view of the historical scope of Needham's account, which carries the story from early China to the modern era. Indeed, each of the topics he discussed merits a monograph. My objective here will be to draw attention to a few of the more general issues concerning linear irreversible time and cyclical recurring patterns, and then to focus on uniquely Chinese perspectives on temporality, causality, and their governing metaphors.

When it comes to Chinese civilization, Needham wrote, "broadly speaking, and in spite of anything that has been said above, linearity . . . dominated":[1]

The apocalyptic, almost the messianic, often the evolutionary and (in its own way) the progressive, certainly the temporally linear, these elements were there, spontaneously and independently developing since the time of the Shang kingdom, and in spite of all that the Chinese found out or imagined about cycles, celestial or terrestrial, these were the elements that dominated the thought of Confucian scholars and the Taoist peasant-farmers. Strange as it may seem to those who still think in terms of the "timeless Orient," on the whole China was a culture more of the Irano-Judaeo-Christian type than the Indo-Hellenic.[2]

For Needham the leading indicator, implicit in the rationale quoted here, was his conviction, echoing Marcel Granet, that the Chinese possessed a more

---

[1] Needham (1981, 133).
[2] Ibid., 135. But see Jack Goody (2006, 18), who cautioned that Needham's characterization is of an "over-generalized kind that wrongly contrasted cultures and their potentialities in an absolute, categorical, even essentialist fashion."

highly developed historical sense than any other civilization.[3] In his essay Needham was generalizing about the entire sweep of Chinese civilization as a whole. If, instead, he had restricted his discussion to the ancient period – that is, ending with the Han Dynasty at the beginning of the Common Era – it is unlikely he would have concluded that linearity dominated Chinese thinking about time. For, as Needham also stresses in discussing historical causation in the early imperial period,

> The conviction that the universe and each of the wholes composing it have a cyclical nature, undergoing alternations, so dominated [Chinese] thought that the idea of succession was always subordinated to that of interdependence. Thus retrospective explanations were not felt to involve any difficulty. "Such and such a lord, in his lifetime, was not able to obtain the hegemony, because, after his death, human victims were sacrificed to him." Both facts were simply part of one timeless pattern.[4]

Clearly, what is implied here is hardly a conventional notion of causality, much less "historical sense" in any ordinary meaning of the term, as Needham is quick to note. Carl G. Jung also realized that in such a conception "a causal event might not be strictly prior in time to its effect, bringing the latter about rather by a kind of absolutely simultaneous resonance," or "synchronicity," as he called it:

> It cannot be a question of cause and effect, but of a falling together in time, a kind of simultaneity. Because of this quality of simultaneity, I have picked on the term "synchronicity" to designate a hypothetical factor equal in rank to causality as a principle of explanation . . . I defined synchronicity as a [psychologically] conditioned relativity of space and time. [J.B.] Rhine's experiments show that in relation to the psyche space and time are, so to speak, "elastic" and can apparently be reduced almost to [the] vanishing point, as though they were dependent on [psychological] conditions and did not exist in themselves but were only "postulated" by the conscious mind.

> Synchronicity is no more baffling or mysterious than the discontinuities of physics. It is only the ingrained belief in the sovereign power of causality that creates intellectual difficulties and makes it appear unthinkable that causeless events exist or could ever occur.

> Synchronistic events as acausal exceptions . . . prove to be relatively independent of space and time; they relativize space and time in so far as space presents in principle no obstacle to their passage and the sequence of events in time is inverted, so that it looks as if an event which has not yet occurred were causing a perception in the present. But

---

[3] Needham (1969, 289). In a perceptive discussion of Granet's account, Paul Ricoeur (1985, 26) draws important distinctions with regard to time consciousness between linear and cyclical conceptions, and between the cyclical and the rhythmic. For a different perspective on Granet, see Saussy (2000, 20).

[4] Needham (1969, 289). The episode referred to concerns Duke Mu of Qin (d. 629 BCE). Nienhauser et al. (1994, 102).

if space and time are relative, then causality too loses its validity, since the sequence of cause and effect is either relativized or abolished.[5]

In China, a significant factor was the consolidation of the imperial institution during the two centuries before the Common Era, and especially its union of convenience with Confucianism. The ascendancy of the latter, with its deep commitment to hierarchy, reverence for the ancestors, and the maintenance of social and political harmony, assured that ancient and more subtle Daoist and Naturalist conceptions of pattern and phenomenological connectedness would be relegated to specialized applications, and with them an intense interest in timeliness, "returning" as the movement of the Dao, and especially "ideas of causality distinctly different from the Indian or Western atomistic picture in which the prior impact of one thing is the cause of the motion of another."[6] In other words, the devaluation of the correlative cosmology of the Naturalists and of Daoist intuitive attunement with the timeless patterns of the cosmos meant that the synthesis of their concepts of causality and temporality, which were, strictly speaking, neither cyclical nor linear, would never be fully elaborated. In any case teleological theorizing was foreign to Chinese historians and philosophers.[7] We shall return to causality and this idea of timeless pattern below.

---

[5] Carl G. Jung, "Synchronicity: An Acausal Connecting Principle," quoted in Ornstein (1973, 450, 456, 457). Wolfram Eberhard (1958, 54) used "parallelism" to refer to the same operative principle in Chinese portentology. In referring to the "discontinuities of physics" Jung may have had in mind "quantum entanglement," the "spooky action at a distance" that Einstein could not accept, or possibly Niels Bohr's principle of complementarity. According to Lawrence Fagg (1985, 168), "The generalized complementarity principle finds an interesting and beautiful comparison with the Chinese *yin-yang* principle . . . Therefore, the notion of complementarity is instinctive and deep-seated in the East and is not new . . . For time comparisons *yin* would obviously be associated with intuitive (subjective) or religious spiritual time and *yang* with rational (objective) or physical time." The principle holds that "wave and particle modes are mutually exclusive but complementary, and each is necessary in rendering a complete description of a microscopic physical phenomenon."

[6] Needham (1981, 97).

[7] As Yü Ying-shih (2002, 168) has pointed out, "The lack of an impulse to speculate on the whole process of history on the part of Chinese historians makes it rather difficult for us to assert with any degree of definiteness whether Chinese historical thought is linear or cyclical." Yü Ying-shih overlooks the early hints of Confucian evolutionary thinking preserved in the "Revolving of the Rites" (*Li yun*) chapter in the *Book of Rites*. Although the title suggests cyclical change, the discussion of the sequence of three epochs – Disorder, Approaching Peace, Great Peace – clearly implies evolutionary progress. This unorthodox theory of history is elaborated further in the exegesis on the *Gongyang Commentary* by He Xiu (129–82 CE) in the Later Han, but never achieved great currency. For further explorations of notions of time and space, the Chinese historical sense, and linear time versus cyclical time in ancient Chinese culture, see Huang and Zürcher (1995), and Loewe (1995, 305–28).

Characteristics of ancient Chinese concepts of time and pattern

Even if it is a truism that no civilization has proceeded from a dominantly linear to a dominantly cyclical conception of time, it is no easy matter to determine how or even whether a transition might have occurred. Indeed, the modern tendency to dichotomize time into "linear" and "cyclical" is itself suspect. As Paul Ricoeur has argued,

> Both cyclical and noncyclical models, when released from the yoke of a comparison with modern linear time . . . display a wide spectrum of modalities, and, more interestingly, numerous overlappings and mixed forms – forms that will look inconsistent only to someone who takes the linearity of chronology as the basic yardstick for the description, analysis, and interpretation of mythical time.[8]

A famous passage from the *Zuozhuan*, the narrative quasi-commentary on the canonical chronicle *Springs and Autumns*, offers a taste of the time-sense in the fourth century BCE:

> In the second month, on day *guiwei*, the [dowager] marchioness Dao of Jin entertained all the men who had been engaged in the walling of Qi. A childless old man from the District of Jiang went and took his place at the feast. Some participants were curious about his age and would have him tell it. He said, "I am a lowly person and do not know how to keep track of the years (*bu zhi ji nian*). Since the year of my birth, on day *jiazi* [first day of the sixty-day cycle], the day of the new Moon in the first month, there have elapsed 445 *jiazi* days, and finally until today one-third of the cycle [of sixty]." The officiants [of the feast] hurried to the court to ask about it. Music-master Kuang said, "it was the year when Shuzhong Hui Bo of Lu had a meeting with Xi Chengzi in Chengkuang. In that year, the *Di* barbarians invaded Lu, and Shusun Zhuangshu defeated them at Xian, capturing their elders Qiaoru, Hui, and Pao, after all of whom he named his sons. [Hence] seventy-three years [have elapsed]."[9]

Several elements are noteworthy in this passage. To begin with, more sophisticated timekeeping methods than counting the cycles of sixty are inaccessible to the freeman commoner, and probably only slightly less so to other non-specialist participants at the feast. The old man says *ji nian*, literally "string the years," to mean "reckon the date," a revealing usage to which we shall return. Second, the specialist who is in charge of record keeping at court, who is also music-master, in the first instance places the timing of the man's birth situationally, as if quoting an annalistic record, and then only secondarily, after

---

[8] Ricoeur (1985, 27). See Jack Goody's (2006, 18) critique of characterizations in terms of a "cyclical East" and "linear West": "any idea of exclusive calculation having to be made in a linear mode rather than a circular one is mistaken and reflects our perception of an advanced forward looking west and a static, backward-looking east."

[9] *Zuozhuan*: Duke Xiang, thirtieth year, 543 BCE; trans. Legge (1972, IX.556, trans. modified). See also Cullen (2001).

an arithmetic operation, is he able to fix the event chronologically.[10] This anecdote is fairly representative and gives a good indication of the relative value attached to different kinds of temporal awareness in daily life. It also points up a central problem at the heart of temporal consciousness to which the *Book of Changes* seemed to offer a solution, as we shall see; that is, how to systematically relate subjective mental states or states of the world to an often contradictory description of the world in terms of events happening.[11]

Here is another, elite perspective on attunement as a vital concern of the ruler who aspired to achieve universal harmony and hegemony, lest his negligence or ineptitude provoke disasters. The admonition is attributed to the chief minister of the first Hegemon, Duke Huan of Qi, in the seventh century BCE:

Since mankind is one entity with the totality, the one who establishes the laws must also make a study of heavenly timeliness and earthly advantages as a basis for devising his laws. Kuan Tzu [Guanzi] said: "Commands have their proper times ... the Sage King strives to adjust to time, and to relate his governmental measures to it." Spring, summer, autumn, and winter each has its activities which should be done at those times. "When man and Heaven are in accord, only then can the perfection of Heaven and Earth come into being." When commands and orders are not appropriate to the season, then "things undertaken will not get accomplished, and there is sure to be a great calamity."[12]

During the last few centuries before the beginning of the present era, old traditions and new speculations about the connectedness of all things were increasingly systematized and elaborated, not least by *yin-yang* correlative cosmologists and propounders of Daoist inspired Lao-Huang theories of rulership, for whom deep understanding of the phenomena was held to be crucial. These ideas drew on ancient roots in divination methods deriving from microcosmic–macrocosmic analogies, numerology, cosmo-political theories of cyclical dominance by the Five Elemental Phases – *Wood, Metal, Fire, Water, Earth* – coupled with a highly developed sensitivity to phenomenological correspondences perceived to exist in nature. The historical sense, such as it was, displayed a strong bias in favor of time and historical events seen not as historical instantiations of human actions and motives per se, but in their timeless, emblematic aspect, as examples of admirable or dishonorable motives, or adherence to protocol and tradition-bound propriety. This quality of annalistic

---

[10] This is, of course, consistent with our own subjective experience of time. As Anthony Aveni (2002, 2) has remarked, "The absolute chronology called history is thrust upon me ... It conflicts with the way I naturally want to think about the past – purely as a sequence of events that happened in a particular order, like knots on a long string that run from my origin to the present. Who cares how long the string is or how many inches between the knots?"

[11] See, for example, the discussion of the philosopher J.E. McTaggert's view of the unreality of time in Davies (2002, 42).

[12] Trans. Hsiao (1979, 337, trans. modified). "Season" and "time" here both translate the same word, *shi*.

history, the beginnings of which are already discernible during Shang and Zhou a millennium earlier, is attributable to the origins of historical record keeping in the context of divination and ritualistic reporting of temporal goings-on to the ancestors.[13] By the early imperial period, the responsibilities of the Grand Scribe-Astrologer, the official post which had evolved out of the diviners of old, encompassed everything from divination to portentology (including sky pattern reading and the interpretation of anomalous events and prodigies), to calendar making, to advising on the relevance of historical precedent, to current events. Rulership demanded mastery of the complex pattern of events and motives, both human and natural, in order to successfully manage their harmonization with the inchoate and constantly evolving complexion of the times. Take for example the characterization of this enterprise by the philosopher Jia Yi (201–169 BCE), in the early second century BCE:

A popular maxim has it: "Prior events, not forgotten, are the guide to what follows" (*qian shi zhi bu wang, hou shi zhi shi ye*). For this reason, in ordering the state, the accomplished ruler observes the events of antiquity, tests them against the present, matches them with human affairs, examines into the principles of flourishing and decline, and looks for what is appropriate according to the expedient and the trends. In this way discarding and adopting measures have their proper sequence, adapting and transforming their due seasons. Thus, his reign is untroubled and enduring, and his altars to the soil and grain are safeguarded.[14]

Here again is the historical sensibility in the service of statecraft, and statecraft according to a paradigm in which the very cycles of the cosmos and the movements of *yin-yang* and the Five Elemental Phases were all implicated. In this worldview, disturbance or disharmony at any point in the fabric of space–time or human affairs could reverberate throughout the whole, with unpredictable consequences; and not merely prospectively, but perhaps even retrospectively, as we saw above.[15] A popular early metaphor employed in this connection is that of a mirror, the aspiring ruler of men being enjoined to seek

---

[13] David Keightley (2004, 3–64) has characterized this as an impersonalization of roles and "flattening out" of the ancestors (e-mail communication of March 5, 2002).

[14] *Shiji*, 6.278; Nienhauser et al. (1994, 165). This maxim calls to mind Shakespeare's *Henry IV, Part II*, Act III, scene i:

> There is a history in all men's lives,
> Figuring the natures of the times deceased,
> The which observed, a man may prophesy,
> With a near aim, of the main chance of things
> As yet not come to life.

[15] John B. Henderson (1995, 227) calls upon the same textile metaphor in a similar context: "anomalous movements were not simply complicating factors for which adequate adjustment could be made in calculation. Rather, they were woven into the fabric of the cosmos."

guidance in the image of his deeds and motives when reflected in the mirror of previous reigns and the lives of the people.

In attempting to grasp what this worldview was like, we need to eschew conventional ideas of causality. Even terms like "reverberate" or "propagate" above tend to evoke conventional ideas of action and reaction and thus to inject a presentist perspective. Rather, what is implied here is a kind of "acausal orderedness," in which, as Needham says, the "idea of correspondence has great significance and replaces the idea of causality, for things are *connected* rather than caused."[16] Or, in Needham's inimitable phrase, "in such a system causality is reticular and hierarchically fluctuating, not particulate and singly catenarian."[17] In explaining the interpretation of the French sinologist Marcel Granet, Needham remarks, "if two objects seemed to them to be connected, it was not by means of a cause and effect relationship, but rather 'paired' like the obverse and reverse of something, or to use a metaphor from the *Book of Changes*, like echo and sound, or shadow and light":[18]

What Granet had in mind were patterns simultaneously appearing in a vast field of force, the dynamic structure of which we do not yet understand . . . The parts, in their organizational relations, whether of a living body or of the universe, were sufficient to account, by a kind of harmony of wills, for the observed phenomena.[19]

Compare this with Anthony Aveni's observation from another cultural context:

This associative way of thinking is, as I have said, an easy way to recall perceived patterns in the universe, a way of fitting patterns and events into a scheme or system covering all the mutual influences that might occur among them. If I assign an entity in a particular position in a list to one of a series of rotating states or qualities, then that entity automatically acquires a relationship to the quality; there is no need to look for a causal connection.[20]

Number too plays a crucial role in this conception:

In China numbers were used as qualitative instruments of order. According to Granet, the Chinese did not use numbers as quantities but as polyvalent emblems or symbols which served to express the quality of certain clusters of facts and their intrinsic hierarchical orderedness. Numbers, in their view, possess a descriptive power and thus serve as an ordering fact for "clusters of concrete objects, which they seem to qualify merely by positioning them in space and time." In Chinese thought there is an equivalence between the essence of a thing and its position in space–time.[21]

---

[16] Needham (1969, 289), quoting Jabłoński (1939); Major (1978, 13).
[17] Needham (1969, 289).
[18] Ibid., 290. The examples cited must have been understood at the time as simultaneous.
[19] Ibid., 302, n. b. Here one immediately thinks of "quantum entanglement."
[20] Aveni (2002, 92).   [21] Needham (1969, 229).

This idealized role of number achieves its highest expression in the elaborate system of the *Book of Changes* already mentioned above, in which the sixty-four individual hexagrams give graphic shape to the symbolic descriptive power of numerical relations, while at the same time embodying, in their dynamic relations, the infinite changeability and creative potential of the cosmos.[22] As Lawrence Fagg observed,

> The *I Ching* [*Book of Changes*] tells us that each moment can be denoted by a number indicative of the quality of that moment. Therefore, while there is in a real sense a value placed on the moments of Chinese linear time, it is not obviously goal-directed or influenced. Hence, this time also may not be easily identified with the physical world's historical arrow.[23]

The figure of "time as speeding arrow," which appears in the epigram at the head of this chapter, evokes time's fleeting nature and does not connote linearity or teleology as in the West. In this regard it is also worth mentioning that some days in the cycle of sixty in continuous use since Shang times were more auspicious than others, sometimes because of punning associations with homophones having lucky significance. Remarkably, the identity of this set of favorable and unfavorable terms persisted largely unchanged from the early Bronze Age through the Han. In discussing this aspect of the day-dates divined about in the late second millennium BC Shang oracle bones, David N. Keightley called to mind Clifford Geertz's telling remark that the Balinese "don't tell you what time it is, they tell you what kind of time it is."[24]

Synchronicity in the *Book of Changes*

Let me offer an illustration of the intersection of timeless pattern and dynamism in the *Book of Changes*, which shows how this distillation of so much early Chinese thinking about change and timeliness can enlighten us about views of temporality and causation, as well as about certain other prefigurative metaphors

---

[22] According to Carl Jung (1969, §870), "number...is a more primitive mental element than concept. Psychologically we could define number as *an archetype of order which has become conscious*... the unconscious often uses number as an ordering factor much in the same way as consciousness does. Thus numerical orders, like all other archetypal structures, can be pre-existent to consciousness and then they rather condition than are conditioned by it. Number forms an ideal *tertium comparationis* between what we usually call psyche and matter, for countable quantity is a characteristic of material phenomena *and* an irreducible *idée force* behind our mathematical reasoning. The latter consists of the 'indisputability' which we experience when contemplating arrangements based on natural numbers. Thus number is a basic element in our thought processes, on the one side, and, on the other, it appears as the objective 'quantity' of material objects which seem to exist independently outside our psyche." It must also be noted that Needham saw the Chinese tendency to rest content with the apparent explanatory power of number in the *Book of Changes* as the chief impediment to further development in philosophy.

[23] Fagg (1985, 155).   [24] Quoted in Keightley (2000, 33, n. 55).

in ancient Chinese thought. In the *Changes*, where the quality of the moment is a function of *shi*, "inherent spatio-temporal advantage," the timeliness of every action (or inaction) is especially prominent and repeatedly stressed.[25] This is the *kairos* of the Greeks, "the right time." Plato contrasted *kairos* ("occasion") with chance:

What happens by "chance" is said to be opaque to human understanding; chance is a coming together of events that, for all we can understand or determine, could have happened at "any time." Occasion, on the contrary, points to a favorable time which makes possible what, under different circumstances, could not come to pass.[26]

Small wonder, then, that a preoccupation with not encountering receptive times or meeting with unfavorable circumstances that frustrate one's ambitions should have loomed large in the minds of Chinese thinkers in the late Warring States and Han periods, especially given the troubling precedent of Confucius' own failure to achieve due recognition in his day. In his famous prose poem on the theme of "Gentlemen of Integrity Unappreciated in Their Time," *Shi bu yu fu*, the most influential Confucian thinker of the Western Han Dynasty, Dong Zhongshu (*c.*179–104 BCE), was deeply influenced by the *Book of Changes*:

> Alas, the whole world goes along with perversity!
> I grieve that we cannot join together in turning back.
> What else can I do but return to the constant task?
> And not let myself be cast about by the times.
> Though all profit be gained by violating the true self,
> Still it is better to straighten the mind and cleave to the good.
> If only the buffeting of urgency causes me to be moved,
> Surely I cannot be said to have an intemperate nature?
> Clearly manifesting "Unifying Men"
> Means "Possession in Great Measure."
> And to brightly show forth the "radiance of modesty,"
> Means to further the cause.

Tong ren, "Fellowship with Men," and *Da you*, "Possession in Great Measure," are hexagrams thirteen and fourteen in the received text of the *Changes*. Their pivotal importance in the Han Confucian interpretation of the *Changes* is second only to that of the first two, *Qian* and *Kun*, in that they are taken to symbolize the means (humanism and self-cultivation) and the end (political unity and social harmony) of the Confucian sociopolitical agenda. In terms of

---

[25] Cf. e.g. Lin Li-chen (1995, 98). In the *Changes*, "the quantitative measurement of duration or change is made in order to gain insight into the proper qualitative moment to take action . . . With such an achievement-in-process worldview, 'time' is not fully distinguishable from spatial circumstances. The world is a spatio-temporal matrix of interrelated changing particulars." Sellman (2002, 193). *Shi*, "inherent spatio-temporal advantage," is not the same word as *shi*, "time, season, timeliness."

[26] Smith (1986, 13).

their structure, there was thought to obtain an intrinsically dynamic relationship among the central ideas and images embodied in these two hexagrams, which is represented graphically in their configuration. These are two of the very few hexagrams that have a complementary pair of *yin* and *yang* lines occupying the two central, mutually interacting and supremely important positions in the hexagram – the second and fifth lines. Traditionally, the second line is associated with the concept "subordinate" and the fifth line with that of "superior" or "ruler." In both cases, then, we have a representation of the ideal situation in which a yielding or receptive line and an assertive or creative line finds its counterpart in precisely the right location. Both hexagrams therefore symbolize the ideal relationship of a wise ruler paired with a sagely adviser, but in two different aspects.

That is not all, however, because the two hexagrams are also mirror images of each other (Figure 12.1), denoted by the term *zong*. Drawn from the craft of weaving, this term originally referred to the threading (or tying on) of the longitudinal warp threads to the heddle-shafts (*zong kuang*) that alternately raise and lower the warp threads in different sequences and configurations to create the patterns in the weave. What this means in the case of *Tong ren* and *Da you* is that the one is immanent in the other, the one simultaneously *is* the other. Through the dynamics of their unique relationship and the changeability of the hexagram lines, the *yin* line in the second place in "Fellowship with Men" advances to the ruling place in "Possession in Great Measure." In terms of the *Changes*, therefore, in a very real sense "Possession in Great Measure" is inherent in "Fellowship with Men." Though perforce portrayed graphically in linear fashion, and elaborated sequentially, in reality the elements and number symbolism of the one are simultaneously their mirror image.

In his prose poem Dong Zhongshu expressed this dynamic relationship linguistically, linking the two emblematic hexagrams by means of the co-ordinating conjunction *er*, thereby grammatically transforming the dynamics embodied in the two hexagrams into contingent verbal processes. In this way Dong was able to convey immanence and complementarity syntactically. In other words, implicit in achieving "Fellowship with Men" is the realization of "Possession in Great Measure," which here refers, of course, to the ascendancy of a superior man to rulership of the empire. The yielding virtue of the superior man in a subordinate position rises to occupy the central and ruling place by virtue of his ability to expand the principles of fellowship from the few to the many. In the language of the commentary: "the yielding finds its place, finds the middle, and the Creative corresponds with it: this means Fellowship with Men... Only the superior man is able to unite the wills of the sub-celestial realm."[27]

[27] Wilhelm (1967, 452).

Tong ren
Fellowship with Men

Da you

*Possession in
Great Measure*

Figure 12.1 The "mirror image" *zong* relation between hexagrams *Tong ren* and *Da you.*

As we saw, the term *zong* for the mirror image relationship between the two hexagrams "Fellowship with Men" and "Possession in Great Measure," like many of the most important metaphors in ancient Chinese philosophy relating to time and order, is drawn from cordage and the art of weaving. By far the most important of these terms from the weaver's craft are *jing* and *wei*. *Jing* is the warp of a piece of woven goods, and by extension pre-established, order-giving principles, canonical texts, and, in our own day, meridians of longitude. *Wei*, in contrast, are the weft threads of a piece of woven goods; the Five (visible) Planets (*wu wei*) which shuttle back and forth across the sky crossing the meridians in succession; apocryphal, unorthodox commentaries on the Confucian canon; and, in modern times, parallels of latitude.[28] In the present context, *zong* is thus evocative of a *fabric of relations*, comprising warp and weft, sequentially linear as well as recursive in the making (like the trigrams themselves and the numerical manipulations used to derive them), but whose full composition, texture, and qualities can only be grasped in the totality of their complex patterning and structure. The fabric of relations and philosophical ideals evokes a complementarity of principle and pattern – *any segment of which invokes the whole tapestry*. Again, it is not that the two hexagrams are linked as cause and effect, or that one brings the other into being. Rather the one *is* simultaneously the other, like the obverse and reverse of a silk brocade.

In early Han, Dong Zhongshu said,

> Your servant has heard that in Heaven's great conferring of responsibilities on the king there is something that human powers of themselves could not achieve, but that comes of itself. This is the sign that the Mandate has been granted. The people of the empire with one heart all turn to him as they would turn to their fathers and mothers. Thus it is that Heaven's auspicious signs respond to [the people's] sincerity and come forth.[29]

Here we see Dong grappling with the problem of finding the figurative language to express a quasi-mystical sense of noncausality, but ultimately resorting to the prevailing notion of mutual resonance (*gan ying*). Consider now the following explanation of dynastic prosperity and decline by the late Han scholar and iconoclast Wang Chong (27–97 CE) a century and a half later: "When the Mandate of Heaven is about to be issued, and a Sage-King is on the point of emerging, before and after the event the *materia vitalis* gives proofs which will be radiantly manifest."[30] Even in Wang's time, the principle of causality invoked here in relation to auspicious portents tended to be understood simplistically by "mere prognosticators" in terms of cause and effect: "The errors of the School

---

[28] The *Huainanzi*, like many classic texts, has explicit recourse to this metaphor: "In the fabric of the earth's shape, east and west are the weft, north and south are the warp." Major (1993, 167). In the "Treatise on the Celestial Offices," Sima Qian explicitly applies the terminology to the sky, showing that his mental image is of longitudinal sections converging on the *Purple Palace* at the Pole, cross-cut by the movements of the Sun and Moon and the shuttle-like motion of the planets, producing a reticulated "woven" scheme: "the *Purple Palace, Chamber* (5) and *Heart* (5), *Weight* and *Balance Beam* (25–6), *Pool of Heaven, Ruins* (11) and *Roof* (12), the *Sectional Asterisms* within the array of lodges, these are the seats of *Heaven's Five Offices*. They are the warp *jing* – immobile, their sizes differ but their separation is constant. *Water, Fire, Metal, Wood*, and Saturn, these five stars are *Heaven's Five Assistants*; they are the weft *wei* – their appearing and disappearing have their seasons." *Shiji*, 27.1349.

[29] *Han shu*, "Dong Zhongshu zhuan," 56.2500.

[30] Trans. Hsiao (1979, 594). Cf. Kalinowski (2011, 194).

of Prognosticators are not in acknowledging the occurrences of calamities and auspicious happenings, but lie in their erroneous belief that the successes and failings of government bear a cause-and-effect relationship to those."[31] According to Wang Chong,

The accession of a worthy ruler happens to occur in an age that is going to be well governed; his virtues are self-evident above, and the people are automatically good below. The world is at peace and the people are secure. The auspicious signs all display themselves and the age speaks of those as induced by the worthy ruler. The immoral ruler happens to be born at a time when chaos is to exist; the empire is thrown into troubles and the people's ways become disorderly, with unending disasters and calamities, leading to the fall of the state, the death of the ruler, and the displacement of his successors. The world all refers to that as having been induced by his evils. Such observations are clear about the external appearance of good and evil, but fail to perceive the internal reality of good and bad fortune.[32]

Hence, in this view, all the actions of an individual or an undertaking which is about to flourish will spontaneously accord with the spatio-temporal config-uration. In the case of an emerging Sage King: "Followers will come to him unsummoned, and auspicious objects will come to him unsignaled. Invisibly moved, they will all arrive in concert *as if they had been sent*."[33]

This is precisely what Granet was referring to when he spoke of "patterns simultaneously appearing in a vast field of force," and what C.G. Jung meant in stressing that the Chinese world-outlook involved a causality principle quite unlike that of Galilean–Newtonian science, which he denoted "synchronistic." According to Lucien Lévy-Bruhl the "primitive mind," "unlike our own notions of causality, is indifferent to 'secondary' causes (or intervening mechanisms); the connection between cause and effect is immediate and intermediate links are not recognized."[34] He said, "primitives do not probe causal connections in the scientific mode, not because of deficiencies in their individual mental structures, but because such examination is precluded or excluded by their social doctrines, or by the parameters of their systems of knowledge."[35]

Classical Chinese is tenseless, so that temporal relations are somewhat fluid and typically indicated contextually by the use of aspect markers and explicit time words. Taken together, these factors seem to militate in favor of a relative devaluation of precision when it comes to temporal indications, and prioritizing

---

[31] Hsiao (1979, 594).   [32] Trans. Hsiao (1979, 594, italics mine). Cf. Kalinowski (2011, 50).
[33] Ibid., 595.   [34] Tambiah (1990, 86).
[35] Quoted in ibid., 88. One is reminded here of the chapters "The Syllogism and the Tao" and "The Social Origins of Mind" in Richard E. Nisbett's *The Geography of Thought* (2003, xvii): "The collective or interdependent nature of Asian society is consistent with Asians' broad, contextual view of the world and their belief that events are highly complex and determined by many factors. The individualistic or independent nature of Western society seems consistent with the Western focus on particular objects in isolation from their context and with Westerners' belief that they can know the rules governing objects and therefore can control the objects' behavior."

relational or situational import. This state of affairs, as in the account of the aged commoner at the feast and in Clifford Geertz's remark about "what kind of time it is" in Bali, brings to mind another suggestive parallel from the anthropological literature, an account of the cognitive devaluation of linear time among the Trobriand islanders first documented by Jacob Malinowski. Consider the following description of Trobriand concepts of time and temporality:

> There is no boundary between past Trobriand experience and the present; he can indicate that an action is completed, but this does not mean that the action is past; it may be completed and present or timeless. Where we would say "Many years ago" and use the past tense, the Trobriander would say, "In my father's childhood" and use non-temporal verbs; he places the event situationally, not temporally. Past, present, and future are presented linguistically as the same, are present in his existence, and sameness with what we call the past and with myth, represents value to the Trobriander.[36]

Perhaps in this description of the cultural devaluation of temporally structured narrative in favor of patterned relations and activity, we can also gain an inkling of what inspired the Chinese metaphorical recourse to the art of weaving. The early Chinese synthesis of the complementary aspects of time and space into an all-embracing fabric of acausal, patterned orderedness, far from being a metaphysical innovation of the immediate pre-imperial period, like many other images in the *Book of Changes*, owes much to concepts harking back to China's remote past. As David Schaberg says of the historiographical perspective of the *Zuo Commentary* and the *Discourses of the States*, "historical progression is not so much a line as a fabric."[37]

Cordage, weaving, and structural metaphors of space–time

From the above discussion and the highlighted terminological peculiarities it should already be clear that the principal evocative metaphors referring to time and space in early China derive from cordage, weaving, and patterned fabric.[38] As in the case of the invention of writing considered above in Part Two, this fruitful source of structural metaphors enabling the ancients to conceptualize time and space has a long history.

The "Appended Commentary" to the *Book of Changes* (third century BCE) says, "In high antiquity they knotted strings and brought order; the Sages of later generations switched to writing with inscribed graphs." Figure 5.2a is an example of an Inka *khipu* illustrating its basic structure. Since none

---

[36] Lee (1973, 139).   [37] Schaberg (2001, 275).

[38] I exclude here from consideration the architecturally inspired term *yuzhou*, "cosmos," which emerged quite late and the currency of which appears to have been limited to certain kinds of specialized Warring States and Han period cosmological and metaphysical discourse. For Zhang Heng in the first century CE, at least, it referred to what lies beyond the physically observable universe (see below).

of these artifacts has survived in China we cannot be certain their ancient recording devices were identical; nevertheless, the nature of the medium and the structural constraints it imposes ensure that the prehistoric Chinese variety would have been analogous.[39] This is also evident from the terminology and symbolic applications we will explore now. What the following discussion will illustrate is the seamless and uniquely successful transformation of an analog device for recording information into an abstract symbol, a "mind tool" for the conceptualization of space–time. As Anthony Aveni has shown, the very structure of a string-counting device like the *khipu* "invokes radial and hierarchical ways of representing space and time."[40] The motif of *gang ji*, "selvedge and threads," probably represents the earliest Chinese figurative conceptualization of the heavens, long antedating the "canopy-sky," *gai tian*, theory of Warring States and Han times. As Merlin Donald has observed,

Quite early in the history of visuographic symbolism, analog devices were invented that served both a measurement and predictive function in representing time. These devices eventually allowed humans to track celestial events, construct accurate calendars, and keep time on a daily basis... Space was also represented in early visuographic symbolism. Some of the earliest analog symbolic devices were primarily geometric and reflected an abstract understanding of spatial relationships.[41]

The first model of space–time

By far the most important figure of speech in this regard is *gang ji*, "bring regular order to; govern."[42] The compound is made up of *gang*, "cord forming the selvedge of a net or textile; and by extension control, maintain in order, direct," and *ji*. The meaning of *ji* comes from the process of silk production. A crucial step in unwinding the silk from the boiled cocoon is first to find the head-end of the thread. Once that is done the thread can be unwound from the softened cocoon and reeled up smoothly and easily. Similarly, the individual warp threads (*ji*) need to be individually tied onto the loom before beginning to weave. From this, *ji*, "thread-end," acquired the extended meanings of "to straighten out, put in order," its most common usage in pre-Qin literature. Later it also came to mean "keep time, temporal record (annals), period of years, to record," etc., which all derive from the basic notion of "linearity, serial order, sequential, stringing."[43] In philosophical literature one encounters

[39] For East Asian examples from the Ryukyu archipelago, which may give an idea of the appearance of ancient Chinese analogues, see Figure 5.2c and Simon (1924).

[40] Aveni (2002, 252).   [41] Donald (1991, 335, 336).

[42] Compare modern Mandarin *gang ling*, "guidelines, principles."

[43] Needham (1969, 555). Defining *ji* as it applies to time concepts in the Han period, Michael Loewe (1995, 312) stresses the linear connotation: "the term *chi* 紀, or thread, suggests the

the combination *gang ji* as a metaphor for "positions and connections with other things in the web of relationships in both nature and human society; implicit structural organization."[44] We saw an example of this verbal usage "to keep track of (reckon, count) time" in the anecdote about the Old Man of Jiang at the banquet. The headings of the twelve divisions of *Master Lü's Springs and Autumns Annals* setting forth the seasonal and symbolic correlations of the months of the year are called *ji*, combining order-giving, temporally sequential, and celestial senses.[45] In the *Book of Odes* we find "govern the four quarters (*gang ji si fang*)," where *gang* is glossed as "to deploy" and *ji* as "to bring order to." In *Mozi*, "Conforming Upward" (*Shang tong*) it says, "Anciently the Sage Kings instituted the Five Punishments for use in governing the people. Just as a skein of silk has a lead-end (*ji*) and a net has a head-rope (*gang*), [the Five Punishments] are what bring into line commoners in the sub-celestial realm who fail to conform with their superiors." Again in *Mozi*, "Heaven's Will" (*Tian zhi, zhong*): "Moreover, I have reason to know that Heaven abundantly cares for the people – by alternately deploying Sun, Moon, stars, and asterisms to illuminate and guide them, and by making the four seasons spring, autumn, winter, summer *to structure* (*ji gang*) [their activity]."[46] In *Xunzi* (*Yao wen*), his followers describe the Master this way: "His wisdom [having attained] ultimate clarity and his according with the Dao in acting aright were sufficient to serve as the *order-giving structure* [*zu yi wei ji gang*]." In the *Book of Rites* (*Liji*), "Vessel of the Rites" (*Li qi*) it says, "Therefore, one cannot but seriously attend to the superior man's practice of the Rites. They are the *guiding threads* [*ji*] of the masses, if the threads are in disarray the masses are disorderly."

In the early imperial glossary *Cang Jie pian*, attributed to the Qin chancellor, Li Si, in a list of semantically associated meteorological graphs one encounters the following series: "clouds, rain; dew, frost; new Moon, season; Sun, Moon;

line . . . formed by a series of successive incidents or segments." However, *ji* also has a special-ized meaning in reference to resonance periods or constants such as planetary periods, especially Jupiter's, in astro-calendrical calculations. Here it is the recursiveness that is significant.

[44] Paraphrasing Needham (1969, 555).

[45] *Pace* Lewis (2007, 46), who renders *ji* as "calendar" in a famous passage from *Zuozhuan* about the model bureaucratic order of the sage kings of antiquity. While overlooking the paradigmatic significance of both *jing*, "warp," and *ji*, Lewis (2006a, 296) claims the geometric terrestrial schema of the *Classic of Mountains and Seas, Shanhaijing*, forms "a complete model incorpo-rating the entirety of space and time. It is in fact the earliest cosmography in China." In point of fact, there is virtually no "cosmo-" in the *Shanhaijing*'s "cosmography." The *Shanhaijing*'s treatment of the celestial (and consequently that of Mark E. Lewis's chapter, "World and Cos-mos") is decidedly earthbound. The focus of the classic is terrestrial organization, topography, and the horizon calendar as they relate to the gods and culture heroes. The early and pervasive cordage and weaving metaphors that concern us here have a much better claim to constitute the earliest cosmographic model. They are certainly more archaic.

[46] Johnston (2010, 253, added emphasis).

star, asterism; *ji*, *gang*; winter, cold; summer, heat."[47] Unless one is aware of the link between the sky and "structuring threads–cords" the appearance of *ji gang* in this series would be mystifying. In the mind of the lexicographer, however, the association was obviously very close. Lu Jia (*c*.228–*c*.140 BCE), in his "New Analects" (*Xin yu*), opens by discussing cosmography, techniques of the Way, and correlative cosmology:

[Heaven] positions them [sc. myriad things] according to the host of stars, regulates them with the *Dipper*, envelops them with the Six Directions, *reticulates them with threads and ropes* (*luo zhi yi ji gang*), reforms them with disasters and disturbances, makes announcements to them with auspicious omens, motivates them with life and death, and makes them aware with patterns and revelations.[48]

Lu Jia refers here to the structuring order of things as conferred by Heaven. All the above passages convey the flavor of *ji gang* (or *gang ji*) as a figure with deep cosmological associations universally understood in Zhou through Han discourse.

The main selvedge or top-cord that holds the shape of the net and permits its proper deployment is a *gang*, while the pendants suspended from it are *ji*. In a net they form a mesh, with its "eyes," *mu* (e.g. in *gang mu*), while in a *khipu*-like recording device *ji* would be the dependent strings on which information is recorded by knots tied at various points. The analogous pair of ancient terms from weaving – *jing*, "warp," and *wei*, "weft" (for surely the two were also named by the late Neolithic) – overlap in meaning and extend the metaphor to the maintenance of moral order in society.[49] Later these ethical and social guidelines which provide guidance for both individual self-cultivation and state governance were laid down by the sages in the timeless "canonical writings," *jing*, whose "unorthodox" commentaries are the *wei*, "weft," or apocryphal texts so influential from the mid-Han Dynasty. Raising and lowering of the warp threads is accomplished by means of treadles that manipulate the heddles (*zong*, already encountered in its technical sense in the *Changes*). Alternately lifting and lowering the heddles in sequence produces the pattern in the fabric, *wen* (with silk signific), essentially the same word as the "patterning" of literate culture, *wen*, which is the hallmark of a well-ordered society.

---

[47] See slip 3228 from *Cang Jie pian* in the recently unearthed corpus of bamboo-slip texts conserved by Beijing University. Cf. Laozi's use of *Dao ji*, "guiding thread of the Dao," in Chapter 14 of the *Laozi*.

[48] Goldin (2008, 14, trans. modified, my emphasis). Note especially Lu Jia's verbal use of the word "net," *luo*, aptly translated by Paul R. Goldin as "reticulates."

[49] As David Schaberg (2001, 298) observes of the *Zuozhuan*: "In the usage of the historiographers, the words *jing* and *jingwei* suggested not a written canon, as they would for later ages, but a body of inherited norms . . . These norms guided social relations without being entirely visible to all classes. They commanded obedience without being fully available for inspection and testing."

As metaphors for the creative ordering of human activity, the processual nature of weaving and the "fabric" of human society are effective and highly evocative metaphors. Indeed, we exploit the same metaphor in English. Returning for a moment to the Trobiand Islands, it is illuminating to compare what Dorothy Lee meant to express by suggesting a knitting analogy for this form of time-consciousness:

There is organization or rather coherence in their acts because Trobriand activity is patterned activity. One act within this pattern brings into existence a pre-ordained cluster of acts. Perhaps one might find a parallel in our culture in the making of a sweater. When I embark on knitting one, the ribbing at the bottom does not *cause* the making of the neckline, nor of the sleeves or the armholes; and it is not part of a lineal series of acts. Rather it is an indispensable part of a patterned activity which includes all these other acts.[50]

The Trobriand Islands are not China, obviously, but one cannot help but be struck by the anthropologist's homely knitting analogy. Perhaps in this description of the cultural devaluation of temporally structured narrative in favor of patterned relations and activity, we can also gain an inkling of what inspired the early Chinese metaphorical recourse to the art of weaving.

The *Book of Changes* is concerned with *shi*, "the potentiality of spatio-temporal position," and above all with *shi*, "timeliness"; that is, with penetrating insight into the quality of the moment so as to know when and how to act (or not to act, as the case may be).[51] As we saw above in Dong Zhongshu's prose poem, in the final analysis the potentiality of one's spatio-temporal position is *prefigured* by the complex combination of elements and influences making up both the structure and the pattern of the fabric of the space–time continuum as it spontaneously weaves itself into being out of *materia vitalis*. Herein lies the tragedy of the situation of the "Gentlemen of Integrity Unappreciated in Their Time," whose ability to significantly alter spatio-temporal circumstances is severely limited.

Metaphors of the astronomical frame of reference

For of Meridians and Parallels
Man hath weav'd out a net, and this net throwne
Upon the Heavens, and now they are his owne.

John Donne (1611)[52]

---

[50] Lee (1973, 135).
[51] Sellman (2002, 193). The two *shi* homophones are written differently. As Sima Qian said, "What I call time is not the passage of time; men inevitably have propitious and impropitious times," quoted in Harbsmeier (1995, 52, trans. modified). As Harbsmeier points out, Greek *kairoi*, "points/moments in time," is close in meaning.
[52] Quoted in Needham (1970, 11).

The traditional Chinese vocabulary of basic features of the cosmos (and also of the political and social order of this world, which derives from it), is full of images from the language of threads, textiles, weaving, cords, and nets. Too little attention has been paid to this subject by philologists.[53]

What are the implications of such metaphors for astronomy in the earliest period? Given the great antiquity of knotting and weaving technology in China, archaeologically attested by 4300 BCE at Hemudu, might such knowledge also have prefigured how the ancient Chinese conceptualized the cosmos and time as early as the Neolithic?[54] The epigram, "*time [flashes by] like an arrow, Sun and Moon [fly back and forth] like the shuttle of the loom, and another year is past*," likens the motions of the Sun and Moon in their paths across the heavens to the reciprocal movement of the shuttle of a loom as it trails the weft thread behind it. The shuttle shoots back and forth through the opening (shed) formed by the sequential lifting and lowering of the gathered warp threads in alternating, predetermined sequences. (Actually, the poet is taking considerable liberties here since the movements of Sun and Moon are not reciprocal, but repetitive, and the two do not actually reverse direction; the five visible planets do, of course.) Calling the *Five Planets* the *Five Wefts* is obviously evocative of the polar-equatorial *jing–wei* grid, which was coming into use in the late Warring States period.[55] The great first-century astronomer Zhang Heng put the matter succinctly:

The Sage, as the essence of all mankind, traces the implicit structural organization of nature and society [*ji gang*], fixing the celestial co-ordinates [*jing wei*] and the eight limits [*ba ji*] . . . what is beyond [the celestial] no one knows, and it is called the "cosmos" [*yu zhou*]. This has no end and no bounds.[56]

But the history of this way of conceptualizing the structure of the heavens and the passage of time as recorded in the sky is much older. We saw above how the poet deploys what must already have been a familiar trope in early Zhou in the *Book of Odes*: "Tireless is our King, [he] brings order

---

[53] Schafer (1977, 262).   [54] Kuhn (1995, 92).

[55] In glossing *jing* in his *Commentary on the Shuowen jiezi*, Duan Yucai (1735–1815) quotes *Dai the Elder's Record of Ritual Matters* (*Da Dai Liji*), "for north–south one says *jing* ['warp' or longitudinal threads] for east–west one says *wei* ['weft' or lateral threads]." We also saw in the *Huainanzi* chapter "Quintessential Spirit," where that cosmogonic passage speaks about the emergence of *yin* and *yang*, it says, "two spirits sprang from obscurity, *ordering in [jing] Heaven and aligning on (ying) Earth*." In the "Treatise on the Celestial Offices" (*Shiji*, 27.1350) Sima Qian deploys these same metaphors and terminology; see Appendix 1.

[56] Trans. Needham and Wang (1959 , 217, trans. modified). Clearly, for Zhang Heng, *yu zhou* denotes the realm of the formless. Needham glosses *ji gang* figuratively here as "nexus of connections in Nature," which may be appropriate for *ji gang* in philosophical contexts, as we saw above. However, the technical astronomy in the passage that follows (elided here) suggests that "celestial guides and threads" would be more apt.

[*gang ji*] to the four quarters."[57] And again: "Abundantly flowing are the Jiang and Han (rivers), they are the structural threads (*gang ji*) of the southern states."[58] Later, similarly referring directly to things celestial, the "Old Text" (*Yin zheng*) chapter of the *Book of Documents* says, "Then it was that Xi and He subverted their virtue and sank into the confusion of drunkenness. They deserted their offices, left their places and *first disrupted the Celestial Pendants* [*tian ji*].*" The negligent astro-calendrical officials literally "lost track of the time," but the text invokes the image of "disrupted *Celestial Pendants*."[59] By the time these cordage metaphors became conventional in Spring and Autumn and Warring States times, their celestial referentiality was already well established.

In the Han Dynasty, the application of such ruling metaphors to social relations reached a climax, as exemplified by the discussion of the Three Major and Six Minor Relationships which occupies an entire chapter in the *Comprehensive Discussions in the White Tiger Hall*, *Bo hu tong* (first century CE). There *ji gang* and its converse *gang ji* are treated as synonymous in their application to social relations. The Three Major Relationships (*gang*) are those between lord and subject, father and son, and husband and wife. The Six Minor Relationships (*ji*) are those with father's brothers, and between male siblings, kinsmen, and friends. Glossing the terms in its classic catechetical format, the *Comprehensive Discussions* says,

What do *gang*, "Major Relationship," and *ji*, "Minor Relationship," mean? *Gang* means *zhang*, "to spread out"; *ji* means *li*, "to put in order." The greater [relationships] form the *gang*, the lesser the *ji*; thereby [the positions of] superior and inferior are spread out and ordered, and the way of man is adjusted and managed. All men harbor the instinct for the Five Constant [Virtues], and possess the disposition to love; they are developed by [the rules for] the Major and Minor Relationships, *as a net which has small and large net ropes [ji gang] spreads out its ten-thousand meshes.* The *Book of Odes* says: "Very zealous was King Wen, [he promulgated the rules for] the Major and Minor Relationships to the Four Quarters" ... The *Three Major Relationships model themselves on Heaven, Earth, and Man. The Six Minor Relationships model themselves on the Six Cardinal Points.* [The relation between] Lord and subject models itself on Heaven, and represents the coming and going of the Sun and the Moon, due to the workings of Heaven. [The relation between] father and son models itself on Earth, and represents the Five Elemental Phases begetting one another. [The relation between

---

57　*Da ya* ode "*Yupu*" (Mao 238). The version quoted in *Discussions in the White Tiger Hall* has *weiwei*, "tirelessly," for *mianmian*, "energetically," which the commentator says is the preferred reading. Som (1952, Vol. 2, 560).

58　Ode "Fourth Month" *Si yue* (Mao 204); trans. Karlgren (1950b, 155). The Mao commentary glossses this: "Their daemons are sufficient to bring order to [*gang ji*] the whole quadrant."

59　The second line is a paraphrase of the original "abandoned the seasons and disordered the [count of the] days," *fei shi luan ri*, in *Shiji*, "Basic Annals of the Xia Dynasty," but it illustrates how *Celestial Pendants* (*tian ji*) was understood in the Han.

husband and wife models itself on Man, and represents the unison of *yin* and *yang* in man, by which he possesses the faculty of propagation.[60]

Here we see the typical Han Confucian modeling of basic social relations in terms of the natural order, and the penchant for reading contemporary aspects of that· order back into the canonical *Book of Odes* (compare the translation of the same ode Mao 204 above). More relevant to our present discussion, however, is the astronomical origin of the metaphor of the net with its ropes and cords that provide shape and confer utility. The significance of the connection between the six *ji* ("Minor Relationships") and the Six Cardinal Points will become clear in a moment.

Celestial ropes and pendants

Before attempting to elucidate what *"Celestial Guidelines"* are, we first need to revisit the time-keeping role of the *Dipper*. It is helpful to realize that the *Dipper* was conventionally known as *dou gang*, or the *Dipper Top-Cord*. Referring to the structure of the *khipu* with added *Dipper* in Figure 12.3 for comparison, this is a general allusion to the *Dipper*'s serving as controlling, stabilizing center and guide rope. We have already seen in Part Two the crucial role of alignment utilizing the Pole from earliest times, which no doubt involved the stretching of cords, just as in Egypt. The figure of the *Dipper Top-Cord* preceded and inspired its extension to other realms of experience – theorizing about the nature of society and governance as well as later metaphysical speculation.[61] Commenting on the Andean *khipus*, William J. Conklin provided an illustration of an early horizontal loom and observed,

The structural form of quipus suggests a relationship to the format of ancient Peruvian looms, in that the main cord of the quipu is similar to a heading cord and the pendant strings are conceptually similar to the warp. In a quipu, the information is added on to the pendant cords: in weaving, the image pattern is added on to the warp.[62]

---

[60] Trans. Som (1952, Vol. 2, 559–60, trans. modified, italics mine).

[61] On this point, see Harper (1999, 831), who comments that Warring States, Qin, and Han manuscripts "have provided abundant data to bear out Joseph Needham's characterization of Chinese astronomy as 'polar and equatorial' . . . polar and equatorial because Heaven was organized around the polestar, which radiated outward to the constellations marking the celestial equator (the sun's path was noted but it was not the central fact informing spatial and temporal schemes) . . . The basic conception of the celestial regions predated the Warring States period. Between the fifth and third centuries B.C., astrologers and calendrical experts gave specificity to this conception with precise observations and theoretical elaboration. Their explanations of the macroscopic operation of Heaven and Earth probably contributed most to the formation of the idea that all phenomena and human activity were linked in microcosmic synchronicity; that is, their role in the emergence of yin-yang and Five Phases correlative cosmology was seminal." See Needham and Wang (1959, 229).

[62] Conklin (1982, 265, and Figure 3). For the circumpacific distribution of *khipu* and the possible diffusion of the technology with Polynesian seafarers, see Chapter Five above.

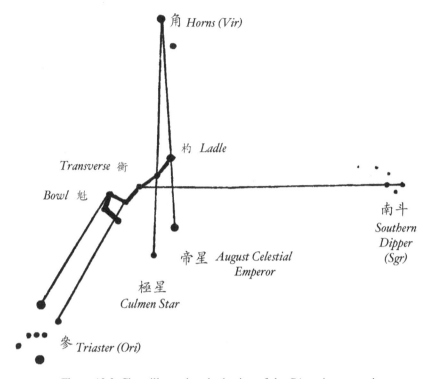

角 *Horns (Vir)*

杓 *Ladle*

Transverse 衡

Bowl 魁

南斗
Southern Dipper (Sgr)

帝星 *August Celestial Emperor*

極星
Culmen Star

參 *Triaster (Ori)*

Figure 12.2 Chart illustrating the keying of the *Dipper*'s stars to important seasonal lodges. Redrawn from Needham and Wang (1959, 233, Figure 88), © Cambridge University Press, reproduced by permission.

Figure 12.2 illustrates the role of the *Dipper* in serving as a pivotal determinant of the locations of three supremely important seasonal lodges at times when they are not visible in the sky. In the *History of the Former Han Dynasty* we are told, "The *Jade Transverse* [Alioth, ε UMa] and the *Ladle Indicator* [Benetnash, η UMa] comprise *Heaven's Top-Cord* [*tian zhi gang*]: [where] Sun and Moon begin their motion, these are the *Pendants of the Stars* [*xing zhi ji*]." The "Treatise on the Celestial Offices" in *The Grand Scribe's Records* explains what is illustrated in Figures 12.2 and 4: "*Ladle* connects to the *Dragon's Horns* [α Vir]; *Transverse* [Alioth, ε UMa] hits the *Southern Dipper* [π Sgr; Winter Solstice]; the *Bowl* rests on the head of *Triaster* [Orion]."[63]

---

[63] *Shiji*, 27.1291; see Appendix 1. The "Monograph on Harmonics and the Calendar" in the *History of the Later Han Dynasty, Hou Han shu*, elaborates: "Anciently, when the sages created the calendrical [i.e. astronomical] system, they observed the rotation of the *Jade Device* [*Dipper*], the motions of the *Three Luminaries*, the movements of the Dao, the length of the [gnomon's]

Figure 12.3 The *Dipper* as *Top-Cord* (*dou gang*). Ancient looms similarly gathered all the warp threads together in a loose knot tied to a head rope or warp beam, allowing the weaver to put tension on the warp (photo DWP).

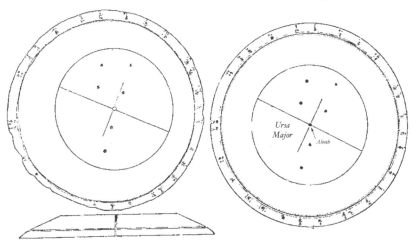

Figure 12.4 Laquer mantic astrolabe from the Han Dynasty tomb of the Marquis of Ru Yin (second century BCE) on which stellar guidelines appear. Redrawn from Zhongguo shehui kexueyuan kaogu yanjiusuo (1980, 115), reproduced by permission.

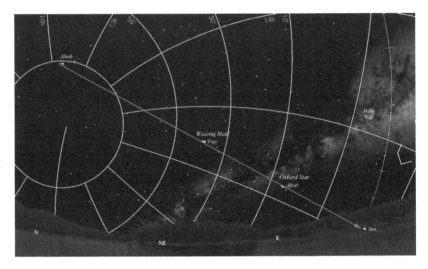

Figure 12.5 Chart showing the alignment on Alioth in the handle of the *Dipper* down through the *Weaving Maid* (Vega) and the *Oxherd* (Altair) stars to the Sun around winter solstice in the mid first millennium BCE (Starry Night Pro 6.4.3).

Thus in its position at the apex of the sky, the *Dipper* is likened in this scheme to the top-cord of a net, and particular lines radiating from the *Dipper* are the guidelines or pendants.[64] *Stellar Pendants* or *Guidelines* (*xing ji*) in particular refer to the alignment between the *Jade Transverse* (Alioth) and the *Southern Dipper* (Figure 12.5) at the point where Sun, Moon, and Planets commence their journeys after fording the *Sky River* (Milky Way). The "Monograph on Harmonics and Calendrics" in the *Han shu* explains the connections:

shadow, where the *Dipper Top-Cord* (*dou gang*) points, and the path the *Cerulean Dragon* treads. 'They investigated their changes, synthesized the numerical constants,' and made of them a system." *Hou Han shu*, "Monograph on Harmonics (B)," *Lü li zhi, xia*, 3.3055: "They investigated . . . constants" is quoted from the first part of the "Appended Commentary" to the *Book of Changes*. In the astronomical systems of the Han Dynasty and later the cordage terminology extends even to cyclical resonance periods. Planetary cycle constants were called *ji mu*, "periodic factors," and day–month-cycle constants were termed *tong mu*, "gathered factors" (*tong* being a technical term from silk weaving for the gathered head-ends of silk threads, or picks); cf. Needham and Wang (1959, 406). The "path the Cerulean Dragon treads" was discussed in Part One above and it will reappear in Chapter 13 below.

[64] Note that, as a practical matter, to establish such stellar alignments one would necessarily have to employ the kinds of sighting techniques described in Part 2. Here, perhaps, we also have the cosmographic explanation for the pairing of the *Northern Dipper* and *Southern Dipper* in the "Dipper Wall" of Chang'an.

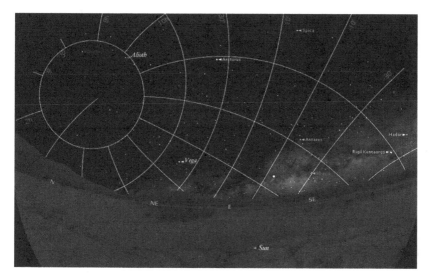

Figure 12.6 Chart showing the *Weaving Maid* star Vega's predawn rising on January 1, 799 BCE (Julian) just ahead of the Sun at winter solstice, at about the epoch of the ode "Great East" in the *Book of Odes* (Starry Night Pro 6.4.3).

Therefore the *Tradition* [*Zuozhuan*] does not say "winter solstice," but says "the Sun reached its southernmost extreme," the limit being the initial [degree] of lodge *Ox-Leader* [β Cap]. At the hour of noon [the Sun's] shadow is longest, and from this one knows it is at its southernmost extreme. The end of the *Dipper Top-Cord* connects through *Align-the-Hall* [Pegasus], and the *Weaving Maid's Pendant* [through Vega, α Lyr] serves to mark off the [periods of] Sun and Moon, hence it is called *Stellar Pendant*. The Five Planets start from its beginning, and the Sun and Moon from its middle. For each of the twelve stations, whenever the Sun reaches the initial [degree of the station] it marks that *qi*-node, and when it reaches the centers, where the *Dipper Indicator* points below are the twelve chronograms [*chen*]. By observing where it indicates one knows the Sun's station.[65]

[65] *Han shu*, "Monograph on Harmonics and the Calendar," 21.984. The end of the *Dipper Top-Cord* refers to the first star of the *Dipper*'s bowl, Dubhe or α UMa, otherwise known as *Celestial Pivot* (*Tianshu*). A line from this star through the celestial Pole leads directly to the Sun's location at the beginning of spring in *Ding* (our Pegasus), the celestial square whose multifaceted role we encountered above. Just ahead of the bowl of the *Dipper* is the asterism *Flourishing of Civic Culture* (*Wenchang*), whose role, according to the apocryphal *Xiao jing yuan shen qi*, is defined as "pattern" (*wen*), wherein quintessence gathers, expansively distributing the *Celestial Pendants* (*tian ji*) [emanating from the *Dipper*]. The inscribed perpendiculars on the astronomical instrument in Figure 12.4 from the tomb of the Marquis of Ru Yin (mid second century BCE) intersect at Alioth (ε UMa) in the handle of the *Dipper* and show the location of the *Celestial Pendants* at the inter-cardinal nodes which divide the sky into the four palaces. Zhongguo shehui kexueyuan kaogu yanjiusuo (1980, 115); Yin Difei (1978, 342); Shi Yunli, Fang Lin, and Han Zhao (2012).

In referring alternately to alignments *Weaving Maid* ↔ *Southern Dipper* (*Han shu*) and *Weaving Maid* ↔ *Ox-leader* (*Shiji*), the two texts seem to be in contradiction. That they are not is explained by the fact that *Stellar Pendant* is actually the name of the so-called "Jupiter station" (*sui ci*) or thirty-degree space which includes both lodges *Southern Dipper* (*8*) and *Ox-Leader* (*9*).[66] The two *Stellar Pendants* identified in the texts thus bracket the space containing the asterism *Establishment Star* (*jian xing*). This asterism is at the edge of the Milky Way in the northeast corner of the sky where Sun, Moon, and planets emerge after "fording" the *Sky River*. This is why the *Han shu* "Monograph" can speak of the *Stellar Pendant*'s "beginning" and its "middle," as *Stellar Pendant* is here a stand-in for *Establishment Star*.[67]

In view of the pervasiveness of the cordage and weaving motifs in cosmology it is surely not happenstance that it is precisely the bright star Vega and not some other that came to be identified as the *Weaving Maid* (*zhi nü*), since it is that very star that plays the pivotal role in marking this all-important alignment – the "*Weaving Maid's Pendant*" (*zhi nü zhi ji*). The myth about the *Weaving Maid* star's tryst arose from this important astro-calendrical role. The ode "Great East" (*Da dong*, Mao 203) invokes stellar imagery to exemplify the mismatch between name and reality: "In the heavens there is the *Sky River*, looking down and bright; slanting is the *Weaving Maid*, all day long she shuttles seven times, but does not complete any woven pattern; brilliant is the *Draught Ox*, but it is not yoked to any wagon."[68] The language is too ambiguous to be certain that the juxtaposition of the *Weaving Maid* and *Draught Ox* actually alludes to a liaison.[69] Indeed, for the metaphor to work the *Draught Ox* here has to be an ox and not the herdboy of later legend. The line (bridge or ford) from the *Weaving Maid* across the Milky Way through the *Oxherd* star (Altair), and down to the ecliptic (Figures 12.5–6),[70] marked the location of winter solstice at the time of the composition of the famous ode in mid first millennium BCE. Moreover, Vega's predawn rising just before the solstice probably conferred on the *Weaving Maid* the status of herald of the approach of

---

[66] This particular *Stellar Thread*, of course, also marks the location of the northeast "node" or "connecting cord," *wei* (lit. "to tie, bind together; drop-cord of a net"), which joins earth and sky.

[67] The four inter-cardinal "cords" (*wei*) terminating at the corner "hooks" (*gou*) are shown on the *Xingde* MS diagram shown in Figure 5.9.

[68] Trans. Karlgren (1950b, 155, trans. modified). For a possible early cosmogonic role for the *Weaving Maid*, see Porter (1996, 97). Chen Banghuai (1959b, 6) identifies a feminine star "declining" in the north in the Shang oracle bone inscriptions as *Serving Girl* (lodge 10 in Aquila), but at this early date more than likely the reference is to a single bright star like Vega.

[69] That the legend was certainly current by the Warring States period is shown by the fact that the poet in the ancient "Nine Songs" (*Jiu ge*) poem from Chu, *Ai sui*, imagines himself wedding the *Weaving Maid* during his celestial journey.

[70] Schafer (1974, 404).

this all-important turning point and may explain why she is called the Daughter of Heaven in the "Treatise on the Celestial Offices" in *The Grand Scribe's Records*.[71]

This dawn role of the *Weaving Maid* is immortalized in a "Rhapsody on the Luminous River," *Ming he fu*, by Xie Yan (seventh century), who refers to her as "made up with rosy dawn-light, dimpled with stars."[72] This and the story of her liaison with the *Oxherd* on summer nights half a year later originally encoded this vital astro-calendrical knowledge, just like the more ancient etiological myth in *Zuozhuan* about the deified feuding brothers Yan Bo and Shi Chen (encountered in Chapter 9) recalls their ancient association with the equinoxes in Scorpius and Orion.[73] The subsequent obsolescence of the Vega–Altair alignment due to precession explains why attention eventually shifted away from the *Draught Ox* star Altair to the solstice's location in lodge *Ox-Leader* (*qian niu*), β Cap.[74]

In early Zhou, the *Weaving Maid*'s first evening appearance would have occurred just after sunset at the beginning of May, the *Draught Ox* Altair about a month later. Then, in late July and early August, the two stars shone brightly, obliquely contemplating one another from opposite banks of the *Sky River* as they slowly traversed the northern sky throughout the night. In the *Lesser Annuary of Xia*, the orientation of the *Weaving Maid*'s three component stars close to the zenith was a principal indicator of the seventh month (Figure 12.7).[75] It is for this reason that the seventh day of the seventh

---

[71] Confirming this, the *Weaving Maid* and *Oxherd* [alt. *Draught Ox*] stars figure prominently as seasonal indicators (called "time's/season's beginning," *shi shou*, and "time's/season's end," *shi wei*) in the astronomy of the Yi-minority of Sichuan, Guizhou, and Yunnan. In Yi astronomy the entire northern quadrant is held to evoke the relationship of the *Weaving Maid* and *Oxherd* that finds an echo in the two lodges *Serving Maid* and *Draught Ox* which supplanted them. Chen Jiujin, Lu Yang, and Liu Yaohan (1984, 102, 106–7). The commentarial consensus is that the current reading "Granddaughter" of Heaven is an error; see Appendix 1. In Sima Qian's own time, in the Shanglin Park Emperor Wu famously "carved out an artificial lake, and erected statues representing the *Weaving Maid* and *Cowherd* stars to make it a replica of the Milky Way." Lewis (2006a, 178).

[72] Trans. Schafer (1977, 260).

[73] Needham and Wang (1959, 248, 282); Yi Shitong (1996, 29). Nathan Sivin (1989, 59) recognized that "mythology may contain the earliest allusions to an oral tradition of ancient astronomy." In a similar vein, Erik Havelock (1983, 14) observed, "as long as preserved communication remained oral, the environment could be described or explained only in the guise of stories which represent it as the work of agents: that is, gods."

[74] On the interversion of the *Draught Ox* and *Weaving Maid* stars due to precession, see Needham and Wang (1959, 251); also Schafer (1977, 144). Karlgren's misidentification of the star *qian niu* as β Cap rather than as the *Draught Ox* α Aquila exemplifies the confusion this redundancy has caused. Karlgren (1950b, 155).

[75] Two millennia earlier at this time of year the *Weaving Maid* and *Oxherd* stars, like *Ding* (Pegasus), would have aligned precisely on the Celestial Pole. For the *Weaving Maid* in the *Lesser Annuary of Xia*, see Hu Tiezhu (2000). For accounts of the legendary romance, see

Figure 12.7 The "summer triangle" of Deneb, Vega, and Altair on July evenings in 800 BCE. Note that given their respective locations, the path across the *Sky River* from the *Weaving Maid*, Vega, to the *Oxherd*, Altair, is always oblique, which may explain the otherwise obscure line of the ode, "slanting (or leaning) is the *Weaving Maid*" (Starry Night Pro 6.4.3).

month traditionally came to be known as the "Pleading for Artfulness" festival (*qi qiao jie*), a day when young girls implored the *Weaving Maid* to bestow on them exceptional talents in needlework and weaving.[76] Trapped house spiders' spinning habits during the night provided a basis for predicting the fortunate bestowal of such talent. In a passage on a theme traditionally composed for this occasion, no less a figure than Liu Zongyuan (773–819) writes,

Schafer (1974, 404–5); Krupp (1991); Birrell (1999, 165). For informative studies of the Milky Way in cultural astronomy, see Urton (1978); Rappenglück (2002); Bertola (2003).
[76] *Quan Tang wen*, 583, 1b.

Heaven's grandchild,
Skillful specialist of the Sky,
Her shuttle the *Jade Device*,
Her warp and woof the starry signs,
Capable of creating the patterned emblems,
To adorn the body of the Supernal Lord.[77]

Liu Zongyuan is echoing Wang Yi's (89–158 CE) "Rhapsody on Women Weavers" (*Jifu fu*) centuries earlier. There, as Dieter Kuhn observed, "the operation of the loom is often rendered in terms of images taken from astronomy. Thus the rhapsody relates the three lights (*san guang*), sun, moon, and stars, to the weaving girl star."[78] In the "Rhapsody," "warp-beam and cloth-beam revolve, like the heavenly bodies"... "the headle-rods (*guang = zong kuang*) [rising and falling] are the Sun and Moon, they evoke the shining [of Sun and Moon as they rise and set]"... "The three round beams [warp-beam, back-roller and cloth-beam] keep the fabric in proper order, modeled on the [*Three*] *Terraces* above."[79] As Edward Schafer remarked on translating Liu Zongyuan's poem, "This is truly cosmic weaving... producing a fabric patterned in... figures that correspond to the numinous asterisms. In effect, the Weaver Maid weaves the sky patterns (*t'ien wen*) that have such a powerful influence on the destiny of men and nations" (Figure 12.8).[80]

In "The Spider's Web, Goddesses of Light and Loom," Justine Snow compiles evidence bearing on the connection of the *Weaving Maid* with pan-Eurasian myths about weaving, spiders, and celestial dawn goddesses as "weavers of light."[81] She overstates the case in asserting a specifically Indo-European origin for the mythic complex, but the comparative evidence she adduces proving the widespread mythological and folkloric associations is unmistakable, including the leitmotif of the spider as the patroness of weaver goddesses and celestial maidens who weave the variegated light of dawn.

---

[77] The *Jade Device* represents the *Dipper*; see Appendix 1. Cf. Schafer (1977, 147).

[78] Kuhn (1995, 99).

[79] Ibid.; trans. Knechtges (1987, 263–77, trans. modified). The *Three Terraces* are formed by the six stars ι, κ, λ, μ, ν, ξ UMa beneath the *Dipper*; Appendix 1. Here again, we find the circumpolar stars playing a prominent role, in this case as the framework of the loom itself, whose shuttle Liu imagined he saw in the *Dipper*.

[80] Schafer (1977, 147). The metaphor continues to inspire, as a recent comment by Diane Ackerman (2011, 5) shows: "Every society has been tantalized by the great loom of the sky with its quilt of stars."

[81] Snow (2002). The curious line "all day long she shuttles seven times" in the ode "Great East" is obscure, but presumably refers to the *Weaving Maid*'s activity while Vega is invisible during the roughly seven double-hours of daylight in the seventh month. No discernible change in the starry patterns results from her labors. At this time the two star-crossed lovers were highest in the sky and the Milky Way was becoming increasingly visible as the atmosphere began to clear after the dust storms of spring and early summer. See also Scheid and Svenbro (2001); and especially Witzel (2013, 155–6).

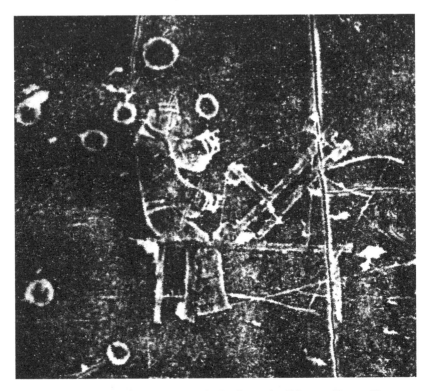

Figure 12.8 The *Weaving Maid* at her loom; detail from an Eastern Han stone relief. The large star above her head is Vega; after Zhongguo shehui kexueyuan kaogu yanjiusuo (1980, 51, Figure 49), reproduced by permission.

Indeed, now that we can affirm that China's *Weaving Maid* also played a seminal role as dawn harbinger of the winter solstice, the link is even stronger. But there is no evidence that the art of weaving itself was introduced to China from the west. On the contrary, by the time Indo-Iranian pastoral nomads arrived in the Tarim Basin, weaving had already been established in China for millennia. Silk was being woven by Liangzhu weavers in the southeast by 3000 BCE. The Shang were already weaving patterned silk fabrics and gauzes using fixed horizontal looms by the mid second millennium BCE.[82] Woolen textiles with their characteristic weaves, like those unearthed in present-day Xinjiang, were not important to China's sedentary agriculturalists, mainly because of major differences in the principal raw materials; that is, silk versus short-staple

[82] Kuhn (1995, 92, 93, n. 39).

fibers like wool or linen.[83] And as we have demonstrated, weaving and cordage images and terminology provided the dominant structural metaphors in cosmology from an early date. The connection between weaving and cordage is self-evident, as is an iconic role for the spider, but the evidence for other borrowing in astrological or cosmological realms is scant.[84] In my view, the mythic complex delineated by Snow must greatly pre-date the second millennium BCE. Given the universality and great age of spinning and weaving technology (dating from the Paleolithic), the omnipresence of a deified patroness of weaving throughout Eurasia is more likely explained by a common origin in archaic traditions connected with spinning, weaving, and netting skills, dating at least from the Mesolithic. Ethnography in the Americas only reinforces the impression. As Cecelia F. Klein observed,

That at least some of Mesoamericans conceived of their universe in terms of cords and fabric should not really surprise us, given the importance of weaving in their daily life. Aztec girls were taught to use the loom from early childhood, and their most important manual task was the production of cloth. The entire life cycle could be visualized in terms of various forms of twisted and netted, or woven, objects . . . [85]

Conclusion

In their seminal study of the *khipu* as a "visual language," M. Ascher and R. Ascher describe that prestigious and highly symbolic object as "special" in a very particular sense:

In every culture, there are special objects that stretch to regions far beyond the particular purpose for which they were intended. These objects sum up areas of meaning for which there may be no verbal counterparts. They usually are common and replaceable, but they are not always inexpensive . . . sometimes they are keys to the culture for the outsider; for the insider, in particular children in the process of learning, they are indispensable cues. They may be fixed and rather permanent as the Medieval cathedral, or they can be impermanent and leave no trace, like an igloo. We think that the quipu is such a special object.[86]

The essential requirement of a "special object" is its ability to give resonant symbolic shape to central cultural ideas. From the above discussion it seems

---

[83] Kuhn (1995, 79–80).

[84] As for the spider, one recalls, of course, the Greek fable of Arachne whose surpassing skill at the loom provoked a fit of jealousy in Athena, goddess of the craft. It is worth noting here that, in a nearby burial close to that of the Puyang cosmo-priest accompanied by cosmogonic mosaics of tiger and dragon, one mosaic is said to represent a spider. Feng Shi (1990b, 117).

[85] Klein (1982, 29). Klein remarks further, "conception and birth were accordingly compared to the acts of spinning and weaving: all Aztec and Maya creation and fertility goddesses were described as great weavers. Today, Tzotzil grandmothers give spinning lessons to young women at winter solstice to encourage them to be good sexual partners to their husbands." Ibid., 15.

[86] Ascher and Ascher (1975, 355).

that the Chinese knotted-string "tool of governance" akin to the *khipu* and the art of weaving jointly fulfilled this cultural function. The metaphorical use of the images and terminology of the two technologies came to be so thoroughly intertwined that one cannot discern any distinction between them in cosmography and cosmology. The early Chinese synthesis of the complementary aspects of time and space into an all-embracing fabric of acausal, patterned orderedness derives its inspiration from the most ancient everyday crafts of cordage and weaving (the latter exclusively women's work in ancient times).[87] The metaphors fundamental to their cosmo-vision, the making of the starry images above, and the socializing patterns of the life-world below – all derive from that earliest, most vital industry.

If we return now to the theme with which we began, ancient Chinese perceptions of time and temporality, rather than thinking in terms of cyclical or linear time, it would seem that the conception is recursive in inspiration. Recursion is a process in which elements (or processes) repeat in a self-similar way. A familiar illustration is when two mirrors are placed facing each other, producing nested images that form an infinite recursion. Recursion is intrinsic to the art of weaving, of course, and at some level also to the idea of generating a cosmos from binary or quinquinary elemental forces, as in the *Book of Changes*.[88] Another illustration of recursion, however, particularly apt in the Chinese case, is the recursive definition of one's ancestors: "One's parents are one's ancestors (*base case*). The parents of one's ancestors are also one's ancestors (*recursion step*) . . . " etc. Here, perhaps, we have a clue to the dominant rearward-looking perspective in Chinese thought, in which it is assumed that the ideal models were all laid down in the past as the work of the ancestors. Therefore the task of descendants is not to innovate but to faithfully carry on those age-old patterns, and in this way contribute to the persistence of the civilization in perpetuity.

---

[87] "In the time of Shen Nong . . . the women wove and the people were clothed." *Book of Lord Shang*, *Shang jun shu*, "Hua ce," 4.9b.

[88] Recursion in practice is not so straightforward. As one everyday source puts it, "Even if properly defined, a recursive procedure is not easy for humans to perform, as it requires distinguishing the new from the old (partially executed) invocation of the procedure; this requires some administration of how far various simultaneous instances of the procedures have progressed. For this reason recursive definitions are very rare in everyday situations." See http://en.wikipedia. org/wiki/Recursion.

13   The Sky River and cosmography

Fu Xi, Nü Wa, and the *Sky River*

As the most obvious large-scale physical structure in dark skies, as one might expect the Milky Way has left a deep imprint in astral lore the world over. It has been seen and is still seen by many people as a flowing river, a link between the world ocean and the sky; as the road or path taken by the spirits of the dead to ascend into the sky; as Γαλαξίας (*Galaxias*) in Greek (Latin *via lactea*), a stream of milk accidentally spilling from the breast of Hera as she suckled Herakles; as the goddess of the firmament Nut who daily gave birth to the Sun in ancient Egypt,[1] or the Egyptian–Greek Ourobouros; as a monstrous, serpent-like Tiamat in ancient Mesopotamia whose tail became the Milky Way;[2] as the Celestial Cayman of the Maya or "Cloud Serpent" (Mixocatl) of ancient Mexico,[3] all representing primieval cosmogonic forces. The worldwide variations on the theme are nearly endless, but they share a number of common attributes – the Milky Way is typically associated with water or clouds and rain, often with half-human reptilians, a means of ascent into the sky to an alternative world of existence, incomprehensibly powerful natural forces, and so on (Figure 13.1).[4]

We saw above in Chapter 2 how the earliest textual reference to the Milky Way in China occurs in the *Book of Odes* where it appears as a *Silvery River* in the sky. The ancient almanac *Lesser Annuary of Xia*, whose astral harbingers of the seasons would have been accurate throughout the Bronze Age, contains the earliest mention of the *Sky River* as a seasonal indicator: "in the seventh month the [*Heavenly*] *Han* [points] straight up; at twilight the *Weaving Maid* faces due east, and when the handle of the *Dipper* points straight down, it is

---

[1] Kozloff and Bryan (1992).
[2] Jacobsen (1968). The Greek "Serpent-Bearer" constellation Serpens-Ophiucus and its likely predecessor, the strange triple-bodied, chthonic demon "Bluebeard" (prominently depicted on the pediment of a sixth-century BCE temple on the Acropolis, now in the Acropolis Museum), may both have derived from the Babylonian constellation of half-human serpent deities in the same location, the "Sitting and Standing Gods." White (2008, 187); Hurwit (2000, 108–9).
[3] Milbrath (1999).   [4] Krupp (1991); Wu (1963, 25); Witzel (2013, 125).

Figure 13.1 The creation myth in the Greek Theogony. Zeus is shown here slaying Typhoeus with thunderbolts; Chalcidian black-figured hydria (*c.*550 BCE): Zeus and Typhon (Inv. 596) (reproduction), Staatliche Antikensammlungen und Glyptothek München, reproduced by permission. As in the case of Marduk, who slew Tiamat in the Babylonian epic *Enuma Elish*, Typhoeus is also slain to form heaven and earth.

dawn."[5] Subsequently, as we saw, the Milky Way reappears in its crucial role as the archetypal *Sky River*, which defines the basic structure of the system of astral–terrestrial correspondences in field-allocation astral prognostication and in the layout of the Qin imperial capital. Then, in the Han Dynasty, the *Sky River* is the essential feature of the reconceptualized binary macro-astrology that made its appearance in the Simas' "Treatise on the Celestial Offices."

Together with the Five Elemental Phases scheme by now firmly linked in correlative cosmology with the seasonal attributes of the palaces of the heavens, *yin* and *yang* re-emerge in the new astrological paradigm as astral–terrestrial south and east versus north and west in reference to the *Sky River*. This appears to contravene the conventional association of *yang* toponyms with the north bank of a river. The reason is obvious. Like the primacy accorded to the astral fields in the field-allocation scheme outlined in Chapter 9, the stars determined what was *yang* and what was *yin*. Simply put, the Sun spends the *yang* portion of the year in the southern and eastern palaces of the heavens, and the *yin* portion of the year in the northern and western. Once again, the celestial skyscape trumped terrestrial topography.

The parallels outlined between certain attributes of cosmic monsters and the Chinese mythic duo Fu Xi and Nü Wa (male and female) will become apparent. Cosmogonic Fu Xi and Nü Wa, representing the conjoining of *yin* and *yang*,

[5] Hu Tiezhu found that the *Lesser Annuary of Xia* (*Xia xiao zheng*) was used in Zhou but may date from the early Bronze Age. The best indicators are the stars said to culminate at dusk in various months, which are all self-consistent. For the Milky Way, see Hu Tiezhu (2000, 234). For the Milky Way in the Vedic literature, see Witzel (1984).

figure prominently in representational tomb art, dressed in contemporary garb and holding the artisanal tools compass (Nü Wa) and square (Fu Xi).[6] They are portrayed as "human" from the waist up and dragon- or serpent-like from the waist down, often with their tails twined in intimate (procreative) embrace. In addition to the obvious cosmogonic role suggested by the tools they brandish, texts attribute to them all manner of civilization-bringing innovations, including the idea of knotting strings to keep records. Perhaps not surprisingly, in view of our discussion in Part Two, the principal invention attributed to Fu Xi in the earliest accounts is the invention of the calendar and record keeping using knotted strings. But their mythic role also includes procreating the human race. So as not to stray too far afield, however, our discussion here will focus on their cosmic attributes.

Depictions of Fu Xi and Nü Wa as cosmogonic figures appear first in late Warring States and Han iconography, probably as a consequence of the growing cultural influence of the south (Figure 13.2).[7] In *Huainanzi*, the pair's attributes are identified with the god Tai Hao (aka Fu Xi) and spirit Gou Mang respectively: "What are the Five Planets? The East is Wood. Its god is Tai Hao. His assistant is Gou Mang. He grasps the compass and governs spring. His spirit is the Year-star [Jupiter]. His animal is the blue-green dragon."[8] Two centuries earlier, *Zuozhuan* (seventeenth year of Duke Zhao) records, "Tai Hao used *Dragon*(s) to keep records [*ji*, "thread; string"], so he made a Dragon Master and named him after the *Dragon*." This jibes with the attribution to Fu Xi (Pao Xi) of the invention of keeping records using knotted strings in the "Appended Commentary" to the *Book of Changes* discussed above.

Some images allude to the pair's procreative role, while others show them in the company of the Queen Mother of the West in her realm of immortals.[9] The most interesting account of their exploits, which captures nearly all of their attributes, appears in a holographic manuscript from Dunhuang entitled "Record of Imperial Rule since the Separation of Heaven and Earth" (*Tian di kaipi yilai diwang ji*), copied in the year 950.[10] The early account that it

---

[6] An account of the careers of Fu Xi and Nü Wa is Lewis (1999, 197–209). For cosmogonic Fu Xi–Nü Wa in the fourth century BCE Chu Silk Manuscript, see especially Kalinowski (2004, 92, 106, 109). Also Wen Yiduo (1982, Vol. 1, 3–68); Liu Huiping (2003); Birrell (1999, 33, 44, 163); Goldin (2008, 11).

[7] Kalinowski (2004, 92, n. 16) provides references.

[8] *Huainanzi*, "Heavenly Patterns," §4: "The Five Planets." Note that here it is Goumang and not Nü Wa who wields the compass. For a comparison of the cosmogonic pair in the Chu Silk Manuscript with their analogues in the "Canon of Yao" and the Guodian text *Tai yi sheng shui*, see Kalinowski (2004). In some accounts, Nü Wa is actually said to be the wife of Yu the Great. Lewis (1999, 203).

[9] Eugene Y. Wang (2011, 58).

[10] There are two versions of the account translated here, in MSs P. 2652 and P. 4016, two of the scrolls acquired by Paul Pelliot now in the Bibliothèque national de France. Others acquired by Sir Aurel Stein also exist (S. 5505 and S. 5785). See the IDP site at http://idp.bl.uk. My translation is principally based on the text established and studied by Su Peng (2009).

(a)

Figure 13.2 (a) Fu Xi and Nü Wa on a stone relief from the Eastern Han (151 CE) Wu Liang Shrine in Shandong (Nü Wa's compass has been lost due to damage). The caption on the left mentions Fu Xi's having "created the institution of kingship, drawn the eight trigrams, and knotted strings to bring order within the seas." After Feng and Feng (1821, 3.7); see Liu Huiping (2008, 297, Figure 5). (b) The same pair holding square and compass carved on a stone sarcophagus from Han tomb no 4 in Hejiang, Sichuan. After Liu Huiping (2008, 296, Figure 1). (c) Fu Xi and Nü Wa on another Wu Liang Shrine relief; here the pair are emerging from the waves. After Chavannes (1913b, Vol. 2, Plate 60).

reproduces in the form of a vernacular dialogue probably dates from before the Sui (589–618), since the last dynasty mentioned in the "Record" is the Jin (265–420).

A.

Question: What did Fu Xi do? About how long did he live?

Answer: Fu Xi had the body of a dragon, his surname was Feng (wind), his given name was Wang ["king"]. He was able to make clothes, to fix in place the Sun, Moon, stars, and planets, establish the myriad things, and motivate *yin* and *yang* so as to bring

(b)

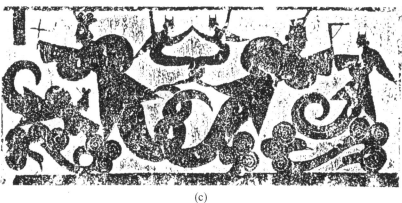

(c)

Figure 13.2 (*cont.*)

winter and summer to completion. At that time the people were stubborn, ignorant, plain and simple, and did not know to differentiate the Rites and Proprieties. As yet there were not Five Grains, clothes, farm holdings, or dwellings, and [people] nested in caves. When [one] met a man he became the husband; when [one] met a woman she became the wife. The myriad things were not yet all in place, [so Fu Xi] looked up at Heaven[11] and fashioned the eight trigrams of the sub-celestial realm. Fu Xi domesticated cattle and broke horses. He lived 18,000 years, whereupon he brought into being Heaven and Earth, the many insects, birds and animals, dragons and snakes, fish and turtles, gold and silver, pearls and jade. [He] governed the myriad things, and promoted procreation without [anyone] knowing his dragon-speech; he was styled Fu Xi.

B.

Question: [You] have enlightened me and relieved my confusion regarding the length of reigns of the Three August Sovereigns, Five Emperors, Xia, Yin, Zhou, Qin, Han, and Jin. But [I] have not yet understood how Fu Xi got to succeed to the rulership of men.

Answer: Fu Xi and Nü Wa were born of a father and mother, but because of a flood the people all died, so brother and sister rode a dragon up to Heaven and their lives were saved. They saw that the sub-celestial realm was all laid waste and chaotic, so they bribed the Heavenly Spirits to teach them how to manipulate *yin* and *yang*, but afterward they felt ashamed and hid themselves in Mt. Kunlun. Fu Xi perambulated to the left and Nü Wa perambulated to the right, but soon they met each other and became man and wife. Heaven blamed them for their union and especially for knowing one another intimately. So Fu Xi used tree leaves to cover his face and Nü Wa used reed catkins to hide her face, and together they lived as husband and wife. That's how it began that people today exchange gifts, wear hats, and adorn themselves. When [Nü Wa's] time was fulfilled she gave birth to 120 sons, recognizing each with a surname. Sixty were respectful, affectionate, and filial, so today *they can be seen in the Sky River*. But sixty were unfilial and unrighteous, so they went into the dense wilds and thus became the cave[-dwelling] Qiang people of Ba and Shu. So it is said [Fu Xi] got to succeed to the rulership of men.

C.

Question: In the time of the Three August Sovereigns, Fu Xi was Heavenly August Sovereign. What was his surname and what was his cognomen? What rules were there?

Answer: Fu Xi was from Luoyang and surnamed Feng. He was the son of the Yellow Emperor of Hanzhong. At that time the people died, and there were only Fu Xi and Nü Wa left. The two of them, brother and sister, rode a dragon up to Heaven and so saved their lives. They were afraid humanity would come to an end, so they became man and wife. Their government was at Yuzhou [Province = Henan], from where they long ruled the sub-celestial realm. They ruled for 18,000 years, after which the Fiery Emperor and Shen Nong succeded [them] as August Sovereigns of Earth. From then on they became [known as] the Three August Sovereigns.

---

[11] The text has "the sub-celestial realm" but the insertion of "under" (*xia*) after Heaven here must result from an eye-skip since *tianxia* appears just a few words later.

Fu Xi and Nü Wa's cosmogonic role is clearly signaled even in the earliest representations by the presence of Sun and Moon, compass and square. Their depiction as hybrid deities symbolizes the combination of cosmic and chthonic attributes, which empowered them to shape both realms. It is their serpentine, earthly aspect that is responsible for their symbolic role as promoters of abundance and procreation. The classic description of their contributions to world formation and civilization familiar from early textual accounts appears in passage A above. In passage B, a long-standing debate about whether Fu Xi and Nü Wa are brother and sister or husband and wife is finessed, albeit by resorting to the unorthodox device of attributing the survival of the human race to the pair's incestuous relations. In passage B this transgression is said to have been the reason for the Supernal Lord's creation of social prescriptions and rites (elsewhere the institution of marriage is attributed to the pair themselves), as well as the use of clothing and adornment.

Representations of Fu Xi and Nü Wa proliferate from the Han Dynasty through the Tang (618–907), so that ultimately a large number of excavated early tombs, including examples as far west as Turfan, have been found with Fu Xi and Nü Wa painted in various locations in the tomb or on funerary banners pegged to the ceiling of the burial vault. One can distinguish two main historical periods in the representation of Fu Xi and Nü Wa. These different styles may be characterized as follows:[12]

(1) An early period represented by the stone reliefs and sacophagi such as in Figure 13.2, in widely distributed Han and Three Kingdoms-period tombs (206 BCE–280 CE). In these Fu Xi's and Nü Wa's tails are loosely crossed or, rarely, entwined snakes appear between their legs at the crotch, suggesting their procreative role. Often accompanied by other reptilian spirits they generally hold compass and square, Sun and Moon, associated with them by gender; i.e. *yin* = female ~ Moon ~ compass, *yang* = male ~ Sun ~ set square.

(2) A middle period represented by Figure 13.3 showing paintings on silk banners or tomb murals from the third through the eighth centuries, in which the pair are vertically elongated, with Sun and Moon, compass and square present, and with tails entwined. They are surrounded by an array of stars, either haphazardly strung together as connected white dots, or clearly recognizable as the twenty-eight lodges. Such representations are especially prevalent in Xinjiang tombs of the period.

(3) Style Two is paralleled in the late Tang by a third stylized variation on the theme, a striking example of which from Astana is shown in Figure 13.4. In these tomb ceiling paintings all trace of Fu Xi and Nü Wa has

---

[12] Liu Huiping (2008, 293–310) studies the styles, distribution, and periodization, and provides copious references to the relevant archaeological reports.

(a)

Figure 13.3 (a) Fu Xi and Nü Wa. After Liu Huiping (2008, 296, Figure 1).
(b) Fu Xi and Nü Wa on a silk tomb banner from Gaochang (Turfan), c.500 CE.
After Zhongguo shehui kexueyuan kaogu yanjiusuo (1980, 58, Figure 56),
reproduced by permission.

now disappeared, the only surviving mythological motifs being the creatures long associated with the Sun and Moon – sunbird, toad, rabbit. More
or less impressionistic depictions of the stars and asterisms are replaced by
relatively accurate depictions of the band of twenty-eight lodges and the
Milky Way, an indication of the spreading of detailed knowledge of the
sky.
    Such stylistic changes of motif in the context of burial customs are no trivial
matter, since mortuary practices were surrounded by all manner of prescribed
rituals, prohibitions, symbolism, and taboos in order to protect the living and

(b)

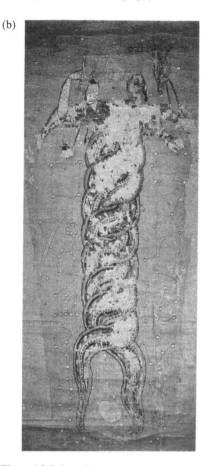

Figure 13.3 (*cont.*)

guarantee a comfortable existence for the deceased in the afterlife. Great effort and expense were obviously invested in reproducing in the vaulted tomb a simulacrum of the sky above.[13]

In some middle-period cases, like the example in Figure 13.3a, the astronomically unschooled artist went overboard in representing the Sun,

[13] Liu Huiping (2008, 299). Compare this depiction with the even more elaborate and detailed star ceiling in the Koguryo era Tokhung-ri tomb (408 CE) in Nampo city, Pyongan South Province, North Korea. More than twenty-four such vaulted star ceilings have been discovered, in some cases showing asterisms distinctively different from their Chinese counterparts. The *Sky River* and distinctively Koguryo representations of Fu Xi and Nü Wa appear in many of the tombs as well. Kim Il-gwon (2005, 25–32).

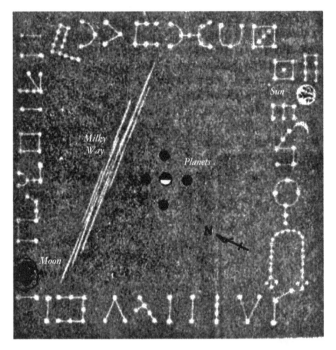

Figure 13.4 Tracing of a celestial ceiling painting from Astana (65TAM38) showing the Sun, Moon, twenty-eight lunar lodges, and Five Planets, with the *Sky River* (Milky Way) represented by the series of white streaks running northeast to southwest. After Zhongguo shehui kexueyuan kaogu yanjiusuo (1980, 66, Figure 69), reproduced by permission.

Moon, and stars, so that the lodges are completely unrecognizable.[14] Instead of showing individual asterisms he has encircled the whole with something like a modern-day string of decorative lighting. In contrast, Figure 13.3b, encountered in the discussion of asterism *Ding* in Pegasus, was executed by an artist who had superior knowledge, if not of the stars themselves, then at least of representational conventions concerning the lodges, which are all faithfully depicted.

Most impressive, however, is the ceiling painting from a tomb in far away Astana in Xinjiang (Figure 13.4), which was already mentioned in the context of the River Diagram. Here the *Sky River* is substituted for Fu Xi and Nü Wa. Remarkably, as is evident from the directional arrow showing compass north, great care was also taken to correctly orient the star ceiling in the tomb so that

[14] For additional examples and references to the relevant archaeological reports, see Liu Huiping (2008, 296–7). See also www.nhu.edu.tw/~NHDH/pdf/dunhung/27/27-1-20.pdf.

it would conform to the true cardinal directions in the upper world. The artist-astrologer who designed this ceiling was obviously an expert in astronomy and geomancy. Star ceilings like this provide invaluable information about how the sky was perceived and about the level of astronomical knowledge at the time.[15] This representation, though stylized, is important for another reason. If we compare Figures 13.3 and 13.4 we find something quite revealing. As already noted, instead of representing Fu Xi and Nü Wa amidst the stars, the ceiling in Figure 13.4 is a square schematic representation of the night sky reminiscent of the famed River Diagram *Hetu*, with the Milky Way correctly oriented flowing southwest to northeast and bisecting the band of twenty-eight lodges. This mural is exceptional in its precision and attention to detail.[16]

In paintings and banners depicting the sky, as *yang*, Fu Xi (who spoke "dragonish") is most often shown leaning to the side occupied by the southern and eastern lunar lodges, while Nü Wa as *yin* occupies the north and west. Their reptilian tails intersect the band of lodges in the northeast, precisely where the tail of the *Dragon* constellation in Scorpius emerges from the *Sky River*. The Moon, a *yin* body, is most often depicted at this location in sky ceilings, this being the ford (or bridge) across the *Sky River* traversed by the Sun, Moon, and planets when they enter the winter sky. Fu Xi's and Nü Wa's upper bodies and implements, representing their cosmogonic aspect, together with the "Great *Yang*" of the Sun, are typically located near the point in Taurus where the paths of the Sun, Moon, and planets again cross the *Sky River*, only now heading into the summer sky.[17]

---

[15] One can only guess at the richness of the celestial imagery to be found in the simulacrum of the sub-celestial realm constructed in the tomb of the First Emperor of Qin, said in the *Records of the Grand Scribe* "to have used mercury for the myriad streams: the Yangtze, Yellow River and the wide sea, with mechanical devices to cause them to flow into one another, and with the heavenly patterns complete above, and all the features of earth complete below." *Shiji*, "*Qin Shihuang benji*," 6.265; Nienhauser et al. (1994, 155).

[16] Zhongguo shehui kexueyuan kaogu yanjiusuo (1980), Figure 69. This and another star ceiling are discussed in Liu Huiping (2008, 303, 305). The same geometrical arrangement of the Five Planets, as with that seen in the center of the River Diagram, appears in a color illustration captioned "Prognostication on the Joining of the Five Planets" (*wu xing ruo he zhan*) in the late Ming astrological compendium, *Tian yuan yu li xiang yi fu* (1628–44). Particularly noteworthy is that since the Sun and Moon are both already represented at top right and bottom left in the Astana tomb painting, the five discs in the center must certainly represent the planets. The half-lighted sphere in the center may actually depict the dichotomy phase of Venus. If so, this is surely the earliest such graphic depiction in history. The ancient Chinese are known to have observed the moons of Jupiter with the naked eye, and observing the Venus crescent requires about the same degree of visual acuity. MacRobert (2005, 116); Hostetter (1990). See also Liu Huiping (2008, 305, Figure 14); Schafer (1977, 79); Miller (1988, 10, Figure 3, image mistakenly reversed).

[17] In this respect the couple may call to mind that other pair of "illicit" lovers with an intimate connection to the Milky Way, the *Oxherd* and *Weaving Maid*, the latter's weaving of the starry patterns perhaps a reflex of Nü Wa's cosmogonic accomplishments. This is speculative, however, for if the *Oxherd* and *Weaving Maid* are a reflex of Fu Xi–Nü Wa, they conspicuously reverse the *yin-yang* associations of their respective sides of the *Sky River*.

(a)

Figure 13.5 (a) Chinese dragon on a Western Han Dynasty stamped tile; redrawn from Liu et al. (2005, 254, Figure 1). (b) Folio 74 from the Dresden Codex showing the Celestial Monster as a sky-band, with the head of a crocodilian instead of a serpent and feet of a deer, its body covered with celestial signs representing Venus, the Sun, the sky, and darkness. A torrent of water gushes from the monster's mouth down past Venus, here represented as God L, while additional floods pour from the glyphs of the Sun and Moon onto the aged Moon Goddess (center). Museo de la Cultura Maya Chetumal, Quintana Roo, Mexico (Photo DWP).

Comparative denizens of the *Sky River*

In addition to the reptilian association with the *Dragon*'s tail in the northeast, it is precisely from the Northern Cross constellation in Cygnus down through Scorpius that the Milky Way unmistakably bifurcates, like the free ends of the demiurges' tails. Their serpentine tails suggest the winding course of the Milky Way, split as it is in places by dark patches of interstellar dust, like islands. The *Sky River*'s gyrations in the sky during the course of the night and through the passing seasons evoke the writhing of a reptilian.[18]

---

[18] Susan Milbrath (1999, 288–91) describes this "writhing" from the Mesoamerican perspective. See also Carlson (1982); Freidel et al. (2001, 85–91); Harris (2011); Isbell (1982, 362); Parpola (2012, 9); Witzel (1984); Wu (1963, 25). In the Babylonian cosmogony *Enūma Elish*, after killing the monstrous, serpentine sea-goddess Tiamat, Marduk (Jupiter) cleaved her body in two and fashioned heaven and earth from the halves: "her tail he bent upwards into the sky to make the Milky Way, and her crotch he used to support the sky." Jacobsen (1968; 1976, 179). This description immediately calls to mind the posture of the Egyptian sky goddess Nut, whose arched body supports the sky. Each day she swallows the setting Ra-sun and he is reborn from her womb at dawn. On the identity of Nut and the Milky Way, see Kozloff and Bryan (1992, 333), especially Figure 11.8, which illustrates the bifurcation of the *Sky River* at the location of Cygnus, identified as Nut's animal familiar, a goose. Egyptian Apep (Apophis), the serpent god (later dragon) who is also associated with the Milky Way, is a malevolent reflex of Nut and a deadly nemesis of Ra.

(b)

Figure 13.5 (*cont.*)

A strikingly similar perception of the Milky Way appears in the iconography of the Cosmic or Plumed Serpent in Mesoamerica (Figure 13.5[19]). An account of the earliest example of this figure in murals recently discovered at San

---

[19] See Baudez and Picasso (1992, 105) for a color reproduction of the original in the Bibliothèque national, Paris. See also Freidel et al. (2001, 106) and Milbrath (1999, 276, Figure 7.4d, 283). Carlson (1982, 153, Figure 10) provides a detailed line drawing.

Bartolo in the Petén, Guatemala (*c*.100 BCE), describes the widespread motif this way:

In the San Bartolo scene, the plumed serpent is the ground-line support for the human figures, constituting the first known instance of this motif, which is widely found in Mesoamerica as well as in ancient and contemporary Puebloan art of the American Southwest. However, the plumed serpent is not just a support or platform; *it is a road or vehicle for supernatural travel... individuals would symbolically ascend or descend atop the back of serpents, which constitute a path or road*... The very symbol of the wind, an Early Classic conch carved in Teotihuacan style, portrays a pair of individuals striding atop the back of a plumed serpent marked with stars. The upwardly turned head of this creature recalls the earlier example from San Bartolo... A series of footprints appears on the San Bartolo serpent, marking it as a *supernatural road*... A mural from Zone 5-A at Teotihuacan features *a plumed serpent body marked with both stars and footprints*, immediately recalling the San Bartolo plumed serpent.[20]

The Maya Cosmic Serpent is bicephalic, precisely like the Shang oracle bone graph ᐳᐸ identified as a rainbow (*hong*).[21] This same bicephalic image occurs in very early curved jade pieces (*huang*) from the Neolithic Hongshan Culture (*c*.4500–2500 BCE) in China's northeast.[22] It is a curious fact that no single pre-Qin word or graph has been identified as standing for that eye-catching *Sky River* arching across the night sky. All the epithets familiar to us – "Heavenly Han," "River of Clouds," "Silvery River" – sound like poetic metaphors for a phenomenon that surely must have deeply impressed Chinese skywatchers from time immemorial and been given a graphic representation. I propose that

[20] Saturno, Taube, and Stuart (2005, 24–5, italics mine), see esp. page 24, Figure 18, and also pages 8–9. The snake-like aspect of the Milky Way is also evident from the Maya name of the starry band, *tamacaz*, which is also the name of the denizen of Mayaland, the deadly "bearded" pit-viper known as the fer-de-lance. Milbrath (1999, 282). A long mural from the tomb of Bo Qianqiu in Luoyang (first century BCE) depicts a man riding a snake to the Queen Mother of the West's realm of the immortals. Wu Hung (1989, 113, Figure 43).

[21] An equivalence between the Milky Way and the rainbow is widespread in South American mythology, for which Lévi-Strauss proposed the structural relation rainbow : Milky Way :: death : life. Their serpentine nature is also well represented. Lévi-Strauss (1969, 246–7). My thanks to Alejandro Martín López for this reference.

[22] In Figure 3.10a we saw the cosmic deity Supreme One grasping a bicephalic rainbow in his right hand and a dragon in his left while standing astride another dragon. Carlson (1982, 145, Figure 6 and passim) provides numerous illustrations of two-headed dragon representations from throughout East and Southeast Asia (a jade *huang* in the case of China) and strikingly similar objects from the Americas. Carlson concludes, "this preliminary investigation examined an assemblage of double-headed forms from cultures that surround the Pacific Basin... The pervasive Asiatic tradition is thought to have migrated into the Western Hemisphere along with those nomadic peoples whose progeny now populate the indigenous New World. The original mythological systems were subsequently modified and adapted to the needs of each specific culture according to the particular ecological circumstances" (ibid., 160). In view of the cosmogonic parallels between the Babylonian Tiamat, the Greek Typhoeus and Python, Fu Xi–Nü Wa, and the Maya Cosmic Monster (among other American parallels), there probably was a pan-Eurasian mythic complex, dating from before the peopling of the Americas in the Holocene; cf. Needham and Lu (1985); Chang (1983, 74–5); Fraser (1968).

the Shang glyph identified as "rainbow" may in fact do double-duty. It may have stood for the arc of the rainbow during the day and the *Sky River* at night, as if the two were a single phenomenon. As Edward H. Schafer observed,

The old un-Indianized *lung* [dragon] . . . as the linguistic evidence shows, was accustomed to display himself – or herself – as the arch of the rainbow . . . [A list of related] words . . . appear to be members of an archaic word family whose meaning combined "serpent" with "arch; vault" . . . Our Chinese dragon, then, is bent and curved like a bow and, like the surface of the sky dome itself, hovers over the aerial hemisphere. The dragon in its rainbow form was widely represented in the early art of south and east Asia. It was the *makara* of India which, like its Chinese counterpart in Han decorative art, appears as a rainbow emblem with a monstrous head at each extremity.[23]

Similarly, David N. Keightley observed that it

seems likely that there was both a semantic and phonetic similarity between 虹 *g'ung/hong*, "rainbow," and 龍 *l'iung/long*, "dragon," and that both words derived from a still earlier word, which may be reconstructed as close to *kliung*, which had the basic meaning of "arched, vaulted" . . . The Shang graphs for dragon . . . were not just *drawings of* a dragon; they recorded the *word for* dragon.[24]

* * *

Fu Xi–Nü Wa's identification with the Milky Way is strongly suggested by passage B above, where their sixty filial and dutiful sons are said to be the personification of that *River of Clouds* (or its distinctive stars and features). In the fourth-century BCE Chu Silk Manuscript their sons are said to be four in number and the implication is that the cosmogonic pair engendered the deities of the four seasons and Four Quarters.[25] Thus the season-defining role of the *Sky River* is alluded to as early as the late Warring States period. Given its prominence as archetypal *River* in the old astral divination and in the new binary macro-astrology promoted by the Simas in the late second century BCE, it would be astonishing if the strikingly prominent *Sky River* dividing the night sky into *yin* and *yang* halves (and the year into spring–summer, autumn–winter) were not also represented in iconography. In the Han Dynasty, as male and female spirits of *yin* and *yang*, Fu Xi and Nü Wa symbolize the original cosmogonic separating out and later systemic and periodic conjoining of *yang* and *yin* (light and dark), first giving shape to the cosmos and then populating it. In this role, though *depicted* in semi-human shape, they are no more human than is Urizen in William Blake's famous print showing him leaning down from

---

[23] Schafer (1973b, 15); quoted in Carlson (1982, 139).
[24] Keightley (1996, 86, original emphasis).
[25] Kalinowski (2004, 106, 109–10) interprets this as an analogical transference of the logic of genealogy into the cosmological realm.

Figure 13.6 Urizen as geometer, William Blake, 1794. Courtesy Library of Congress, Lessing J. Rosenwald Collection (acc. no. 2003rosen1806).

the clouds with fingers extended to form a compass with which to frame the cosmos (Figure 13.6).

Fu Xi–Nü Wa did not continue to serve as guarantors of order in the world or as "tomb guardians."[26] Rather, their original cosmogonic, flood-quelling, and procreative functions portrayed in early accounts having been accomplished, they withdrew into the background like other otiose deities and mythic sovereigns, to be principally invoked in their subsequent mortuary role as tutelary gods of the means to immortality. The conceptually linked modalities of procreation and immortality proceeded through them.[27] Here we have an explanation for their apparently exclusive appearance in mortuary contexts, as pointed out by Hayashi Minao.[28] Rather than being merely "liminal" spirits or simple tomb guardians, in their "dragonitic" watery aspect they personify the cosmic pathway to be taken to the ethereal realm of the immortals. This explains the dynamic linkage of the terrestrial and celestial realms by twining dragons on early tomb banners like the famous example from Mawangdui. As in Fu Xi and Nü Wa's escape from the cataclysmic flood, dragons are the means of transport from here below up to the Gates of Heaven.

The persistence of their chthonic aspect shown by their serpentine tails (sometimes in a variation on the "bicephalic" theme the two torsos share just one tail) evokes their androgynous "two aspects in one."[29] The substitution of the Milky Way for Fu Xi–Nü Wa demonstrates the equivalence between the cosmic pair and the *Sky River*, so that in the course of progress in representing knowledge of the heavens, especially in more expert renderings, it became possible to replace them with the actual physical feature in the sky. The exact correspondence between Fu Xi–Nü Wa and the *Sky River* in position and orientation among the circuit of lodges shows that, whatever other cosmic attributes they may have possessed, they also personified the Milky Way.[30]

---

[26] *Pace* Lewis (2006a, 125–6; 2006b, 129).

[27] Eugene Y. Wang (2011, 58, 84, n. 108).   [28] Lewis (2009, 579).

[29] A long mural from the tomb of Bo Qianqiu in Luoyang (first century BCE) makes their identities abundantly clear by depicting Fu Xi (*yang*) with the Sun at the right end and Nü Wa (*yin*) with the Moon at the left end. Wu Hung (1989, 113, Figure 43); cf. Lewis (1999, 203).

[30] As Edward H. Schafer (1973a, 102) remarked, "After Han times, Nü Kua faded away to become a mere fairy tale being, neglected by the upper classes and ignored in the state religion." For the Maya and Andean bicephalic skyband reptilians, their rain-bringing attributes, and identification with the Milky Way, see Carlson (1982, 146 ff.); Grofe (2011, 72, 75); van Stone (2011, 13–4). Ethnographic evidence from Quechua communities in the Andes provides strong support for these celestial associations: "*Machácuay* is the Quechua name for the dark-cloud 'serpent' constellation lying in the Milky Way – the *Mayu* or 'River'" (ibid., 152–3). Carlson concludes, "these data strongly suggest that the arched, double-headed dragon forms of the Andes are celestial symbols probably related to serpents, rainbows, water, the Milky Way, and the yearly cycles of fertility. The similarities in these celestial representations are far more striking than their differences" (ibid., 159, 149).

Consider the descriptions of the elaborate cosmic inventory of Han tomb paintings. One example, at Guitoushan, helpfully provides captions to the images, including "the 'Gate of Heaven,' the two cosmic deities Fu Xi and Nü Gua, the four directional animals, the Sun and the Moon, numerous immortals as well as the Queen Mother of the West."[31] Another stone relief from Nanyang, Henan, "depicts a snake-bodied creature holding the moon aloft . . . the exact equivalent of Nu Gua on the Sichuan coffins . . . paralleled by a dragon lodged within a cluster of stars."[32] One cannot ignore the essential cosmic nature of the imagery, which is dynamic and evocative of the spirit journey to be undertaken by the spirit on departing the physical body, as depicted on the silk mortuary banner from the Mawangdui tomb of Lady Dai. The implication is not that the tomb conveys finality, representing the furnishings of a static, eternal resting place, however elegantly appointed. Such an interpretation would leave unreconciled the tension between the fervent desire to be transported to the realm of the immortals and the notion of the tomb as the deceased's otherworldly dwelling.

Lai Guolong shows that

this idea of tombs as underground houses is at best an incomplete picture of the conception of tombs in early China . . . Instead, the tomb in part defines the nature of the otherworldly journey: whether conceived of as waystation or starting point, its accumulated chariots and other travel paraphernalia will orient the soul in time and space as it undertakes its journey; and the tomb texts are provided as travel documents so as to orient the deceased within a social, political, and cosmic order.[33]

The tomb is the lodging of the corporeal component or *po* of the deceased, which has the potential to become a malevolent hungry ghost. It is this component that must be enticed by creature comforts to remain quiescent in the tomb. The iconography, in contrast, has the magical potency to conjure the guiding spirits of Fu Xi and Nü Wa. Like the Egyptian *Book of the Dead*, the figures depicted in tombs and on funerary banners are an elaborate visuographic code perceptible in that form to the *hun*, "spirit."[34] Providing guidance to the spirit of the deceased,

---

[31] Lewis (2006a, 126).

[32] Lewis (2006b, 129). Cf. Von Falkenhausen (2006, 312–18).

[33] Lai (2005, 2, 42 and passim). For the Milky Way as the path of souls to the afterlife in legends and myths around the world, see Lebeuf (1996).

[34] Hulsewé (1965, 87, 89); Lai (2005, 31). Mark E. Lewis (1999, 195, 199) questions the existence of this *hun-po* distinction, instead portraying the tomb as the permanent "household" of the dead, while also serving as a "paradise of immortals." Cf. K. E. Brashier (1996), but also Chang Kwang-chih (1990). Mark E. Lewis recognizes (1999, 196), however, that the imagery may have served as "charts for cosmic journeys," and that (ibid., 197) second-century texts testify that sacrifices at ancestral shrines and gravesides caused "the spirits to all descend." Descend from where, one wonders, if they had been thought to be hermetically confined in the tomb? This distinction between corporeal and spiritual components of the soul is, of course, consistent with traditions such as portrayed in the Chu funereal ode "Summons to the Soul" (*Zhao hun*) as well as in other facets of contemporary thought. Pankenier (1986). For the interred body as the locus of transformation whose aim is to "exit from death and enter into life," see Eugene

they portray sights and signs encountered on the cosmic journey westward to the Queen Mother's realm of the immortals and the delights the spirit will ultimately find there.[35] Here we might only add that it is possible to draw a direct connection from this later mortuary imagery, through the cosmic simulacrum of Qin Shihuang's tomb and the elaborate astral decoration of Marquis Yi of Zeng's magnificent coffin, all the way back to the cosmological shell mosaics of the Neolithic cosmo-priest's burial at Puyang. The fervent desire to be transported in style to that supra-visible cosmic realm is symbolically encoded throughout. The principal difference is that now the imagery has transformed from abstract cosmological symbols and emblems into anthropomorphized cosmic deities, and the goal of the spirit journey has been localized in the Queen Mother's western paradise, a realm of immortals whose attributes are celebrated in suitably mythic imagery and song.

This journey of the spirit is mirrored in Tang period tales of poets and others inadvertently straying up the *Sky River*. In one, an errant fisherman magically floats up the *Sky River* and temporarily appears to the earthbound as a mysterious new star before inevitably returning to his former prosaic existence, in some versions even carrying one of the *Weaving Maid*'s loom weights. About this trope Schafer remarks, "this is not *just* poetry – it is accepted belief expressed in poetic language."[36] The flooding motif in the Fu Xi–Nü Wa myths derives from the belief that the downward-flowing *Sky River* joins its earthly namesake and world-encompassing sea at the horizon, producing a continuous circulatory communication between the two realms that sometimes resulted in an overabundance of water here below. As Li Po said, "Han water has from the first communicated with the flow of *Starry Han*."[37] One could hardly ask for a more evocative description of *Sky River*-related images than these lines reminiscent of the "Nine Songs" of Chu from the "Dragon Pool" lyric (*Long chi pian*) by the early Tang poet Shen Quanqi (650–729):

> The Pool is open to the *Sky River*,
> Where it bisects the Yellow Road;
> The Dragon heads for the Gate of the Sky,
> And enters the Palace of Purple Tenuity.[38]

Y. Wang (2011, 75). According to Wang, "for the transformation to take place, a cosmic arena encompassing both heaven and earth is required . . . No matter how cosmic the setting or the scale may appear to be, the dramatic action remains within the bodily theater."

[35] Jiang Xiaoyuan (1992, 214–7) is particularly eloquent in showing how the impulse to "communicate with Heaven" (*tong tian*) distinguishes the politico-religious role of sky pattern reading/astrology (*tian wen*) from the earliest times.

[36] Schafer (1974, 404, original emphasis).

[37] Ibid. For an Inka parallel, see Urton (1981, 111, 113).

[38] "To put it plainly, the divine reptile makes for the intersection of the Milky Way and the ecliptic, and then heads by way of the 'Gate of Heaven' (in our constellation Virgo) to the palace of the greatest of the celestial deities, the Palace of Purple Tenuity." Schafer (1974, 405). Of the "Nine Songs," see especially the poems *Da Siming, Ai sui, Wei Jun*.

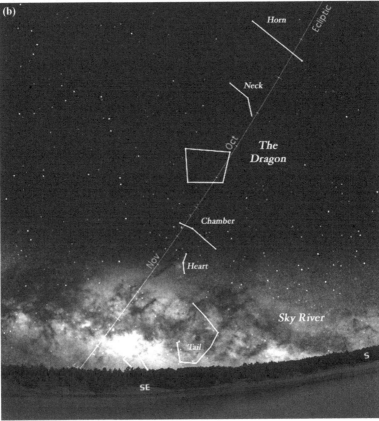

Figure 13.7 (a) The intersection of the *Sky River* and *Yellow Road* at the *Mineral Spring* in the sky; (b) the *Dragon* climbing into the sky out of the *Sky River* or *River of Clouds* (Starry Night Pro 6.4.3).

In Qu Yuan's lament "Encountering Sorrow," *Li sao* (fourth century BCE), in the "Nine Songs," and in *Huainanzi*, the mythic *Mineral Spring* (*Xianchi*) is in the west, where the Sun god bathes after his daily journey across the sky. In the "Treatise on the Celestial Offices" and later, the *Mineral Spring* is in the constellation Auriga in the Milky Way, a source of the *Sky River* (Figure 13.7a). The *Yellow Road* is the ecliptic, the path of the Sun, Moon, and planets across the sky, which intersects the Milky Way in the northeast and southwest. Diametrically opposite Auriga's *Mineral Spring* at the other end of the *Sky River* is the *Dragon* constellation, whose tail is immersed in the *River* where it meets Scorpius, while the *Dragon's* right horn (α Vir) lies just above the *Gate of Heaven* (*Tian men*) in Virgo (Figure 13.7b). The poet's evocative image is of the *Dragon* taking flight from its abode in the *Sky River*, climbing northward through the *Gate of Heaven* to reach the circumpolar *Palace of Purple Tenuity*, seat of the *Supernal Lord*. Can there be any doubt that this is the same *Dragon* ridden by Fu Xi and Nü Wa as they fled up to Heaven to escape the rising flood caused by the overflowing *Sky River*?

14   Planetary portentology East and West

> But when the planets,
> In evil mixture, to disorder wander,
> What plagues, and what portents? what mutiny?
> What raging of the sea? shaking of earth?
> Commotion in the winds? frights, changes, horrors,
> Divert and crack, rend and deracinate
> The unity and married calm of states.
>
> William Shakespeare, Troilus and Cressida, Act I, scene iii

Introduction

As we have seen, the five visible planets have attracted attention for millennia. Some are brighter than the brightest stars, and this, together with their unique ability to "wander" independently in a direction contrary to the diurnal revolution of the normally placid stellar background placed them in a category of their own. Freedom of movement led to their being invested with divine power and the ability to influence temporal affairs, especially when they clustered together in a joint display of supra-sensible influence. Several physical factors determine how impressive such a grouping will be, including proximity to the Sun and how closely the Five Planets crowd together. Unlike eclipses, where weather conditions are a crucial factor, in the case of the planets weather plays a minor role, as groupings may persist for days or weeks as the planets converge and then disperse again. In previous chapters we saw how the Chinese witnessed and recorded two of the three densest such groupings during the first two millennia BCE.[1]

---

[1] A recent study found only 102 cases between 3000 BCE and 2750 CE when the Five Planets could be circumscribed by a circle less than twenty-five degrees in diameter. Of this number, in only twenty-six cases was the circle less than 10.5° in diameter. Eighteen of those occurring between 2000 BCE and 1700 CE had enough of a visual impact to leave a lasting impression. Finally, three out of only four five- to seven-degree groupings, which could easily be covered by a fist held at arm's length, manifestly influenced political events on the ground. De Meis and Meeus (1994) The term "conjunction," strictly speaking, refers to when two planets are located at the same degree of longitude, so that when applied to groupings of multiple planets conjunction is used in the broader sense of "grouping."

Detailed study of individual circumstances is the only way to gauge the impact of these events on contemporary observers at different times and places. Judging from the historical record, circumstances on the ground play a major role in determining whether the event is deemed astrologically significant. There are times when such events may be eagerly anticipated; there are times when occurrences may be overlooked due to social disorder or careless observation; and there are times when inconvenient astrological prognostications may be "massaged" to render them more acceptable.

Examples of these different motives occurred in the early years of the Han Dynasty when a so-called "linked pearls" (*lian zhu*) alignment of planets in May 205 BCE extending over thirty degrees was pressed into service as a sign of heavenly approbation of Han founder Liu Bang's (206–195 BCE) rise to the dignity of emperor. In contrast, only twenty years later, in March 185 BCE, one of only four truly spectacular planetary groupings of less than seven degrees in the past 4,000 years was overlooked (or ignored), when it occurred during the reign of the Empress Dowager Lü (187–180 BCE), the first woman to rule the empire – not a precedent deserving commemoration in imperial China.[2]

Planets, periodicity, and divination

Probably the most ancient form of legitimation is the conception of the institutional order as directly reflecting or manifesting the divine structure of the cosmos, that is, the conception of the relationship between society and cosmos as one between microcosm and macrocosm. Everything "here below" has its analogue "up above."[3]

The conjunctionist thesis leads imperceptibly and unconsciously to the idea of variability and pluralism in religious and political regimes. If changes depend on the movements and conjunctions of the upper planets with certain signs of the zodiac, then major historical events can only be considered "providential" in the metaphorical sense.[4]

In the above quotations Peter Berger and Karl Löwith allude to fundamental ideas that lent cogency to planetary astrology as traditionally practiced in the West and China. The belief in antiquity that future events "here below" could be foretold by special means implies a presupposition that the future was in some sense preordained. Divination and portent astrology were methods

---

[2] For further examples, see Schafer (1977, 211–19).

[3] Berger (1990, 34). The maxim Berger is paraphrasing was popularized by Ptolemy (90–168 CE). It derives from the hermetic tradition of *The Emerald Tablet of Hermes Trismegistus*: "That which is Below corresponds to that which is Above, and that which is Above corresponds to that which is Below, to accomplish the miracles of the One Thing."

[4] Löwith (1949, 10, 20).

of "reading" the dispositions of supra-visible forces as if they were accomplished facts or knowledge, hence the revealed nature of omens from an early date.[5] At the same time, Karl Löwith also points to the inherently periodic nature of planetary motions, a fact whose recognition militated in favor of the removal of astral omens from the realm of the unpredictable and worrisome, gradually leading to their being seen as in principle calculable and ultimately harmless.

In China and the West the gradual recognition of the regularity of planetary phenomena did not proceed apace. How the periodicity of planetary cycles was exploited in astral prognostication diverged even more. Classical and medieval astrological theory in the West long indulged in grand speculations about the future of humanity being either mechanistically "determined" or merely "influenced" by the movements of the celestial bodies. The role of the divine varied, correspondingly, from merely indifferent to interventionist – disposed to regularly alter the course of history. When properly understood, from one perspective astrology could provide insight into the inevitable course of events, from the other it could indicate trends and directions that might still be influenced by divine or human will. In China, imperial astrologers, in contrast, never evinced much interest in speculation about long-term future trends, nor in planetary conjunction cycles, even after persistent exposure to Indian and Persian astrology's Great Years (not to mention the vast cycles of the *yugas*) and their associated conjunctionist theories.[6] This is entirely in keeping with the scribe-astrologers' time-honored role as archivists, who made "no systematic attempt... to theorize about 'ultimate causes' or search for 'general laws' in history as such."[7]

The Japanese historian of Chinese science Yabuuchi Kiyoshi characterized the Chinese approach to things celestial in the imperial period this way:

There are two sorts of celestial phenomena. One was cyclical in a simple way, and its regularity or periodicity could be discovered with relative ease; the other could not be predicted by human effort, but only observed. The former was systematized within the framework of calendrical science, while the latter became the object of astrological interpretation. Since they were complementary, they were equally important to Chinese administrators... The breadth of the Chinese ephemerides reflected the grave concern of Chinese rulers constantly to expand the demonstrable order of the sky, while reducing

---

[5] For a brief survey of the theoretical underpinnings of Chinese astrology and receptivity toward resonance theories, see Xu Fengxian (1994). For a historical survey of Chinese cosmology in the imperial period more generally, including changing attitudes toward general astrology, see Henderson (1984).

[6] Paul Tillich remarked on this aspect of Chinese time-consciousness: "The past is predominant over the future. The present is a consequence of the past, but not at all an anticipation of the future. In Chinese literature there are fine records of the past but no expectations of the future." Quoted in Needham (1981, 235).

[7] Yü (2002, 169).

the irregular and ominous. The parallel with the ruler's responsibility in the political realm is obvious.[8]

Chinese astral prognostication, with its view that Heaven's intentions are made manifest by the stars (among other things), held fast to a motivation similar to that of the ancient Mesopotamian astrologers who compiled collections of astral omens with a view to keeping the king informed of potential disasters or successes in the near term.[9] After the Han Dynasty, Chinese "portentology" at the official level remained conservative and resistant to change, despite a persistent undercurrent of skepticism. Path-dependency assured that throughout the history of imperial China, celestial omen-reading was practiced according to essentially the same ancient principles and astral–terrestrial correspondences, which assigned virtually the entire sky to the Chinese empire.[10] For the Chinese imperial court it remained an article of faith that threatening presages could, at least in theory, be deflected or "defused" if appropriate actions were taken or policies changed to mitigate the misgovernment that had called forth the ominous signs.

For the Latin West, Krzysztof Pomian summarized the essential principles of astrological practice as follows:

Different celestial bodies have different qualities, exert different influences and are connected with different individuals, peoples, institutions, etc. Each celestial event is therefore a peculiar combination of qualities and influences and as such it makes intelligible the peculiar character of a terrestrial event it is considered to be the cause of. Hence, an astrologer who knows what celestial events will happen in the future and when, is able to forecast the terrestrial events which will necessarily follow them, in other words, to make a horoscope of an individual, a dynasty, a city, etc. Likewise the knowledge of dates of past terrestrial events opens up the possibility of identifying their celestial causes and thus of setting up of a historical horoscope. The description of qualities of celestial bodies, of the range of influence of each one of them, of the peculiarities of different events which happen in the sky and of their supposed terrestrial effects furnishes the greatest part of the content of astrological books, as one can easily ascertain turning over the pages of the *Tetrabiblos* of Ptolemy or of the *De magnis coniunctionibus* of Albumasar (Abū-Ma'shar), two specimens of astrological literature referred to with the greatest frequency during the Middle Ages.[11]

From the outset, astrology in the West was no less eurocentric than the Chinese version was sinocentric, at least in the sense that it was exclusively

---

[8] Yabuuchi (1973, 93–4); also Nakayama (1966).

[9] I avoid the term "divination" for the Chinese practice because, unlike in Babylonia, in China there was no divinization of celestial bodies in the context of omen reading.

[10] The late imperial encyclopedia *Gujin tushu jicheng* (1725) routinely assigns essentially the same astral–terrestrial correspondences for the Chinese territory as were current in late Warring States "field-allocation" astrology. By then the scheme was already being criticized, however, not least for its sinocentrism. Henderson (1984, 214–15).

[11] Pomian (1986, 33).

Judeo-Christian and centered on the European experience, like the philosophy of history. Astrologers strove in their own way to identify the principle of intelligibility in history, producing an astrology characterized as a "naturalistic theology of history." In China, by contrast, one searches in vain for evidence of similarly grandiose theoretical ambitions or formulations. Instead, like other forms of divination, Chinese astral prognostication was deployed opportunistically.

Planetary astrology in China and Germany

In late February and early March of 1524 there occurred an impressively close grouping (less than 10.5°) of all five naked-eye planets in Aquarius–Pisces (lodge 13 *Align-the-Hall*), the very same location as the grouping of February 1953 BCE discussed in Parts One and Two. This was the densest such gathering in centuries. In both China and the West such planetary phenomena had long loomed large in astrology because of their presumed association with world-changing events on the grandest scale: the rise and fall of empires, changes of dynasties, or (in the West) the appearance of great prophets, even though reliance on reason as the best guide for belief and action gradually made inroads on the mental territory occupied by rival habits of thought, directly challenging those systems claiming esoteric knowledge, such as astrology. But in early sixteenth-century Europe the Scientific Revolution still lay decades in the future, although this was also the age of Copernicus (1473–1543), the first to formulate a heliocentric theory of the solar system, and the father of modern Western astronomy. In China, Guo Shoujing (1231–1316) had already devised high-precision instruments for use in positional astronomy and produced the "Granting the Seasons," *Shoushi*, astronomical system (1280), whose calculation of the length of the year of 365.2425 days anticipated the Gregorian calendar by some three centuries.[12] But despite the growing intelligibility of the cosmos and the predictability of celestial phenomena, belief in astrology was still pervasive. This was especially true of the popular imagination in Europe where age-old religiously and astrologically inspired millenarian ideas held powerful sway.[13]

Events of 1524 in China and Europe in response to the planetary phenomenon provide insight into divergent Chinese and Western responses to such

---

[12] Sivin (2009).

[13] C. Scott Dixon (1999, 406–7) describes the spirit of the times this way: "The sixteenth century was an anxious age. Knowledge creates anxiety, as does uncertainty or a sense of dissociation, and the century of Reformation gave rise to its share of novel and divisive ideas. Yet whereas medieval cosmology offered the anxious thinker 'a fully articulated system of boundaries' for understanding the world, the onset of the early modern age saw the disintegration of this order. What replaced it, in the first instance, was not an alternative cosmology, but an anxious scramble to reassociate the culture's disparate parts."

"millennial" events. To explore the topic fully would require a lengthy mono-graph, not least because Ming China and Reformation Europe were such vastly different places. In what follows I propose simply to portray, with a broad brush, the conceptual background and contemporary impact of the planetary grouping of 1524 in early Reformation Europe and Ming China. I will sketch the sociopolitical comparisons and contrasts, which are brought into especially clear focus by reactions to the planetary phenomenon. In Europe the event had long been predicted, and, consequently, widely and anxiously anticipated, but in China to all appearances it was not – knowledge of the event was mainly confined to the imperial court, its disquieting implications closely identified with court politics.

Planets, periods, and prophecy in the West

In the *Timaeus* (39D), Plato (427–348 BCE) asserted that the "perfect year" began with a conjunction of all the planets. Soon after, the Babylonian Berosus (fl. *c*.300 BCE) introduced the theory that the world, having begun with a conjunction of planets, would end with another.[14] In Qin and Han China, in a remarkable echo of the ancient massing of 1953 BCE, the astronomical system of the Zhuan Xu calendar of Warring States date (*c.* fourth century BCE) began with a conjunction of the Five Planets, Sun, and Moon at a precise degree of longitude in lodge *Align-the-Hall* (Pegasus). The contrast is instructive, in that the same astronomical phenomenon is redolent of eschatology in the West, while it commemorates the initiation of a bureaucratic instrument for time management on the other.

In China, pretensions to universality provided a powerful ideological impetus for the foundation of empire, and after unification, with no serious challengers to China's status as the supremely dominant civilization in East Asia, the idea of pluralism in religious and political regimes could emerge only with difficulty, and then only within the context of the dynastic cycle. This may be part of the explanation why Löwith's conjunctionist thesis failed to take root, in spite of nascent late Warring States speculation along those lines. The potential that planetary resonance periods offered for the formulation of an astrological history was ignored, despite the theory of Zou Yan (third century BCE) of historical change based on the alternation of Five Elemental Phases, which was soon diverted into more harmless pursuits than dynastic politics, such as divination and the interpretation of the natural world. Part of the explanation may also have been that as late as the Tang Dynasty it was still believed that the regularity of planetary periods could not be counted on, because it was contingent and subject to manipulation by Heaven. In the eastern Mediterranean world the situation was rather different.

---

[14] Campion (1994).

Given the plurality of ancient civilizations contesting for imperial supremacy there, and under the influence of Babylonian, Indian, and Persian astrology, theories of astrological history based on planetary resonance periods came to dominate. In Indian astrology the *Kaliyuga* began with a planetary conjunction in 3102 BCE, initiating a period of 432,000 years at the end of which the world would end and begin anew.[15] Expanding on this astronomically derived date, Greek and Persian traditions held that this was the date of the Deluge. With this the stage was set for the development of contending theories of astrological history, one theocentric and informed by Judeo-Christian mysticism and eschatology, the other naturalistic and concerned with the "natural science" of astrology and with discovering the principles of its functioning. The two are logically incompatible.

According to Krzysztof Pomian,

Until the sixteenth century the existence of a connexion between celestial and terrestrial bodies and between celestial and terrestrial events was admitted by everyone as self-evident. But on the nature of this connexion there was no agreement. The augustinian current considered celestial events as signs of terrestrial ones. The former announce the latter, because God conferred upon them such a meaning. And this meaning can be truly understood only by those who are looking on the skies guided by the divinely inspired scriptures. To such an attitude . . . the aristotelians opposed their conviction that celestial events are causes of terrestrial ones. In order to understand their action one has therefore to inquire into their powers in conformity with the principles of natural science. Between these two poles lay an entire spectrum of intermediate positions which tried to reconcile or synthesize Augustine with Aristotle, theology with physics and astronomy, significance with causality, prophecy with prediction.[16]

Whatever one's position about causality in relation to the celestial bodies, there was a fundamental distinction between the classical and Christian perspectives with regard to forecasting the future: "the fulfillment of prophecies as understood by the Old and New Testament writers is entirely different from the verification of prognostications concerning historico-natural events."[17] Nevertheless, as has long been observed, both are integral aspects of the Judeo-Christian and Europe-centered orientation characteristic of the philosophy of history in the West.

With regard to planetary conjunction periods in particular, the most influential contribution after Ptolemy[18] was made by a ninth-century Persian Christian

---

[15] Kennedy (1963).   [16] Pomian (1986, 32).   [17] Zambelli (1986, 13, n. 35).

[18] According to John D. North (1986, 50, original emphasis), "Ptolemy falls comfortably into the Aristotelian–Stoic meterological tradition. A certain power (δύναμις) emanates from the aether, causing changes in the sublunar elements and in plants and animals. Effluence from the Sun and Moon – especially the Moon, *by virtue of her proximity* – affects things animate and inanimate, while the planets and stars also have their effects. If a man knows accurately the *movements* of the celestial bodies, and their *natures* (perhaps not their essential but at least their potentially

astrologer in Balkh, Abū Ma'shar (Ja'far ben Mohammed ben Omar El Balki Abū Ma'shar, 786–866), held to be the single most important transmitter of Aristotle's theories of nature to the Latin West.[19] The astrological history of Abū Ma'shar (Albumasar) was largely derivative, being inspired by Indian astrology transmitted via third-century Sāsānian (Persian) intermediaries. The formulation of the theory of astrological history, the writing of past and future history on the basis of planetary conjunction periods, Hellenistic horoscopy, and Zoroastrian millenarianism, was a characteristically Sāsānian innovation. Having made a study of the planetary conjunction periods, especially Jupiter–Saturn and their sublunar influences, in his *On The Great Conjunctions and On Revolutions Of the World, Kitāb al-qirānāt*, and *The [Book of the] Thousands, Kitāb al-ulūf*, Abū Ma'shar set forth a cycle of historical events, adapting Berosus and based on an Indian *yuga* of 180,000 years, according to which the world was created when the Sun, Moon, and the Five Planets gathered in the sign of Aries (in 183,102 BCE), and would end when another such conjunction occurred in the sign of Pisces (in 176,899 CE).[20]

Another crucial contribution was that of the eighth-century Jewish astrologer Māshā'allāh, who likewise transmitted the Sāsānian theory that history represents the playing out of the consequences of the twenty-year Jupiter–Saturn conjunctions.[21] Māshā'allāh's astrological world history, *On Conjunctions, Religions, and Peoples*, survives only in fragments. According to Māshā'allāh, political and religious developments are signaled by the planets' periodic meetings in particular signs of the zodiac. A horoscope cast for the day of the vernal equinox will predict the ordinary course of events during a year when a twenty-year conjunction occurs. Twelve successive such conjunctions tend to recur within the same "triplicity" before moving on to the next after about 240 years (Figure 14.1).[22] A shift from one triplicity to the next indicates that a higher-order change is in the offing, like the rise of a new nation or dynasty. The most

---

effective qualities), and if he can deduce scientifically the qualities resulting from a *combination* of these factors, why should he not judge both weather and human character?"

[19] In the Latin West, "without any doubt, the most influential Islamic writer was Abū Ma'shar (786–866). Fashions came and went in astrology, but he seems to have been read and quoted constantly from the time of the translations of John of Seville and Hermann of Carinthia in the twelfth century to the decline (or at least turning native) of the subject in the seventeenth." North (1986, 52).

[20] Pingree (1968); De Meis and Meeus (1994, 293–4).

[21] Roy A. Rosenberg (1972, 108) shows how in the ancient world "conjunctions of Saturn and Jupiter signified the transfer of power from one planetary *daemon* to another," as from Kronos (Saturn) to Zeus (Jupiter).

[22] A triangular configuration of three zodiacal signs belonging to the same classical element, each sign being separated by 120 degrees. Aveni (2002, 115). The four triplicities are:
  Fire – Leo, Aries, Sagittarius
  Earth – Taurus, Virgo, Capricorn
  Air – Libra, Gemini, Aquarius
  Water – Cancer, Pisces, Scorpio

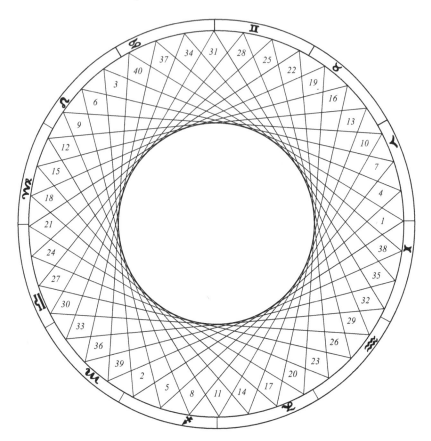

Figure 14.1 The "Great Trigon" of Johannes Kepler, from his *Mysterium Cosmographicum* (1606), showing successive Jupiter–Saturn conjunctions. The diagram illustrates how every third meeting of the planets occurs at nearly the same location, and how after 820 years the forty-first conjunction occurs at the original starting point. This plot shows at a glance how different the Western geometrical conception was from the Chinese.

portentous of all shifts, occurring every 960 years after cycling through all four triplicities, heralds truly epoch-making events like the coming of a major prophet.[23] In Māshā'allāh's astrological history, one date no doubt stood out as especially relevant in the sixteenth century – the 19 March 571 shift indicating the rise of Islam. If one adds another 960 years to arrive at the next "millennial"

[23] Kennedy (1971); also Kennedy (1963, 245).

shift, the result is 1531, and the concomitant revolutionary development should be the coming of a major prophet.

The third, and last, Western astrological history we need briefly to consider is that of Don Isaac Abrabanel (1437–1508), statesman, philosopher, theologian, and renowned biblical commentator, scion of one of the most prominent and venerable Ibero-Jewish families. Abrabanel served Ferdinand and Isabella for eight years as finance minister until 1492, when the Jews were banished from Spain at the instigation of the Inquisition. He moved with his family to Italy, eventually settling in Venice, where he held office as a minister of state until his death in 1508. Abrabanel is famous as a prolific biblical exegete, and his commentary on the prophetic passages in the Book of Daniel was highly influential in the sixteenth to seventeenth centuries, not least among Christian theologians.[24] In his commentary on Daniel in *The Wells of Salvation* (*Ma'ayney hayyešu'ah*, 1497), Abrabanel explains in detail the significance in Jewish astrology of the Jupiter–Saturn conjunctions as portents of earthly events. In his scheme, the "watery" sign of Pisces figures importantly as the location of "mighty conjunctions" at intervals of 2,860 years, and on the basis of his astrological chronology he assigns to the conjunction of 1465 unique significance as a portent of the Messiah.[25]

In the first chapter (or "Gate") of *The Wells of Salvation*, Abrabanel shows that six of the prophetic passages in Daniel have Messianic forecasting interpretations and that the dates forecast in these prophecies for the appearance of the Messiah converge on the decade of the 1530s, more specifically 1531 (a date from Māshā'allāh's theory encountered above). Immediately following this in the second chapter, Abrabanel sets forth his refinement of Māshā'allāh's astrological history with an explication of how Jupiter–Saturn conjunctions, and minor (240 years) and major shifts (2,860 years), combined with the influence of the four elements of Nature, profoundly influence historical events on the grandest scale. Israel's destiny is especially closely linked with the sign of Pisces: "if we begin with the First Redemption of Israel, from Egypt, we

[24] The second chapter of Daniel recounts how Daniel was able to establish his bona fides as a true prophet by offering a convincing interpretation of a recurring dream of King Nebuchadnezzar II (*c*.605–562 BCE) about a giant statue fashioned from head to foot from five layers of gold, silver, brass, iron, and clay. With the benefit of divine prompting, Daniel succeeds where all the king's seers and advisers failed, even though he was not told the content of the king's dream beforehand. In his prophetic interpretation, Daniel explained that the dream symbolized five successive world empires, extending far into the future. These prophecies, together with the teachings of Augustine, make up the standard paradigm of what Krzysztof Pomian (1986, 30) calls the "theocentric theology of history."

[25] Abrabanel (1960, 12.2); Rosenberg (1972, 105–7). Abrabanel's 2,860-year period for the mighty conjunctions represents a refinement of Māshā'allāh's (3 × 960 =) 2,880-year period. Abrabanel's 2,860 years represents almost exactly 144 Jupiter–Saturn cycles if one uses the modern figure for the conjunction period of 19.859 years (144 × 19.859 = 2859.696).

are led to conclude that the next and decisive redemption of Israel and the transformation of the world will begin around 1534" (Figure 14.1).[26]

A convergence of eastern, Islamic, Christian, and Jewish astrological speculation thus pointed to a historical culmination on the grandest scale in the early sixteenth century, so it is hardly to be wondered at that the turn of that century brought with it a heightened sense of anticipation that epoch-making events were in the offing. But the development that turned these somewhat esoteric astrological facts into common knowledge and a matter of widespread popular concern was the publication in 1499 of ephemerides calculated by two German astrologers, Johannes Stöffler and Jacob Pflaum, showing that a great conjunction of all five planets would occur in the "watery" sign of Pisces in February 1524 – Pisces being the sign of the biblical Flood. In the years immediately following the turn of the sixteenth century, as political tensions and dissension within the Church escalated, predictions of a second Deluge provoked widespread collective fear and heated debates among astrologers and theologians, Catholic and later Lutheran. In 1515, George Tanstätter, a Viennese astrologer, added the prediction that a schism between the people and the Church would give evidence of the approach of the universal disaster.[27] By 1519, two years after Luther initiated the Protestant Reformation by publicly denouncing Church abuses in his Ninety-Five Theses, these animated debates and profound apprehensions reached unprecedented scope and pitch in both Germany and Italy.[28]

---

[26] Prof. Stephen L. Goldman, personal communication; I am grateful to Prof. Goldman for an English synopsis of the relevant Hebrew passages. Abrabanel's *The Wells of Salvation* contains the most authoritative exposition of Jupiter–Saturn conjunctions in Jewish sources and was consulted by Johannes Kepler (1571–1630) in attempting to establish the date of Christ's birth. Kepler published several works on this subject beginning in 1606 and culminating in his *De anno natali Christi* (1614), which advances his mature theory that a triple conjunction of Jupiter, Saturn, and Mars in Pisces in 7 BCE was the Star of Bethlehem. Rosenberg (1972, 105); Hughes (1979, 96); Jenkins (2004).

[27] Gritsch (1967, 123).

[28] In C. Scott Dixon's (1999, 405) account, "In 1499 two German stargazers, Johannes Stöffler and Jacob Pflaum, warned that in February 1524 a planetary conjunction would occur in the sign of Pisces. Although the authors themselves made no mention of a flood, astrological speculation soon appropriated the notion of a second deluge, a washing away of sinful man (*Sintflut*). Before it was over, the debate had engaged fifty-six authors – Germans, Frenchmen, Spaniards, Netherlanders and Italians. In total, 133 works dealing with the 1524 prediction were published. Although some writers, such as the Spaniard Augustino Nifo, doubted the truth of the prognostication, most authors were willing to concede the possibility of a flood. Even Johann Carion, Philip Melanchthon's classmate at Tübingen and later Brandenburg's court astrologer, granted this much. But, in a sense, the reception of the theory in the astrological community was strictly academic, for by the year 1517 the prediction had spread to all social levels. As witness to its popular profile, the episcopal secretary to Würtzburg, Lorenz Freiss, blamed the run of horrifying title-page woodcuts for the scare. It had caused such disquiet that Martin Luther felt obliged to address the portent in his 1522 Advent sermon in Wittenburg." See also Barillà (n.d.).

Pamphlets, portents, and prophecy

It is already ten years now since I predicted those terrible wars that have wracked Italy. Those who laughed at me then now see what great things God has brought about through the stars. Even the ruin of the Papacy, the change of the Law, Emperor Franz's capture, the Peasants' War, all have I predicted, and not simply more or less, but based on astrological principles. (W. Pirckheimer (c.1525))[29]

The literature on these "portentous" developments in the early years of the Reformation is vast, certainly vaster than my modest comparative ambition here allows me to consider. I will focus simply on the role of apocalyptic astrological prognostications in Germany leading up to the radicalization of the reform movement and the Peasants' War or *Bauernkrieg* of 1525.

There were many influential actors in the controversy over the doom-saying prognostications that so agitated the popular mind prior to the predicted Deluge in 1524. On the one side were the astrologers whose predictions became increasingly comprehensive, as the quote by Pirckheimer about his 1515 prognostications shows. They served to reinforce the conviction that events were building toward an imminent climax. Astrologers began to assume an increasingly public function, representing the "analytical" approach to the interpretation of events – "based on astrological principles" – "urging science against the Bible," as it were.[30] In this they were aided and abetted by the rapid development of printing and publishing, which they lost no time in exploiting to expand their audience.

Even more influential, however, were the creators and publishers of the many pamphlets (*Flugschrift*) already alluded to, which were in great demand and widely circulated. For many, especially the illiterate peasantry, the garish woodcuts on the title pages of these pamphlets had a more unsettling impact than the astrological prognostics in the text. Notwithstanding the frightening images, however, the message of many of these pamphlets was a positive one, looking forward with anticipation to the new beginning following the cataclysm. Shown in Figure 14.2 is a famous example of a pamphlet from 1525 depicting the grouping of the Five Planets inside Pisces the fish, from which a deluge is pouring down, inundating the town below. Confronting each other across the impassable flood are the militant common folk on the left and representatives of secular and ecclesiastical authority on the right. Also shown is the comet of 1525. The genius of the artist speaks for itself.

---

[29] "Es sind nun schon zehn Jahre, daß ich jene schreklichen Kriege, welchen Italien zerütten, vorhergesagt habe. Wer mir damals auslachte, siht nun doch welche grosse Ding Gott durch die Gestirne ausrichtet. Auch dem Ruin des Papstes, die Veränderung der Gesetz, die Gefangenschaft des K. Franz, den Bauernkrieg habe ich vorausgesagt und zwar nichts aufs ohnegefähr, sondern auf astrologische Principien gestutzt," quoted in Zambelli (1986, 9, n. 24).

[30] Hammerstein (1986, 140).

Figure 14.2 Nürnberg practica of 1525. The upper caption reads, "A practica concerning the major and extensive interaction of the planets which will appear in the year 1524 and without a doubt will bring many wonderful things." Collection of Owen Gingerich, reproduced by permission.[31]

---

[31] See Laube, Steinmetz, and Vogler (1974, 205). Artist Werner Tübke incorporated a representation of this woodcut in his huge panorama mural *Frühbürgerliche Revolution in Deutschland* (1987) in Bad Frankenhausen, commemorating the Battle of Frankenhausen in 1525.

Perhaps surprisingly, many of the "doomsayers and stargazers" were them-selves clergymen.[32] There were others who delivered consoling messages in their sermons; fewer who railed against astrology as baseless nonsense, pre-senting no credible challenge to Scripture; and a very few who were moved enough by the spirit of the times and the rising tide of civil and religious dis-content to seize the moment and offer themselves as instruments of God's will in bringing the revolution to pass. One such radical cleric was Thomas Müntzer (1490–1525), Luther's nemesis and the spiritual motivator of the Peasants' War, the mass uprising of the peasantry in Saxony and Thuringia in the summer of 1525 that led to the massacre of 5,000 peasants as well as Müntzer's own beheading. Müntzer, it has been said,

distinguished himself from other "revolutionary Spiritualists" by introducing into the Radical Reformation the idea . . . based upon the Danielic–Hieronymic concept of the four world periods or monarchies – that a fifth historical period, namely that of Christ's direct rule in the saints, had begun . . . Such a concept, developing into an intense convic-tion of the imminent end of the world, encouraged numerous representatives of the three major movements (Anabaptism, Spiritualism, and Evengelical Rationalism) to attempt a Radical Reformation which, in its entirety, represented an "abortive constitutional revolution."[33]

Although Müntzer is not known to have embraced astrology, indeed he "disassociated 'spiritual experience' from all external media,"[34] nevertheless, as a noted Hebrew scholar he would certainly have been aware of Isaac Abra-banel's influential astrological reading of the Danielic prophecy.[35] Significantly, Müntzer's final sermon aimed at persuading the secular authorities to support his reform movement, the Sermon to the Princes (1524), takes the second

---

[32] Regarding the role of astrology in Christianity, C. Scott Dixon (1999, 408, 411) remarks, "This was not such an odd pairing, for astrology had always been central to the Christian faith, especially in the middle ages. Granted, the two schools of thought, Aristotelian and Augustinian, could not agree on the relationship between celestial cause and terrestrial effect, but most religious thinkers yielded to the notion that God might reveal man's destiny in the stars. Luther granted that certain heavenly signs might herald the judgments of God, but he gave the practice of astrology short shrift. In his eyes, it was not a predictable science; moreover, he rejected it on theological grounds: it placed limitations on the powers of God." Further, "Although Luther himself balked at the notion that astrology amounted to a science, he was willing enough to concede that certain astral signs might portend divine intelligence. 'For it is incredible,' he said, 'that they [the planets] be observed to move without inquiring whether there isn't somebody who moves them.'"
[33] Williams (1962, 858, 865); see also Baylor (1993, 30) and Gritsch (1967, 103).
[34] Gritsch (1967, 187).
[35] Müntzer's name appears in the 1506 registration rolls of Leipzig University. In 1512 he attended the University of Frankfurt-an-der-Oder where he earned the degrees of master of arts and bachelor of theology. In 1519–20 he continued his literary studies in the monastery of Beuditz at Weissenfels. Müntzer was a noted scholar of classical and humanistic literature and thoroughly conversant with Latin and Greek, as well as Hebrew. One or more of these institutions would no doubt have had a copy of Abrabanel's important and highly regarded commentary.

chapter of Daniel as its theme. In that crucial sermon, Müntzer develops his scriptural foundation for a theory of revolution and offers himself as the new Daniel who, as he sees it, has been chosen to lead the vanguard in overturning the corrupt established order. Even if Müntzer did not personally subscribe to ordinary horoscopic astrology, astrological history on the millennial level is something else again. It is very doubtful whether he could have remained completely aloof from the feverish anticipation and growing militancy among the common folk as the February 1524 date of the predicted Deluge approached. It could be coincidental that he chose March 1524 to set in motion the militant phase of his radical movement by leading his secret League of the Elect in attacking and putting to the torch a local symbol of Church authority – but that it was precisely in 1524–5 that mass insurrection erupted is in no small part due to the apocalyptic nature of the astrological predictions in anticipation of the planetary grouping. The spirit of those years is eloquently captured in a saying current in Germany in 1525:

> He who does not die in 1523,
> In 1524 does not drown,
> And is not beaten to death in 1525,
> Can truly claim miracles in his life.[36]

China

Earlier we saw how, in ancient China, long-established tradition held that rare groupings of all five visible planets within the space of a single lodge carried dynastic implications. The astrological significance of the pre-dynastic precedents, as well as the one pertaining to the Han Dynasty founder Gaozu, are spelled out very clearly in *The Grand Scribe's Records*, as well as in *Han shu*.[37] By the time Sima Tan (d. 110 BCE) and his son Sima Qian (d. 86 BCE) were compiling the summary of astronomical–astrological knowledge contained in the "Treatise on the Celestial Offices," the imperial system was entrenched. A growing recognition of the "national security" implications of unpredictable celestial phenomena contributed to a subtle reformulation of the astrological prognostications associated with the planets – if the mandate-conferring omens could not be predicted, at the very least they would have to be managed.

One indication: alongside the traditional field-allocation scheme of astral–terrestrial correspondences described in Chapter 9 one also finds in the Simas' "Treatise" the new corollary doctrine previously discussed, according to which, "when the Five Planets are disposed in mid-heaven and gather in the east, China

---

[36] Quoted in Gritsch (1967, 123).
[37] *Shiji*, 27.1348 and 89.2581; *Han shu*, 26.1301 and 36.1964.

benefits, when they gather in the west, foreign kingdoms using weapons gain."[38] In this new scheme, which for the first time actually took account of non-Hua–Xia peoples by allocating to them the *yin* half of the sky, a strung-out display of the Five Planets in the *yang* sky merely bestowed tactical advantage on imperial China, rather than heralding something so drastic as a dynastic transition. At the same time, by comparison with the situation in the pre-imperial period, we saw how astrological prognostication had also moved decidedly "downscale," so that what had begun as a hermetic science of astral omenology enjoyed such widespread popularity by the second to third centuries of the Common Era that the maxim "when the Five Planets appear in the east, it is beneficial for China" could even figure as an epigrammatic motif in silk brocades which found their way to the remotest outpost of Niya on the northwest frontier.

Although rough approximations for the synodic periods of the planets were known by the early imperial period (i.e. twelve years for Jupiter, and twenty years for Jupiter–Saturn conjunctions), with the exception of the astronomical computations typical of comprehensive calendrical systems like Liu Xin's "Triple Concordance System" (*Santong li*),[39] little attention seems to have been paid in China to resonance periods in planetary astrology. Indeed, at the popular level astromantic divination had become thoroughly mathematized and divorced from observation. Although the ancient names of the lunar lodges and Jupiter stations continued to be used, their arithmetic combination with the sexagenary sequence, binary *yin-yang* principles, and Five Elemental Phases correlations in ever more intricate numerological schemes meant that any meaningful connection with the movements of actual heavenly bodies had long since been abandoned.[40]

Astronomical observations were regularly carried out largely in order to account for political developments after the fact. In this, the compilation of precedents played an important role, leading to the creation of an entire genre of *zhan jing* "canonical prognostication texts." Astrological prognostics deriving from such works were frequently adduced, based on familiar analogical principles:

During the Shiyuan reign period of Emperor Xiao Zhao [of Han; r. 87–74 BCE], the Han eunuch Liang Chenghui and the King of Yan's stargazer Wu Moru saw a fuzzy star [tailless comet] emerge in the west at the east gate of *Celestial Market* [in Oph; March 84 BCE]; it traveled past *River Drum* [Aql] and entered *Align the Hall* [Aqr-Psc]. [Liang Cheng-]hui said, "when a fuzzy star appears for sixty days, before three years have passed there will be rebellious officials executed in the marketplace below."[41]

---

[38] *Shiji*, 27.1328.   [39] Xi Zezong (1989, 46–58); Cullen (2001).
[40] Kalinowski (1996, 55–81).   [41] *Han shu*, 26.1306.

The significance of otherwise unremarkable movements of Venus through the *Celestial Court* (roughly Leo) only subsequently became clear after the plot to assassinate imperial regent Huo Guang (d. 68 BCE) was discovered in 80 BCE, leading to the public execution of a king, two generals, and the senior princess. Consequently, the following prognostic was pronounced:

Taiwei is the *Celestial Court*. When Venus moves into it, the gates of the palace are closed, great generals don armor and weapons, and evil officials submit to punishment.[42]

From the Eastern Han Dynasty (23–220 CE) on, planetary resonance periods were not a preoccupation of the officials charged with observing the skies, who continued the retrospective interpretive approach historically applied to astral omens. This is true even though it is apparent that Jupiter's nominal twelve-year period had been used since the time of compilation of the *Zuozhuan* in the fourth century BCE, in an effort to retrospectively compute the astrological circumstances of historical events centuries earlier. From Liu Xin's discussion of the Zhou conquest of Shang (1046 BCE) in his "Canon of the Generations" (*Shijing*) in the *History of the Former Han Dynasty*, it is clear that he also sought unsuccessfully to derive the date of that epoch-making event by this means.[43] That astro-calendrical officials did not prognosticate on the basis of longer planetary resonance periods seems surprising, since the principle of least-common multiples was exploited in the construction of ambitious calendrical systems like Liu Xin's "Triple Concordance System" (*Santong li*). Although some had absorbed the implication of Zou Yan's (third century BCE) phenomenological interpretation of historical change that periodicity "leads unconsciously to the idea of variability in political regimes," astrologers evinced little, if any, interest in the predictive potential of grander cycles. Seeking to discover the astrological principles underlying past events in order to interpret the present, they ignored the far future. This is somewhat surprising since remarks by Warring States philosophers like Mencius (372–289 BCE) point to the currency of early speculation about a 500-year "dynastic" cycle, whose inspiration may be traceable to the memorable planetary groupings associated with the founding of the Three Dynasties described in Part Two. At about the same time that Plato was talking about the "perfect year" beginning with a conjunction of all the planets, Mencius implicitly alluded to the predictive potential of astronomical cycles:

All who speak about the nature [of things] have only former instances [*gu*] to reason from, and nothing else. Fundamental to precedents [*gu*] is their ready application. What I dislike in your learned men is the way they drill into things. If they would do as Yü the Great did when he conveyed away the waters, there would be nothing to disapprove of in their learning. Yü's method of draining the flood was to convey the water in a

---

[42] *Han shu*, 26.1306.   [43] *Han shu*, 21B.1011–24; Cullen (2001, 27–60).

way consistent with its natural inclination. If your learned men would do likewise, their knowledge would be great indeed. Consider the heavens so high and the stars so distant. If we seek out former instances [*gu*] we may, while sitting still, have command of a thousand years' of solstices.[44]

Mencius' real point has to do with the historical precedents of moral order, but even the astrologers seem not to have grasped the forecasting implications of his analogy of inductive reasoning drawn from calendrical science. Part of the explanation lies in the fact that the astrologers were also scribes and historiographers, and, as Yü Ying-shih has explained,

Chinese historians... recognized the existence of "historical trends" or "patterns of change" in the past. However, when they ventured generalizations, these generalizations are invariably limited in time and confined to a particular aspect. It seems never to have occurred to them that it was their business to establish "universal historical laws" or theorize about the entire process of human history.[45]

The other part of the explanation, of course, is the profound influence of the *Book of Changes*. As we saw in Chapters 2 and 12, the *Changes* established the parameters for interpreting the potentiality of a given configuration of circumstances in space–time. Once the *Changes* became enshrined during the Han Dynasty as the original wisdom text of the canon, this virtually guaranteed that the primary divinatory interest would henceforth be focused on the immediate present, or at most on short-term trends. The Simas' unprecedented ambition to study the *longue durée* – "to investigate the boundary between the celestial and human and thereby to comprehend historical change, past and present, in order to formulate a theory of our own" – is the exception to the rule.

Astral divination in the later empire

The tension between the regular and the anomalous, already making itself felt in Warring States- and Han-period Five Elemental Phases metaphysical speculation (witness the skepticism of Xunzi (310–237 BCE) and his disciple Hanfei (*c*.280–233 BCE) with regard to the supra-visible), came increasingly to the fore in relation to astral omens, leading by the Northern Song Dynasty (960–1127) to a significant shift in thinking about the implications of such phenomena for imperial policy. Astrology gradually became domesticated, the

---

[44] *Mencius, IV*. B (26); Needham and Wang (1959, 196); and Lau (1970, 133, trans. modified). D.C. Lau's and others' rendering of Mencius' statement about solstices as prospective (rectified here) may be mistaken. Nathan Sivin's translation goes to the heart of the last line of the passage (*gou qiu qi gu, qian sui zhi ri zhi, ke zuo er zhi ye*): "if we seek how they were before, as we sit there we can call before us [i.e. imagine] the winter solstices of a thousand years." Sivin (2000, 128).

[45] Yü (2002, 167). Yü Ying-shih overlooks the Simas' astrological theory of history, for which see Chapter 10 above and Appendix.

emperor and scholar-officials at court eventually adopting a posture denoted "pragmatic agnosticism."[46]

This process and the jockeying for power and influence between scholar-officials and bureaucrats at court, for which the eleventh-century astrological debates served as proxies, have been analyzed by Wu Yiyi, who focused a study on precisely the half-century when some of the most impressive and unsettling astral anomalies occurred – the sudden appearance of two spectacular "guest-stars" (supernovae) in 1006 and 1054, as well as comets, meteors, and an impressive planetary grouping in 1007. The supernovae each took many months to fade away, occasioning great disquiet and debate at court in the meantime, not least because of the loss of the northwest and northeast of the country to the Xi Xia (1032–1227) and Liao (907–1125) kingdoms. An uneasy peace along the frontier, bought with heavy payments of tribute, meant the future of the Northern Song Dynasty was far from secure. The ominous external situation provided ample fodder for the manipulation of the interpretations of the astral omens (though not the phenomena themselves) by one court faction or another in an effort to advance their political fortunes, so it is noteworthy that it was at precisely this juncture that a pragmatic agnosticism was able to assert itself.

How perturbed the popular imagination was by these eleventh-century omens is impossible to say since the sources are silent on popular sentiment. Dabbling in astronomy and the calendar had often been proscribed activity (though up to now this was seldom rigorously enforced), but prognostication texts were still closely held within the Directorate of Astronomy, so that popular speculation on these matters may well have been subdued, certainly by comparison with the situation in Europe a few centuries later. Ordinary people had plenty of other insecurities to worry about; in any case, there were no chiliastic *fin de dynastie* mass movements reminiscent of the Yellow Turbans rebellion of 184, which precipitated the ultimate fall of the Han Dynasty in 220.

But Wu Yiyi's is really only one side of the story. There were influential individuals from Han through Song who were very well informed about the link between significant planetary events and the conferral of Heaven's Mandate. In Chapter 3 we saw how exceptionally close encounters of the Five Planets must have captivated astrologers for several days. Both clusters were certainly witnessed, and more importantly remembered, by the ancient Chinese, who

---

[46] The term "pragmatic agnosticism" was coined by Wu Yiyi (1990, 252). A representative example of this posture is the statement of the historian Sima Guang (1019–86): "From ancient dynasties on, historians kept records of meteors as weird and anomalous events . . . but meteors appear every night and are uncountable in number, without regard to the blameworthiness or prosperity of the Empire. Even regular observations cannot provide a complete record; it is a meaningless waste of effort." The guidance regarding anomalies which Sima Guang offered his collaborators in compiling the *Comprehensive Mirror to Aid in Government, Zizhi tongjian*, was, "if the anomalies were nothing more than weird prodigies, feel free to omit them." For these and other examples, see Wu (1990, 256 ff.); see also Xu Fengxian (1994, 13).

must have gazed in amazement as they strove to comprehend their significance. Still another curious planetary "dance" in 1576 BCE, at the beginning of the Shang Dynasty, when "the planets moved criss-cross," was remembered for its association with the transformation of the political landscape of north China by the Shang. This phenomenon, in addition to the astrological circumstances of the Zhou conquest itself, was cited by advisers to Cao Pi (187–226 CE), first emperor of Wei, as a precedent in connection with the dynastic transition from Han to Wei two millennia later, documented in Chapter 6. There is no doubt that these milestones in astrological history were recorded and firmly established in traditions about the celestial signs that signaled the heavenly conferral of legitimacy.

In addition to the passage quoted above from the official correspondence between the courts of Han and Wei preserved in the "Account of the Proferring Emperor" (*Xiandi zhuan*), other documents also explicitly invoke the behavior of the planets in connection with the "abdication" in 220 CE of the Han emperor Xiandi in favor of Cao Pi.[47] The two concluding documents in the exchange capture so well the historic resonances and symbolic import of the event that they are worth quoting in full:

Document 26: On day *xinyou* [December 1, 220] Consulting Erudites Su Lin and Dong Ba submitted a memorial saying, "The twelve stations of heaven are astral fields; each principality and domain belongs to such an allotment. Zhou is in *Quail Fire*; Wei is in *Great Span.* Jupiter progresses in succession through each territory's station, the Son of Heaven receives the Mandate and the Lords of the States are enfoeffed accordingly. When King Wen of Zhou first received the Mandate, Jupiter was in *Quail Fire*; down to King Wu's attack on [Shang] Zhòu it was thirteen years and Jupiter was again in *Quail Fire*, therefore the *Springs and Autumns Commentary* [i.e. *Discourses of the States, Guoyu*] says, '*When King Wu attacked [Shang] Zhòu, Jupiter was in Quail Fire; where Jupiter was located is none other than the astrological field allotted to us, the Zhou.*'[48] Formerly, in the seventh year of [the reign period] 'Brilliant Harmony' [184 CE, year

---

[47] *San guo zhi*, 2:75.
[48] Liu Ciyuan et al. and Zhou, Xiaolu (2001, 1217–18); also Cullen (2001, 47–51); Pankenier (1981–2, 3). Zheng Xuan (127–200) cites the detailed astral omens in *Discourses of the States* when commenting on the line '["Heaven] protected, helped, [and gave] you its Mandate; [you] harmonized [with Heaven] in attacking Great Shang" in the ode "Da ming" in the *Book of Odes*. Zheng explains that '"harmonized' harmonized" is an allusion to King Wu's taking care to conform with the multiple astronomical indications reported in *Guoyu*, denoted the "Three Locations and the Five Positions." Kong Yingda (574–648), supports Zheng in this. Further strong support comes from *Han shu*, "Xu zhuan shang," which reproduces Ban Gu's (32–92 CE) prose poem "You tong zhi fu," which has the line '["when his] 'Neighbor to the East' tortured and killed the humane; King [Wu] conformed his positions to the Three and the Five.'" Explication by Ying Shao (140–206) has survived in rthe commentary of Yan Shigu's (581–645) commentary, which again quotes the *Discourses of the States* astrological passage. Ying Shao was a high official, eminent historian, and eyewitness throughout Emperor Xian's reign, so we can be quite certain of the currency of these astrological notions at the time.

jiazi], Jupiter was in *Great Span* and the Martial King [of Wei; i.e. Cao Cao (155–220)] first received the Mandate when he suppressed the Yellow Turbans. That year was changed to become the first year of the 'Pacified Middle' reign period. [Subsequently,] in the first year of 'Established Tranquility' [196 CE], Jupiter was again in *Great Span* and [Cao Cao] rose to become General-in-Chief. Thirteen years later [Jupiter] was once more in *Great Span* and [Cao Cao] was elevated to the post of Chancellor. Now twenty-five more years have elapsed, Jupiter is again in *Great Span*, and Your Majesty has received the Mandate. *Thus Wei's obtaining Jupiter's [confirmation] corresponds to King Wen of Zhou's receipt of the Mandate.*"

Ten days later came the enthronement ceremony and Cao Pi's announcement to Heaven:

Document 42: The "Account of the Proferring Emperor" says, "On day *xinwei* [December 11, 220] the King of Wei [Cao Pi] mounted the podium to receive the abdication. The high officials, Lords of the Principalities, Generals of the Armies, the Shanyu [Chief] of the Hsiungnu, and representatives of the barbarians of the Four Quarters, all numbering in the tens of thousands, attended the ceremony. Burnt sacrifices were offered to Heaven and Earth, to the Five Sacred Peaks, and to the Four Waterways [as the King] announced, 'Your servant, Emperor Pi, has the temerity to employ dark sacrificial offerings and to announce to the Very August Sovereign Lord that the generations of the [Lords of] Han amounted to twenty and four, in all four-hundred and twenty-six years. [All within] the Four Seas are impoverished and beset, the Three Mainstays [of society] are not upheld, *the Five Planets move criss-cross*, and anomalous omens appear in concert. Those proficient in the mantic arts, reflecting on the ways of antiquity, all take these signs to mean that the sequence of Heavenly numbers has come round to conclude in this generation, [and that] all the portentous omens as well as the will of the people and spirits signal the final culmination of the numbers of the Han and the receipt of the Mandate by the House of Wei. The Lord of Han has appropriately bestowed the sacred insignia [of state] on your servant, and emulating Shun has passed the rulership to Pi. Pi trembles in trepidation before the Mandate of Heaven [and recalls the maxim], "though flattered, be not distracted."[49]

The many dignitaries, counsellors, and heads of the Six Bureaus [within the court administration], and the Generals and Officers without, including the Princes and Chiefs of the Man and Yi barbarians, all say, "The Mandate of Heaven may not be declined; the sacred insignia cannot be long vacant; the numerous subjects cannot be without a master; and the myriad affairs of state cannot be left ungoverned." Pi reverently receives the august emblems, not daring not to recover them all. The venerable turtle oracle has been consulted and the cracks indicated great unexpected [developments]; the milfoil oracle was consulted thrice and displayed the signs of change of the Mandate. A day of Origination was carefully selected and together with the assembled dignitaries [I, Pi] ascend the dais to receive the imperial cord and seal, to make sacrifice and announce it to you the Great Spirits: "May you the Spirits perpetually receive the sacrificial repasts and grant favor. May you, the hope of the masses, bless the Wei with generations of enjoyment [of the Mandate]." Thereafter [Cao Pi] issued a decree to the Three Excellencies: "From the first, rulers of highest antiquity esteemed the transforming

---

[49] Quoting from the *"Lü xing"* chapter of the *Book of Documents*.

influence of benevolence in ameliorating the customs and mores of the Empire, and thus the people conformed with their instruction, and punishments and regulations were established. Now, I continue the legacy of those sovereigns by letting the first year of the current 'Prolong Well-Being' reign period [220 CE] become the origin year of the 'Yellow Inception' reign period. I propose to reform the calendar; to change the costume and colors of the court; to differentiate the emblems and appellations; to unify the musical tones, weights, and measures; to acknowledge the rule of the element Earth; and to declare a general amnesty throughout the Empire. From the various capital sentences on down, all those who should not otherwise obtain clemency, shall be granted a pardon and remission of their offences. The *Springs and Autumns* of the Wei says: "When the Emperor had ascended the throne and the ceremony was completed, he turned to the assembled company of officials and said, 'The matter of Shun and Yu the Great,[50] I know about it.'""

Centuries later, in the Southern Song Dynasty (1127–1279), no less an authority than the eminent Neo-Confucian Zhu Xi (1130–1200) invoked the same millennial tradition concerning the linkage between planetary omens and the receipt of Heaven's Mandate:

In general, after the decline of the Zhou and Mencius' passing, the transmission of this Dao was not entrusted [to anyone], skipping Qin to reach the Han Dynasty. [Then] bypassing Jin, Sui, and Tang [the transmission] arrived at the "Sage Ancestor" [Taizu (964–97)] of our Song Dynasty. The receipt of the Mandate [signaled by] the gathering of the Five Planets in [lodge] *Kui* [Psc–And], [confirmed that] the course of enlightened [teaching of the Dao] had truly begun.[51]

Zhu Xi is obviously alluding to the Mandate tradition and to Mencius' remarks (IV.B.26) about the periodicity of sages being "slightly more than 500 years." From the Zhou founding in 1046 BCE to Confucius' birth in 551 BCE was 500 years, so by Mencius' day it was evident a sage was long overdue. The Han Dynasty legitimately received the Mandate because, as we saw, there was a Five Planet alignment for Han founder Liu Bang in lodge *Eastern Well* in 205 BCE. Bypassing the centuries of disunity and the intervening three dynasties of Jin, Sui, and Tang, one finally arrives at the Song Dynasty founding in the mid tenth century. In the almost offhand way Zhu Xi invokes the significance of planetary groupings it is apparent that by his time the tradition was long held to be established fact.

---

[50] An allusion to Shun's abdication in favor of Yu, rather than his heir. Yu founded the Xia, ostensibly the first hereditary dynasty.

[51] Zhu Xi (1130–1200); 1434.2–1435.1. The record of the planetary event in the *History of the Song Dynasty* (*Song shi*, 9.56) reads, "In the fifth year of the *Qiande* reign period of Emperor Taizu, third month (April 13–May 11); the Five Planets gathered like linked pearls in *Kui* [Psc–And] and *Lou* [Aries]." For more on Zhu Xi's interpretation of the planetary precedent of 967 CE and its subsequent enduring legacy as the sign of the revivifying of the "legacy of the Dao," see Wei Bing (2005, 27–34). I am indebted to Lionel Jensen for bringing Zhu Xi's remarks to my attention.

But there is more to this passage than that. As is well known, Zhu was no great admirer of the Song regime. In reflecting on Zhou Dunyi (1017–1073) in his essay, Zhu Xi ostensibly pays homage to the Song founder Taizu's sagely status, signaled by an alignment of the Five Planets in the year 967. But the real import of the passage, and its epochal significance for Neo-Confucianism, lies in Zhu Xi's identifying the re-emergence of traditional Confucian learning, the "legacy of the Dao" (*Dao tong*), with the conferral of the Mandate of Heaven. In so doing he fundamentally redefines the meaning of the doctrine of the Mandate. Henceforth the transmission of the tradition of the Dao came to be felt by adherents as transcending in importance the continuity of dynastic rule (*zheng tong*) itself. In other words, the supremely auspicious planetary omen was not a sign of the restoration of legitimate temporal rule, but of something of a different order entirely, the renaissance of the long neglected learning of the Dao. As we will shortly see, this idea continued to energize Neo-Confucian thinkers through the Ming Dynasty.

Events of 1524 in Ming China

Following the cataclysm of the Mongol conquest, it is not surprising that the earlier Song agnosticism with regard to astrology was less in evidence. The Yuan court was much more receptive to unorthodox and esoteric teachings, astrology among them, and in the Ming Dynasty portent astrology still found favor at court.[52] Indeed, even in the late sixteenth century, Zhu Guozhen (1558–1632; *jinshi* 1579) matter-of-factly asserted that "the court placed great emphasis on sky pattern reading."[53]

This is no doubt true of Emperor Shizong, Zhu Houcong (r. 1521–67), eleventh (*Jiajing*) emperor of the Ming Dynasty, a tyrant who, besides his cruelty, also came to be noted for devotion to Daoist rites, the search for immortality through magic and alchemy, and an interest in omens. Zhu Houcong was still

[52] Evidenced by the fact that there was a report of a "simultaneous appearance" of the five planets during the *Hongwu* reign period (1368–98) of the Ming founder; see *infra.*
[53] See Zhu Guozhen (1998, 15.329). According to John B. Henderson (1984, 132), "Cosmological skepticism did not flourish in early and mid-Ming China. On the contrary, the various divinatory and magical arts, based ultimately on correlative cosmology, enjoyed almost unprecedented support from both state and literati during this era." On astral and other omens, see Mote and Twitchett (1988, 136, 326, 479). Strong confirmation of the attention paid to astral omens in the Ming is found in the astrological compendium *Tianyuan yu li xiang yi fu*, which had been circulated among his high officials by Emperor Renzong (1425–6). It contains hundreds of hand-painted multicolor illustrations of astral and meteorological omens, with accompanying prognostications attributed to Zhu Xi (Zhu Wengong). Now rare, the book was among those banned in the literary inquisition connected with the compilation of the *Complete Writings of the Four Treasuries* (*Siku quanshu*) in the last decade of the eighteenth century. It survives in an "Orchid Cabinet" (*Lan ge*) hand-copied edition printed in late Ming between 1628 and 1644. Zhu Guozhen was no doubt well familiar with the book.

a minor of fourteen when he ascended the throne, and his adamant refusal to accede to protocol and allow himself to be adopted into the direct line of succession precipitated what is known to history as the Great Rites Controversy, which culminated precisely in 1524.[54] This struggle for unrestrained autocratic power in the face of strenuous opposition by officialdom, led by the imperious Grand Secretary Yang Tinghe (1459–1529; *jinshi* "presented scholar" degree 1478), consumed the court for the first few years of his reign, but Zhu Houcong refused to budge in his insistence on elevating his father posthumously to the status of emperor, with attendant rites. In August 1524 a protest demonstration by 200 officials kneeling and wailing outside the court gates brought down the emperor's wrath on over a hundred of them, who were flogged and imprisoned, many dying from the beatings or lengthy imprisonment. In the end, following the retirement of Grand Secretary Yang, the emperor's rigid refusal to give ground eventually carried the day.[55] An authoritative account of Zhu Houcong's reign characterizes the period this way:

In his time the rich grew richer and the poor became impoverished, particularly in the lower Yangtze area. Wealth bred leisure, which demanded luxuries and entertainment: it also encouraged the development of theatre, art, literature, and printing. The political vigor of the empire, however, began to decline, and the house of Ming showed signs of senescence.[56]

Outside the court the empire was in turmoil. Before the climax of the Rites Controversy that summer ground tremors were felt across a wide area: "In a period of ten months, from July 1523 to May 1524, as many as thirty-eight were recorded. Nanking, which has not been known in recent years to have earthquakes, reported fifteen in one month in 1524, and six in a single day."[57] The external affairs of the *Jiajing* period were no less fraught, with Mongol raiders pillaging and killing virtually at will all along the northern frontier from the northwest to the northeast, especially during the later years. The Datong border garrison on the northern frontier staged two revolts within the same year of 1524. Along the southeast coast, attacks by the marauding *wokou* ("dwarf") pirates were frequent and devastating. If ever there was a time to attach significance to celestial signs of Heaven's intentions, 1524 should have been it.

---

[54] For the succession controversy, see Fisher (1990).
[55] The *Cambridge History of China* sums up his reign this way: "The Chia-ching [*Jiajing*] emperor continued to rule in the brutal and despotic style of his cousin. He overrode all counsel and precedent to get what he wanted; he tolerated no interference, no criticism of his person or his policies. His officials retained their positions so long as they carried out his will without question and quickly lost them when they did not or could not." Mote and Twitchett (1988, 440–50, 479).
[56] Goodrich and Fang (1976, 315).   [57] Ibid., 320.

A memorial to the emperor by the official Wu Yipeng (*jinshi* 1493) enu-
merates the disasters and anomalies that occurred between the late summer
of 1523 and mid-spring of 1524 and entreats the emperor to act. At the time,
Wu was a high-ranking Attendant Gentleman in the Bureau of Rites, who had
previously remonstrated with the emperor concerning the latter's ritual excess.
In late spring of 1524, Wu memorialized,

> Yipeng earnestly reports there are disasters and anomalies in all quarters. It is said that
> since the sixth month of last year through the second month of this year, in that space of
> time the sky has rung three times; the earth has quaked thirty-eight times; in autumn and
> winter there was thunder, lightning, and hail eighteen times; one each of windstorms,
> fog, fissures in the ground, landslides, and monstrous births; and on two occasions the
> starving turned to cannibalism: there are twice as many extraordinary incidents as in
> former times. I wish that your Majesty would be first to lead your masses of workers
> in rescuing the sick and suffering, cease construction projects, place your trust in your
> officials, accept loyal remonstrance, and in this way return to Heaven's intention. "The
> Emperor was concerned and responded with a decree."[58]

This was before the confrontation between the emperor and his high officials in the summer of 1524. As an old and experienced official, Wu enjoyed
Emperor Shizong's respect, and his memorial was circumspect. As we shall see,
a memorial along virtually the same lines submitted later by a less experienced
official met with harsh retribution.[59]

Reaction to the planetary omen

The 1524 planetary portent is recorded in typically minimalist fashion in the
"Treatise on the Heavenly Patterns" in the *History of the Ming Dynasty*, though
the date of closest approach of the planets is accurately reported: "In the first
month of the third year of the *Jiajing* reign period, on day *renwu* [February 20],
the Five Planets gathered in *Yingshi* [*Align-the-Hall*]."[60] That there is a good
deal more to this episode than meets the eye may be gleaned from the private
musings of literati observers of court affairs, like Lang Ying (1487–*c*.1566),[61]
who held no official position, and the aforementioned Zhu Guozhen. Lang

---

[58] Wu Yipeng's biography and a several of his memorials are in the *History of the Ming Dynasty*,
Ming shi, 191, 5061–3. Because of the political sensitivity of astrological prognostication, the
first Ming emperor, Zhu Yuanzhang, reinstituted the ban on the private study of astrology and
strictly prohibited the transfer of officials from the Astronomical Bureau to other positions in
the bureaucracy. As will become quite evident, however, although the prohibition in the *Great
Ming Code* remained in force throughout the dynasty, it was not strictly enforced after Zhu
Yuanzhang's reign. For a recent study of the ban and the reaction to it by officialdom, see Shin
(2007). Jiang Xiaoyuan (1992a, 217–21) succinctly surveys the history of the prohibition and
its rationale from the early imperial period through the Qing Dynasty.

[59] For Emperor Shizong's decree, see *infra*.   [60] *Ming shi*, 26, 377.

[61] His biography is found in Goodrich and Fang (1976, 791).

Ying lived through the entire *Jiajing* reign period and witnessed the events of 1524. He was a bibliophile and connoisseur who devoted his whole life to scholarship. He was a perceptive observer of state affairs and the author of a famous miscellany, *Qi xiu lei gao*, containing observations on diverse subjects ranging from the history of the early Ming Dynasty to contemporary affairs. In contrast, Zhu Guozhen, writing a century later, was a high-ranking official who served concurrently from 1623 as minister in the Ministry of Rites and grand secretary in the Hall of Literary Profundity, as well as sometime tutor to the heir apparent. In addition, Zhu was a noted historian and author of a substantial history of the Ming Dynasty.

A benign prognostication

The first and most extensive account of the 1524 planetary portent, Lang Ying's, reads as follows:

When Zhou was about to attack Yin, the Five Planets gathered in *Chamber* [Sco]. When Duke Huan of Qi was about to become Hegemon, the Five Planets gathered in *Basket* [Sgr]. In the first year of Emperor Gao[-zu of Han], the *Five Planets* gathered in *Eastern Well* [Gem]. The retainer Zhang Er [King of Zhao] said, "*Eastern Well* is the allocated field of Qin, The King of Han [i.e. Liu Bang, Emperor Gaozu] should enter Qin and take all under Heaven"; and so it turned out before long. In the eighth month of the third year of the *Kaiyuan* reign period [715 CE] of Emperor Xuanzong of the Tang Dynasty, the Five Planets gathered in *Basket* [Sgr] and *Tail* [Sco]. The prognostication said, "the virtuous one will be fortunate, while the one lacking virtue will suffer."[62] As expected, the *Kaiyuan* [713–41] reign period was well ordered, but the *Tianbao* reign period [742–55] was chaotic. In the eleventh month of the third year of the *Jianlong* reign period [962] of Emperor Taizu of the Song Dynasty, the Five Planets gathered in *Stride* [Psc]. The prognosticator said, "a virtuous one will receive the Mandate, possess the Four Quarters, and his posterity will flourish."[63] Afterward, [Taizu's] reign endured many years, as expected. To my mind the start of the Lian, Luo, Guan, and Min teachings all spring from this.[64] During the *Hongwu* reign period [1368–98], the Five Planets also gathered in *Stride*. I believe the prognostication must come to pass as in the Song. In the second year of the *Jiajing* reign period [*sic*] [1523], the Five Planets gathered in [*Ying-*]*shi*. I said to someone, "*Shi* [lit. "Hall"] is *Align-the-Hall* [Pegasus]. Gan De and Shi Shen [fourth-century BCE astrologers] both indicated that 'Hall' stands for the Great Ancestral Temple. I know from this that the country is bound to have occasion

---

[62] This grouping also figures in Zhu Guozhen's discussion below. The record is in the *Xin Tang shu*, 33.865; Xu, Pankenier, and Jiang (2000, 247). More impressive groupings on April 15, 730, and September 30, 748, apparently went unnoticed, as neither is recorded in *Tang shu*. De Meis and Meeus (1994, 295).

[63] Paraphrasing *Shiji*, 27.1321. However, the prognostication in *The Grand Scribe's Records* continues, "but the one lacking in virtue suffers calamities or destruction."

[64] These are the four major schools of Song dynasty Neo-Confucian "School of Principle" (*lixue*) philosophy.

to celebrate some event in the Great Ancestral Temple and in this way bring glory and greatness to the state." By the fifteenth year of the *Jiajing* reign period [1536], great construction projects were set in motion and the Nine Temples were renovated [in response to the omen]. When the sub-celestial realm is cultured, the Way of Heaven is manifest. Alas! From the Zhou Dynasty to the present is more than 2,800 years, and yet the Five Planets have only gathered as rarely as this, while cases of a single planet trespassing on an astral lodge are numerous indeed! Alas! Alas! From this it is evident that peace and good order are the exception, while unrest and disorder are the rule![65]

If we overlook for the moment the apparent sycophancy of the passage, several things are worthy of note. First, Lang Ying cites eight historical occasions when gatherings of the Five Planets carried dynastic implications. One of these, that supposed to have occurred during the reign of the Ming founder Zhu Yuanzhang, is tendentious at best.[66] Lang Ying's account of the official prognostication predicting a replication of the glorious Song precedent – "I believe the prognostication must come to pass as in the Song" – is a good illustration of the kind of interpretive spin designed to win approval. Oddly enough, Lang Ying gets the date of the *Jiajing* planetary event wrong, since he places it in 1523, though this could well be the result of a common copyist's error of "second year" for "third year." Space does not permit thorough study of the other precedents, but it will be instructive to briefly describe the circumstances.

In Part Two I discussed the spectacular planetary groupings in the Three Dynasties period, and in Chapter 9 the much less impressive alignment of 205 BCE that inaugurated the Han Dynasty. Unlike the earlier precedents which are amply documented during the Han period, the account of the grouping presaging the ascendancy of Duke Huan of Qi in the seventh century BCE, which makes its first appearance in the *History of the Song Dynasty* (420–497), "Treatise on the Heavenly Patterns," over a thousand years after the fact, is somewhat suspect.[67] Lang Ying is confused about the Tang precedent. There

---

[65] Lang Ying (1961, Vol. 1, 25).

[66] The record is dated *Hongwu* eighteenth year, second month (March 24, 1385), and states only that "the Five Planets appeared together," *wuxing bing jian*. In fact, they were separated by some fifty-four degrees. At the time of a subsequent report of their "all appearing" two years later (February 19, 1387) they were spread across the entire sky (over 150 degrees). *Ming shi*, 26.376.

[67] *Song shu*, 25.735. Recently, Salvo de Meis (2006, 18) focused on an impressive grouping on January 11, 661 BCE, as a possible candidate, at which time the Five Planets were located in Sagittarius–Capricorn with a maximum separation of just over seventeen degrees. It is worth mentioning in this context a comment attributed to Emperor Taizong (r. 627–49), cofounder of the Tang Dynasty. Though considered a no-nonsense rationalist, the following exchange (quoted in Wang Fangqing (1983–6, 4.13b)) recorded by a high Tang official gives an impression of contemporary views on planetary portents:

Taizong asked a retainer, "For an emperor to rise he needs must have Heaven's Mandate; [his position] is not achieved by luck." Fang Xuanling replied, "One who would rule as King must have Heaven's Mandate." Taizong said, "What you say is right. When I observe the Kings of

was no grouping in 715, nor is there any such record in the *History of the Tang Dynasty* (*Tang shu*). There was, however, an alignment observed in *Tail* (Sco) near the end of Xuanzong's reign (712–56; see infra).[68] On the date given by Lang Ying in the year 715, only Jupiter was located in lunar lodge *Winnowing Basket* in Sagittarius. Lang Ying also misdates the Northern Song Dynasty precedent. As we saw, this is a reference to a fortuitous early Song planetary cluster in mid-April of 967, during the reign of Taizu, the first emperor of the Northern Song Dynasty.[69] Coming at the beginning of the dynasty it should have been interpreted as auspicious.

Lang Ying was not the only influential literatus to put a positive spin on the 1524 planetary phenomenon. The *History of the Ming Dynasty* also records celebratory and commemorative hymns and music. Two of these congratulatory hymns on the theme "Heaven Mandates the Virtuous" were composed specifically for performance at court during the *Jiajing* period and probably contemporaneous with the events in question since they refer explicitly to the auspicious implications of the 1524 planetary massing. They offer an example of the kind of predictably diversionary phraseology one might expect. The first of these, *Wan sui yue* (lit. "10,000 Year's Longevity Song"),[70] suffices to give a taste of the extravagantly allusive and flattering language employed:

old who possessed Heaven's Mandate, their compelling influence was as if divinely inspired, as if they arrived at their objective without acting. Those who lacked Heaven's Mandate in the end only met with destruction. Anciently, King Wen of Zhou and Emperor Gaozu of Han initiated grand sacrifices and first received the Mandate, whereupon the Red Sparrow [augury] came; they first made their reputation, and then the Five Planets gathered. Thus when combined with what Heaven displays above, the verification is never empty. If not preordained by Heaven, the true course is never inappropriately achieved. If I had served the Sui Dynasty, I would not have risen beyond [the rank of] Capital Guard; indeed, as I am lazy and slow to act, I would not have acted as time required." The Duke (Fang Xuanling) said, "In the *Changes* it says, 'concealed dragon, do not act,' which is to say that at a time when sagely virtue is in concealment, one does not act in a manner known to one and all, and so when Gaozu of Han served the Qin, he did not rise above the post of headman of a township."

[68] It is noteworthy that there was a spectacular six-degree grouping in Cancer–Leo slightly before the beginning of Xuanzong's reign in late June of the year 710 CE, the planets reaching minimum separation on June 25. This occurred just weeks before the death of Emperor Zhongzong and the accession of Ruizong, and may have been overlooked for that reason. It is not recorded in *Tang shu*. This particular grouping of planets was, however, a focus of intense interest by the Maya, who conducted a major ritual on that day in Naranjo, and used the movements of Jupiter and Saturn to time a military attack on Yaxha. Olson and White (1997, 63–4). For military associations of Jupiter, Saturn, and Venus in Maya astrology, see Schele and Freidel (1990) and Milbrath (1999). Lang Ying also overlooked a thirty-four-degree alignment of all Five Planets in Cancer–Leo in the third year of Xuanzong's successor, Daizong (766–79), carrying the prognostication "beneficial for China." *Xin Tang shu*, 33.866.

[69] The Five Planets lined up in Pisces within about twenty degrees of each other. Wen Ying (1991, Chapter 2); Xu et al. (2000, 249).

[70] *Ming shi*, 63.1577.

> A Son of Heaven arises in an Age of Great Peace –
> The days are flourishing;
> At the start of [Hexagram] Lü increasing *yang* –
> Returning to Primal Auspiciousness;
> The sweet wine springs and the Spirit Fungus –
> Don't collect them!
> Of the Five Planets it's been said –
> They gathered in (lodge) *Align-the-Hall*.

Perhaps more interesting was the later positive interpretation, popular among some devotees of the philosopher Wang Yangming (1472–1529) that the planetary portent signaled the ascendancy of Wang's intuitive School of Mind over the orthodox rationalism of the Song Dynasty Cheng-Zhu School of Principle. None other than the eminent Huang Zongxi (1610–95) made this assertion more than once in his *Case Studies of Ming Confucians* (*Ming Ru xue an*), China's first intellectual history.[71] Even though court astrology typically concerned itself with affairs of state, Huang Zongxi's report of such an opinion, ending with a rhetorical "How could it not be Heaven's doing?", is in keeping with the venerable tradition that Confucius' own symbolic elevation to the status of "uncrowned king" was likewise heralded by an auspicious sign – the capture of a *qilin* ("unicorn").

That knowledge of the celestial phenomenon was widespread at least among the literate elite is undeniable, since a period obituary inscribed on a stone stele specifically mentions it, in one case actually dating the birth of the individual being eulogized to the very day of the planetary massing: "born in the third year *jiashen* of the *Jiajing* reign period, first month, on the seventeenth day; on this day the Five Planets gathered in Jupiter station *Juzi* (*Align-the-Hall*, Aqr–Psc)."[72] Clearly, in the mind of the author of that memorial, Guo Fei (1529–1605), this must have been a propitious sign or he could not have alluded to it this way.

An ominous prognostication

Now, let us consider what Zhu Guozhen has to say about these same events, nearly a century later. In a section in his miscellany *Yong chuang xiao pin* (1622), entitled "Gatherings of the Five Planets," Zhu reports a prognostication rather different from Lang Ying's. According to Zhu,

---

[71] Huang Zongxi (1974, 14.28, 62.36). Huang asserts that a planetary portent of 1133 signaled the rise of the Cheng-Zhu School and that "those who know" say that the planetary massing of 1524 presaged the flourishing of Wang Yangming's thought. For the epochal importance for the "legacy of the Dao" that Zhu Xi attached to the alignment of 967, as well as its enduring motivational significance for Neo-Confucians, see Wei Bing (2005).

[72] See *Ming wen hai*, 450.38b.

In the third year of the *Jiajing* reign period [1524], the Five Planets gathered in *Align-the-Hall*. The Director of Astronomy, Le Huo, submitted an opinion: "When the planets gather, either there is great good fortune, or there is great calamity. When they gathered in *Chamber* [Sco], Zhou flourished; when they gathered in *Winnowing Basket* [Sgr], Qi became Hegemon; when the Han arose, they gathered in *Eastern Well* [Gem]; Song prospered when they gathered in *Stride* [Psc]. In the *Tianbao* reign period [of Tang, 742–56], they gathered in *Tail* [Sco–Sgr], and the An Lushan rebellion erupted.[73] The pertinent prognostication is: 'when armies throughout the realm are plotting, the planets gather in *Align-the-Hall*.'"[74]

Zhu Guozhen leaves it at that, no doubt confident that his readers will be able to draw the appropriate conclusions. However, an entry in the *Veritable Records of the Ming Dynasty* (*Ming shilu*), based on court diaries and daily administrative records, quotes at length from Le Huo's admonition to the emperor immediately following his prognostication above. Le Huo recited a litany of difficulties confronting the empire – unrest along the inland frontiers, banditry in the Central Plain, piracy along the southeast coast, economic and social decline, exploitation of the peasantry, neglect of infrastructure and husbandry. From their juxtaposition with the planetary portent, the linkage in his mind is patently clear. Le Huo then says, "I hope that His Majesty will promote the worthy, accept remonstrance, cultivate himself, settle the populace, desist from construction projects, and screen himself off from pleasures and indulgences, in order externally to cut off the marauders at the gates, and to put an end to strife within."[75] But the *Veritable Records* omitted a crucial part of the memorial. A quotation in the local Jiangxi gazetteer, *Jiangxi tong zhi*, records that Le Huo also said this:

The prognostication texts say that when the Five Planets gather, this means a change of kingship: the virtuous one is fortunate but the one lacking in virtue is destroyed. [The Five Planets] gathered in *Chamber* and the Zhou sacrifices flourished; when they gathered in *Winnowing Basket*, Duke Huan of Qi became Hegemon; when Han arose they gathered in *Well*; and when Song flourished they gathered in *Stride*. In all these

---

[73] The Five Planets drew to within about thirty degrees of each other, actually no more impressive a grouping than Gaozu's in 205 BCE. The An Lushan rebellion, which nearly toppled the dynasty, lasted from 755 to 763. The planetary alignment occurred in the eighth month (early October) of the ninth year of the *Tianbao* reign period, or 750. It is reported in *Xin Tang shu*, 33.865; Xu et al. (2000, 247).

[74] Zhu Guozhen (1998, Chapter 15). The other source that quotes the precise wording of Le Huo's memorial with its five historical precedents and ominous portent is the great Qing Dynasty scholar Yan Ruoqu's (1983–6, 1.24b). Also worth mentioning is the opinion of Zhang Xuan (1558–1641; *juren*, "selected scholar," 1582), who also cites the five historical precedents. Zhang expresses his bafflement that such figures as the Spring and Autumn period hegemon Duke Huan of Qi and the Tang rebel An Lushan were deemed worthy of celestial omens, implying that he too saw the signs as portending a reassignment of Heaven's Mandate to rule. Zhang Xuan (1983–6, 3.9b). Zhang Xuan's biography is in Goodrich and Fang (1976, 78).

[75] See *Ming shilu* (n.d., 36, 319) for March 19, 1524 (third year, *Jiajing* reign period, second month, day *jiyou*).

four cases, at a turning point in time, the one prospered and the other met with disaster –
there is clearly no mistaking it. Only when they gathered in *Tail* and *Winnowing Basket*
in the *Tianbao* reign period [742–56], the virtue of Tang was unenlightened, and so
in the end there was the [An] Lushan rebellion. Your Imperial Majesty has arisen in
mid-dynasty, and the Five Planets have appropriately gathered. How could one not all
the more burnish sagely virtue in order to be granted this great beneficence?[76]

Recalling the highly charged atmosphere at court in 1524, one must admire
Le Huo's temerity in proffering the above opinion to the emperor. Having
only recently been promoted in 1521 to the post of vice director in the Directorate of Astronomy, Le Huo had already earned a reputation as a competent
astronomer.[77] Not surprisingly, the *Jiangxi* gazetteer goes on to report that
after the memorial was submitted Le Huo was thrown into prison and was only
spared through the intercession of high officials. He was subsequently banished
to an inconsequential post far from the court.

Le Huo is otherwise seldom mentioned in the standard histories of the period,
but he does figure prominently in a memorial preserved in the brief biography of
another freshman court official, Wei Shangchen,[78] a newly minted "presented
scholar" (*jinshi* 1522) serving in his first posting as case reviewer in the Court
of Judicial Review. It was Wei's unhappy lot to be assigned to review the cases
of the officials who were condemned as the succession controversy reached its
climax in the late summer of 1524. According to Wei Shangchen, since he took
up his post, not a day had passed without new cases of official misconduct being
referred to him. In his memorial, Wei wrote in defense of forty-five accused and
imprisoned high officials, mentioning many by name, including Le Huo. Wei
considered all of them wrongly indicted and deserving of pardon. In closing his
plea to the emperor, Wei cited the numerous natural calamities causing great
distress throughout the empire, as well as the heavenly signs (including comets),
all of which "chill the heart" of any who are aware of what is happening. The
reference to signs in the heavens is another clear indication that, as an apparent
response to current affairs, the ominous planetary grouping provoked more
official consternation than the official record would otherwise seem to suggest.
Wei concluded his memorial by proposing that pardons or exoneration of the
accused and compensation for the families of the dead were the means to relieve
the suffering caused by the many disasters. In making policy suggestions, Wei
too was imprudent. The emperor's response to Wei Shangchen's "seeking to
make his reputation peddling frankness" was to have Wei banished to a remote
prefecture in Qingjiang, Jiangxi, to serve as a lowly jailer.

---

[76] *Jiangxi tongzhi* (1983–8, 81.47b).
[77] For a critical evaluation of the performance of the Directorate of Astronomy during the Ming
Dynasty, see Deane (1994).
[78] His biography, which quotes the memorial, is in *Ming shi*, 208.5500.

This contrast between divergent interpretations offers a good illustration of the context of court astrology in late imperial China. Even the most ominous phenomena were duly reported; not to do so would have invited execution. Laudatory prognostications illustrate the kind of spin that could be placed on the phenomena to render the standard prognostication less ominous.[79] But even under the harshest regimes there were individuals who took their professional responsibilities most seriously at considerable risk to themselves. Le Huo was evidently such a person. Looking carefully at the five recognized portents he cites, the implications of his unadorned conclusions are, as he suggests, unmistakable:

when the planets gather, either there is great good fortune or there is great calamity, and when armies throughout the realm are plotting, the planets gather in *Align-the-Hall*... The prognostication texts say that when the Five Planets gather, this means a change of kingship: the virtuous one has occasion to celebrate and the one lacking in virtue is destroyed... When the Five Planets gathered at the beginning of a regime they heralded the rise of three prestigious dynasties, Zhou, Han, and Song, and a Hegemon, Duke Huan of Qi. When they gathered in mid-dynasty, on the other hand, as in the case of Tang emperor Xuanzong, they foretold imminent disaster in the form of armed rebellion.[80]

Le Huo's interpretation was suppressed, however, and after the fact the officially sanctioned interpretation of the planetary massing became "*Shi* [*Align-the-Hall*] stands for the Great Ancestral Temple ... the country is bound to have occasion to celebrate some event in the Great Temple and in this way bring glory and greatness to the state." It was not until twelve years later (!), after the appearance of Halley's Comet in 1533, that the Nine Temples were belatedly renovated in fulfillment of the prediction.

Well before that, however, Emperor Shizong was taking no chances. A senior imperial censor, Jin Xianmin (*jinshi* 1484), on learning of the inauspicious implications of the planetary portent, recommended to the emperor that inspectors of the various defense commands make preparations for military engagement. Emperor Shizong agreed.[81] Despite the auspicious prognostication and congratulatory hymns that reflected the officially sanctioned view, the emperor was attentive enough to ritual form to follow the prescribed protocol. For a ruler in the late imperial period this meant issuing a decree assuming personal responsibility for the disequilibrium of the cosmic forces arising from misrule and for the resulting natural disasters and calamities visited on the empire. While one might assume that this was a mere formality by this late

---

[79] For another interpretation of the planetary portent along much the same lines as Lang Ying's, see Wang Yingdian (1983–6, 269.29a).

[80] Le Huo's prognostication and admonition are cited approvingly by his near contemporary, the classical scholar Zhang Huang (1983–6, 25.45a). For Zhang Huang's biography, see Goodrich and Fang (1976, 83).

[81] *Ming shi*, 194.5141.

date, a ceremonial nod in the direction of tradition, in the case of an individual as superstitious as Emperor Shizong, no matter how tyrannical and vindictive, he may not have been wholly immune to disquiet.[82] The imperial decree circulated throughout the administration read,

Heaven displays warnings and disasters, and anomalies are frequent. With regard to our person, vigilantly and fearfully we will reform our behavior and, together with the senior and junior officials and workers of the inner and outer courts, make extra efforts to mend our ways and examine ourselves critically, in order to revert to the will of Heaven.[83]

In view of the anxious responses to the 1524 planetary grouping in China and Europe, it is especially instructive to contemplate the fact that the event was, in fact, *invisible everywhere* – the astral omen transpired in daylight, about eleven degrees from the Sun. It was calculated and predicted, but impossible to observe.

Solar and Planetary Longitudes on 19 February 1524

Sun	Mercury	Venus	Mars	Jupiter	Saturn	Sign	Separation
339°	340°	351°	350°	343°	341°	Aqr–Psc	10.51°

A reflection on a cultural intersection of astrological ideas

When it comes to planetary astrology in particular, the outcome in China is surprising in view of the crucial role played by Sāsānians as intermediaries in transmitting to the Mediterranean world the theory of world ages punctuated by Jupiter–Saturn conjunctions. China's direct contact and

---

[82] Indeed, in the nineteenth century the Xianfeng emperor (1831–61) still issued similar *mea culpas*.
[83] Emperor Shizong's decree of the third year of *Jiajing* survives in *Nan gong zou gao* (c.1535), a collection of memorials authored by Xia Yan (Xia Wanchun, 1631–47, *jinshi* 1517). Comparison with other similar imperial decrees preserved in the same section "Memorials on Disasters and Anomalies" shows that their form and substance varied only minimally. After quoting the emperor's order, Xia Yan helpfully continues in his own words, "checking this against earlier historical records, only the [Tang Dynasty] *Tianbao* reign period [planetary massing] was inauspicious, for Emperor Xuanzong's rule was dissipated." This confirms that this 1524 decree was Emperor Shizong's official response to the planetary omen. It also suggests why, since four of the five historical precedents were seen as auspicious, some were able to put forward a more sanguine view based merely on the percentages. In 1524, Xia Yan was a supervising secretary in the Ministry of War in Nanjing. His talent and ambition ultimately led to his elevation to the highest office of Chief Grand Secretary in 1539, concentrating great power and authority in his hands. Xia Yan (1983–6, 5.19b). Xia Yan's biography is in Goodrich and Fang (1976, 527).

involvement with the Sāsānid Empire (224–651) was, if anything, even more extensive than that of the Latin West from the Eastern Han through the Tang Dynasty. Besides centuries-long trade contacts with Persia via the Sogdians in what is now Xinjiang, who were themselves Iranian, Persian seaborne trade with Southeast Asia was also extensive. Sāsānid merchants maintained settlements in Canton and other ports during the Tang Dynasty. Sogdians assumed Chinese surnames (collectively denoted "the nine families"), filled important military posts, and held public office.[a] The appearance of the seven-day week in Chinese almanacs beginning in this period is attributable to the Sogdians, as is the introduction of the Western zodiac and several well-known compendia of planetary ephemerides and star lore.[b] After the destruction of the Sāsānid Empire by the Arabs in 651, much of the nobility were exiled at the Tang court. Yet Sāsānian astrological history, especially integral numbers of Jupiter–Saturn conjunctions, their migration through the triplicities of the zodiac, and the epoch-making political and religious implications of "mighty conjunctions," seemingly had no discernible impact in China.

This is all the more remarkable because in the first half of the eighth century the work of the Imperial Bureau of Astrology and the Calendar was actually in the capable hands of renowned astronomers and mathematicians deeply knowledgeable about both Chinese and Indian astrological theory and methods. The first was the Buddhist monk Yi Xing (683–727), and the second the Indian Qutan Xida (aka Gautama Siddha, fl. c.720). It was Qutan Xida who oversaw the compilation of the famous *Prognostication Canon of the Kaiyuan Reign Period* (*Kaiyuan zhanjing*) completed in 729. In this comprehensive manual were collected and collated all the surviving ancient astrological text passages and prognostications, including the most complete versions of the *Canons of Stars* attributed to the famous fourth-century BCE astrologers Shi Shen and Gan De. This compilation is the most important astrological compendium since Sima Qian's "Treatise on the Celestial Offices" compiled nearly a millennium earlier – and it is devoted to traditional Chinese astral prognostication.

That said, it is intriguing that it was the son of a prominent Sogdian military family and court favorite, the schemer An Lushan (703–57), whose mutinous rebellion almost succeeded in bringing down the Tang Dynasty.[c] An Lushan was a Zoroastrian, the religion whose astrologer-priests, the Magi (or Chaldeans), are well known in the history of astrology. An's Turkish mother was reputedly a sorceress herself. After a long and checkered military career General An was able to insinuate himself into the emperor's good graces. He was doted on by the emperor's favorite concubine Yang Guifei (719–756) to such an extent that he even managed to convince her to adopt him. As a result General An rose to the highest military rank and

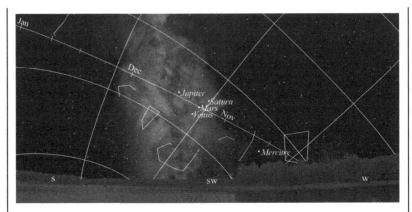

Figure 14.3 The Tang Dynasty planetary grouping of October 750 CE in Scorpius–Sagittarius (Starry Night Pro 6.4.3).

influence at court and was able to create his own powerful frontier defense force commanded largely by non-Chinese Khitan and Turkic officers personally beholden to him. His allegiance to the Tang Dynasty was opportunistic and decidedly tenuous.[d]

When in early October 750 the five visible planets gathered in Scorpius–Sagittarius the event could be thought to carry dynastic implications.[e] Jupiter, Saturn, Mars, and Venus were merely a few degrees apart, and only Mercury was far away (see Figure 14.3).[f]

In Chinese thinking, as a possible sign of a shift of Heaven's Mandate, a mid-dynasty planetary grouping like this was particularly ominous for the ruling house, and as we saw, the verdict of later history in this instance was that the planetary portent certainly foretold the fall of the Tang Dynasty. In seeming confirmation of the inauspiciousness of the omen, in the following year, 751, Chinese forces were decisively defeated by Abassid Arab and Turkic armies at the Battle of the Talas River, marking the end of Tang expansion and the beginning of withdrawal from Central Asia. Shortly after this, General An Lushan set in motion the devastating rebellion that he had apparently been planning all along. In 756, after capturing the ancient capital, Luoyang, An Lushan declared himself emperor. The ensuing fighting, which cost tens of millions of lives, lasted from 755 to 763 and nearly toppled the Tang Dynasty.

In Māshā'allāh's astrological world history, *On Conjunctions, Religions, and Peoples*, written around the year 800 in Baghdad, it comes as no surprise that one event stood out prominently the March 19, 571, shift of triplicity signaling the rise of Islam.[g] The Tang Dynasty planetary conjunction of

750 came 179 years afterward, a mere sixty years from the next shift of triplicity, which theoretically should signal another world-changing event like the rise of a new nation or dynasty. What might An Lushan's knowledge of Sāsānian and Chinese planetary astrology have led him to conclude about the impressive planetary omen he was witnessing? It is a safe bet that there was rampant astrological speculation in knowledgeable circles about the omen's significance, based on both Chinese and Western precedents. I submit it is likely that An Lushan knew of the dynastic implications in the two astrological traditions when he decided to seize the opportunity to usurp the throne. Of course, this is not to claim this was the only, or even the principal, factor prompting the General to launch his bid, just that astrology could well have played an important role in his thinking. The case is admittedly circumstantial, but this may have been the one and only instance when an otherwise impervious Chinese astrology coalesced with its Western counterpart, at a watershed moment and in the mind of a uniquely well-placed actor.

[a] A famous case in point is that of Yu Hong (d. 592), "a high-ranking member of a community of Sogdians who had settled on the northern border of China at the beginning of the fourth century. While barely in his teens, Yu Hong began his career in the service of the most powerful nomadic tribe at the time, known as the Ruru, and was posted as an emissary to several countries, including Iran. During the Northern Qi [550–77], Northern Zhou [557–88], and subsequent Sui Dynasty [581–617], he served as a *sabao* (Sogdian *sartpaw*, originally 'caravan leader'). an appointed leader of a foreign enclave on Chinese soil. Iwami (2008, 53). Yu Hong, his father, and his grandfather were all respected leaders of the Sogdian community. His wife (d. 597) was interred along with him in the same tomb six years later in 598. They lived their later years and were buried in Taiyuan, Shanxi Province, the Northern Qi Dynasty's second capital city." For a full description of Yu Hong's well-preserved tomb, see Shanxisheng kaogu yanjiusuo (2005). DNA analysis has shown that Yu Hong belonged to one of the oldest western Eurasian lineages; his are the easternmost Europoid remains thus far discovered in ancient China; see Xie et al. (2007).

[b] Edward H. Schafer (1977, 10–11) discusses some of these cross-cultural contacts, also Sen (2010), and especially Jao and Vandermeersch (2006). Needham and Wang (1959, 258) quote a revealing anecdote that underscores the cosmopolitanism of the time. In the *Xiu yao jing* commentary (764 CE) the seven planets are given their Sanskrit, Sogdian, and Persian names, and linked to the days of the week. The Chinese reader unfamiliar with the names of those days is advised to "ask a Sogdian or a Persian, or the people of the Five Indies, who all know them."

[c] Pulleyblank (1955). An Lushan's original name was Aluoshan; his sinicized given name Lushan transcribes the Sogdian word "light," *rokhsh*, which was also the name of Alexander the Great's Sogdian wife, Roxanne.

[d] Iwami (2008, 64).

[e] The planetary alignment occurred in the eighth month (early October) of the ninth year of the *Tianbao* reign period, or 750. It is reported in the *New History of the Tang Dynasty* (*Xin Tang shu*), 33.865; Xu Zhentao, Pankenier, and Jiang (2000, 247).

^f After sunset on October 8, 750, the locations of the planets were as follows:
Jupiter: 16h 39' δ –22° 6.7'
Saturn: 16h 1' δ –19° 19'
Mars: 15h 59.6' δ –21° 48'
Venus: 16h 5.8' δ –23° 52'
Mercury: 14h 26' δ –17° 40'

^g This was the year after the birth of Mohammed, although the Muslim calendar takes its beginning from the Hegira in 622 CE, fifty-one years later.

Conclusion

But how can the elemental rhythm of *yin* and *yang* and the cycle of growth and decay be adjusted to the belief in a meaningful goal and a "progressive revelation" of divine truth in history?[84]

Löwith's rhetorical question suggests that it was chiefly the non-teleological nature of Chinese concepts of time and temporality, Tennyson's "cycles of Cathay," that constituted the major obstacle to the development in China of a philosophy of history, or, in the case of astrology, a naturalistic theology of history. That is indeed part of the story, since, as we saw in Chapter 2, there was always an interplay between the circularity implicit in Chinese correlative cosmology and the linearity (as in "lineage") at the heart of the Great Tradition, the well-developed historical sense, and the perennial dynastic system. As Yü Ying-shih has remarked, it is significant that

by and large no systematic attempt was made to theorize about "ultimate causes" or search for "general laws" in history as such. In contrast to Western theorists of history, nor was the Chinese historian disposed to develop a systematic theory out of an important historical observation, due perhaps to his rather underdeveloped "theoretical reason."[85]

Unlike Indian astrologers or Maya calendar keepers, Chinese thinkers evinced no interest in eschatology, philosophical ruminations about "deep time," or about cycles of destruction and regeneration. There is no reconciling the "rhythm of *yin* and *yang*" in China with "the belief in a meaningful goal," nor should we expect there to be. Chinese history "was written by bureaucrats for bureaucrats. Its purpose was to provide collections of the necessary information and precedents required to educate officials in the art of governing."[86] Seen in this light, the close affinity between history and astral prognostication is clear, at least through the Ming period. Both were handmaidens of the imperial system, responding exclusively to its institutional demands and operating entirely within the conceptual parameters of that system.

It is abundantly clear, as the above discussion has shown, that cultural responses reflected in the reading of planetary signs differed markedly in China

---

[84] Löwith (1949, 16).   [85] Yü (2002, 169).   [86] Beasley and Pulleyblank (1961, 3, 5).

and the West. Though there were religiously inspired quasi-millennial movements that roused the masses of the Chinese peasantry to concerted action, the potent combination of teleology and astrological history witnessed in the West did not materialize in China. The traditional inertia and conservatism of sky pattern reading (*tian wen*), together with the imperial system's ability to co-opt and domesticate its potentially troublesome ideas, conspired by the Song Dynasty in the eleventh century to marginalize portent astrology, albeit at times with difficulty. By the *Jiajing* reign period of the Ming Dynasty 500 years later, even with a superstitious emperor on the throne and the empire beset with natural calamities and worrisome anomalies, a potentially ominous planetary omen apparently occasioned only modest disquiet. Alarmist prognostications by responsible officials were not met with complete indifference, since those responsible were severely punished. But any dynastic implications of the portentous planetary massing seem to have been deflected in a harmless direction with comparative ease.

Epilogue

It is a curious fact that the planetary clusters observed by the Chinese during the second millennium BCE and associated by them with the founding of the first three dynasties began with an event in Aquarius–Pisces, because the latest, in 1524, also occurred in Aquarius–Pisces. When the optimal period of 516.33 years for such groupings is extrapolated forward from 1576 BCE (1953 BCE having been an outlier), it becomes apparent that the Ming Dynasty cluster of February 1524 belongs to the same remarkable series of "conjunctions" – 1576, 1059, 543, 27 BCE, 491, 1007, 1524. All but the event in 543 BCE were observed and remarked upon by the responsible astrological officials in China, and five of the seven involved all five planets.

In China these early observations provided the empirical basis for the theory that Heaven was disposed to intervene in history at periodic intervals, as well as for later naturalistic speculation based on Five Elemental Phases correlative cosmology. Both theories were fundamental aspects of traditional Chinese cosmo-political ideology, and the tension they engendered at the heart of the Great Tradition about the agency underlying dynastic change has endured until the present. In times of stress the popular imagination continues to have recourse to a cosmic authority that transcends the temporal power, nourishing a hope in intervention by supra-visible forces on behalf of those disenfranchised by the historical process.

The seductive appeal of Zou Yan's (third century BCE) Five Elemental Phases cycle notwithstanding, historical change remained inexplicable by means either of sheerly mechanical causes or of inscrutable divine agency, hence Confucius' (and his followers') dismay at being passed over, in view of the ripeness of the time and their own self-evident qualifications to receive Heaven's Mandate (or at least decide who should). With the Han Dynasty (206 BCE–220 CE) and the advent of the imperial era, the co-optation of Confucianism and incorporation of the doctrine of Heaven's Mandate as elements of imperial ideology was fully realized. Henceforth, Five Elemental Phases process theory declined in political significance, not least because it was fundamentally inimical to the dynastic system. As national security concerns, astral prognostication and calendrics remained within the domain of imperially

442

authorized and controlled activity. Throughout the history of imperial China, astrology, wedded as it was to this-worldly, near-term concerns, remained non-teleological and never transcended the cyclical paradigm of the growth and decay of dynasties.

In more recent times, despite decades of indoctrination in materialist causation and Marxism-Leninism–Mao Zedong Thought, one still encounters ordinary Chinese quite willing to opine about political legitimacy in terms of portentous omens and the traditional concept of the Mandate of Heaven. The year 1976 was a particularly striking instance, when Zhou Enlai, Chiang Kai-shek, and Mao Zedong all died within months of each other and the catastrophic Tangshan earthquake struck north China, killing hundreds of thousands. Rumors proliferated about the occurrence of anomalous heavenly signs. Given the pace of change in China and the sometime turbulence of the sea of people that keeps the current regime afloat, there are no doubt still those of a millenarian bent both inside and outside the Communist Party who, when apprised of its approach, will look forward with trepidation to the next exceptionally dense planetary grouping in the 517-year sequence.

The Planetary Massing of 8 September 2040 in Virgo

Sun	Mercury	Venus	Mars	Jupiter	Saturn
166°	186.5°	192.9°	195.8°	190.2°	191.4°

Appendix
Astrology for an empire: the "Treatise on the Celestial Offices" in *The Grand Scribe's Records* (*c*.100 BCE)

More than fifty years ago Joseph Needham called the "Treatise on the Celestial Offices" in Sima Qian's *Shiji, The Grand Scribe's Records* (*c*.100 BCE) "a text of the highest importance for ancient Chinese astronomy."[1] It is that and more. Here are highlights of what the "Treatise" does:

- It "imperializes" astral nomenclature, identifying by name and relative location eighty-nine asterisms and some 412 individual stars, in some cases providing rudimentary descriptions of apparent magnitude, color, and variability.
- It preseves a number of regional variations in the names of stars, asterisms, and prognostics by fourth-century BCE scribe-astrologers Shi Shen and Gan De.
- It shows detailed knowedge of planetary motions and is the first to claim that all five visible planets regularly exhibit retrograde motion (*ni xing*).
- It focuses entirely on state-level astral portentology and eschews popular astromancy, hemerology, and numerology (e.g. calendrical spirits, *xing-de*, etc.).
- It provides the most systematic description of the practice of observing the *materia vitalis* (winds, vapors, clouds, etc.) and of harvest prognostications at the New Year.
- It provides an account of numerous anomalous astral phenomena and their prognostics – comets, meteors, guest stars, auroras – including the earliest description of the zodiacal light.
- It adapts age-old astrological schema to meet the requirements of the new universal empire.
- It displays self-conscious theory formation in advancing an astrological theory of historical change.

The Simas' "Treatise" is clearly much more than merely a compendium of star lore. Sima Qian's references to Sima Tan in the text as *Tai shi gong*, his filial

---

[1] Needham and Wang (1959, 200). As Needham remarks, any earlier speculation about the authenticity of the "Treatise" has been laid to rest by the work of Liu Zhaoyang and others, e.g. Wang Shumin (1982, 1089).

444

promise to fulfill his father's dying wish that he complete the work, and his characterizing the whole as a family affair, make it clear that the "Treatise" is the work of both father and son. A certain derangement and sketchy coverage of some topics, especially by comparison with the specialized Mawangdui silk MS "Prognostications of the Five Planets," *Wu xing zhan*, and "Diverse Prognostications on Heavenly Patterns and Formations of *Materia Vitalis*," *Tianwen qixiang za zhan*, might suggest that the "Treatise" was originally longer.[2] This is particularly true of the final section on astrological history, which barely hints at the deeper meaning of the concluding line: "[Having explored events] from beginning to end, from ancient times to the present, [we have] looked deeply into the vicissitudes of the times, examining the minute and the large-scale." Of course, it is also possible that the "Treatise" was never designed to be more than merely an "executive summary" for Emperor Wu's personal benefit because the details of the theory of prognostication were too sensitive.[3]

The title "Celestial Offices" evokes the direct linkage between the stellar patterns and the analogous imperial offices and departments of the Celestial Empire. Astronomy in Sima Qian's time was, of course, not the dispassionate scientific pursuit that term implies. Rather, it was the prevailing conception that celestial events had ramifications on Earth, and conversely, that temporal affairs could perturb the cosmos, mutual resonance (*ganying*) influences being accomplished by means of the long-established system of astral–terrestrial correspondences and the medium of *materia vitalis*. It was thought that the signs of this reciprocal interaction between Heaven and Man could be discerned at an incipient stage, that study of historical precedents and the maintenance of careful records could aid in the interpretation of such omens, and that it was crucial to the welfare of the state, then as later, that the emperor be kept regularly and accurately informed of celestial events, especially unanticipated or anomalous phenomena. This is why the Simas' official title *Taishi gong* is rendered in English as "Prefect Grand Scribe-Astrologer," the two responsibilities being functionally inseparable. But it is the astrological motive that especially informs the "Treatise" and lends it its special character.

The Grand Scribe's Records was written and compiled during an epoch-making period in China's early history. By Sima Qian's time (*c*.145–86 BCE), the Han Dynasty had come into its own after nearly a century of consolidation following the debacle of the Qin Dynasty's extinction in 206 BCE. The early Han emperors' successful stewardship of the Mandate of Heaven and the growing self-confidence that it engendered prompted the Han rulers increasingly to distance themselves from the excesses of the failed Qin Dynasty

---

[2] Cullen (2011b); Harper (1999, 831–43).

[3] For an overview of Sima Qian's achievements as an astronomer, see Bo Shuren (1981).

(221–206 BCE). By the reign of Emperor Wu (r. 141–87 BCE), in particular, this ambition manifested itself in concerted efforts to reform political, legal, and financial administration, to re-create state religious rites and ceremonial, to reinvent the imperial ideology, and to project Han military power abroad. Sima Tan and Sima Qian, who together occupied the office of Prefect Grand Scribe-Astrologer for over half a century, witnessed these transformative developments, and the enduring importance of their life's work is a testament both to the intrinsic value of their comprehensive account of Chinese history and to the normative influence *The Grand Scribe's Records* exerted on later Chinese historiography. What makes the "Treatise on the Celestial Offices" so important from the perspective of the history of ideas, apart from its clear exposition of the theory and practice of official portentology in the Western Han (202 BCE– 8 CE), is that it also marks an unmistakable transition from ancient traditions to a new interpretive paradigm.

Theory of astrology in the early imperial period

Although field-allocation (*fenye*) astrology is hardly mentioned in Western Han philosophical works, numerous passages in fourth- to second-century BCE narratives like *Zuo's Tradition* (*Zuozhuan*), *Discourses of the States* (*Guoyu*), the *Springs and Autumns of Master Lü* (*Lüshi chunqiu*), and the encyclopedic *Huainanzi* (*Book of Master Huainan*), as well as an abundance of archaeologically excavated Han Dynasty artifacts ranging from diviners' mantic astrolabes (*shi*) to the Mawangdui silk manuscripts, all make it clear that both a theory of astral–terrestrial correspondences and the metaphysical notion of mutual resonance between Heaven/Earth and Man were pervasive in the cosmological thought of the Qin and Han periods.[4] Unlike Hellenistic astrology, the Chinese did not stress the unidirectional influence expressed by Ptolemy's axiom "as above, so here below," so much as the reciprocal "as here below, so above" reflected in the belief that celestial anomalies and other ominous manifestations of Heaven's displeasure could be seen as an index of temporal misrule. For this reason, astral omenology and calendrical astronomy bore directly on the security of the state and hence were closely controlled activities. This explains the hereditary, even hermetic, character of the scribe-astrologer's profession from the earliest times.[5]

The ancient Chinese were acute observers of celestial phenomena. Such observation was not the result of disinterested stargazing. Original records

---

[4] For a general discussion of these and other aspects of pre-imperial and Han dynasty astrology and cosmology, see Harper (1999, 831 ff.).
[5] Needham and Wang (1959, 193, 359).

of regular astronomical observation ranging from the mundane (sunrise and sunset, solstices, individual stars and planets) to the exceptional (lunar and solar eclipses, sunspots, supernovae, etc.) appear as early as writing itself in the Shang oracle bone inscriptions beginning in the thirteenth century BCE. The very fact that astronomical omens first appear in the late Shang Dynasty (1554–1046 BCE) inscriptions, and that the *Springs and Autumns Annals* accurately reports thirty-six solar eclipses and several comets during the period 722–479 BCE, should be sufficient to excite curiosity, but the role of astral–terrestrial correspondences in the earliest period has previously not been adequately explored.[6] For years the conventional view has been that prognostication based on the stars played little role in the history of ideas in China before the late Warring States period; in fact, however, by that time it already had a history reaching back more than a millennium.[7]

In the Western Han Dynasty (206 BCE–8 CE), it was the duty of the Prefect Grand Scribe-Astrologer, *Taishi gong*, to know the historical precedents, to follow the movements of the heavenly bodies, and to advise the emperor on the implications of developments, especially unanticipated changes or anomalies. The Simas' "Treatise on the Celestial Offices" provides a comprehensive survey of the cosmological and astronomical knowledge in the keeping of their office, as well as its practical application. This included tracking the locations, movements, and changes affecting the stars and planets and interpreting their significance based on the well-established system of astral–terrestrial correspondences. According to Sima Qian, this preoccupation with following the movements of the heavenly bodies could be traced all the way back to the dawn of civilization: "For as long as the people have existed, when have successive rulers not calendared the movements of the Sun, Moon, stars and asterisms?"[8]

The astrological portents described in the "Treatise" had implications for the ruler, for high dignitaries, and for major affairs of state – warfare, the harvest, natural disasters, etc. No astral omens applied to ordinary individuals. The form that astral omens and prognostics took conveyed a sense of inevitability; in other words, given the formula "if *x* occurs/occurred, then *y* will occur," the

---

[6] For recent studies of later Chinese eclipse records, see Steele (2000); also Pankenier (2012).

[7] *Pace* Sivin (1995b). In contrast, Peter Berger (1990, 34) observed, "Probably the most ancient form of legitimation is the conception of the institutional order as directly reflecting or manifesting the divine structure of the cosmos, that is, the conception of the relationship as one between microcosm and macrocosm. Everything 'here below' has its analogue 'up above.'"

[8] *Shiji*, 27.1342. Whenever the "Treatise" cites the historical or theoretical judgments of the "Prefect Grand Scribe-Astrologer" one must presume that Sima Qian is referring to his father, Sima Tan. Sima Qian emphasizes that the completion of the work represents the fulfillment of his father's dying wish, hence we assume that a substantial portion of the account of astrology in the "Treatise" forms part of his father's legacy.

implication is that there is no possibility of dispelling the evil consequences of baleful astral omens. In contrast to Mesopotamian astral divination, where personalized offerings and appeals were routinely directed to the deities deemed individually responsible, in the "Treatise" no appeal for relief is admitted to the impersonal cosmic forces of *yin* and *yang* and the Five Elemental Phases. The prevailing view in the early empire, promoted by prominent thinkers like Dong Zhongshu (*c*.179–104 BCE), was that ominous signs were warnings to the ruler from a beneficent Heaven that his governance and conditions in the realm were not in harmony with the cosmic order and needed to be corrected, lest greater harm befall the state.[9] No apotropaic rituals or procedures are hinted at, even though we know that propitiatory offerings to "Heaven" and natural spirits occurred at other levels, for example sympathetic magic involving effigies of dragons coupled with invocations for rain.[10] The term "evil consequences" must be used advisedly, since there is no implication of a causal relationship per se between omens and disasters; instead, the connection appears to be associative, correlative, even analogical, consistent with the cosmological conceptions of the time. This is also the reason why the term "divination" is best avoided in connection with Chinese astral omenology – the celestial bodies were not divinized as in other astro-mantic practices. In contrast to the Mesopotamian and Buddhist pantheons, astral bodies were not divine but rather a manifestation of the *materia vitalis* that gives shape to and animates the cosmos. It is this concept that underlies the Chinese understanding of the fabric of space–time in which everything relates to everything else, all partaking of the same *qi* whose operations Joseph Needham memorably characterized:

[*Materia vitalis* manifests as] patterns simultaneously appearing in a vast field of force, the dynamic structure of which we do not yet understand ... The parts, in their organizational relations, whether of a living body or of the universe, were sufficient to account, by a kind of harmony of wills, for the observed phenomena.[11]

---

[9] Henderson (1984, 5).

[10] Referring to similar practices of an earlier epoch, at the very beginning of his chapter on "*Luan long*," the late Han sceptical thinker Wang Chong (27–*c*.97 CE) wrote, "Emulating the [Great] *Yu* sacrifice in the *Springs and Autumns Chronicle*, Dong Zhongshu fashioned clay dragons to bring down the rain, thinking that clouds and dragons go together. The *Changes* says, 'clouds accompany the Dragon, wind accompanies the Tiger' ... Duke She of Chu [late sixth century BCE) was fond of dragons – wall panels, basins, goblets – he had dragons painted on all of them. [He] caused the real thing [to materialize] by means of images of a similar kind, and so it regularly rained in the Duke's state." "Discourses Weighed in the Balance," *Lunheng*, fascicle 16. Even more noteworthy is that Emperor Wu inaugurated a new regnal era in 100 BCE called "*Sky River*," *Tian Han*: "the title had a double significance, since the Han empire had been named after its founder's fief on the Han River. But more significantly the proclamation was a magical act intended to alleviate a long period of severe drought. It was expected that the heavenly river would respond to the honor by sending down some of its excess." Schafer (1974, 406).

[11] Needham (1969, 302).

The very fact that astral and meteorological omens were classified and cataloged implies that, however rare the phenomenon (or improbable, from a present-day perspective), recurrence was always a possiblity.[12]

Because of their rarity and precedent-setting historical occurrences, multiple planetary conjunctions, especially dense clusters involving all five naked-eye planets (Mercury, Venus, Mars, Jupiter, Saturn), ranked as the most portentous of all celestial phenomena and as such carried dynastic implications. This pre-eminence was based on the historical association of planetary massings with epochal dynastic transitions during the Three Dynasties of the Bronze Age, and culminating in the most recent such alignment, in 205 BCE. This sign in the heavens was officially recognized and later memorialized in *The Grand Scribe's Records* as the astral omen signaling the imminent founding of the Han Dynasty: "When Han arose, the Five Planets gathered in *Dongjing* [*Eastern Well* or lodge 22, Gemini]."[13] A later account in the *History of the Former Han* is even more explicit:

First Year of Emperor Gaozu of Han, tenth month, the Five Planets gathered in *Dongjing* [lodge 22, *Eastern Well*]. Extrapolation based on the calendar [indicates] they followed [the lead of] Jupiter. This was the sign that August Emperor Gao had received the Mandate. Therefore a retainer said to Zhang Er, "*Dongjing* is the territory of Qin. When the King of Han [i.e. Liu Bang, soon to be Emperor Gao] entered Qin, the Five Planets followed Jupiter in gathering together, signifying [he] ought to gain the sub-celestial realm by means of Rectitude."[14]

From Sima Qian's account of the significance of planetary massings, it is evident that by the beginning of the Han Dynasty heavenly endorsement of the transfer of the Mandate to a new dynasty in the form of a cluster of all five planets had become ideologically de rigueur. In a conservative "science" dependent on precedents like astral prognostication this kind of axiomatic premise could not take shape and win general acceptance overnight, it must have the sanction of tradition behind it. Clearly, by Sima Qian's time the connection was self-evident. His concluding summary of the astrological knowledge of his day in

---

[12] For a brief summary of the numerous texts in this category in the imperial library at the end of the Former Han, see Raphals (2008–9, 83).

[13] *Shiji*, 27.1348.

[14] See also *Han shu*, 26.1301. Sima Qian is more circumspect, saying only "when Han arose." For his theoretical statement that conjunctions of the Five Planets in which Jupiter takes the lead portend the rise of a righteous dynastic founder, see *Shiji* 27.1312. The date "tenth month of the First Year of Emperor Gaozu" for the planetary event is physically impossible and an obvious fiction evidently designed to make the timing coincide with Qin heir Wangzi Ying's surrender to Liu Bang at Xianyang, the Qin capital. The actual planetary alignment occurred the following year, in May 205 BCE. The faulty record was ridiculed early on by Gao Yun (390–487), who pointed out that in the tenth month the Sun should be in *Tail – Winnowing Basket* (lodges 6–7, Sco–Sgr). Given that Mercury and Venus could not possibly be found so far from the Sun as *Eastern Well* (lodge 22, Gem), Gao concludes that the date cannot be right. *History of the [Northern] Wei Dynasty*, "Gao Yun zhuan," *Wei shu*, 48.1068.

the "Treatise" displays both ancient conceptual roots and the Han theoretical reformulation based on the prevailing *yin-yang* and Five Elemental Phases correlative cosmology:

Ever since the people have existed, when have successive rulers not systematically calendared the movements of the Sun, Moon, stars and asterisms? Coming to the Five Houses [Huang Di, Gao Yang, Gao Xin, Yao-Shun] and the Three Dynasties [Xia, Shang, Zhou], they continued by making this [knowledge] clear, they distinguished wearers of cap and sash from the barbarian peoples as inner is to outer, and they divided the Middle Kingdoms into twelve regions. Looking up they observed the figures in Heavens, looking down they modeled themselves on the categories of the Earth. Therefore, in Heaven there are Sun and Moon, on Earth there are *yin* and *yang*; in Heaven there are the Five Planets, on Earth there are the Five Elemental Phases [Wood, Fire, Metal, Water, Earth]; in Heaven are arrayed the lunar lodges, and on Earth there are the terrestrial regions.[15]

Therefore, Sima Qian says,

When the Five Planets gather, this is a *change of Elemental Phase*: the possessor of [fitting] virtue is celebrated, a new Great Man is set up to possess the four quarters, and his descendants flourish and multiply. But the one lacking in virtue suffers calamities or extinction.[16]

That the Simas' was the conventional conception is confirmed by Shen Yue's (441–513) reiteration of this principle in the *History of the Song Dynasty*, *Song shu*, where he paraphrases Sima Qian, "the one lacking in virtue suffers punishment, is separated from his household and kingdom, and devastates his ancestral temple," and then says,

Now, in my judgment, based on surviving texts, there have been three clusters of the Five Planets. The Zhou and Han [dynasties, each] relied [on one] to rule as king, as did Qi as Hegemon. [Duke Huan of] Qi [685–643 BCE] finally ended up as Hegemon, but in the end there was no epochal change. Therefore, there has never occurred such a thing as a cluster of planets with no [concomitant] change of Elemental Phase.[17]

Similarly, arguing that dynastic change may be portended by clusters of only four planets, Shen Yue later points to the record of the transfer of the Mandate from the Han Dynasty to the Wei in 220 CE: "Emperor Xian of Han, twenty-fifth year of the *Jian'an* reign period: Emperor Wen of Wei [Cao Pi; r. 221–237 CE] received his [Han Emperor Xian's] abdication. This constitutes the change of Phase [portended by] the Four Planets' three [recent] gatherings."

---

[15] *Shiji*, 27.1342.   [16] *Shiji* 27.1321. "History of the Song Dynasty," *Song shu*, 25.736.
[17] *Song shu*, 25.735.

A new astrological paradigm

A noteworthy attribute of the earlier field-allocation system of the Warring States period (403–221 BCE) is that it was unabashedly sinocentric in conception. The Chinese world was all that mattered, so that the identity between the Milky Way (the *Sky River*) and the Yellow River provided the paradigm for the entire scheme of correlations between the starry sky and the terrestrial provinces below (Map 9.2). No accommodation was made in Heaven for non-Chinese peoples. By the Western Han Dynasty (206 BCE–8 CE), however, some concession had to be made to the new political reality. Leaving no room for prognostication concerning non-Chinese peoples was an anachronistic bias that portentology could no longer afford if it was to have a claim to relevance in the imperial period.[18] Therefore, in view of the increasingly ominous threat to the unified Han Empire posed by aggressive non-Chinese peoples on the periphery like the Xiongnu, the Simas assert in the "Treatise" that in macro-astrological terms the warlike nomadic peoples are *yin* with respect to the *yang* of the Chinese world. As such, they correspond to the northern and western quadrants of the heavens, while the Chinese world corresponds to the south and east. By way of theoretical support, Sima Qian adduces the historically powerful Chinese border states of Jin and Qin as cases in point of hybrid Chinese polities whose martial proclivities clearly reflected the influence of non-Chinese peoples with whom they had been in intimate contact for generations:

Coming to Qin's swallowing up and annexing the Three Jin [i.e. Wei, Zhao, Han], Yan, and Dai [Shandong], from the [Yellow] River and Mount [Hua] southward is China [*zhongguo*]. [With respect to] the area within the Four Seas, China therefore occupies the south and east as *yang* – *yang* is the Sun, Jupiter, Mars, and Saturn.[19] Prognostications [about China are based on astral locations] situated south of *Celestial Street* [κ Tau], and *Hunting Net* [lodge 19, ε Tau] governs them. To the north and west are the Hu, Mo, Yuezhi and other peoples who wear felt and furs and draw the bow

---

[18] See also the discussion in Di Cosmo (2002, 305–11). Angus C. Graham (1989a, 314–15) also stressed that for both political and philosophical reasons the early empire needed "a unified world-view to back a politically unified world . . . [a problem] solved at the start by the rapid adoption of the cosmology of the court astronomers and diviners."

[19] In the early Han *Mawangdui* silk MS "Prognostications of the Five Planets," *yin* and *yang* are commonly used as directional terms in the sense of "north and west" and "south and east" respectively, but only with respect to relative locations of the kingdoms of the Warring States. For example, in discussing prognostications based on the position of Venus in particular astral fields, the "Prognostications" says, "Yue, Qi, Han, Zhao, and Wei, are *yang* with respect to Jing (Chu) and Qin. Qi is *yang* with respect to Yan, Zhao, and Wei. Wei is *yang* with respect to Han and Zhao. Han is *yang* with respect to Qin and Zhao. Qin is *yang* with respect to the *Di* barbarians. They are prognosticated on the basis of [Venus's lying] north or south, advancing or retreating." See Liu Lexian (2004, 86). Sima Qian's use of *yin* and *yang* here with respect to the geography of the empire as a whole is new.

as *yin* –*yin* is the Moon, Venus, and Mercury. Prognostications [about them are based on astral locations] situated north of *Celestial Street*, while *Topknot* [Pleiades, lodge 18, 7 Tau] governs them . . . On the whole, China's mountain ranges and watercourses run north and east, their head[-waters] in [Mount] Long and Shu [Gansu and Sichuan] and tail at the Bo[-hai Gulf] and [Mount] Jie[-shi] [Shanhaiguan]. For this reason, Qin and Jin are fond of using weapons; furthermore, their prognostications [depend on] Venus, governor of China, while the Hu and Mo, who have repeatedly invaded and despoiled, are uniquely prognosticated [based on] Mercury. Mercury's appearances and disappearances are swift and sudden, so as a rule it governs the *Yi-Di* barbarians. These are the guiding principles. They are modified according to who acts as the guest and who the host. Mars means order, externally [north and west], the army should be mobilized, but internally [south and east], the government should be put in order. Therefore it is said, "Though there may be a perspicacious Son of Heaven, one must still look to where Mars is located."[20]

This new paradigm reflects the animosity between the Chinese and frontier peoples which became a major preoccupation of the imperial court from the Qin Dynasty on, and intensified greatly during Emperor Wu's reign (140–87 BCE), under whom the Simas held office. This antagonism also surfaces in the interpretation of important planetary phenomena, with its suggestive modern terminology, for as we read in the "Treatise," "when the Five Planets are disposed in mid-heaven and gather in the east, China benefits, when they gather in the west, foreign kingdoms using weapons gain."[21] Remarkably, given astrology's resistance to change, and in spite of a pious nod in the direction of their esteemed predecessors, the Simas' account bears witness to a major reformulation of astral divination theory and practice, in which the former preoccupation with a multivalent sinocentric world is adapted to the circumstances of the Han Empire with its "us-versus-them" view of contemporary power relations.

By referring again to the map of the sky (Figure A.1) it becomes apparent that this broad generalization invokes the parallelism between the celestial topography and the geopolitical realities of north central China in the pre-imperial and early Han periods.[22] Essentially, the wintry or *yin* celestial fields north and west of the Milky Way as archetypal *Sky River* (i.e. lodges 10–18,

---

[20] *Shiji*, 27.1347. For the portrayal of Qin (and its self-identification) as "the other," see Lewis (1999, 40 ff.).

[21] *Shiji*, 27.1328. *Kaiyuan zhanjing* quotes a near parallel from Shi Shen: "When the Five Planets divide the sky in the middle and gather in the east, the Middle Kingdoms (*zhongguo*) greatly benefit; when they gather in the west, kingdoms on the seacoast making war benefit." Sima Qian has apparently adapted the older prognostication to conform to the reality of his own time. Precisely because of the new macro-astrological context, I do not think it is anachronistic to translate *zhongguo*, "Middle Kingdoms," as "China" in this part of the "Treatise."

[22] It is noteworthy that even in this revised scheme Sima Qian's geographical focus is still on north China and the Yellow River watershed. He gives surprisingly short shrift to the Yangtze region and southward despite the Han dynasty's aggressive expansion into this area as far south as present-day northern Vietnam.

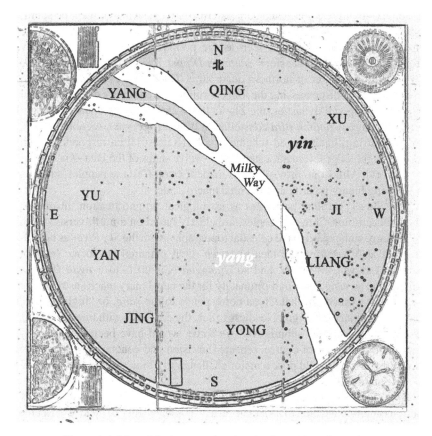

Figure A.1 Map of the heavens showing the disposition of the Nine Provinces surrounding the Milky Way in field-allocation astrology. The Milky Way, as archetypal *Sky River*, divides the celestial skyscape into *yin* (north and west) and *yang* (south and east) halves.

provinces *Yang* through *Liang*, roughly Capricorn through Taurus) correspond to the historical fields of activity of the peripheral "barbarian" peoples. In contrast, the *yang* celestial fields south and east of the Milky Way (i.e. lodges 15–19, provinces *Yong* through *Yu*, or Taurus through Scorpius) correspond to the Chinese heartland. Sima Qian singles out the powerful northern Chinese border states of Jin and Qin as emblematic of those who favored martial exploits over the purported Chinese ideological emphasis on civil pursuits and ethical self-cultivation.

Astrological prognostication on this new binary macro level clearly departs in important respects from the earlier astral omen watching. To cite one telling example, linking the powerful southeastern coastal state of Wu with the southern lodge *Bird* (23; α Hya) represents a radical break with the earlier scheme in which Wu and Yue, then seen as the peripheral south, were originally associated with lunar mansions *Southern Dipper* 8 [φ Sgr] and *Ox-Leader* 9 [β Cap] in the winter or northern quadrant.[23] This move also had the radical consequence of dispossessing the longest-ruling and culturally most influential of the pre-imperial dynasties, the Zhou (1046–256 BCE), by disassociating it from the huge *Vermilion Bird* constellation (Cancer–Corvus). As noted above, in the Warring States-period scheme of astral–terrestrial correspondences all twenty-eight lodges were allocated to the twelve states of the Hua–Xia or Chinese world. Almost no provision was made for non-Chinese peoples, nor was a *yin-yang* binary polarity applied on a macro scale.

The Simas' cardinal principles of astrological prognostication, insofar as they concern non-Chinese peoples, are clearly based on a north-versus-south dichotomy with regard to the celestial topography, with the *Sky River* as boundary, but also based on the idea that *wen* (civil pursuits) are to *wu* (martial pursuits) as *yang* is to *yin*, and so *zhongguo* (China) is to *waiguo* (foreign kingdoms) as *yang* is to *yin*. Fortunately for the new binary macro-astrological scheme, it worked out that China corresponds to the *yang*, or "high-culture" half of the binary paradigm. Needless to say, the converse, with imperial China corresponding to the "submissive" *yin* force, would have been ideologically unacceptable. Later, of course, repeated invasion and conquest by aggressive nomadic peoples would pose a major challenge to China's traditional devaluation of martial pursuits.

This "Treatise" of the Simas, father and son, represents the first and only serious attempt in China to articulate an astrological theory of history.[24] Let me

---

[23] *Ox-Leader* is my preferred translation here for *qian niu*, "*Draught-ox*," to distinguish Han and later references to lodge 9 from the older use of the same name to refer to the star Altair (α Aquila), the "*Oxherd*" of legend, who pines for the *Weaving Maid* sitting on the opposite bank of the *Sky River*. In the ode "Great East" (Mao 203), where we find earliest mention of the pair, *qian niu* seems to refer to the ox and not the oxherd, so it remains uncertain whether the romantic legend actually dates from early Zhou.

[24] For discussion of the Simas' unique contributions, see Liu Shaojun (1993, 100–6). The Simas' innovative binary macro-astrology formulated in the "Treatise" had a negligible impact on later astral prognostication. The sinocentrism of the field-allocation scheme was still being questioned in the Tang Dynasty (618–907), but in response the famous astronomer Li Chunfeng (602–70), in a blatant expression of Han chauvinism, denigrated peripheral peoples in his astrological treatise *Yisi Prognostications* (*Yisi zhan*) of 645 and baldly reasserted the validity of the ancient scheme. Jiang Xiaoyuan (1992a, 70). In the Yuan Dynasty, Wang Zhen (1271–1368), in a chapter on land utilization (*dili*) in his influential *Treatise on Agriculture* (*Nong shu*), proposed a new scheme classifying the suitability of the entire country's land and soils for agriculture and stock-raising in accordance with the twenty-eight lodges and the twelve astral fields. He includes a circular graphic called the "Diagram of the Bestowal of the Seasons and Easy Method of [Timing Farm] Work" (*Shoushi zhizhang huo fa zhi tu*). At its center is the

end this introduction with Sima Qian's own words, which conclude his brief account of his investigation of the interface between celestial phenomena and human events:

The revolutions of Heaven are: thirty years make one Small Revolution, one hundred years a Medium Revolution, and five hundred years a Major Revolution; three Major Revolutions make one Era, and three Eras make a Grand Consummation. These are the great cycles. Those who manage the affairs of states must respect the Three and the Five. [By looking] forward and backward a thousand years before and after [events], the continuity of the interaction between Heaven and Man will be fully in evidence... Now, changes in the constant stars are rarely seen, but prognostications concerning the Three Luminaries [Sun, Moon, stars and planets] are frequently applied. The Sun and Moon with their halos and blemishes, clouds and winds, these are the transient [effects] of celestial *materia vitalis*; their production and appearance also have their major cycles. Moreover, with regard to the vicissitudes of political affairs, they are the most proximate tallies of [the interaction] between Heaven and Man. These Five are the responsive movements of Heaven. Those who deal with the Regularities of Heaven must comprehend the Three and the Five... [Having explored events] from beginning to end, from ancient times to the present, [we have] looked deeply into the vicissitudes of the times, examining the minute and the large-scale, [so that exposition of] the *Celestial Offices* is now complete.[25]

The overarching, unifying principle in the Simas' astrological theory of history is the role of Heaven in human history, apart from that they say little in specific about observational methods or prognosticatory principles.[26] As in the exegetical traditions of the *Springs and Autumns Annals* one must draw one's own inferences from historical instantiations. There is good reason to credit Sima Qian's claim in the "Postface" to have "investigated the interaction between the celestial and the human, thereby to comprehend historical change, past and present, in order to formulate our own thesis [*yi jia zhi yan*]." The "Treatise

Dipper, surrounded by concentric bands containing the sexagenary signs, four seasons, twelve chronograms, and seventy-two weather conditions. Around the outer circumference are spelled out numerous agricultural works appropriate to the seasonal divisions and weather conditions; see Mitsukuni (1979, 60–1). The enormous Qing Dynasty encyclopedia of 1725, the *Complete Collection of Illustrations and Writings from Ancient Times to the Present (Gujin tushu jicheng)*, continues to identify geographic locations in terms of the traditional field-allocation scheme of astral–terrestrial correspondences.

[25] *Shiji*, 27.1344, 27.1350. Analysis of the numerology in this discussion seems to suggest that the Simas' implicitly recognized the Qin unification as a Grand Consummation. Van Ess (2006, 103–4).

[26] Mark Edward Lewis (2007, 310) avers that "the single ruler advocated in late Warring States polemics as the ground of intellectual unity... [in *Shiji*] reappears as the unifying principle of history." On the contrary, the unifying principle in the Simas' history is the first-order continuity of celestial influence (*tian ren zhi ji*) as reflected in the history of astral omens. As Sima Qian says in the "Treatise," "it has never happened that an event was not first preceded by some visible manifestation."

on the Celestial Offices" is thus a crucial part of their project to "create a new 'expertise' of considerable value" deserving of imperial recognition.[27]

Notes on the text and translation

The source text for the translation is the 1959 punctuated edition published by Zhonghua shuju and *pinyin* transcription is used throughout.[28] That standard text was collated with the *Bona ben ershisishi* edition from the *Sibu congkan* series. In addition, the following have been consulted: Takigawa Kametarō, *Shiji huizhu kaozheng*; Wang Shumin, *Shiji jiaozheng*; Wang Liqi, *Shiji zhuyi*. The portions of the "Treatise" which are reproduced in the *Han shu* "Monograph on Astrology" were also compared with the source text. In every case where significant variation was found this is indicated in the footnotes. Emendations, omissions, etc., are identified as to source and marked by curly brackets { } in the translation.[29] Words not present in the original and inserted into the translation to facilitate comprehension are placed in square brackets [].

The "Monograph on Astrology" in the *Jin shu* was consulted, as was Ho Peng-yoke's *The Astronomical Chapters of the Chin shu* and the "Kaiyuan reign period Classic of Divination," *Kaiyuan zhanjing*; however, I have avoided introducing later material so that the translation will accurately reflect the Simas' original text. For the same reason, for stellar identifications I have tended to rely on Sun Xiaochun and Jakob Kistemaker's *The Chinese Sky during the Han: Constellating Stars and Society*. Édouard Chavannes's century-old translation "Les Gouverneurs du Ciel" was also consulted. Although an impressive accomplishment in its time, Chavannes's rendering is now out of

---

[27] Although Sima Tan's use of the term *jia*, "household, family, school of thought," elsewhere to denote a more or less coherent teaching is open to interpretation, given the hereditary nature of the office of Scribe-Astrologer both historically and in the present instance, the connotation "one family's tradition" in this case is literally true; see Csikszentmihalyi and Nylan (2003, 69, n. 24).

[28] Sima Zhen (eighth century), *Shiji suoyin*; Pei Yin (fifth century), *Shiji jijie*; Zhang Shoujie (eighth century), *Shiji zhengyi*.

[29] This does not include the very frequent deletions of redundant characters in *Han shu*'s extensive quotation from the "Treatise." In the "Treatise" the same word often ends a clause and also begins the next, and in virtually every instance the second occurrence of the character will be elided in *Han shu*. The question as to the direction of borrowing between the "Treatise" and the *Han shu* "Monograph on the Heavenly Patterns" is well established in favor of the primacy of the "Treatise." This and other traces of editing on the part of Ma Xu (70–141 CE), the presumed author of the *Han shu* "Monograph," simply reinforces the point. Wang Shumin, *Shiji jiaozheng*, 1089. Discrepancies between the compositions and locations of numerous asterisms in their "Treatise" and those given in surviving fragments of earlier texts show also that the Simas did not simply reproduce the stellar identifications handed down from their predecessors Shi Shen and Gan De; see Bo Shuren (1981, 686). Many of the stars and asterisms in the "Treatise" appear in the *Chuci*, "Songs of the South," centuries earlier.

date and contains errors. Presented here is an entirely new translation, which I believe supersedes previous versions in accuracy and comprehensiveness.[30] The translation is provided with subheadings identifying these divisions of the text:

> The Offices, comprising an account of the stars of the Central (Polar), Eastern, Southern, Western, Northern Palaces of the sky;
> The Planets: Jupiter, Mars, Saturn, Venus, Mercury, together with a discussion of general planetary theory (between the accounts of Saturn and Venus);
> Field allocations (a simple listing of the astral–terrestrial correlations);
> Solar halos, the Moon;
> Auspicious and inauspicious stars, clouds and *materia vitalis*;
> The harvest;
> History and resposibilities of the office of Grand Scribe-Astrologer;
> Theory of astrology (astrological precedents, rationale for writing the "Treatise").

Standard abbreviations for modern astronomical constellations have been used (e.g. Sco for Scorpius, Vir for Virgo), together with conventional Greek letters for their constituent stars, as these are common to contemporary Chinese and Western usage. For ease of recognition, all names of stars, asterisms, and heavenly bodies mentioned in the text are placed in *italics*. The words *she*, "lodging," and *xiu*, "lodge for the night," are used interchangeably in the "Treatise" and are rendered as "lodge" or "lodging."[31] Inevitably, there is the problem of how to translate the term *qi*, the more or less energetic matter of which everything in the world is constituted. The forms and manifestations of *qi* in the text all refer to vaporous emanations and their various forms and I have opted for the expedient rendering *materia vitalis*.

---

[30] Derek Walters's (2005) rendering of the "Treatise" into English is not a scholarly translation and has numerous deficiencies.

[31] The translation "lodge" is consistent with conventional usage in the late Warring States and Han; see Cullen (2011a). However, it is clear from their number (originally twenty-seven, then twenty-eight), their locations, and the basic meaning of the interchangeable terms *xiu* and *she*, "lodge for the night; lodging," that the "lunar lodges" originated as astral mark-points used to designate the Moon's nightly progress (and at full Moon the position of the Sun opposite); see Feng Shi (1996, 113–15). They comprise an odd assortment of asterisms with greatly varying angular dimensions. Some are archaic, like the components of the *Dragon* and the *Vermilion Bird, Horned Owl, Determiner* (*Ding* or Pegasus), and some are recent, like *Ox-leader* and *Eastern Wall*. In later regular application the twenty-eight are allocated among to the "chronograms," *chen*, that divide the celestial equator into twelve equal segments, corresponding to solar months. When applied to Jupiter's position in a given year of the planet's cycle, however, the twelve divisions are purely an idealization. Both terms *xiu* and *chen* come into regular use in this sense about the same time as the mantic-astrolabe *shi*, suggesting that the device and the idealization of the cosmos it represents may have prompted their invention.

I am particularly grateful to Christopher Cullen for reading the translation and for comments, suggestions, and criticisms, which obliged me to revisit my interpretation in several instances.

The treatise on the celestial offices

From Sima Qian's "Postface" to The Grand Scribe's Records

The writings concerning the astral bodies and materia vitalis [contain] many and varied inauspicious and propitious omens, but are unorthodox. [We have] inferred their patterns, investigated their correspondences, apart from the outliers, collating them all [to permit] discussion [in reference to their corresponding] deeds and events, verifying them in terms of paths and measures, and ordering them sequentially in compiling the "Treatise on the Celestial Offices." (*Shiji*, 130.3306)

[1289] *The offices*

The brightest among the *Celestial Culmen* stars in the *Central Palace*[32] is the constant abode of the *Supreme One*.[33] The three stars beside it are the *Three Eminences*, otherwise called the *Crown Prince* and *Cadet Princes*.[34] The large star at the end of the curve formed by the four stars in back is

[32] Received versions of the *Tianguanshu* (hereafter *TGS*) write "*Central Palace*" 中宮 here; however there is an argument for reading 宮 as 官; see Takigawa Kametarō, *Shiji huizhu kaozheng* (1955, hereafter *KZ*, Vol. 4, 27.3–4). Also Wang Shumin, *Shiji jiaozheng* (1982, 1090–1) (hereafter *JZ*).

[33] Sima Zhen's (eighth century) *Suoyin* commentary (hereafter *SY*) quotes from the apocryphal *Chunqiu he cheng tu*, which says, "the *Purple Tenuity Palace* is the Hall of the *Great Lord*, and the essence of the *Supreme One*." "The brightest" can only refer to β UMi, while the name *Supreme One* has also been associated with δ Dra. Sun and Kistemaker (1997, 96, 167) (hereafter *SK*); Maeyama (2002, 3–18), and Ho (1966, 71). Zhang Shoujie's (fl. *c*.737) *Zhengyi* commentary (hereafter *ZY*) says *Supreme One* is another name for the *Celestial Lord*, by which epithet β UMi is conventionally known. *TGS* does not identify β UMi as the Pole Star, nor was it understood to be so in the Han. *Celestial Culmen* refers to the five stars of the asterism as a whole; see Maspero (1929, 327 ff.). For a survey of views concerning the identity of polar deities *Supreme One* and *Celestial One*, see Teboul (1985). Teboul also recognizes their change of identity over time. For a description of the central deity in the Chu Silk Manuscript identified as *Celestial One (Tianyi)* or *Supreme One (Taiyi)*, see Harper (1999, 851, 870), and also Li Ling (1995–6, 13–4). Discussion of the institution of the cult of *Supreme One* by Emperor Wu (r. 140–87 BCE) may also be found in Twitchett and Loewe (1986, 661–8).

[34] This describes the *Celestial Culmen* array of five stars. Alongside the *Celestial Lord*, β UMi is the *Crown Prince*, γ UMi; the *Cadet Princes* (sons of concubines) would then be *a*, *b* UMi. Later accounts identify the *Three Eminences* (*Grand Commandant*, *Grand Overseer of Works*, *Grand Overseer of the Masses*) with stars near the *Dipper*, marked by 24 CVn. Here, however, we are clearly concerned with stars within the *Central Palace* itself. The third, unnamed star is otherwise identified as the *Celestial Pivot* (or *Knot*), the Pole Star of Han time, 4339 Cam. Polaris, α UMi, was belatedly identified as the Pole Star in the Ming dynasty, well after the Pole had moved away from the *Knot Star*; cf. Maspero (1929, 324), Needham and Wang (1959, 259) [hereafter *SCC*]; *SK*, 154; Teboul (1985, 17).

the *Official Consort*;[35] the other three stars are [concubines] belonging to the *Inner Palace*.[36] The twelve stars framing and guarding [the celestial Pole] are *Screening Ministers*.[37] Collectively, they are all referred to as the *Purple [Tenuity] Palace*. [1290] In front of [the *Purple Tenuity Palace*], directly ahead of the [*Northern*] *Dipper*'s opening, is a triangle of three stars with pointed end to the north, sometimes visible, sometimes not, called *Virtue of Yin*, or some say *Celestial One*.[38] The three stars to the left of the *Purple Palace* are called *Heavenly Lance*.[39] The five[40] stars to the right are called *Celestial Flail*.[41] The six[42] stars in back, which cross the [*Sky*] *Han* [*River*] (i.e. Milky Way) toward *Align-the-Hall* [13],[43] are called the *Screened Causeway*.[44] [1291]

[35] In contrast to later accounts, the identification of the three stars alongside β UMi as the *Crown Prince* and *Cadet Princes* means that the *Official Consort* and [*Ladies of the*] *Inner Palace* must belong to our "handle" of UMi instead; but see SCC, 261. Sima Qian employs a rudimentary descriptive scale in characterizing the apparent brightness of stars, "large," *da*, 大, being the brightest, and "sometimes visible, sometimes not," *ruo jian ruo bu*, 若見若不, being the dimmest (possibly also designating variable stars). Bo (1981, 688).

[36] The unnamed "large star at the end of the curve" belongs to the here unnamed formation lying behind the *Celestial Culmen* asterism, made up of stars in the handle of UMi, and known as the *Angular Array*, *Gou Chen*. Later it comprises six stars, not four; see Ho (1966, 67). The "large star" to which Sima Qian is referring must, therefore, be Polaris, α UMi. The other three stars, the concubines, are ζ, ε, δ UMi.

[37] Two arcs of stars surrounding the *Pole*, made up of the *Right Purple Tenuity Enclosure* and the *Left Purple Tenuity Enclosure*, each star being identified with a court official. The two enclosures comprise fifteen stars: α, κ, λ Dra, δ, 24 UMa, 43, α, 3947 Cam, on the right; ι, θ, η, ζ, ν, 73 Dra, π Cep, 23 Cas, on the left. SK 134, however, says that in Han times the gap separating the left and right enclosures was between κ Dra on the left and Thuban, α Dra the third millennium BCE Pole Star, on the right (looking north).

[38] There is some uncertainty as to the exact translation of the line rendered here as "a triangle of three stars with pointed end to the north," and I have followed the commentarial consensus. Maeyama (2003) identified 6, κ, 4 Dra as the three stars in question. Sun and Kistemaker (*SK*, 166–7) opt for 7, 8, 9 Dra. This same ambiguity with regard to *Tianyi* and *Taiyin* is also present in *Huainanzi*: "of the most honored spirits of Heaven none is more revered that *Qing long* [*Cerulean Dragon*], who is sometimes called *Tian Yi* [*Celestial One*] and sometimes called *Tai Yin* [Grand *Yin*]." Major et al. (2010, 143).

[39] 9, κ, ι Boo.

[40] The "Monograph on the Heavenly Patterns" in the *History of the Former Han Dynasty*, *Han shu*: *Tianwen zhi* (hereafter *HS*), has "four." Although Ban Gu (32–92 CE) compiled the *History*, Ma Xu (70–141 CE) was the author of the "Monograph on the Heavenly Patterns," *Tianwen zhi*.

[41] ξ Dra, five stars. *KZ* (Vol. 4, 27.6) points out that the perspective in the *TGS*, here as elsewhere, assumes an observer at the Pole. Below a comet-like object by the same name – *Celestial Flail* – is said to be caused by irregularities in Jupiter's motion. In the Chu Silk Manuscript (*c*.300 BCE) and the *Kaiyuan Reign Period* (713–41 CE) *Canon of Prognostications* (*Kaiyuan zhanjing*, hereafter *KYZJ*, Chapter 85) *Celestial Flail* is identified as a variety of comet. In *KYZJ* its body is said to resemble a star, but it is four spans (*c*.2.3 m) long and pointed at the end, which parallels the description of the phenomenon below produced by Jupiter; Li Xueqin (2001, 38, 44–5).

[42] *HS* writes "seventeen" for "six."   [43] α, β Peg.

[44] ι, ε, δ, φ, θ, μ, ν, o Cas, an enclosed passageway to permit movement about in secret.

The seven stars of the *Northern Dipper*, the so-called *Jade Device* [with its] *Jade Transverse*, are what equilibrate the *Seven Regulators*.[45] The *Ladle*[46] connects to the *Dragon's Horns* [1]; *Transverse* hits the *Southern Dipper* [8];[47] the *Bowl*[48] rests on the head of *Triaster* [21].[49] The dusk indicator is the *Ladle; Ladle* [governs] from Mount Hua southwestward. The midnight indicator is the *Transverse*; *Transverse* corresponds to the Middle Province, the area between the Yellow and Ji Rivers. The dawn indicator is the *Bowl; Bowl* [governs] from the [Eastern] Sea and Mount Tai northeastward.[50] The *Dipper* is the [*Supernal*] *Lord's Carriage*. Revolving in the center it oversees and controls the four directions, separates *yin* from *yang*, establishes the four seasons, equalizes the Five Elemental Phases, shifts the seasonal nodes and angular measures, and fixes the various cycles – all are tied to the *Dipper*. [1293]

Ahead of the *Dipper's Bowl* is a frame of six stars called the *Palace for Promotion of Civic Virtue;*[51] the first is called *General-in-Chief*, the second is called *Lieutenant General*, the third is called *Chancellor*, the fourth is called *Overseer of Fate*, the fifth is called *Overseer of the Middle Way*, the sixth is called *Overseer of Ranks and Emoluments*. Within the *Bowl* of the *Dipper* is a *Prison* for the nobility.[52] Beneath the *Bowl* the six stars paired two by two are

---

[45] The passage is a clear allusion to the "Canon of Shun" in the *Book of Documents*. The *Seven Regulators* are Sun, Moon, and Five Planets. The *Jade Transverse* is ε UMa. Later, the *Jade Device* was sometimes taken specifically to denote the star β UMa, but early texts make it clear that the term originally referred to the *Dipper* as a whole; cf the first sentence in *Hou Han shu*, "Monograph on Harmonics and Calendrics": "Anciently, when the sages created the calendrical system, *they observed the rotation of the Jade Device*, the travels of the *Three Luminaries*, the movements of the Dao, the length of the [gnomon's] shadow, what the *Dipper Top-cord* indicates, and the path the *Cerulean Dragon* treads. 'They investigated their changes, synthesized their numerical constants,' and made of them a system." ("They synthesized . . . constants" is quoted from the first part of the "Appended Commentary" to the *Changes*.) On the erroneous identification of the "*Jade Device*" with ancient serrated *bi*-discs supposedly used as templates to locate the Celestial Pole, see Cullen and Farrer (1983). The *Zhoubi suanjing* says the true Celestial Pole (*zheng ji*) "is in the center of the *xuanji*." That this statement was taken literally is shown by the placement of the pivot of the revolving heaven plate on mantic astrolabes in the Han. It need not be taken as saying precisely that, but could simply refer to the fact that the Pole is in the middle of the space defined by the *Dipper*, aligned with the star Alioth. Positions of celestial bodies within the lodges are recorded in much the same way.

[46] δ, ε, ζ, η UMa. For an explanation of the system linking the stars of the *Dipper* with specific lodges, allowing their position to be determined even when invisible, see Chapter 12, 232 ff. Two different words are conventionally used to refer to the *Dipper*, biao, 杓, *Ladle* and bing, 柄, *Handle*. It is not always clear whether the two are exact synonyms, and since *Ladle* is distinguished from *Bowl* here, in the present context the two are not synonymous.

[47] π, λ, μ, σ, τ, ζ Sgr.   [48] α, β, γ, UMa.   [49] ζ, ε, δ, α, γ, κ, β Ori.

[50] For an interpretation of these correlations as a remnant of an early system of astral–terrestrial correspondences, see Feng Shi (2007, 372).

[51] *Wénchāng*: the semi-circle formed by o, 23, ν, π, θ, 15 UMa.

[52] Called *tiānlǐ*, 天理, meaning *Celestial Judge*; other sources identify the dungeon with the small ring of stars called *Celestial Prison* just outside the bowl of the *Dipper*, see *SK*, 153.

called the *Three Terraces*.[53] When the aspects of the *Three Terraces* are alike the Lord and his ministers are in harmony, when dissimilar, there is disharmony. When the *Assistant Star*[54] is bright and close by, the *Associate Ministers* are on intimate terms [with their Lord] and strong, when it is distant and small they are alienated and weak.

[1294] At the end of the *Ladle* there are two stars; the one nearest is the *Spear*, or *Twinkling Indicator*,[55] the one farthest away is the *Shield*, or *Celestial Spearhead*.[56] There is an encircling curve of fifteen stars associated with the *Ladle* called the *Commoners' Prison*.[57] When the *Prison* is [visibly] full of stars many are incarcerated, when empty it is open and they are let out.[58] [1295] When *Celestial Unity, Lance, Flail, Spear,* and *Shield* scintillate with large rays, fighting breaks out.

The Eastern Palace

In the Eastern Palace of the *Cerulean Dragon*, [the principal asterisms are] *Chamber* [4][59] and *Heart* [5] (Figure A.2).[60] *Heart* is the *Luminous Hall*. Its large star is the *Celestial King*,[61] the stars in front and behind it are the *Heir Apparent* and *Cadet Princes*. Their being straight is undesirable, for if straight

---

[53] ι, κ, λ, μ, ν, ξ UMa; discussed in the context of the *Grand Tenuity Palace* in *Jin shu*; see Ho (1966, 80).

[54] Pei Yin's fifth-century commentary *Jijie* (hereafter *JJ*) quotes Meng Kang (fl. *c*.220–54), who states explicitly that Alcor is meant. *HS* has "[Dipper] Handle's Assistant Star." The ZY commentary provides detailed prognostications based on Alcor's appearance. Alcor is a member of a multiple star system and an eclipsing variable, potentially varying in brightness.

[55] γ Boo.

[56] The *JJ* says that *Tianfeng* refers to *Sombre Dagger-Axe, Xuange* (λ Boo), but that star is actually nearer the *Dipper Handle* than *Twinkling Indicator*, γ Boo. Both *ZY* and *SY*, based on the *Canon of Stars*, say the two stars in question comprise the *Moat* asterism (ε, σ, ρ Boo); *Tianfeng* would then be one of the three. *Twinkling Indicator*, γ Boo, is identified with the *Hu* barbarians, and fluctuations in the *Moat*'s visibility foretell border troubles. For the Supreme One Spear (*Taiyi Feng*) wielded by the Grand Scribe-Astrologer in an apotropaic ritual prior to invading Southern Yue in 212 BCE, see Nienhauser, Cao, and Galer (2002, 239).

[57] Wang Yuanqi (1714–86) distinguishes the seven stars of the "hook," or *Seven Eminences* (δ, μ, ν, χ, φ, τ, 42 *Boo*), from the eight stars of the "ring," or *Fetters* (π, θ, β, α, γ, δ, ε, ι, ρ CrB).

[58] That is, having been amnestied. *ZY* cites detailed prognostications based on the *Prison* stars' visibility.

[59] π, ρ, δ, β₁ Sco, four stars. For discussion of the details of the Xi'an tomb ceiling painting in connection with the "Treatise," see Hu Lin'gui (1989).

[60] σ, α, τ Sco, three stars. *Chamber* (4) and *Heart* (5) are said to be the brightest of the components of the *Cerulean Dragon*.

[61] This is Antares, α Sco, otherwise known as *Great Fire*. Antares is the focus of the famous ode 154 in the *Book of Odes*, "In the Seventh Month the *Fire* [Star] Declines," which describes the seasonal activities correlated with the *Fire Star*. Antares is the uniquely red-colored star depicted next to the left rear claw of the dragon in the detail above from the Western Han tomb painting of the twenty-eight lodges discovered on the campus of Xi'an Jiaotong University in 1987. See Feng Shi (2007, Plate 5.2).

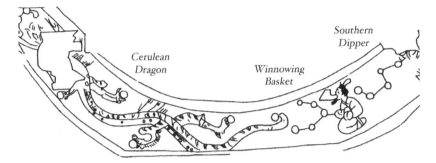

Figure A.2 Western Han depiction of the (antlered) *Cerulean Dragon* of the East, from a star map on the ceiling of a Western Han tomb discovered at Xi'an Jiaotong University. Detail redrawn from Luo Qikun (1991, 237, Figure 1).

the *Celestial King*'s plans go awry.[62] *Chamber* is the {*Heavenly*} *Directorate*, called *Celestial Quadriga*, its northern star is the *Right Outer Trace-Horse*.[63] The two stars alongside are called *Bolt* and *Latch*;[64] the star to the north is called *Linchpin*.[65] To the northeast, the curves of [1296] {twenty-two} stars are called *Banners*.[66] The four stars in the middle of *Banners* are called *Celestial Market*,[67] the six stars in the middle are called *Market Tower*.[68] When the stars within *Market* are numerous [the grain harvest] is abundant, when they are sparse there is scarcity. The many stars to the south of *Chamber* [4] are *Cavalry Officers*.[69] [1297]

---

62 *HS* has *zhi ze tian wang* for *TGS zhi wang*.
63 Lodge *Chamber* (4), the *Heavenly Quadriga*, plays an important role in prognostication by the Moon and Five Planets based on which of the three paths they take through its row of four stars (ιι, ρ, δ, β₁ Sco) that straddle the ecliptic. For characterization of those pathways, see below and Ho (1966, 96). To locate the next star *TGS* identifies only the *Right Outer Trace-Horse* (β₁ Sco), leaving the others unmentioned. The Moon's location in *Heavenly Quadriga* is cited in *Discourses of the States* as one of the particular observations noted at the time of the Zhou campaign of conquest against Shang in 1047–1046 BCE; see *Guoyu*, 3.18a.
64 ω Sco, two stars, the lock securing the entrance to the *Mansion*.
65 ν Sco; *ZY* says (Shi Shen's) *Canon of Stars* gives the name *Gate-Bolt* for this star.
66 The *TGS* has "twelve" stars, but it is likely the initial "two" has dropped out of the text; see *KZ*, 27.12. The direction indicated points to the curves of stars identified as the *Celestial Market's Right* and *Left Walls* (β, γ, κ Her; γ, β, δ, α, ε Ser; δ, ε, ζ Oph and δ, λ, μ, ο, 112 Her, ζ Aql, 9, η Ser, ν Oph, ξ Ser, η Oph). Moreover, in what follows, *Market Tower* is described as lying within *Celestial Market*, and a prognostication is added based on the visibility of stars within the *Market*, which is consistent with an enclosed open space. Later this will be identified as the third major "Court," the *Great Market Enclosure*; *TGS* gives it short shrift.
67 That is, the four stars marking the north and south gates to the *Market*: δ, β Her, η, ζ Oph.
68 μ, ν, 47, 30 Oph and ο, ν Ser, six stars. *HS* 26.1276 has an eye-skip here, eliding *zhong liu xing wei Shilou*, 中六星曰市樓, "the six stars in the middle are called *Market Tower*."
69 The twenty-seven stars of the *Celestial King*'s Tiger Vanguard, marked by α Lup.

The *Left Horn* [1] is the *Justice*, the *Right Horn* is the *General.*[70] *Great Horn*[71] is the *Celestial King's Court.*[72] To either side are three stars in a curve like the legs of a tripod, they are the [*Left* and *Right*] *Assistant Conductors.*[73] The *Assistant Conductors* are right where the *Dipper's Handle* is pointing, thereby determining the seasons and nodes [of the year], hence they are called *Assistant Conductors. Neck* [2][74] is the *Detached Court*[75] and governs disease. The two large stars to the {north and} south of it [*Neck*] are called the *South Gate* [of the *Celestial Arsenal*].[76] *Base* [3][77] is *Celestial Base* and governs epidemics. [1298] *Tail* [6][78] is *Nine Cadet Princes*, and is called *Lord* and *Ministers* [and concerns] estrangement and disaccord.[79] *Winnowing Basket*

---

[70] The *Left Horn* is ζ Vir; *HS* (26.1276) has *li*, 理, "Legal Officer"; The *Right Horn* is α Vir. Several early sources (e.g. the "Treatise on the Feng and Shan Sacrifices") identify the *Left Horn* as *Numinous Star, Ling xing*, 靈星 (or *Heavenly Field, Tiantian*, 天田) and note that its first dawn appearance in spring signalled that the time had arrived to begin sowing the crops, so it was greeted with a sacrifice at the shrine *Lingxingci* specifically dedicated to that star; see Jao Tsung-yi (1998, 37). Lodge *Chamber*, or *Quadriga* in Scorpius performed the same function and hence was denoted *Farmer's Auspice*. See *Guoyu*, 1.6b-7a; cf. *Han shu* 25A.1211– 12 and 25B.1242. The enumeration of the lodges begins with *Horns* either because the first full Moon of the year appears there, or because the *Dipper's* handle permanently points to it. SCC, 251.

[71] Arcturus, α Boo, the third brightest star in the sky.

[72] *HS* has *Tian Wang Di zuo ting*, 天王帝坐廷, awkwardly inserting *zuo*, "Throne." *TGS Di* should be elided. KZ, 27.13.

[73] *Left Assistant Conductor* is ο, π, ζ Boo; *Right Assistant Conductor* is η, τ, ν Boo. The idea is that they "convey" an extension of the *Dipper* past Arcturus to the first lodge, *Horns*. According to the *Chunqiu yuanming bao* ("Mystical Diagrams of Cosmic Destiny"), "[the *Assistant Conductors*] lead by the hand *Great Horn* [Arcturus] on the one side and the *Dipper* on the other." SCC, 251. For the obsolescence of the *Assistant Conductors* due to precession once Arcturus left the zone of perpetual visibility around 1200 BCE, see SCC, 252. In another reference to the same phenomenon, Shi Shen's *Canon of Stars* states that the *Dipper* originally comprised nine stars but that two of them had "been lost sight of." SCC, 250. For the same phenomenon noted in *Hesiod*, see Worthen (1991, 210). Lai Guolong (personal communication) identifies *sheti*, 攝 提, as the god *Sheshi* (or *Nieshi*) 聶氏 in the Mawangdui silk MS *Xingde*. Thus *Shetige* might perhaps be rendered "*Sheshi* arrives."

[74] λ, χ, ι, π Vir.

[75] Song Jun (first century CE) says *miao*, "temple," should be read as *chao*, "court".

[76] The *South Gate* of the *Arsenal* is conventionally identified as α, ε Cen. Sun and Kistemaker (*SK*, 154) propose ξ Cen (two stars), which can hardly be described as "large." "North" is here placed in curly brackets as an interpolation. Wang Ximing's (aka Dan Yuanzi) *Pacing the Heavens Poem, Bu tian ge*, (sixth century), the lengthy mnemonic recitation of stars and asterisms, takes up *South Gate* and *Arsenal* (see below) under lodge *Horns* (1) rather than *Neck* (2). SCC (201) calls Dan Yuanzi the "Aratus of China." In the *Lesser Annuary of Xia*, *South Gate* is said to be due south on the meridian at dusk in the fourth month.

[77] α, ι, γ, β Lib. ZY quotes the *Canon of Stars*, which says the fours stars represent the traveling court.

[78] ε, μ1, ζ2, η, ϑ, ι1, κ, λ, υ Sco.

[79] KZ (27.13), which departs from the interpretation "*Nine Cadet Princes*, about whom the Lord and Ministers are in disaccord," by placing a full stop after "*Ministers*." In contrast, ZY says the nine stars of *Tail* are cadet princes, or, alternatively, palace ladies ranged in order from the *Official Consort* nearest *Heart* through concubines at the end. If the nine are equally bright,

[7][80] is a devious retainer,[81] and is called *Mouth* and *Tongue*, {the Residence of the Official Consort and Concubines}.[82]

If the *Fire Star* (Mars) trespasses against and guards *Horns* [1] there will be war. If [the *Fire Star* trespasses against and guards] *Chamber* [4] and *Heart* [5], it is hateful to the king.

The Southern Palace

[1299] The *Vermilion Bird* of the Southern Palace [corresponds to] the *Sliding Weight*[83] and *Balance Beam*. The {The *Balance Beam*} *Grand Tenuity* [*Enclosure*], is the court of the *Three Luminaries* [i.e. Sun, Moon and *Planets*].[84] The twelve stars framing and guarding [the *Grand Tenuity Enclosure*] are *Screening Ministers*. On the west are the *Generals*,[85] on the east the *Chancellors*. The {four} stars on the south are the *Enforcers of the Law*,[86] in between is the *Main Gate*.[87] To the [*Main*] *Gate*'s left and right are the [*Left* and *Right*] *Side Gates*;[88] the {six} stars within the *Gates* are the *Vassal Lords*;[89] the five stars within are the *Five Emperors' Thrones*.[90] The luxuriant cluster of fifteen stars behind is

good order prevails in the Inner Palace, if not, there is intrigue and conflict, especially if Mars or Venus stays there.

[80]   γ, δ, ε, η Sgr.
[81]   Lodge *Winnowing Basket* (7) figures twice in the *Book of Odes* as a simile for scheming or good-for-nothing subordinates. For example, in Ode 200, "Xiang Bo": "Great and large is truly that southern *Winnowing Basket*; those slanderers, who likes to consult with them?" Again in "Da Dong" (Mao 203), *Winnowing Basket* is named along with other asterisms as a simile for superficially appealing, but ultimately worthless officials. Karlgren (1950b, 150, 153).
[82]   *HS* appends *hou fei zhi fu*, "mansion of the concubines."
[83]   An alternative name for the constellation *Chariot Pole*, *Xianyuan*, a sinuous array of some fifteen stars extending to the north of Regulus, α Leo, otherwise known as the *Yellow Dragon*, alter ego of the Yellow Emperor (see below).
[84]   The repetition of "*Balance Beam*" here is redundant.
[85]   The *Jin shu* "Monograph on the Heavenly Patterns" mixes the generals and ministers, two of each, senior and subordinate, to the east and west. The four stars on the east are γ, δ, ε Vir and α Com, and the four on the west are σ, ι, θ, δ Leo.
[86]   η, β Vir, *Left* and *Right Enforcers of the Law*. Here again there is disagreement between *TGS* and the *Jin shu* "Monograph on the Heavenly Patterns," since the latter identifies only two stars as *Enforcers* on left (west) and right (east). In any case, the stars comprising the enclosures do not add up to twelve. It is likely, according to *KZ* (27.15), that that total includes the two stars *Captain of the Gentlemen of the Palace* (31 Com) and *Tiger Vanguard* (72 Leo), just to the north of the eastern and western enclosures. *Pacing the Heavens Poem* says that the stars of the enclosure add up to ten.
[87]   The gap between β–η Vir.
[88]   *Right Side-gate*, β–σ Vir and *Left Side-gate*, η–γ Vir. Five more gates are named in the *ZY* commentary.
[89]   39, 35, 26, 20, 5 Com; Sun and Kistemaker (*SK*, 176) suggest 6, 11, 24, 27, 36 Com. These five stars are denoted the *Inner Five Vassal Lords* to distinguish from the *Five Vassal Lords* in *Eastern Well* (22) marked by θ, τ, ι, ν, κ (or φ) Gem. Commentators offer no plausible explanation for the figure "six" in *TGS*.
[90]   Marked by the *Yellow Emperor of the Center*, β Leo. The others are: the *Green Emperor of the East*, the *Red Emperor of the South*, the *White Emperor of the West*, the *Black Emperor of the*

called the [*Palace*] *Gentlemens' Seats*;[91] the large star alongside is the *Captain of the* [*Palace*] *Gentlemens' Seats*.[92] If the Moon and *Five Planets* enter the [*Grand Tenuity Enclosure*] in direct motion via the [proper] path, watch their departure and observe what they guard [to discern] whom the Son of Heaven will execute. If [the Moon and *Five Planets*] enter in retrograde motion, or not by the proper path, [the outcome is] dictated by [the place] trespassed on. The [*Five Emperors'*] *Thrones* in the middle [when trespassed on] manifest [the threat]; always it is on account of the crowd of subordinates conspiring together. [Involvement of] the *Metal Star* (Venus) or the *Fire Star* (Mars) is especially serious. Suspended to the west of the *Court* (*Enclosure*) *Palisade* there are {five} stars called *Ranking Officials* representing commissioned officers, officials, and grandees.[93] The *Sliding Weight* is *Xuan yuan, Chariot Pole; Xuan yuan*[94] is the body of the *Yellow Dragon*. The large star in front is the image of the *Female Ruler*.[95] The small stars alongside are attending women of the Inner Palace. Prognostication based on trespassing or guarding by the Moon and *Five Planets* is as in the case of *Balance Beam*. [1302] *Eastern Well* [22] concerns watery affairs.[96] The curved asterism to its west is called *Broadaxe*.[97] North [of *Eastern Well*] is *North River*, south of it is *South River*.[98] Between the two *Rivers* and the *Celestial Gate Towers* is the *Gate Span*.[99]

Ghost in the Conveyance [23][100] governs sacrificial matters. The whiteness in the middle is *Headsman's Block*.[101] If the *Fire Star* [Mars] guards

---

North; not to be confused with Gan De's *Five Emperors' Inner Seats* in the *Palace of Purple Tenuity* marked by γ Cep, according to SK, 164.

[91] Fifteen stars marked by γ Com. *HS* writes *aiwu*, 哀烏, for *weiran*, 蔚然, "luxuriant."

[92] Probably 31 Com; α CVn suggested by SK (153) is much too far away. The *Pacing the Heavens Poem* also places it at the end of the left enclosure, at the same level as *Tiger Vanguard*, 72 Leo.

[93] SK, 152, proposes 53, 52, 51, 54 Leo. *HS* writes "four" instead of "five," as does the *Pacing the Heavens Poem*.

[94] Clan name of the legendary Yellow Emperor; comprising the sixteen stars in a crooked line stretching northward like a Yellow Dragon from Regulus, α Leo.

[95] I.e. the Official Consort or Empress, consistently identified as "large" or the "largest," is Regulus. The small star to its south, 31 Leo with apparent magnitude 4.37, is *Lady-in-Waiting, nüyu*, 女御 (or *yunü*, 御女).

[96] Eight stars, μ, ν, γ, ξ, ε, δ, ζ, λ Gem.   [97] η Gem.

[98] *North River* is three or four stars marked by α, β Gem; similarly, *South River* is marked by α, β CMi.

[99] The terminology reflects the fact that the ecliptic crosses the Milky Way at the *Celestial Gate*, ζ Tau. Consequently, the prognostications in *ZY* all concern the precise path taken by the Sun, Moon, and Planets as they pass this point.

[100] θ, η, γ, δ Cnc. The name of lodge 23, *Yugui*, 輿鬼 is conventionally rendered "*Carriage Ghost*"; however, *yu*, 輿, "conveyance," is not necessarily a wheeled vehicle. In the Western Han dynasty tomb painting discovered at Xi'an Jiaotong University the image representing this lodge (above) shows two men carrying a sedan chair with a spectral figure seated in it, hence the more generic translation here. Hu Lin'gui (1989, 87).

[101] This is Praesepe, the spectral Beehive Cluster or M44 in Cancer. *TGS zhi*, 質, in the *Jin shu* "Monograph on the Heavenly Patterns" is written *fuzhi*, 鈇鑕, the broadaxe and execution block

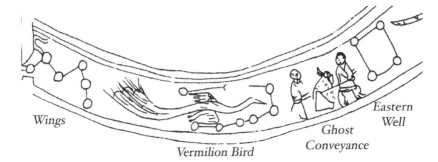

Figure A.3 Western Han depiction of lodge 23, *Ghost in the Conveyance*, through lodge 27, *Wings*, from the Xi'an Jiaotong University tomb ceiling. Detail redrawn from Luo Qikun (1991, 237, Figure 1).

the *South* or *North Rivers*, fighting breaks out and grains fail to grow. Therefore, *Balance Beam* accomplishes virtue,[102] [*Celestial*] *Lakes* accomplishes prognostication,[103] *Broadaxe* accomplishes injury, [*Eastern*] *Well* accomplishes disaster, and *Headsman's Block* accomplishes punishment.

[1303] *Willow* [24] is the [*Vermilion*] *Bird's Beak*[104] and governs trees and grass (Figure A.3). *Seven Stars* [25], the *Gullet*, is the *Round Office* and governs urgent affairs. *Spread* [26], or *Crop*, is the *Kitchen* and governs hosting guests.[105] *Wings* [27] is *Plumage* and governs guests [sc. ambassadors] from afar.[106] [1304] *Chariot Platform* [28] is the *Carriage* and governs the wind.[107] Beside it is a small star called *Long Sandbar*,[108] [which is] disinclined to shine brightly. If it is as bright as the four stars [of *Chariot Platform*] and the *Five Planets* enter *Chariot Platform*, general warfare breaks out. The group of stars south of *Chariot Platform* is called *Celestial Arsenal* and *Tower*.[109] [Within it] there are *Five Chariots*;[110] if the *Chariot* stars' rays increase greatly or are absent, chariots and horses are unavailable.

for cutting in two at the waist; see *KZ*, 27.20. The *Pacing the Heavens Poem* characterizes the ghostly Praesepe as the "*materia vitalis* of piled corpses," which explains those associations of *zhi*, 質, in commentaries.

[102] A reference to the *Grand Tenuity Enclosure*.   [103] 14, 16, 19, φ, σ, μ Aur.

[104] δ, σ, η, ρ, ε, ζ, ω, θ Hya. *HS* writes *zhuo*, 啄 for *zhu*, 注.

[105] κ, ν₁, ν₂, λ, μ, κ, φ Hya and GC13839.

[106] Twenty-two stars including χ₁, χ₂, β, λ, α, γ, ζ, η, δ, κ, ι, ε, ϑ Crt.   [107] γ, ε, δ, β Crv.

[108] ζ Crv, a double and suspected variable star, possibly explaining the comment.

[109] α, β, δ, γ Cru, γ, ι, ϑ, η, α, ε Cen.

[110] Not to be confused with the *Five Emperors' Chariot Sheds* in Auriga (next note), which also has five *Pillars*. The *Pacing the Heavens Poem* places the fifteen stars of "*Five Pillars*" within the *Arsenal*, though conventionally only one *Pillar* (ν₁, ν₂ Cen) is actually inside the walls, while the others surround it.

The Western Palace

The *Heavenly Mineral Spring* of the *Western Palace* are said to be *Heaven's Five Lakes*.[111] The *Five Lakes* are the *Five Emperors' Chariot Sheds*.[112] If the *Fire* [*Star*, Mars] enters there is drought; if the *Metal* [*Star*, Venus], there is armed conflict; if the *Water* [*Star*, Mercury], there are floods. In the middle there are *Three Pillars*.[113] If any of the *Pillars* is not in evidence, armed conflict breaks out. [1305] *Stride* [15][114] is said to be a *Great Boar*, and is *Canals* and *Gullies*. *Pasture* [16] is *Gathered Herd*.[115] *Stomach* [17][116] is the *Celestial Granary*.[117] The group of stars to its south is the *Fodder Barn*.[118] *Topknot* [18][119] is long flowing hair and the star of the *Hu* barbarians [of the west]; it stands for gatherings of white clothes [i.e. mourning rites].

---

[111] Sun and Kistemaker (*SK*, 179) separately assign λ, μ, σ Aur to the *Mineral Spring*, which better agrees with the *Pacing the Heavens Poem*. For some commentators the *Mineral Spring* stands for the entire Western Palace, not just three stars as in *ZY*. In the *Huainanzi*, "Heavenly Patterns" chapter, and in the older Chu poem "Encountering Sorrow," *Lisao*, the *Mineral Spring* is where the Sun bathes after its daily journey across the sky from the Fusang Tree in the east. Major (1993, 81, 94, 102).

[112] *TGS* equates the *Five Celestial Lakes* with the *Five Emperors' Chariot Sheds*, ι, α, β, Ꝺ Aur, β Tau (γ Aur), perhaps reflecting the replacement of the older name by the imperial nomenclature. "Chariot" here is interpreted by some as "grain cart."

[113] Later, τ, ν, υ Aur, ζ, ε, η Aur, and χ, 26 Aur. Sun and Kistemaker (SK, 179), however, identify only one *Pillar* (ζ, ε, η Aur) made up of three stars, which is also an acceptable reading of both *TGS* and the *Pacing the Heavens*.

[114] η, ζ And, 65 Psc, ε, δ, π, ν, μ, β And, 82, τ, u, j, χ, ψ1, ψ2Psc. For a description of the running boar representing this lodge on the Western Han tomb painting from Xi'an Jiaotong University, see Hu Lin'gui (1989, 86).

[115] γ, β, α Ari; livestock for the sacrifices.   [116] 35, 39, 41 Ari.   [117] ι, η, ζ, Ꝺ, τ, ν Cet.

[118] Also known as *chugao*, 芻藁; Sun and Kistemaker (SK, 179) identify six stars west of *Celestial Parkland* (γ, π, δ, ε, ζ, η, Eri, ρ Cet, τ1, τ9 Eri), marked by ρ Cet. Ho Peng-yoke (1966, 110) is in basic agreement. This would place *Fodder Barn* north of *Celestial Granary* rather than south as stated in *TGS*.

[119] The Pleiades, 17, 19, 21, 20, 23, η, 27 Tau. If early commentaries are to be credited, the Zhou Chinese may only have considered the five brightest stars (17, 20, 23, η, 27 Tau) to comprise *Topknot*. In the *Book of Odes*, "Little Stars," they are called *The Five* and are paired with *The Three* (aka *Triaster*) in Orion's belt: "tiny are those little stars, *The Three* and *The Five* are in the east." Sima Qian refers to the cluster of multiple stars as *maotou*, 髦頭, "tassel-headed," which apparently refers to dangling locks of hair; Karlgren (1964b, 1137e): "long tufts of hair, worn uncut (one on each side of the head) by children or youths." *Maotou* (aka *maofa* 髦髮) has been translated as "mane"; however, the connection is simply that horses' manes were often tufted in a similar fashion. The hairstyle (with or without topknots) was an identifying characteristic of nomadic frontier peoples, which explains the association with the *Hu* barbarians. An equivalent hairstyle in Japan is known as *chonmage* and in Korea as *sangtu*, where it was worn by married men in the Joseon Dynasty. It was also commonly worn by adult Mongol men. Notwithstanding Roy A. Miller's assertion (1988, 5) that "Ssu-ma Ch'ien's *mao-t'ou* is transparently little more than a Chinese calque upon Assyrian *zappu* 'bristle, animal hair; comb; the Pleiades, "conceived as the 'mane' of Taurus,"'" there is in fact no connection with Mesopotamian astral nomenclature. Miller's phonological argument fares no better.

Net (19)[120] is called *Netting Carriage* and stands for frontier troops, it governs fowl hunting with bow [and tethered arrows]. The small star next to its large star is *Beside the [1306] Ear*.[121] If *Beside the Ear* scintillates there are slandering and rebellious ministers [by the ruler's] side. Between *Topknot* and *Net* is the *Celestial Street*;[122] north of it are the *yin* kingdoms, south of it are the *yang* kingdoms. *Triaster* (21)[123] is the *White Tiger*, its three stars in a straight line make *Balance* and *Stone* [*Weight*].[124] Beneath it the three stars that come to a point are called *Attack*,[125] signifying affairs [involving] *Decapitation* and *Termination*. The four stars outside it are the left and right *Shoulders* and *Haunches* [of the *Warrior*].[126] The three small stars in an angle are called *Horned Owl* (20) (Figure A.4),[127] constituting the *Tiger's Head*, which governs matters concerning wild herbs and grains. South of it are four stars called the *Celestial Latrine*.[128] Beneath the *Latrine* is one star called *Celestial Feces*.[129] If *Feces* is yellow it is auspicious, if green, white, or black it is baleful. West of it three curves comprising nine stars each are arrayed, the first is called *Celestial*

[120] The Hyades, ε, 68, δ, γ, α, θ₁, 71, λ Tau. *Celestial Fork* appears to have been another ancient name for this asterism incorporating the Hyades and other stars in Taurus; see Karlgren trans., "Da Dong," (Mao 203), and Karlgren (1964a, no 203; 1964b, 124). The Western Han tomb painting from Xi'an Jiaotong University (Figure A.5) depicts a running man casting a net while chasing a rabbit; see Hu Lin'gui (1989, 86). *Pacing the Heavens* calls the asterism "*Talons*."

[121] σ Tau; as in *fu er di yan*, 附耳低言, "speaking softly close to someone's ear."

[122] ω, κ (or υ) Tau; so-called because the Sun, Moon, and Five Planets pass through here. The *yin* kingdoms (north and west of *Celestial Street*) are identified below as non-Chinese and *yang* kingdoms (south and east) as the Middle Kingdoms, *zhongguo*, or "China."

[123] *Triaster* refers to the three stars in Orion's belt ζ, ε, δ Ori. *Triaster* is easily identifiable and an ancient asterism since it already appears in the Shang oracle bone inscriptions; see Jao Tsung-yi (1998, 44). Subsequently it appears in the ode "Little Stars," in the *Book of Odes*. The *Lesser Annuary of Xia* says, "in the first month at dusk, *Triaster* culminates on the meridian, and it is the start of the year." Orion and Scorpius, lying diametrically opposite each other near where the Sun, Moon and planets ford the Milky Way, must have served as harbingers of the seasons from a very early date, since they anciently marked the equinoxes and were immortalized in astral myth as a pair of feuding brothers whom the Supernal Lord banished to opposite ends of the earth to have charge of observations of Orion and Antares.

[124] δ, ε, ζ Ori. "*Stone*" here is *dan*, a standard unit of measure, ten pecks. The expression is a metaphor for the power to weigh in the balance, to establish laws and standards.

[125] c, θ₂, ι Ori. Variously written either *fa*, 罰, as here, or *fa*, 伐. Sometimes Orion is referred to as *shen-fa*, 參伐, and sometimes *fa* stands for the whole; cf., e.g., *Chunqiu fanlu*, Chapter 34, "Feng ben," where the stars of *Attack* are said to number thirteen, probably corresponding to most of our constellation of Orion.

[126] β, γ, α, κ Ori.

[127] λ, φ₁, φ₂ Ori. The discovery in 1987 of the Xi'an Western Han tomb ceiling with a horned owl representing this asterism resolved the confusion about the interpretation of the name *zuixi*, 觜觿, for this lodge; see Hu Lin'gui (1989, 87). Its dual identity as both "Tiger's head" and *Horned Owl* presumably reflects a superposition of the iconic Tiger constellation on the earlier individual asterisms of the western palace. Unlike the *Dragon* of the eastern palace, none of the western asterisms depict any part of the tiger's body.

[128] α, β, γ, δ Lep.

[129] Either μ or ν₂ Col; *shi*, 矢 "arrow," in *TGS* is a euphemism for *shi*, 屎, "excrement."

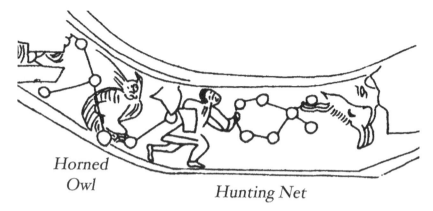

Horned Owl

Hunting Net

Figure A.4 Western Han depiction of lodge 19, *Net*, from the Xi'an Jiao-tong University tomb ceiling. Detail redrawn from Luo Qikun (1991, 237, Figure 1).

Banner,[130] the second is called *Celestial Parkland*,[131] and the third is called *Nine Pennants*.[132] To their east is a large star called the *Wolf*.[133] When the *Wolf*'s rays change color there is an abundance of robbers and bandits. Below it are four stars called the *Bow*[134] pointing at the *Wolf*. Nearer the ground is a large star called the *Old Man of the Southern Culmen*.[135] When the *Old Man* appears there is peace and order, when [the *Old Man*] does not appear, armed conflict breaks out. As a rule one watches for it in the southern suburbs at the autumnal equinox. [1308] {When *Near Ear* enters *Net* [19] fighting breaks out.}[136]

The Northern Palace

The Northern Palace, the *Dark Warrior*, is *Ruins* (11)[137] and *Roof* (12).[138] *Roof* governs coverings of halls;[139] *Ruins* is weeping and tearful affairs. [1309] South of them is a group of stars called *Celestial Army of the Forest of*

---

[130] Aka *Shenqi*, 參旗, o₁, o₂, 6, π₁, π₆ Ori.
[131] Above *Celestial Parkland* was identified as being a ring of sixteen stars. Evidently, it only comprised an arc of nine in Sima Qian's time.
[132] With *JJ* and *ZY*, reading *yu*, 瘀 (斿) as *liu*, 旒.
[133] Sirius, α CMa, the brightest star in the sky.
[134] Initially, ζ, ε, δ CMa, ξ Pup, marking the bow, later expanded to nine stars including the arrow and bowstring.
[135] α Car, or Canopus. "Southern Culmen" signifies the southern limit of visibility, not the South Pole. Canopus is the *Longevity Star*.
[136] *HS* lacks this sentence, which is both out of place and corrupt.
[137] Reading *xu*, 虛, "void," as *xu*, 墟, "ruins." β Aqr, α Equ.   [138] α Aqr, θ, ε Peg.
[139] *SY* follows *TGS* in equating *Roof* with "covering halls" here, however, *ZY* (following *Pacing the Heavens*?) points out that *Covering the Room, gai wu*, 蓋屋, is a separate asterism comprising

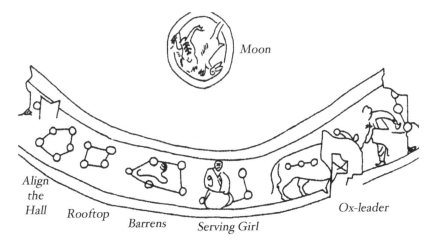

Moon

Align the Hall

Rooftop

Barrens

Serving Girl

Ox-leader

Figure A.5 Lodges 9, *Ox-Leader*, through 14, *Eastern Wall*, of the winter palace of the heavens, from the Xi'an Jiaotong University tomb ceiling. Detail redrawn from Luo Qikun (1991, 237, Figure 1).

Feathers.[140] {West of the} *Army* are its *Fortifications*,[141] which are otherwise called *Broadaxe*.[142] Next to them there is a large star, *Northern Territories* [*Army Gate*].[143] If it fades away and sets,[144] or if the star [*Northern Territories*] scintillates, its rays increasing and then growing fewer, or if the *Five Planets* trespass on *Northern Territories* and then enter the *Army*, armies will be raised. If [the trespasser] is the *Fire* [*Star*, Mars], the *Metal* [*Star*, Venus], or the *Water* [*Star*, Mercury], the consequences will be especially severe. If the *Fire* [*Star*, Mars], the army will be disaffected, if the *Water* [*Star*, Mercury], it will be stricken by floods. If it is the *Wood* [*Star*, Jupiter] or *Earth* [*Star*, Saturn] it will be auspicious for the army. East of *Roof* (12) there are six stars paired two by two, called the *Overseer* {*of Works*}.[145] *Align-the-Hall* (13) is the *Pure Temple*, called the *Detached Palace*.[146] The four stars {of the *Screened Causeway*}[147] in the middle of the Milky Way are called *Celestial Four-in-Hand*.[148] The single star alongside is [Charioteer] *Wang Liang*. When *Wang Liang* whips his team, chariots and horses fill the open country.[149] Alongside are eight stars that intersect the [*Sky*] *Han* [*River*] called *Celestial Ford*.[150] Beside the *Celestial Ford* is the *River Star*;[151] when the *River Star* scintillates, people ford streams.

and considers the text defective here; but see *SK*, 169–70 for the *Overseers*. Both *Supreme One* and the *Overseer of Lifespans* are mentioned in the Baoshan divination record (*c*.316 BCE). Li Ling (1995–96, 13–4) speculates that these are the same deities named in the "Nine Songs" in the *Songs of the South* (*Chu ci*).

[146] In later accounts, the *Detached Palace* comprises six stars radiating from the northernmost star (β Peg) of *Align-the-Hall* (*13*); that is, λ, μ, o, ε, τ, υ Peg. In describing the shifting balance between *yin* and *yang* forces throughout the year, the *Shiji* "Treatise on Harmonics" (25.1243) begins with *Eastern Wall* (*14*) and *Align-the-Hall* (*13*), where the Sun is located at the Beginning of Spring: "*Eastern Wall* dwells to the east of the wind of [Mount] Buzhou, and governs the development of the embryonic *materia vitalis*, moving it eastward. As for *Align-the-Hall*, [it] governs giving foetal form to the *yang* force and gives birth to it." Glossing "*Align-the-Hall*," *SY* comments, "it is asterism *Ding* (the Square of Pegasus). When *Ding* culminates halls may be built, therefore it is called *Align-the-Hall*. The asterism has the [square] shape of a hall, therefore it is called *Align-the-Hall*. Its stars present the image of a hall, so in the 'Treatise on the Celestial Offices' it governs temples." In *TGS*, *SY* quotes the pre-Han glossary *Literary Expositor*, *Erya*, and a comment by Guo Pu (276–324 CE), both identifying *Align-the-Hall* as *Ding*, the Square of Pegasus. Guo comments on the two cognates: "*Ding* is 'correct [*zheng*, 正]'. All palaces and halls built in the world take the culmination of *Align-the-Hall* [due south on the meridian] as correct." Note the *TGS* omission here of any mention of *Eastern Wall* (*14*) (α And/δ Peg and γ Peg), presumably because the two lodges were subsumed under the single asterism *Ding*, as in the commentary just cited. That is how they are represented in the detail from the Xi'an tomb painting from the Western Han (although the northernmost star of *Align-the-Hall* β Peg has largely flaked off in the tomb painting). Liang Yusheng notes the omission of *Eastern Wall* as unusual (*KZ*, 27.29). *Temporary Palace*, as a later development, is not represented at all in the painting.

[147] *Screened Causeway* actually comprises o, ν, μ, θ, δ, ε, ι Cas straddling the galactic equator (Milky Way). In the parallel passage in the *Jin shu* "Monograph on the Heavenly Patterns," *Screened Causeway* is never mentioned, so it is an obvious intrusion here.

[148] *Celestial Four-in-Hand* here actually refers to Charioteer *Wang Liang*'s team of horses. The Zhonghua editors of *Shiji* erred in not placing a full stop after *Palace*, instead treating *Temporary Palace* and *Screened Causeway* as terms in apposition, implying that *Screened Causeway* is an alternate name for *Align-the-Hall*. In fact, the *Screened Causeway* is a different asterism in Cassiopeia well to the north. In the "Treatise on Harmonics" (25.1244), the *ZY* commentary makes the same mistake, which suggests that the text Zhang Shoujie saw already contained the intrusion. According to every other Han or earlier source, *Heavenly Quadriga* is actually the name of the four stars comprising lodge *Chamber* (*4*) in Scorpius (also identified as such in *Guoyu*, 3.18a), so I translate *sì*, 駟, "team of four," as *Quadriga* to distinguish it from *Wang Liang*'s team, *Celestial Four-in-Hand*. *SY* quotes the apocryphal *Chunqiu yuan ming bao* as writing *qi*, 騎, "cavalry," here, so *si* is quite possibly simply an early copyist's error.

[149] β Cas is identified with *Wang Liang* and λ, α, η, ν Cas are doubtless his chariot horses, while γ Cas is his *Whip*, *ce*, 策. This phrase is quoted verbatim in Wang Chong's *Lunheng*, "Biandong" chapter.

[150] Song Jun, reading *huang*, 潢 as *jin*, 津, says this is a reference to the *Celestial Ford* (δ, o, α, ν, τ, υ, ζ, ε, γ Cyg), whose location agrees. This interpretation is endorsed by *KZ*, 27.30.

[151] *ZY* identifies this *River* with the *Celestial River* asterism in Ophiucus (36, θ, 44 Oph) lying at the intersection of the ecliptic with the Milky Way on the opposite end of the *Celestial*

[1310] The four stars of *Mortar* and *Pestle*[152] are {south} of *Roof*. When a green or black star guards *Calabash*,[153] fish and salt are expensive. *Southern Dipper* (8)[154] is the *Court*. North of it is *Fundamental Star*; *Fundamental Star* is [*Celestial*] *Flag*.[155] *Ox-Leader* (9)[156] represents the sacrificial offerings. North of it is *Drum-Bearer*.[157] The large star of *Drum-Bearer* [1311] is the *General-in-Chief*, to its left and right are the *Generals of the Left and Right*. To the north of *Serving Girl* (LM 10)[158] is the *Weaving Maid*.[159] *Weaving Maid* is the daughter {granddaughter} of Heaven.[160]

Ford from *Wang Liang*. This is also where *Pacing the Heavens* places *Tianjiang* 天江. Sun and Kistemaker (*SK*, 150) also take this to be the *Celestial River* "star" in Ophiucus, but then equate that asterism with the Milky Way as a whole, as do some commentaries. *Jin shu*, however, identifies *Celestial River* as a variant name for *Celestial Ford*; see Ho (1966, 88).

[152] *Mortar* is μ Cyg, κ, ι, 32 Peg, while *Pestle* is π₂, 23 Peg, 1 Lac. In the *Canon of Stars Mortar* and *Pestle* are said to comprise seven stars, so "three stars" may have dropped out of the text after *chu*, 杵, "pestle." The direction indicated is also incorrect: "south" should certainly be "north."

[153] ζ, β, δ, γ, α Del.   [154] μ, λ, j, σ, τ, ζ Sgr.

[155] ξ₂, ο, π, 43, ρ, υ Sgr. Also known as the *Gate of the Celestial Capital*, it is fundamental because it marks the point where the Seven Luminaries ford the Milky Way.

[156] ξ₂, α₂, ν, β, ρ, ο, π Cap.

[157] γ, α, β, Aql, conventionally "*River Drum*," but here reading *he*, 河 as *he*, 荷. The brightest star, Altair (α Aql), is mentioned in the *Book of Odes*, 203 "Da Dong": "Brilliant is the *Draught-ox*, but one does not yoke it to any carriage"; Karlgren (1950b, 155). It is this *qianniu xing*, 牽牛星, *Draught-ox* star (often mistakenly called *Oxherd*), which has an annual tryst with the *Weaving Maid* on the seventh day of the seventh month, a bit of star lore long pre-dating the Han Dynasty. The *Celestial Ford* they use to cross the *Sky River* lies just to the northwest in Cygnus. The name *Ox-Leader*, later applied to lodge 9 in Capricorn, originated with this *Draught-ox* asterism in Aquila, the redundancy of the name *qian niu* subsequently causing endless confusion. In *TGS* Altair is transformed into the imperial *General-in-Chief*. The *Weaving Maid* and *Draught-ox* stars figure prominently as seasonal indicators (called "season's beginning," *shishou*, 時首, and "season's end," *shiwei*, 時尾) in the astronomy of the Yi-minority of Sichuan, Guizhou, and Yunnan. The antiquity of their conceptions and influence on mainstream Hua–Xia astronomy is well established. In Yi astronomy the entire northern quadrant is held to evoke the relationship of the *Weaving Maid* and *Oxherd*, which finds an echo in the two lodges *Serving Girl* (*10*) and *Ox-leader* that supplanted them; see Chen Jiujin, Lu Yang and Liu Yaohan (1984, 102, 106–7).

[158] ε, μ, 4, 3 Aql. A *Serving Girl* or concubine in the lowest status female occupation, depicted in a kneeling posture in Figure A.6 below.

[159] α, ε, ζ Lyr. The *Weaving Maid*, Vega (α Lyr), is a celestial maiden who supposedly governs fruits and vegetables, silks, and precious things, the daughter of the Son of Heaven. She enjoys a tryst with the *Oxherd* once a year on the seventh day of the seventh month. See the *Book of Odes*, "Great East" (Mao 203): "in the heavens there is the [*Cloudy*] *Han* [*Sky River* = Milky Way], it looks down and is bright; triangular (oblique?) is the *Weaving Maid*, throughout the day she moves seven times. Although she moves seven times, she does not achieve any pattern in recompense"; trans. Karlgren (1950b, 155, trans. modified). In other words, although she labors throughout the seven double-hours of daylight, she has nothing to show for it when she reappears. The identity of this triangular asterism has remained stable for close to three millennia. *TGS* "granddaughter" is probably an error for "daughter."

[160] This concludes the *TGS* overview of the stellar offices. Zhang Shoujie quotes from the *Jin shu* "Monograph on the Heavenly Patterns" review of the same material: "During the reign of Emperor Wu [265–90 CE], the *Jin* dynasty Prefect Grand Scribe-Astrologer Chen Zhuo

The Five Planets

Jupiter

[1312] Study the movements of Sun and Moon to discern the planet Jupiter's direct and retrograde motion. [Jupiter] is the East, Wood, and governs Spring; its stem days are *jia* and *yi*. When righteousness is deficient, punishment emanates from the planet Jupiter. When Jupiter gains or regresses, the state's fate is determined by the lodge [the planet occupies].[161] The state wherein Jupiter is located may not be attacked, but may attack others.[162] If Jupiter advances prematurely into[163] a lodge and gets ahead of itself, it is called "gaining." Premature withdrawal from a lodge it is called regressing.[164] If Jupiter gains, that state's troops will not return. If [Jupiter] regresses, the state [from which it withdrew] will be beleaguered, its general lost, and the state overthrown and defeated. If the *Five Planets* all follow [Jupiter] and gather in the one lodge [wherein Jupiter is located], that state below will be able to attract the entire sub-celestial realm through Rectitude.

[1313] In a *Shetige* year Jupiter's *yin*-double moves leftward and is in chronogram *yin*,[165] while Jupiter moves to the right and occupies chronogram *chou*.

[陳卓] consolidated the star charts compiled by the three traditions of Shi [Shen], Gan [De] and Wuxian, in all amounting to 283 offices (asterisms) and 1,464 individual stars, to serve as the definitive record. Now, we survey the most prominent among them to complete the [following section on the] 'Celestial Offices'"; cf. Ho (1966, 67). On his chart, Chen Zhuo famously distinguished the three traditions by representing the stars in three different colors. His color coding still appears on the star maps found at Dunhuang dating from the Tang dynasty, several centuries later; see the color plates in *SK*, 28–9.

[161] *TGS* often uses *she*, 舍, "guest house; lodging," as a synonym for the more technical term *xiu*, 宿. Originally, *xiu*, "lodge for the night," specifically denoted the Moon's nightly lodging, since 365^d divided by twenty-eight lodges yields a close approximation of the Moon's daily progress. By the Han this sense was considerably attenuated. Prior to the system of twenty-eight lodges, there were only twenty-seven, yielding an even better approximation of the Moon's sidereal period of 27.3 days.

[162] In *KYZJ* this axiom is attributed to Shi Shen.   [163] Reading *qu*, 趨, as 超; *JZ*, 1112.

[164] "Gaining" and "regressing" refer to a planet's unexpectedly rapid advance or retrogradation (followed by resumption of direct motion); SCC, 399. See the section below dealing with planetary conjunctions after the discussion of Saturn; see also Maspero (1929, 295). In the Chu Silk Manuscript and the *KYZJ* (Chapter 11) these two terms were even applied to the Moon, however, where there can be no question of retrograde movement; Li Ling (1985, 39); Maspero (1929, 295); Jao (1972 118–19). For a concise explanation of Jupiter's motion and its tendency to "gain," see Cullen (2001, 39).

[165] *Sui yin*, 歲陰, rendered as "Jupiter's *yin*-double." However, as Marc Kalinowski (1998–99, 154 and n. 69) has observed, "far from being a function derived from the Jovian year count (a 'counter-Jupiter' cycle, as it is often called), it is rather the *Taiyin* motion that requires that the Jupiter rotations be fitted to the framework of the sexagenary norm of the calendar . . . This does not exclude the possibility, supported by a number of ancient texts, that the *Taiyin* cycle was originally conceived on the model of the revolutions of Jupiter to give the sexagenary year count astronomical legitimacy." The term *Shetige* is the first of twelve distinctly un-Chinese terms given to the successive year-long periods (*sui*) of Jupiter's visibility. Their origin is obscure. Lai Guolong (private communication) has pointed out that one of the monthy spirits

In the first month it appears at dawn in the east with [lodges *Southern*] *Dipper* (*8*) and *Ox*[-*leader*] [9], and is called "Overseer of Virtue."[166] Its color is pale green and gleaming. Should it miss its [proper] station, a response will appear in *Willow* [24] [on the opposite horizon]. If the *Year*[-*star*] is early there will be floods, if late there will be drought. [After] Jupiter appears in the east it travels 12^d in one hundred days and then stops.[167] It retrogrades 8^d in one hundred days and then resumes eastward motion. In one year Jupiter moves 30 7/16^d, traveling about 1/12^d per day. In twelve years it completes a circuit of the sky.[168] As a rule it appears at dawn in the east, and sets at dusk in the west.

In a *Shanyan* year,[169] Jupiter's *yin*-double is located in chronogram *mao*; the planet occupies chronogram *zi*. In the second month, it appears at dawn in the east with lodges *Serving Girl* [10], *Ruins* [11], and *Roof* 12], and is called *Descend to Enter*.[170] [Jupiter] is large and gleaming. Should it miss its [proper] station a response will appear in *Spread* [26]. {Its name is called "Descending to Enter"} In this year there will be major flooding.

In a *Zhixu* year, Jupiter's *yin*-double is located in chronogram *chen*; Jupiter occupies chronogram *hai*. In the third month it [appears] at dawn with lodges *Align-the-Hall* [13], and *Eastern Wall* [14], and is called *Green Emblem;* [Jupiter] is bright green and very splendid. Should it miss its [proper] station a response will appear in *Chariot Platform* [28]. If the *Year* [*Star*] is early there will be drought, if late there will be floods.

named in the Chu Silk Manuscript is called *Nieshi* 聶氏, using the same phonetic as in *she* of *sheti*. Equally intriguing a clue is found in the calendrical astronomy of the Yi-minority of Sichuan, Guizhou, and Yunnan. Yi astrononers–astrologers exercised a strong influence on Hua–Xia astronomy from the Western Zhou through the Simas' own time, then in the person of astronomer Luoxia Hong (*c*.130–70 BCE). Yi-minority calendrical astronomy also prominently featured seasonal timekeeping using the orientation of the *Dipper*'s handle, as documented in the *Heguanzi* by a Warring States-period compatriot of Luoxia Hong; see Chen Jiujin, Lu Yang, and Liu Yaohan (1984). In the Yi language, the *Dipper* in its time-keeping function is called *Shanie*, 沙聶 (in Chinese transcription), once again employing the same phonetic element to render the Yi word; see ibid., 83, 97. Others of the twelve terms also bear a strong resemblance to astral naming conventions in the Yi language, so further research along these lines may ultimately settle the question of their origin.

[166] As noted in *SY*, these indications derive from Shi Shen's *Canon of Stars*, while those in the *HS* "Monograph on the Heavenly Patterns" differ in following Gan De and the Han *Taichu xingli*, "Taichu Reign-Period Astronomical System." On the whole, year names are the same as given in *Erya*.

[167] Superscript "^{d}" stands for the Chinese angular measure *du*, 度, 1/365 of a circle or fractionally less than one degree.

[168] More precisely, Jupiter's sidereal period is 11.86 years, its synodic period is 399 days.

[169] The term *Shanyan* ordinarily denotes the principal wife of the Shanyu, leader of the Xiongnu confederation of frontier nomads who constantly harassed the northwest border areas. Its significance here is unknown, as are the meanings of the other names for Jupiter years, here left untranslated. It has long been thought that they are transliterations of terms borrowed from another language (see n. 165 above).

[170] The *Bona ben* edition (*BNB*) of the *TGS* inserts the name of the year farther on (in curly brackets { } here). Only this and the next entry follow this pattern.

[1314] In a *Dahuangluo* year, Jupiter's *yin*-double is located in chronogram *si*; Jupiter occupies chronogram *xu*. In the fourth month it appears at dawn with lodges *Stride* [15], *Pasture* [16], {*Stomach* [17], and *Topknot* [18],}[171] and is called *Rapid Follower*. [Jupiter] is bright red and gleaming. Should it miss its [proper] station a response will appear in *Neck* (2).

In a *Dunzang* year, Jupiter's *yin*-double is located in chronogram *wu*; Jupiter occupies chronogram *you*. In the fifth month it appears at dawn with lodges *Stomach* [17], *Topknot* [18], and *Net* [19], and is called *Emergent Brightness*.[172] [If Jupiter] flames and gleams military hostilities are halted; [the year] is only favorable for high officials and kings, and not for the conduct of war. Should [Jupiter] miss its [proper] station a response will appear in *Chamber* [4]. If the *Year* [*Star*] is early there will be drought, if late there will be floods.

In a *Xieqia* year, Jupiter's *yin*-double is located in chronogram *wei*; Jupiter occupies chronogram *shen*. In the sixth month it appears at dawn with lodges *Horned Owl* [20] and *Triaster* [21], and is called *Long Brilliance*. [Jupiter] shines and gleams, and [the year] is favorable for the military to take to the field. Should [Jupiter] miss its [proper] station a response will appear in *Winnowing Basket* [7].

In a *Tuntan* year, Jupiter's *yin*-double is located in chronogram *shen*; Jupiter occupies chronogram *wei*. In the seventh month it appears at dawn with lodges *Eastern Well* [22] and *Ghostly Conveyance* [23], and is called *Great Riser*.[173] [Jupiter] shines white. Should [Jupiter] miss its [proper] station a response will appear in *Ox-Leader* [9].

[1315] In a *Zuo'e* year, Jupiter's *yin*-double is located in chronogram *you*; Jupiter occupies chronogram *wu*. In the eighth month it appears at dawn with lodges *Willow* [24], *Seven Stars* [25], and *Spread* [26] and is called *Chief King*. [Jupiter] shoots rays in all directions, and [in this year] its state flourishes[174] and the grain ripens. Should [Jupiter] miss its [proper] station a response will appear in *Roof* [12]. There is prosperity despite drought, obsequies for women, and epidemics among the people.

In a *Yanmao* year, Jupiter's *yin*-double is located in chronogram *xu*; Jupiter occupies chronogram *si*. In the ninth month it appears at dawn with lodges *Wings* [27] and *Chariot Platform* [28] and is called *Heavenly Eye-opener*. [Jupiter] is white and very bright. Should [Jupiter] miss its [proper] station a response will appear in *Eastern Wall* [14]. This year there will be floods and obsequies for women.

---

[171] *KZ* (27.36) and *JZ* (1113) both consider the two lodges an interpolation. The locations are also astronomically inaccurate.

[172] *HS* writes *qiming*, 啓明, for *kaiming*, 開明. In the *Book of Odes*, "Da Dong" (Mao 203), *Qiming* denotes Venus as the morning star, Lucifer.

[173] Reading *Jin*, 晉, for *yin*, 音; see *JZ* (1113).

[174] Reading *qi guo chang*, 其國昌, for *guo qi chang*, 國其昌.

In a *Dayuanxian* year, Jupiter's *yin*-double is located in chronogram *hai*; Jupiter occupies chronogram *chen*. In the tenth month it appears at dawn with lodges *Horns* [1] and *Neck* [2] and is called *Heavenly Augustness*.[175] [Jupiter] is greenish-white. If it speeds ahead and emerges faintly at dawn, this is called *Correct Peace*. If armies and companies are raised, their commanders must be aggressive. [Jupiter's] state is virtuous and will possess [all within] the Four Seas. Should [Jupiter] miss its [proper] station a response will appear in *Pasture* [16].

In a *Kundun* year, Jupiter's *yin*-double is located in chronogram *zi*; Jupiter occupies chronogram *mao*. In the eleventh month it appears at dawn with lodges *Base* [3], *Chamber* [4], and *Heart* [5] and is called *Heaven's Wellspring*. [Jupiter's] somber hue is very bright. Rivers and lakes [1316] flourish; it is not advantageous to raise troops. Should [Jupiter] miss its [proper] station a response will [appear] in *Topknot* [18].

In a *Chifenruo* year, Jupiter's *yin*-double is located in chronogram *chou*; Jupiter occupies chronogram *yin*. In the twelfth month it appears at dawn with lodges *Tail* [6] and *Winnowing Basket* [7] and is called *Celestial Vastness*. [Jupiter's] darkling somber color is very luminous.[176] Should [Jupiter] miss its [proper] station a response will appear in *Triaster* [21].

If [Jupiter] ought to occupy [a place] but does not, or occupies it but wavers to left and right; or if it ought not yet depart but leaves and meets up with another planet, it is malefic for that state. When it occupies a location for a long time, that state enjoys ample virtuous governance. If Jupiter's rays are agitated, now shrinking, now growing, and if its color changes multiple times, the ruler will be afflicted. Should [Jupiter] miss its [proper] station by one lodge or less and advance northwestward, within three months it will produce a *Celestial Flail* four *zhang* long, with pointed tip. If it advances southeastward, within three months it will produce a *Broom Star* two spans [*zhang* = 4.5 m] long, like a sweep.[177] If it retreats northwestward, within three months it will produce a *Heavenly Sickle* [?][178] four *zhang* long, with pointed tip. If it retreats southwestward, within three months it will produce a *Celestial Lance* several spans long, pointed at both ends. Carefully watch the state in which [these phenomena] appear, [for that state] cannot initiate major undertakings or

---

[175] *HS* writes *Tianhuang*, 天皇, "Heavenly Augustness," for *Dazhang*, 大章; see *JZ*, 1114.

[176] What "black" signifies in reference to the color of celestial bodies is difficult to determine. Bo Shuren (1981, 687) contends that *hei*, 黑, in such contexts often just means "dark," but elsewhere in *TGS shen*, 深, "deep," is used in that sense. How light can simultaneously be both "dark" and "luminous" as here is unclear.

[177] The *ZY* comment states that a comet's tail points away from the Sun, pointing to the east when it appears in the evening and to the west at dawn, and that the locations swept by the tail suffer disasters.

[178] *Chan*, 欃, "magnolia" (Karlgren, 1964b, 612b; modern "sandalwood") makes no sense. Dictionary definitions of *chan xing*, 欃星, as a "comet" all simply cite this passage.

use troops. If [Jupiter] appears to float and then sink, that state will undertake construction; [if Jupiter appears] to sink and then float, that state's countryside will be lost. If Jupiter's color is red and rayed, the state it occupies will flourish. Any who confront its rays and do battle with it will not be victorious. If Jupiter's color is reddish-yellow and dark, the astral field it occupies will reap a great harvest. [If Jupiter's color] is greenish-white to reddish-gray, the astral field it occupies will be afflicted. If Jupiter is occulted by the Moon, that field's Chancellor will be exiled. If it duels with Venus, that field's army will be shattered. [1317] One name of Jupiter is *Sheti*, another is *Doubly Resplendent*, or *Responsive Star*, or *Periodic Star*. *Align-the-Hall* (13) is the *Pure Temple*, the temple of Jupiter.

Mars

One observes the punishing[179] *materia vitalis* to locate *Sparkling Deluder* [Mars]. [Mars] is the South, Fire, and governs summer; its stem days are *bing* and *ding*. When propriety is lost, punishment emanates from Mars {and Mars moves anomalously}.[180] When [Mars] appears there is armed conflict, when it disappears troops disperse. One identifies the subject state based on the lodge [Mars occupies]. Mars is rebellion, brigandage, plague, bereavement, famine, war.[181] If it retraces its path for two lodges [1318] or more and then dwells there, within three months there will be calamities, within five months there will be armed invasion, within seven months half the territory will be lost, within nine months more than half the territory will be lost. Accordingly, if [Mars] both appears and disappears together with [a single lodge], that state's sacrifices will be terminated. If [Mars] occupies a place and calamity promptly befalls it, though [anticipated to be] great, it ought to be small; [if the calamity is] long in coming, though it ought to be small, on the contrary, it will be great. If [Mars] is south [of a lodge] there will be male obsequies, if north, female obsequies. If scintillating rays encircle it, reaching now in front, now behind, now to the left, and now to the right, the calamity will be even greater. [If Mars] duels with other planets, their gleams touching each other, it is injurious; if [their gleams] do not touch, it is not injurious.[182] If all *Five Planets* follow

---

[179] *TGS* writes *gang*, 剛, "unyielding," but commentators agree this should be *fa*, 罰, which also better accords with Mars's role as enforcer.

[180] The phrase in curly brackets is thought to be an interpolated annotation; *KZ* 27.40.

[181] Alluding to the tradition that unrest among the people manifests itself in ominous popular sayings, the *Jin shu* "Monograph on the Heavenly Patterns" (*KZ*, 27.41) adds, "Mars comes down as ominous juvenile ditties and sayings, games and skits."

[182] *ZY* says that whenever planets duel it signifies war, if the troops are not abroad it means civil war.

[Mars] and gather in a single lodge, its state below will be able to attract the entire sub-celestial realm through Propriety. [1319]

As a general rule, [Mars] appears in the east and travels through sixteen lodges before halting, then it retrogrades through two lodges; after six ten-day weeks, it resumes eastward travel, [to?] ten lodges from where it halted.[183] After ten months it disappears in the west, then travels for five months in obscurity before appearing again in the east.[184] When it appears in the west it is called *Returning Brightness*, and rulers hate it. Its eastward motion is quick, each day traveling 1½^d. Its motion to the east, west, south, and north is rapid. In each case troops gather beneath it. In war those who comport with it are victorious, those who defy it are defeated.[185] If Mars follows Venus, the army is beset; [if Mars] departs from it, the army retreats. If [Mars] emerges northwest of Venus, the army will split; if [Mars] moves southeast of it, generals on the flanks do battle. If during [Mars's] travel Venus overtakes it, the army will be shattered and its general killed. If Mars enters and guards or trespasses against the *Grand Tenuity* [*Palace*], *Chariot Pole*, or *Align-the-Hall* (13), those in command hate it. *Heart* (5) is the *Hall of Brilliance*, the *Temple of* Mars – carefully watch this.[186]

Saturn

Track its meetings with the [*Southern*] *Dipper* to determine the location of Saturn.[187] Saturn is the Center, Earth, and governs the last month of summer;

---

[183]   The passage literally reads "for several tens of lodges from where it halted," which is so egregious an error that the text must be defective here. Mars's retrogradation lasts some seventy-five to eighty days and covers only about twenty degrees. I suspect the "ten lodges" has been transposed from the preceding lines, because "for ten months" appears to be missing from the first line.

[184]   This implies a synodic period of twenty-seven months or some 797 days, compared to the modern figure of 780 days. As late as the monograph on astrology in the *Jin shu* (648 CE), Mars's movements were still held to be problematical. *Jin shu*, 12.318.

[185]   The same terms, *shun*, 順, and *ni*, 逆, also refer to direct and retrograde motion, which may also be implied here. In other words, if Mars retrogrades away from a lodge, that state is disadvantaged.

[186]   *KZ* (27.43) comments that the pithy conclusion concerning Mars from near the end [*TGS*, 1347] actually belongs here: "Mars causes fuzzy stars [tailless comets]. Externally it governs [the use of] military force, and internally it governs [the conduct of] government. Therefore, it is said: 'though there be a perspicacious Son of Heaven, one must still look to where Mars is located.'" *Bo*, 孛, "fuzzy, bushy," as a description of comets refers to the indistinct head of a tailless comet whose short, fuzzy rays point evenly in all directions. The term *xing bo*, 星孛, "a star became fuzzy," already occurs in the *Springs and Autumns Chronicle* (e.g. Duke Ai, thirteenth and fourteenth years, 482–481 BCE) and suggests that it was thought that comets were stars that spontaneously changed aspect. "Fuzzy" comets are also mentioned in the Chu Silk Manuscript and explained in detail in the *KYZJ*. Li Ling (1985, 44).

[187]   That is, when Saturn passes the ford of the Milky Way in Sagittarius; hence the name of the Jupiter station at this location, *Stellar Guide-thread*. See Chapter 12.

its [stem] days are *wu* and *ji*. [Saturn], the Yellow Emperor, governs virtue and is the image of the female ruler. [Saturn] weighs on one lodge annually and is auspicious for the state it occupies. If [Saturn] ought not to occupy [a place] but does, or if it has already departed but turns back again and returns to occupy it, that state will gain territory, if not then it will obtain women. If [Saturn] ought to occupy a place but does not, after having already occupied [a lodge], but again departs to the west or east, that state will lose territory; if not it will lose women and cannot initiate undertakings or use troops. If [Saturn] dwells for a long time, that state will have abundant good fortune; if only briefly, then the good fortune will be meager. [1320] One name for Saturn is *Lord of the Earth*, governor of the harvest. In a year it travels 13 and 5/112^d, in one day traveling one 1/28^d, making the circuit of the sky in twenty-eight years. Where it stops, if all *Five Planets* follow Saturn and gather in the one lodge, that state below will be able to attract the entire sub-celestial realm through Weighty {Virtue}.[188] If propriety, {virtue,} rectitude, killing, and punishments are entirely neglected, then Saturn will be agitated by it. [If Saturn] gains, the king will not rest easy; if it regresses an army will not return. Saturn is yellow colored, with nine rays; its sound is said to be Yellow Bell, and its note *gong*.[189] If it skips two or three lodges ahead [of its proper station] it is said to "gain" and the ruler's commands are unfulfilled, if not, there are great floods. If it falls behind [its proper station] by two or three lodges, it is said to regress, the Consort will suffer distress and the year will not be seasonable; if not, then the sky will split open and the ground will shake. [*Southern*] *Dipper* (8) is the *Great Hall of Culture* and the temple of Saturn, the star of the Son of Heaven.

General planetary theory[190]

If the *Wood Star* [Jupiter] and *Earth Star* [Saturn] join,[191] it means internal disorder and famine; the ruler must not go to war or [he will suffer] defeat.

---

[188] A plausible argument is put forward in *JZ* (at 1119) that *de*, 德, has been transposed from its proper location here to the following line.

[189] As befits Saturn's status (and color), these are the fundamental notes in the musical and mathematical harmonics of the time. They correspond to the first note C in the pentatonic (and duodecatonic) scales. Major et al. (2010, 927–31).

[190] Based on the *Canon of Stars*, the placement of this section in the middle of the discussion of the individual planets is curious. In the *HS Monograph* it appears after the discussion of the fifth planet, Mercury. The section bears comparison with the parallel passage in *Jin shu*; see Ho (1966, 126 ff.). For a detailed review of knowledge of the periods and movements of all Five Planets in the Han dynasty, including the data in *TGS*, see Maspero (1929, 295–318); also Cullen (2011b).

[191] Strictly speaking, *he*, 合, "join," refers to a conjunction, defined below as "occupying the same lodge." Here for the first time the planets begin to be identified by their associated

If [Jupiter and] the *Water* [*Star*, Mercury join] then change [i.e. overthrow] is being plotted, and usurpation. If [Jupiter and] the *Fire* [*Star*, Mars, join] there will be drought.[192] If [Jupiter and] the *Metal* [*Star*, Venus, join] there will be gatherings of white garments [obsequies] and floods. If the *Metal* [*Star*, Venus is located] to the south [of Jupiter], it is called female and male and the grain will ripen. If the *Metal* [*Star*, Venus, is located] to the north the harvest will tend to fail. If the *Fire* [*Star*, Mars] and Mercury join, it is quenching.[193] If [Mars] and the *Metal* [*Star*, Venus] join, it is fusing and [there are] obsequies; in no case may one begin undertakings {or use troops}. [If Mars and the] *Earth* [*Star*, Saturn, join] it means distress and governs disloyal ministers of state; [there are] great famine, defeat in war, revolt of the army, the army surrounded, and great undoing if affairs are undertaken.[194] {If Mars and Mercury join it means revolt of the army, and if troops are used any undertakings will be greatly undone. If Saturn and Venus join there will be epidemics and armed conflict internally. If Mercury and Venus join it means change is being plotted, and the distress of armed conflict.} If the *Earth* [*Star*, Saturn] and Mercury join, an abundant harvest will be hindered, the army destroyed, and that state may not undertake major affairs.[195] Where it appears [that state] will lose territory; where it disappears [that state] will gain territory. The *Metal* [*Star*, Venus] is disease, internal armed conflict, [1321] and the loss of territory. If *Three Planets* join, the territory and state of that lodge will suffer armed conflict internally and externally, along with obsequies, and kings and eminences will be replaced. If *Four Planets* join, war and obsequies occur together, men of quality are in distress, and mean men run rampant. If the *Five Planets* join together, this means a change of elemental phase – the virtuous benefit, change brings a new great man to power to possess the Four Quarters and his posterity will flourish and multiply. Those lacking in virtue will suffer calamities or extinction.

If the *Five Planets* are all large, the affairs are also great; if [they are] small, the affairs are also small. The earliest one to appear is the gainer, the [state of the] gainer is the guest; the latest one to appear is the regresser,

---

Elemental Phases, a practice which later became conventional, gradually supplanting the older designations.

[192] According to *ZY*, this and much of what follows is from Shi Shen's *Canon of Stars*, from which Zhang Shoujie cites the relevant passages.

[193] The metallurgical terminology evokes the merging of the opposing Elemental Phases; cf. SCC, 408.

[194] The text is defective at this point, probably due to an eye-skip, which is restored following the *HS* version.

[195] *HS* has "If Saturn and Mercury join then a general's army will be overthrown and a legion taken down."

the [state of the] regresser is the host.[196] Heavenly responses must appear at the [*Dipper*] *Handle*.[197] If [two or more] planets [occupy] the same lodge it means to join, if there is mutual encroachment it is dueling;[198] [if the planets approach each other] within seven *cun* [dueling] must occur. [1322] If in aspect the *Five Planets* are white and round, it means obsequies and drought; if red and round, then the center is unsettled, and it means armed conflict; if green and round, it means affliction by floods, if black and round, it means epidemic disease and many deaths; if yellow and round, then it is auspicious. Red rays [mean] invasion of our cities; yellow rays [mean] territorial conflict; white rays [mean] the sound of weeping and lamentation; green rays [mean] there will be the affliction of war; black rays then [mean] deluge.[199] If the *Five Planets* are all the same color, in the sub-celestial realm weapons are put away and the common people are at peace and prospering. There are breezes in spring, rain in autumn, cold in winter, and heat in summer; {if the planets scintillate, it is usually due to this.}[200] After appearing for one hundred twenty days Saturn retrogrades westward, traveling west for one-hundred-twenty {thirty} days before reversing [direction] and traveling eastward. It is visible for three hundred thirty days and then sets, thirty days after setting it emerges again in the east. When *Taisui* is at *jiayin*, Saturn is in *Eastern Wall* (14), therefore in *Align-the-Hall* (13).[201]

Venus

Study the Sun's motion in order to locate *Supreme White*'s [Venus's] {position}.[202] Venus is the West and Autumn,[203] its [stem] days are *geng* and *xin*, and it governs killing. When killing is inappropriate, punishment emanates

---

[196] This implies a slightly different meaning for the two terms than we encountered above in the discussion of Jupiter where "regressing" clearly referred to retrograde motion. The usage here may provide an explanation for the term when applied to the Moon.

[197] The Zhonghua editors of *TGS* erred in placing the full stop after "star."

[198] Defined as their rays touching each other.

[199] An unintelligible phrase, *yi xing qiong bing zhi suo zhong* {意行窮兵之所終}, follows.

[200] It is not perfectly clear how this last follows from what precedes. Planets do not normally twinkle, and scintillation in other contexts is usually not an auspicious attribute. *KZ* (27.48) treats this clause as an interpolation.

[201] *KZ* (27.49) plausibly considers this passage ("After appearing…*Align-the-Hall*") to be a misplaced slip that belongs in the introduction to Saturn above (at 1320) following "governor of the harvest."

[202] Strictly speaking, commentators distinguish between the morning star denoted *Emergent Brightness*, "Lucifer," and the evening star, *Supreme White* (aka *changgeng*, 長庚, "Hesperus," in the *Book of Odes*, "Da Dong," Mao 203).

[203] At this point there is an unintelligible interpolation, *Sibing yue xing ji tian shi*, 司兵月行及天矢.

from Venus. When Venus's motion is anomalous, the lodge [it occupies] identifies the affected state. After it appears, it traverses eighteen lodges in two hundred forty days and then disappears. After disappearing in the east it invisibly traverses eleven lodges in one hundred thirty days; after disappearing in the west it invisibly traverses three lodges in sixteen days and reappears.[204] If it ought to appear but does not appear, or if it ought to disappear but does not disappear, this is called missing its lodge – if the army is not shattered the rulership of the state must be usurped. [1323] Starting at the Superior Epoch, in a *Shetige* year [Venus] appears in the east at dawn together with *Align-the-Hall* [13], and when it reaches *Horns* [1] it disappears; then it appears again with *Align-the-Hall* [13] in the west in the evening, and when it reaches *Horns* [1] it disappears; it appears again with *Horns* [1] at dawn, and then disappears with *Net* [19]; it appears again in the evening with *Horns* [1] and then disappears with *Net* [19]; it appears again at dawn with *Net* [19] and then disappears with *Winnowing Basket* [7]; it appears again with *Net* [19] in the evening, and disappears with *Winnowing Basket* [7]; it appears again with *Winnowing Basket* [7] at dawn, and then disappears with *Willow* [24]; it appears again with *Winnowing Basket* [7] in the evening, and then disappears with *Willow* [24]; it appears again with *Willow* [24] at dawn, and then disappears with *Align-the-Hall* [13]; it appears again in the evening with *Willow* [24], and then disappears with *Align-the-Hall* [13]. In all, it appears in the east and disappears in the west five times in eight years {and two hundred twenty days}, before appearing once again in the east at dawn with *Align-the-Hall* [13].[205] In sum, it makes the circuit of the sky in one year. When Venus first appears in the east it travels slowly, about $\frac{1}{2}^d$ a day, and after one hundred twenty days it must retrograde one or two lodges. Rising to its highest point, it reverses and travels eastward $1\frac{1}{2}^d$ a day, and after one hundred twenty days it disappears. When low and near the Sun it is called *Bright Star* and is yielding; when high and far from the Sun it is called *Great Imperious* and is unyielding. When Venus first appears in the west it travels rapidly, about $1\frac{1}{2}^d$ a day for one hundred twenty days; when it reaches its highest point it travels slowly, $\frac{1}{2}^d$ a day, and after one hundred twenty days it disappears at dawn; it must retrograde one or two lodges and then disappear. When at its lowest, near the Sun, Venus is called *Supreme White* and is yielding. When high and far from the Sun it is

[204] These figures for Venus's periods of invisibility are inaccurate, since they imply a synodic period of 626 days, as opposed to the correct figure of 584 days. *Huainanzi*, Chapter 3, "Heavenly Patterns," gives 120 and thirty-five days respectively; see Major (1993, 76). Sima Qian's most likely source, "Prognostications of the Five Planets," implies 120 and seventeen days respectively (Xi, "Wu xing zhan," 50), so it seems certain that "130" in *TGS* is a copyist's error for "120." Under ordinary conditions, Venus is only invisible for about ninety days in all.

[205] "Two hundred twenty days" is impossibly wrong and almost certainly the result of an early copyist's error – five synodic periods of 584 days equals almost exactly eight solar years.

called *Grand Chancellor*, and is unyielding. It appears in chronograms *chen* [lodges 1–2] and *xu* [15–16], and disappears in chronograms *chou* [8–9] and *wei* [22–3].[206] [1324] If [Venus] appears but should not, or should not yet disappear but does, all the sub-celestial realm ceases hostilities and troops abroad return. If it should not yet appear but does, or if it should disappear but does not, all in the sub-celestial realm take up weapons, and states will be destroyed. If Venus appears on schedule its state will flourish. When it appears in the east, it concerns the east; when it disappears in the east, it concerns the north. When it appears in the west, it concerns the west; when it disappears in the·west, it concerns the south. Where it tarries for a long time, that country benefits; where it passes quickly, that country suffers harm. When [Venus] appears in the west and moves eastward, it is auspicious for the state due west. When it appears in the east and moves westward, it is auspicious for the state due east. When [Venus] appears it does not cross the meridian [in daylight];[207] if it crosses the meridian [in daylight], the sub-celestial realm will change government.[208] When [Venus] is small and its rays scintillate, fighting breaks out. If when it first appears [Venus] is large and then later small, the troops are weak; if it is small when it appears and then [becomes] larger, the troops are strong. If [Venus] appears high, sending troops deep is auspicious, but [dispatching troops] to skirmish is inauspicious; if [Venus appears] low, [dispatching troops] to skirmish is auspicious, but [sending troops] deep is inauspicious. If the Sun is [heading] south and the *Metal* [*Star*, Venus] stays south of it, or if the Sun is [heading] north and the *Metal* [*Star*, Venus] stays north of it, it is called "gaining"; vassal lords and kings are uneasy, advancing troops is auspicious, retreating inauspicious. If the Sun has just reached the south[209] and the *Metal* [*Star*, Venus] stays north of it, or if the Sun has just reached the north and the *Metal* [*Star*, Venus] stays south of it, it is called "regressing";[210] vassal lords and kings are beset, withdrawing troops is auspicious and advancing inauspicious. In employing military force emulate *Supreme White*: if *Supreme White*'s motion is rapid, march quickly; if slow, march slowly. If [Venus] is

---

[206] Édouard Chavannes (1895–1904, Vol. 3, 380) interprets these as double-hours of the day, although identical indications given for Mercury below would then imply observations of the planet in daylight or several hours before sunrise, which is physically impossible. The implication that Mercury exhibits such seasonal regularity on an annual basis is a fiction. But as Jean Meeus (1997b, 223–5) has noted, after exactly thirteen years Mercury's elongations, rising times, conjunctions with the Sun, and the like all repeat, so there may be some basis to this idealized account.

[207] A phenomenon observable only in daylight; Ho (1966, 40). The drastic prognostications associated with this phenomenon are explained by the fact that Venus is actually "contesting with the Sun in brilliance."

[208] *HS* writes *ge min geng wang*, 革民更王, "change the people (?) and replace the king."

[209] That is, among the seven lodges of the Southern Palace.

[210] "Gaining" and "regressing" here appear to mean something other than premature or late appearance in a lodge as previously.

rayed, dare to do battle; if it scintillates vigorously, be vigorous; if it is round and quiescent, be quiescent. If one attacks[211] in the direction its rays point it is auspicious, the contrary is invariably inauspicious. When [Venus] appears [1325], send forth troops, when it disappears, recall troops. If its rays are red, there will be war; if white, there will be obsequies; if it is dark and round then rayed, [it means] affliction and matters of a Watery kind; if it is green and round with smallish rays, [it means] affliction and matters of a Woody kind; if it is yellow and round and mildly rayed, there are matters of an Earthy kind, and there will be a harvest. {If Venus disappears for seven days and then reappears, a general will be killed in war. If it disappears for ten days and then reappears, a chancellor will die from it. Whenever it disappears and then appears again, rulers hate it.}[212] If [Venus] has already appeared for three days and once again fades and disappears, and then three days after disappearing it reappears full, this is called feebleness; its state below will see its army defeated and general beaten. If [Venus] has already disappeared for three days and then appears again fainter, and then having appeared again for three days sets full, its state below will be afflicted, its armies will surrender their provisions and arms to the enemy to use, and though its soldiers be many, they will be captured. If on appearing in the west [Venus] moves anomalously, foreign kingdoms will be defeated; if it appears in the east and moves anomalously, China will be defeated.[213] When its aspect is large and encircled with glossy yellow, good things may be accomplished; when it is ringed with bright red, military forces are plentiful, but will not do battle. The whiteness of *Supreme White* compares with the *Wolf [Star]*;[214] when red, it compares with *Heart [5]*;[215] when yellow, it compares with the left shoulder of *Triaster*;[216] when it is green, it compares with the right shoulder of *Triaster*;[217] when black, it compares with the big star of *Stride [15]*.[218] If all *Five Planets* follow [Venus] and gather in a single lodge, its state below will be able to bring the entire sub-celestial realm into submission through force of arms. If where they stop is substantial, there will be gain; if where they stop is vacuous, there will be no gain.[219] Motion supersedes

---

[211] Reading *shun jiao suo zhi ji zhi*, 順角所指擊之; *JZ*, 1124.

[212] Following *JZ*, 1124, in restoring the missing text.

[213] Sima Qian is introducing a new macro-astrology with an imperial perspective in contrast to the former multivalent world of the Warring States, so it seems appropriate to translate *zhongguo* and *waiguo* accordingly.

[214] Sirius, α CMa, the brightest star in the sky.

[215] Antares, α Sco, whose apparent visual magnitude varies between +0.9 and +1.1.

[216] Betelgeuse, α Ori, whose apparent visual magnitude varies between +0.3 and +1.2.

[217] Bellatrix, γ Ori, whose apparent visual magnitude varies between +1.59 and +1.64.

[218] Mirach, β And, whose apparent visual magnitude varies between +2.01 and +2.10. This suggests that "black" (or "dark") may refer to dimming.

[219] According to *SY*, "substantial" refers to dwelling there in the usual course of travel; "vacuous" refers to anomalous appearance in a place, whether "gaining" or "regressing". The *Jin shu*

aspect; aspect supersedes position; being in position supersedes being out of position; colored supersedes being colorless; motion supersedes all of them. When [Venus] appears and tarries among the mulberries and elms [i.e. low on the horizon], it afflicts the state below. If [Venus] rises [1326] rapidly and crosses a third of the sky before its days [of travel] are up, that afflicts the state opposite. If [Venus] rises and descends again, or if it descends and rises again, a general will be deposed; if occulted by the Moon, a general will be humiliated. If the radiances of the *Metal* [*Star*, Venus] and the *Water* [*Star*, Mercury][220] unite, the [state] below will join battle;[221] if [the planets] do not join, though weapons are raised, there will be no fighting. If they join [and their radiances] obliterate one another, there will be shattered armies in the countryside. When [Venus] appears in the west, if it appears at dusk it is *yin*, the troops to the north are strong; if it appears during the evening meal, they will be slightly weaker; if it appears at midnight, they will be moderately weaker; if it appears at cock's crow, they will be much weaker; this is called *yin* submitting to *yang*. When [Venus] is in the east, if it rides high in full daylight it is *yang*, the troops to the south are strong; if it appears at cock's crow, they will be slightly weaker; if it appears at midnight, they will be moderately weaker; if it appears at dusk, they will be much weaker; this is called *yang* submitting to *yin*. If one fields forces when *Supreme White* has gone in, the troops will meet with reverses. If [Venus] appears south of *mao* [due east], the south will defeat the north; if it appears north of *mao*, the north will defeat the south; if it appears exactly in *mao*, eastern states will benefit. If [Venus] appears in the northwest, the north will defeat the south; if it appears in the southwest, the south will defeat the north; if it appears due west, western states will be victorious. [1327] When [Venus] trespasses on the arrayed [fixed] stars, there are small battles; [when it trespasses on] the *Five Planets*, there are major wars. When a trespass occurs, if *Supreme White* appears from the south, the southern state will be defeated; if it appears from the north, the northern state will be defeated. If [Venus's] motion is rapid, it means militancy; if it does not move, it means civility. If its color is white with five rays and if it appears early, it means a lunar eclipse, if late it means celestial anomalies and a *Broom-star* [comet] will emanate toward states {below which have lost the Way}.[222] When [Venus] appears in the east it is virtuous, and if in starting undertakings one keeps it on the left or facing,

---

"Monograph on the Heavenly Patterns" has a clearer exposition of the following rules, possibly closer to the original; *TGS*, 27.1326, n. 4.

[220] Following modern commentators in emending *mu*, "Wood," to *shui*, "*Water*"; Wang Liqi (1988, 945, n. 18); *KZ*, 27.56.

[221] The Zhonghua editors mispunctuate here.

[222] Following the *HS* version.

it will mean accretion. When it appears in the west it means recission,[223] so if one keeps it on the right or to the rear, it will be auspicious; the opposite will invariably be detrimental. When *Supreme White* is so bright that shadows are seen, it means victory in war. If it is seen in daylight and crosses the meridian, contesting {with the Sun} in brilliance,[224] strong states will be weakened, small states strengthened, and female rulers will flourish. *Neck* [2] is the *Detached Temple*, the temple of *Supreme White. Supreme White* is a Grand Minister, its epithet is *Senior Eminence*. Other names are *Expansive Star, Great Regulator, Controlling Star, Governing Star, Palace Star, Bright Star, Great Enfeebler, Great Glossy, Terminal Star, Grand Chancellor, Celestial Vastness, Sequencing Star, Moon's Weft*. The office of Marshal of State assiduously watches this [planet].[225]

Mercury

Examine the encounters of the Sun with its chronograms to determine Mercury's position.[226] [Mercury] is said to be the North, *Water*, and the essence of Supreme *Yin*; it governs winter, and its stem days are *ren* and *gui*. When penalties are inappropriate, punishment emanates from Mercury; its lodge identifies the affected state. [1328] Thus [Mercury] regulates the four seasons. In the second month of spring at the vernal equinox it appears in the evening with *Stride* [15], *Pasture* [16], and *Stomach* [17] [chronogram *xu*], and [travels] eastward five lodges [to 10–12, that astral field] being the state of *Qi*. In the second month of summer at the summer solstice it appears in the evening with *Eastern Well* [22], *Ghostly Conveyance* [23], and *Willow* [24] [chronogram *wei*], and [travels] eastward seven lodges [to 27–8, that astral field being] the state of *Chu*. In the second month of autumn at the autumnal equinox it appears in the

---

223   For the technical terms "accretion" and "recission," see Major (1987, 281–91); but see also Kalinowski (1998–99, 125–202). Here the terms appear not to denote spirit entities, but rather processes.

224   Reading *yu ri zheng ming*, 與日爭明, for *shi wei zheng ming*, 是謂爭明; see *JZ*, 1127. This is the earliest recorded observation that Venus can be seen to transit the meridian in daylight.

225   The earliest reference to Venus as morning and evening star is in the *Book of Odes*, "Da Dong" (Mao 203), where the planet is called *Opening Brightness* when it appears as a morning star in the east, and *Long Continuer* when it appears as the evening star in the west; see Karlgren (1950b, 155).

226   The implication seems to be that Mercury is called the *Chronogram Star, chen xing*, 辰 星, because it serves as an adjunct of the Sun in marking monthly progress through the chronograms. This is so because Mercury's maximum elongation cannot exceed twenty-nine degrees, which means it can never leave the same chronogram that the Sun occupies. For this reason, Joseph Needham (SCC, 399) chose to call it the *Hour Star*. In a small way, Mercury partakes of the season-indicating function of the *Northern Chronogram (Dipper)* and so shares the epithet *chen*, 辰, "celestial season indicator," with it, the *Dipper, Heart*, and *Triaster* being denoted the *Great Chen*.

evening {morning}[227] with *Horns* [1], *Neck* [2], *Base* [3], and *Chamber* [4] [chronogram *chen*], and [travels] eastward four lodges [to 5–7, that territory] being the Han [Dynasty].[228] In the second month of winter it appears at dawn with *Tail* [6], *Winnowing Basket* [7], *[Southern] Dipper* [8], and *Ox-Leader* [9] [chronogram *chou*], representing the Central States. As a rule, [Mercury] appears and disappears in chronograms *chen, xu, chou,* and *wei*.[229] When [Mercury] is early it means a lunar eclipse; when late, it means a comet and celestial anomalies. When it should appear in season, but does not, this is a lapse; troops will be pursued abroad but without fighting. If [Mercury] fails to appear for one season, that season will be disharmonious; if it fails to appear for four seasons, the whole sub-celestial realm will suffer a great famine. If [Mercury] ought to appear and does, if its aspect is white there will be drought; if it is yellow, the five grains will ripen; if it is red, there will be armed conflict; if it is black, there will be floods. If [Mercury] appears in the east, large and white and there are troops abroad, they will withdraw. Regularly, when in the east, if [Mercury] is red, China will gain the victory; if in the west and red, foreign kingdoms will benefit. If there are no troops abroad and [Mercury] is red, armed conflict will arise. When [Mercury] and Venus appear together in the east, both red and rayed, foreign kingdoms will be utterly defeated and China will gain the victory. When [Mercury] and Venus appear together in the west, both red and rayed, external states benefit. when the Five Planets are disposed in mid-heaven

---

[227] To be consistent with the other indications "evening" here must be emended to "dawn."

[228] Sima Qian here reassigns chronogram *yin*, the province of You (former astral field of the state of Yan), to the Han dynasty.

[229] This statement encapsulates the foregoing account of Mercury's movements but confuses the sequence. Chronogram *chen* corresponds to lodges 1–2 (east-southeast); *xu* corresponds to lodges 15–16 (west-northwest); *chou* corresponds to lodges 8–9 (north-northeast); *wei* corresponds to lodges 22–3 (south-southeast). When Mercury is at greatest elongation either east or west of the Sun in spring or autumn it presents the best opportunity to see the planet due to the high inclination of its ecliptic path with respect to the horizon at that time of year. Mercury averages six such elongations a year, but never gets far enough from the Sun to leave the dusk or dawn twilight. Possibly we have here an attempt to characterize ideal observation times. Christopher Cullen (personal communication) comments, "Earlier on, we were told the same thing about Venus in slightly different words . . . The most suggestive parallel seems to come from a lengthy section dealing with Venus in the *Wu xing zhan*, where the planet is said, for instance, to . . . 'appear in *xu* . . . appear in *hai* . . . appear on the northwest diagonal,' making it clear that the only possible interpretation is that the cyclical characters are used to indicate horizon directions, in which of course the 'northwest diagonal' is midway between the directions conventionally labelled as *xu* and *hai*. If we are to interpret the *Shi ji* and *Huainanzi* in the same way, that means we can reject the idea of following Chavannes and others in taking the cyclical characters in their other sense, i.e. as labels for the twelve double-hours of the day." Chavannes's interpretation would mean that Mercury could be observed late at night (hour of *chou*, LT 1–3) and in broad daylight (hour of *wei*, LT 13–15), which is plainly impossible. Portions of the "Treatise" read like a synopsis of the "Prognostications of the Five Planets"; see Christopher Cullen's translation of the *Wu xing zhan*, Cullen (2011b).

and gather in the east, China benefits, when they gather in the west, foreign kingdoms using weapons gain. If the *Five Planets* all follow [Mercury] and gather in one lodge, the state in which they lodge will be able to attract all the sub-celestial realm through Law. If [Mercury] does not appear, *Supreme White* stands for the guest; if it appears, *Supreme White* stands for the host. If [Mercury] appears but does not accompany *Supreme White*, though there be armies in the field, they will not fight. If [Mercury] appears in the east and *Supreme White* in the west, or if [Mercury] appears in the west and *Supreme White* in the east, this stands for being at odds; though there are troops abroad in the field they will not fight. If [Mercury] misses its season and appears, if [the weather] ought to be cold, on the contrary it will be warm, if it ought to be warm, on the contrary it will be cold. If [Mercury] ought to appear but does not, this is called "attacking soldiers" and war will break out. If [Mercury] enters [i.e. is occulted by] *Supreme White* and emerges above it, an army will be shattered, its general killed, and the invading army will be victorious; if it emerges below [*Supreme White*] the invaders will lose territory. If [Mercury] comes to interfere with *Supreme White* and *Supreme White* does not depart, a general will be killed. If erect banners [rays] emerge upward, the army will be shattered, its general killed, and the invading army will be victorious; if [banners] emerge downward, the invading army will lose territory. Watch the direction in which they point to name whose army will be destroyed. If [Mercury] circles *Supreme White* [without departing] [1329] or if they duel, there will be a major war, the invader will gain the victory, {and the host's officers will die}.[230] If the *Rabbit [Star]* passes by *Supreme White* by the width of a sword blade, there will be a minor battle and the invader will win. If the *Rabbit [Star]* lingers in front of *Supreme White* {for three days more than a ten-day week},[231] the army will withdraw; if it appears to the left of *Supreme White*, there will be a minor battle; if it brushes *Supreme White*, several tens of thousands of men will do battle and the host's officers will die; if it appears to the right of *Supreme White* at a distance of three feet [about seventy centimeters], the armies will speedily agree to do battle. If [Mercury] has green rays, [it means] the distress of war, if black rays, there will be floods.[232] The *Rabbit [Star's]* seven names are *Little Rectifier*, *Chronogram Star*, *Celestial Rabbit*, *Pacify Zhou Star*, *Tiny Lively*, *Clever Star*, *Hook Star*. If [Mercury] is yellow and small, and when it appears abruptly changes place, the cultured civility of the sub-celestial realm will change for the worse. The *Rabbit [Star's]* five aspects are: green and round[233] meaning affliction, white and round meaning obsequies, red and round meaning the center is uneasy,

---

[230] Emendation following the *HS*; *JZ*, 1130.   [231] Emendation following the *HS*; *JZ*, 1130.

[232] An unintelligible string *chi xing qiong bing zhi suo zhong*, 赤行穷兵之所终, intervenes here.

[233] "Ringed" translates *huan/yuan*, 圜, here, though the contrast is with "rayed" so the implication is a round, smooth appearance.

black and round being auspicious. Red rays mean invasion of our state, yellow rays mean [1330] contesting for land, white rays mean sounds of weeping and lamentation. On appearing in the east [Mercury] traverses four lodges in forty-eight days. After about twenty days [of visibility] it backs [toward the Sun] before disappearing in the east. When [Mercury] appears in the west it traverses four lodges in forty-eight days; after about twenty days [of visibility] it moves backward before disappearing in the west.[234] Initially one watches for it in *Align-the-Hall* [13], *Horns* [1], *Net* [19], *Winnowing Basket* [7], and *Willow* [24]. If it rises between *Chamber* [4] and *Heart* [5] the earth will quake. Mercury's colors are: in spring, greenish-yellow; in summer, reddish-white; in autumn, greenish-white, and the crops ripen; in winter, it is yellow but not bright. If [Mercury] changes color the season will not be prosperous. If [Mercury] does not appear in spring, it will be very windy; if autumn, [the crops] will not fruit; if summer, there will be sixty days of drought and the Moon will be eclipsed. If [Mercury] is not seen in autumn, there will be armed combat; if in spring, things will not grow. If it is not seen in winter, it will be overcast and rainy for sixty days and towns will be washed away, if it is not seen in summer, [the crops] will not mature.

Field allocations of the provinces

[Lodges] *Horns* [1], *Neck* [2], and *Base* [3] are [the province of] Yanzhou. *Chamber* [4], *Heart* [5] are Yuzhou. *Tail* [6] and *Winnowing Basket* [7] are Youzhou. [*Southern*] *Dipper* [8] is Jiang [Yangtze] and Hu [the Lakes].[235] *Ox-Leader* [9] and *Serving Girl* [10] are Yangzhou. *Ruins* [11] and *Roof* [12] are Qingzhou. *Align-the-Hall* [13] and *Eastern Wall* [14] are Bingzhou. *Stride* [15], *Pasture* [16], and *Stomach* [17] are Xuzhou. *Topknot* [18] and *Net* [19] are Jizhou. *Horned Owl* [20] and *Triaster* [21] are Yizhou. *Eastern Well* [22] and *Ghostly Conveyance* [23] are Yongzhou. *Willow* [24], *Seven Stars* [25], and *Spread* [26] are Sanhe. *Wings* [27] and *Chariot Platform* [28] are Jingzhou.[236]

---

[234] This implies a synodic period of about ninety-six days; the modern figure is 116 days.
[235] Lakes Dongting and Taihu in the lower Yangtze drainage.
[236] These astral–terrestrial correlations reflect some reapportionment by comparison with the earlier scheme of Shi Shen (closely followed by the *Huainanzi*; Major et al., 2010, 140). Shi Shen's allocations as given in ZY (*Shiji*, 27.1346) are: "*Horns* (1), *Neck* (2) or Yanzhou are the state of Zheng; *Base* (3), *Chamber* (4), *Heart* (5) or Yuzhou are the state of Song; *Tail* (6), *Winnowing Basket* (7) or Youzhou, are the state of Yan; [*Southern*] *Dipper* (8), *Ox-Leader* (9) or Yangzhou are Wu and Yue; *Serving Girl* (10), *Ruins* (11) or Qingzhou are Qi; *Roof* (12), *Align-the-Hall* (13), *Eastern Wall* (14) or Bingzhou are Wey; *Stride* (15), *Pasture* (16) or Xuzhou are Lu; *Stomach* (17), *Topknot* (18) or Jizhou are Zhao; *Net* (19), *Horned Owl* (20), *Triaster* (21) or Yizhou are Wei; *Eastern Well* (22), *Ghostly Conveyance* (23) or Yongzhou are Qin; *Willow* (24), *Seven Stars* (25), *Spread* (26) or San-he are Zhou; *Wings* (27),

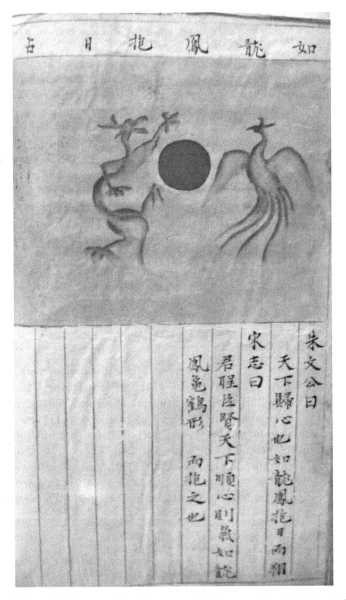

Figure A.6 Image of the Sun surrounded by *materia vitalis* shaped like a dragon and a phoenix, said to appear when the ruler is enlightened, ministers are worthy, and the whole realm is in accord. After the Ming Dynasty astral prognostication manual *Tianyuan yu li xiang yi fu* (1425?). Courtesy Yale University, SML East Asia Library Special Collections, Fv1742 +6223; see http://neworbexpress.library.yale.edu/vwebv/holdingsInfo?bibId=2948643.

{*Seven Stars* [25] is the *Throat*, the temple of Mercury, star of the Man and Yi [barbarians].}[237]

[1331] Solar halos[238]

When two armies confront each other, the Sun will have halos. If the halos are equal, the armies are evenly matched. [The one whose halo of *materia vitalis*] is thick, long, and large will gain the victory. [For the one whose halo] is thin, short, and small there will be no victory. Concentric and enveloping [halos mean] complete destruction; without their enveloping it signifies harmony. [Radiating] back [away from the Sun] means disharmony, and represents separating and departing from each other. [Radiating] straight out means self-appointing, setting up [on one's own authority] marquises and kings. A penetrating *materia vitalis* means a general will be killed. [If the halo] embraces [the Sun] and also crowns it, there will be joy. If the encircling is in the middle, the Middle [Kingdom] will be victorious. If outside, then outsiders will be victorious. If [the halo] is green outside and red within, [opponents] will leave each other in harmony. If [the halo] is red outside and green within, [opponents] will leave each other in enmity. If the halo arrives first and departs afterward, the resident [i.e. defending] army will be victorious. [If the halo] arrives first and departs first, then the earlier [force to arrive] benefits and the later is assailed. [If the halo] arrives later and departs later, the earlier [force to arrive] is assailed and the later benefits. [If the halo] arrives later and departs first, the earlier and the later will both be assailed, and the resident army will not be victorious. [If the halo] appears and then departs, sallies will be quick, but though victorious there will be no merit in it. If it appears longer than half a day, the merit is great. [If the halo is like] a short white arc, with pointed ends at the top and bottom, then [its state] below will suffer great effusions of blood. The Sun's

---

Chariot Platform (28) or Jingzhou are Chu. Traces of an earlier, more fine-grained scheme, in which state subdivisions are correlated with individual stars, are preserved in the Mawangdui text "Prognostications of the Five Planets"; see Liu Lexian (2004, 189 ff.). For discrepancies between the various field allocation schemes, see Chen Songchang (2001).

[237] Some commentators think that this sentence belongs at the end of the preceding section rather than here; see *JZ*, 1131.

[238] Compare the detailed exposition in Ho (1966, 139). Among the recently discovered Shanghai bamboo slips (nos 28–9) containing the text of "*Fan wu liu xing*," there is a generically worded rhymed couplet concerning the interpretation of solar and lunar halos. The phenomena described in the following three sections (and many more besides) are among the hundreds depicted in the Ming Dynasty illustrated compendium of astral prognostications, *Tian yun yu li xiang yi fu* (1628–44). This extraordinary hand-painted manual, said to have been produced at the direction of Ming emperor Renzong (1425–6) for circulation among high officials, was evidently so sensitive that it was included among the works proscribed in the late eighteenth century Qing literary inquisition associated with the compilation of the *Complete Library of the Four Treasuries* (*Siku quanshu*).

halo brings victory [or defeat] at the earliest in thirty days, at the latest in sixty days. [If the Sun] is eclipsed, it is disadvantageous for the locale wherein it is eclipsed; when it recovers, it is beneficial for the locale wherein it recovers. If the eclipse is total, it reflects on the position of the ruler; {if not total it reflects on the senior ministers.}[239] Based on what it [the eclipse] indicates, together with the lodge the Sun [occupies], and adding in the date and time, these serve to identify the affected state.

The Moon

When the Moon travels the Middle Path,[240] it is quiet and peaceful. [When the Moon occupies] the *yin* space[241] there is an abundance of water and intrigue. Beyond [the Middle Path] three feet [sixty-nine centimeters] to the north is the *Yin Star*. Beyond that three *chi* to the north it is the *Great Yin*, [when the Moon is there it portends] great floods and armed conflict. [When the Moon occupies] the *yang*-space [the ruler] is arrogant and willful. [Beyond it three feet to the south it is the] *Yang Star*, [when the Moon is there it portends] great violence and many imprisonments. [Beyond that three feet to the south is] *Great Yang*, [when the Moon is there it portends] great drought and obsequies. [If the Moon passes through] *Heaven's Portal* in *Horns* (1) in the tenth month, it signals [consequences the following] fourth month; if in the eleventh month, it signals [consequences the following] fifth month, if in the twelfth month, it signals [consequences the following] sixth month; water will pour forth at least three feet [deep], at most five feet [deep]. [If the Moon] trespasses on the *Four Counsellors* [of lodge *Chamber*], Minister Counsellors will be executed. [If the Moon] travels past the *North* and *South Rivers*, speaking in terms of *yin* or *yang* [in relation to them],[242] [1332] [one can expect either] drought, flood, armed conflict, or obsequies. If the Moon occults Jupiter, the land [corresponding to] its lodge will suffer famine or catastrophic loss. [If the Moon occults] Mars, there will be rebellion; if Saturn, then the lower orders will trespass on the upper; if Venus, then a powerful state will be defeated in war; if Mercury, then [palace] women will be rebellious. [If the Moon] occults *Great Horn*, rulers hate it;[243] [if the Moon occults] *Heart* (5), it means seditious

---

[239] Following the suggested emendation in *JZ*, 1133.
[240] That is, between the two middle stars of lodge *Chamber*, ρ and δ Sco.
[241] That is, north of the Middle Path.
[242] In fact, the ecliptic passes between *North* (α Gem) and *South Rivers* (β CMi), too distant from both for the Moon to pass either to the north of *North River* or to the south of *South River*.
[243] Arcturus is too far from the ecliptic for the Moon ever to occult it, so "*Great*" here may be an intrusion caused by the "*Great*" of *Supreme White* in the previous sentence. Occultation of the stars of lodge *Horns* (Vir) by the Moon, on the other hand, will occur with some frequency.

Figure A.7 Image of the eclipsing Sun with multiple lobes, captioned with the associated omens. After the a late Ming copy of the astral prognostication manual *Tianyuan yu li xiang yi fu*, Courtesy Library of Congress, East Asian Special Collections.

rebellion within; [if the Moon occults] the arrayed [fixed] stars, the land [corresponding to] those lodges will suffer affliction. The lunar eclipse [cycle's] beginning days are: every fifth month six times; every sixth month five times; every fifth month again six times; then every six months once; then every fifth month five times;[244] in all one hundred thirteen months [elapse and the cycle] begins again. Therefore, [with regard to] lunar eclipses, there is a regular [pattern]; solar eclipses are not good. [For days] *jia* and *yi* [corresponding to lands beyond the Four Seas], Sun and Moon are not prognosticated; [prognosticate on] *bing* and *ding* [corresponding to the regions] Jiang, Huai, and Haidai;[245] [1333] on *wu* and *ji* [corresponding to the regions] Zhongzhou, He, and Ji;[246] on *geng* and *xin* [corresponding to the region] from Mt. Hua westward; on *ren* and *gui* [corresponding to the region of] Mt. Heng and northward. Solar eclipses refer to the lord of the state; [as for] lunar eclipses, Generals and Chancellors are indicated.

Auspicious and inauspicious stars

The *State Majesty Star* is large and red,[247] in aspect resembling the [*Old Man of the*] *Southern Culmen*.[248] Where it appears, in its [state] below armed conflict will arise and the troops will be strong; [the location] opposite will be disadvantaged.

The *Brightly Shining Star* is large and white,[249] lacking rays, now rising now falling [in brightness]. The state wherein it appears will see armed conflict arise and many anomalies.

The *Five Cruelties Star* appears due east in the eastern astral field. In aspect it resembles Mercury, but at a distance of about six spans from the ground {large and yellow}.[250] [1334] The *Great Ruiner Star*[251] appears due south in

---

[244] The implication seems to be that in one-half of a Metonic cycle of 235 synodic months or almost exactly nineteen solar years there are twenty-three occasions in 113 months, at five- to six-month intervals, when an eclipse could potentially occur. For discussion of the technical problems with this passage see *JZ*, 1135.

[245] That is, Mt. Tai in Shandong.

[246] That is, the north China plain, the lower course of the Yellow River, and the Ji River area in the west of Shandong.

[247] According to Meng Kang, quoted in *JJ*, this star results from the dispersal of Jupiter's essence.

[248] That is, Canopus, α Car (apparent visual magnitude −0.72), second-brightest star in the sky after Sirius; *JJ* says this star is also an emanation of Jupiter.

[249] According to Meng Kang, quoted in *JJ*, this is a manifestation of the essence of Mars.

[250] Following *HS*; see *JZ*, 1136. According to Meng Kang, quoted in *SY*, this is a manifestation of the essence of Saturn. This star is mentioned in the *Classic of Mountains and Seas*, "Western Mountains Classic," where it is said to be controlled by the Queen Mother of the West.

[251] *HS* writes *liu*, 六, "six," for *da*, 大, "great," and *xing*, 星 "star" for *ye*, 野, "astral field" – both scribal errors. According to Meng Kang, quoted in *JJ*, it resembles a broom star and is a manifestation of the essence of Venus.

the southern astral field. The star is about six spans from the ground, large and red, moves repeatedly, and gleams.

The *Overseer of Deviance Star* appears due west in the western astral field. The star is about six spans from the ground, large and white, resembling Venus.

The *Imprisoned Han Star*[252] appears due north in the northern astral field. The star is about six spans from the ground, large and red, and moves repeatedly, on inspection it is green in the middle.

Where these [preceding] four stars of the astral fields appear, if it is not in the appropriate direction, [in that field's correlated state] below there will be armed combat, and [the location] opposite will be disadvantaged.

The *Four Pacifing Stars* appear in the four inter-cardinal directions, about four spans from the ground.

The *Splendid Gleam of the Terrestrial Nodes*[253] also appears at the four inter-cardinal points, about three spans from the ground, and is like the Moon when it first appears. Where it is seen [if the region] below experiences rebellion, the rebellious will be extinguished while the virtuous will prosper.[254]

The *Torch Star* is like Venus in aspect, on appearing it also does not move. It appears, then is extinguished. Cities and towns on which it shines will be disordered.

If [something] seems like a star but is not a star, seems like a cloud but is not a cloud, it is denoted *Reverting to Perversity*. Where *Reverting to Perversity* appears, there must be one who is restored to [power in] the state. [1335] Stars are the dispersed *materia vitalis* of Metal, their origin is Fire. When stars are myriad, the state is fortunate, when few, it is inauspicious. The Han (Milky Way) is also the dispersed *materia vitalis* of Metal, its origin is Water. When the Milky Way's stars are many, there will be many inundations, if few, there will be drought – this is the general rule.

A *Celestial Drum* has a sound like thunder but is not thunder; the sound is {in the sky} and reaches the ground.[255] Where it goes, troops are sent forth below.

---

[252] *HS* has *xian han*, 咸漢 for *yu han*, 獄漢, and *xing*, "star," for *ye*, "astral field". *Chunqiu he cheng tu* (*JZ*, 1136) says this "star" governs deposed kings, which supports the reading *yu*, "prison." The northerly direction suggests a connection with the aurora borealis.

[253] *HS* writes *zang*, 臧, for *xian*, 咸. Here the phenomenon appears only tangentially related to the previous one, but in *HS* it is portrayed as a further elaboration on the appearance of the *Pacifying Stars*.

[254] *HS* has an eye-skip here, writing 所見下, 有亂者亡, 有德者昌 where *TGS* writes 所見, 下有亂; 亂者亡, 有德者昌. According to Meng Kang, quoted in *JJ*, this star is a manifestation of the essence of Saturn.

[255] Following *JZ* (1138) in emending to *yin zai tian er xia ji di*, 音在天而下及地.

A *Dog of Heaven*'s shape is like a large *Fleeting Star* [i.e. meteor][256] in appearance and makes a sound, on falling to the ground it resembles a dog. Observing [from below] it is like a fiery gleam flaming across the sky. The bottom is round like a paddy-field several *qing* [*c*.6.5 hectares] in size and it comes to a point above, when it appears it is yellow; for a thousand *li* [500 kilometers] around armies will be shattered and generals killed.

An *Arriving Beneficence Star* has an appearance like flame. Yellowish-white, it rises up from the ground; large on the bottom, its apex is pointed.[257] When it appears, there is a harvest without sowing; if no earthworks are performed, there must be great harm [done to the crops].

Chi You's Banner is like a broom star but curved behind, like a banner. When it appears the King conducts punitive campaigns in the Four Directions.[258] [1336]

A *Decade's Beginning Star*, shaped like a rooster, appears beside the *Northern Dipper*. Its tail is greenish-black, resembling a crouching turtle.[259]

Crooked Arrows resemble large grayish-dark meteors moving sinuously; watching from a distance they appear feathery.

A *Long-Lived* [*Star*][260] is like a bolt of cloth hanging in the sky. When this star appears, armed combat arises.

When stars fall and reach the ground they [are found to be] stones. {Between the Yellow and Ji Rivers there have occasionally been star falls.}[261]

A *Resplendent Star* is seen in bright, clear skies. A *Resplendent Star* is a *Star of Virtue*. Its shape is not constant, but it normally appears in a state possessing the Way.[262]

---

[256] *HS* writes *liu xing*, 流星, "streaming star," for *ben xing*, 奔星, "speeding star."

[257] Bo Shuren (1981, 687) identifies this as the zodiacal light.

[258] According to *Lüshi chunqiu*, "Ming li" chapter, *Chi You's Banner* is long, multi-branched, and splendiferous, yellow above and white below (*Lü Buwei*, 1974, 62). For the oldest account of *Chi You*'s battle with the Yellow Emperor from a Mawangdui silk manuscript, see Lewis (1990, 148). For description of the Mawangdui cometary atlas whose illustration no 28 depicts *Chi You's Banner* (the MS caption reads, "*Chi You's Banner*, troops afield return"), see Xi Zezong (1989a, 31). A photographic enlargement of the atlas is in Zhongguo shehui kexueyuan kaogu yanjiusuo (1980, 23, Plate 21). Comets identified as *Chi You's Banner* appeared twice in Emperor Wu's reign (see *infra*).

[259] "*Decade's Beginning*" figures as an astral deity in the *Chuci* poem "Wandering Far," so it has a long history. "Black" here and elsewhere may just mean "dark, dim."

[260] Presumably the aurora borealis.

[261] *HS* lacks this last sentence, which appears to be a comment inadvertently copied into the text.

[262] Descriptions of this most auspicious "star" vary as to color and shape. Wang Chong (1974, 174) observed, "anciently people could not predict the movements and appearance of planets like Jupiter and Venus, so when they saw a large star they called it a *Resplendent Star*." Wang goes on to quote from *Erya*, "Explicating the Four Seasons," which states that *Resplendent Star* refers to the temperateness of the four seasons. Wang also mentions that the term could have been applied to sightings of Venus in broad daylight, once notably during Wang Mang's reign (9–25 CE). In *Jin shu* it is somewhat improbably identified as "earthshine," seemingly

Clouds and *materia vitalis*[263]

Whenever one observes cloud-like[264] *materia vitalis* at a distance, seen from below it is three to four hundred *li* [distant]; on viewing from level ground [when it is] above the mulberries and elms [i.e. low on the horizon], it is two thousand *li* [distant];[265] climbing high to view it, where it hugs the ground it is three thousand *li* [distant]. Among cloud-like *materia vitalis*, those atop which wild animals dwell are victorious.[266] [1337]

From Mt. Hua southward, the *materia vitalis* is dark below and red above. Around Mt. Songgao and Three Rivers, the *materia vitalis* is pure red. From Mt. Heng northward, the *materia vitalis* is dark below and green above. Within the Bohai Gulf, Mt. Jieshi, the sea, and Mt. Tai, all the *materia vitalis* is dark. Between the Yangtze and the Huai River, the *materia vitalis* is all white. The *materia vitalis* of conscripts is white; the *materia vitalis* of earthworks is yellow; the *materia vitalis* of chariots now rises, now falls, always gathering.[267] The *materia vitalis* of cavalry is low and wide; the *materia vitalis* of infantry clumps together. [If the *materia vitalis* of a force is] low in front and high in back it is quick [to advance], [if] blunt in front and high in back,[268] low and pointed in the rear, [it is] retreating. Whichever [force's] *materia vitalis* is level, its movements are slow. [If a force's *materia vitalis* is] high in front and low in back, it turns about without stopping. When two [forces'] *materia vitalis* meet, the low overcomes the high, and the sharp overcomes the blunt. When a [force's] *materia vitalis* comes in low and follows the ruts of

too commonplace a phenomenon; Ho (1966, 129). For a study of an identification with the comet of 110 BCE in Emperor Wu's reign, see Kern (1997, 251).

263 For a parallel in *Jin shu*, see Ho (1966, 147). For an extraordinary parallel with *TGS* demonstrating the persistence of such techniques of prognostication based on clouds and *materia vitalis* right down through the Ming dynasty, see Yates (2005, 24–31). For Ming dynasty illustrations, see *Tian yuan yu li xiang yi fu*, which contains a vast number of hand-painted pictures of cloud formations, haloes, and the like, as well as astronomical phenomena of other kinds. The Simas' treatment of the subject may be compared with the illustrated Mawangdui MS "Diverse Prognostications on Heavenly Patterns and Formations of *Materia Vitalis*," *Tian wen qi xiang za zhan*; Wang Shujin (2007; 2011). Earlier studies on the ancient practice of "watching for the *materia vitalis*" (*wang qi*) include Bodde (1981); Hulsewé (1979); He and He (1985); Loewe (1994b); Huang and Chang (1996); Harper (1999).

264 Evidently condensations of *materia vitalis* like fog, mist, low-hanging clouds, etc.

265 Following *JZ*, 1142.

266 See the Mawangdui MS "Diverse Prognostications on Heavenly Forms of *Materia Vitalis*" for illustrations of such creatures; also Liu Lexian (2004, 100, and Plate 2). *JZ* (27.1143) quotes a parallel passage from the *Book of War*: "when a cloud [forms shaped like a] cock and approaches a state, that state must surrender"; *ZY*, 27.1337.

267 Hereafter the focus is exclusively on prognostication in military contexts, which appears to have been the principal application of the technique, according to Yates (2005, 34).

268 *BNB* has an eye-skip here, writing 後高兌而卑者 for 後高者兌, 後兌而卑者; see also *JZ*, 1143.

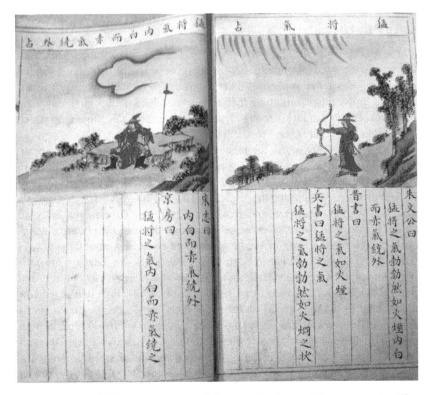

Figure A.8 The appearance of the *materia vitalis* of fierce generals. After the Ming Dynasty astral prognostication manual *Tianyuan yu li xiang yi fu* (1425?). Courtesy Yale University, SML East Asia Library Special Collections, Fv1742 + 6223; see http://neworbexpress.library.yale.edu/vwebv/holdingsInfo?bibId=2948643.

the chariot wheels,[269] for up to three or four days [what this presages] may be witnessed as far as five or six *li* away. When a [force's] *materia vitalis* comes in seven or eight feet high, for up to five or six days [what this presages] may be witnessed more than ten {to twenty} *li* away. When a [force's] *materia vitalis* comes in one to two spans high, for up to thirty or forty days [what this presages] may be witnessed fifty to sixty *li* away. A moving mist that is greenish-white [signifies] the general is courageous but the soldiery is cowardly. [A *materia*

---

[269] *HS* writes *dao*, 道, for *tong*, 通; according to *JJ* the substitution *tong*, 通, for *che*, 轍, was in observance of the taboo on Emperor Wu's personal name *che*, 徹, though this seems doubtful.

vitalis] whose base is broad but trails off in the distance [signifies that] there will be an even contest.[270] [A *materia vitalis*] that is greenish-white, but whose front is low, [signifies] victory in battle; if it is red in front and uplifting, [it signifies] defeat in battle.

Battle-Array Clouds are like standing walls. Shuttle Clouds are like the shuttle of a loom. Axle Clouds roll up with two tapered ends. Stretched Clouds are like a cord, lying ahead of one they stretch across the whole sky, if only half a one, it [stretches across] half the sky. Arc Clouds [are so called] because they resemble battle flags. Hooked Clouds are curved. If these various clouds appear, then prognosticate based on their five colors. If they are glossy, rolling, dense, or their appearance excites people, then they should be prognosticated – armed combat must occur and battle will be joined where they indicate. [1338]

Wang Shuo's [practice of] watching for the *materia vitalis* relied on those next to the Sun. Cloud-like *materia vitalis* next to the Sun is the image of the ruler of men; in all cases [Wang] prognosticated based on their appearance. Therefore, the *materia vitalis* of the northern Yi-peoples resembles livestock herds and yurts; the *materia vitalis* of the southern Yi-peoples resembles watercraft and pennants. Wherever there have been major floods, on the battlefields of defeated armies, or in the wastes of ruined states, and wherever there is hoarded cash below,[271] and above golden treasure, in every case there will be *materia vitalis* – one cannot but look into this. The *materia vitalis* of shellfish[272] by the seaside resembles multistoried buildings and terraces. The *materia vitalis* of broad expanses of wilderness takes the shape of palaces and gate towers. Cloud-like *materia vitalis* is the image of places wherein mountains, streams, and the populace gather and accumulate.[273] [1339] Therefore, those who watch for [signs of] growth or decline on entering a state or a town will look closely at the regularity and orderliness of boundary mounds, paddies and fields, and at how well-appointed appear the inner and outer walls, mansions, houses and doorways of the humble, and after that [they look at] how exquisite are the vehicles, garments, livestock, and goods. What is fruitful and growing is auspicious, what is empty and dissipated is inauspicious.

---

[270] *JZ*, 1144.

[271] *HS* writes *quan*, 泉, "spring," for *qian*, 錢, "cash"; "an accumulation of springs" may be an alternative reading.

[272] This belief has given rise to the expression *hai shi shen lou*, 海市蜃楼, "sea markets and shellfish pagodas," for a mirage.

[273] *ZY* quotes a long passage from *Huainanzi*, Chapter 4, "Terrestrial Forms," which amplifies on this theme; Major et al. (2010, 160).

What looks like smoke [*ʔen; **qiin], but is not smoke [*·iən, *ʔen; **qiin],[274]
what looks like cloud [*ɦiun; **ɢun] but is not cloud [*ɦiun;**ɢun],
luminous and plenteous [*phiun; **phuɯn]; [but] cool and distant [*khʷin;
    **khrun],
is called a *Propitious Cloud* [*ɦiun; **ɢun].
Propitious Clouds are joyous *materia vitalis* [*khɨi; **khuɯds].
What looks like fog [mio; **mos], but is not fog [*mio; **mos],
and does not dampen cap and clothes [*ɳio; **njo],
where it appears that region [*ɦɨk; **ɢʷruɯg],[275]
by armored troops is overrun [*tsʰio; **tsho].

Now, lightning, rainbows, thunderclaps, gleams in the night, these are [due to] the action of *yang materia vitalis*; they occur in spring and summer and vanish in autumn and winter; therefore, {writings} about watching [for the *materia vitalis*] all monitor them.[276]

Heaven parts and displays images [*ɱiut; **sɢlaŋʔ],
Earth moves and cracks open [*khəi; **khuɯɯl],
mountains collapse and slide [*siɛ; **selʔ],
watercourses are blocked and streams choked off [*biŭk, *biu; **buɯgs],
waters ebb and flow, the land expands [*ʈiɐŋ **krlaŋ],
marshes dry up – revealing signs [*ziɐŋ; **sɢlaŋʔ]:
outer and inner walls, doors and village gates [*ɭiɤ; **g-ra],
where the gloss is faded and the wood decayed [*khụo; **khaa],
on palaces and temples, mansions and residences [*dei; **liils],
the places wherein people dwell [*tshi **snʰis];
rhymes and common usages, carts and clothes [*biŭk, *buk; **buɯg];
– observe what the people drink and eat [*zɨk; **ɢljuɯg],
the Five Grains, grass and trees – observe their kinds [*diuk, *dʑiok; **djog];
granaries and offices [*pio; **poʔ],
stables and storehouses [*khụo; **khlaas],
and the roads in all directions [*lụo; **g-raags];
the six kinds of livestock, fowl and game [*ɕiu; **qjus],
the goods produced, their goings and comings [*dzɨu; **dzugs];
fish and turtles, birds and rats [*ɕiɤ; **qhjaʔ],
– observe where they are found [*tɕhiɤ; **khljaʔ].
ghosts weep and cry out [*ɦuọ; **qhaa]
and people meeting them are gripped with fear [*ŋụo; **ŋaa];
– so these words[277] can be factual [*ɳiɛn; **njen]. [1340]

[274] Old and Middle Chinese reconstructions are taken from *TLS, Thesaurus Linguae Sericae.*
[275] *HS* writes *cheng*, 城, for *yu*, 域; neither fits the rhyme in *-o.
[276] Following *HS*, which has *hou shu zhe*, 候書者. The quoted lines that follow are a rhymed litany of the kinds of indication to be observed in prognosticating the good or ill fortune of a place.
[277] Following *KZ*, 27.78.

The harvest

Whenever watching for whether the harvest will be fine or poor, pay particular attention to the beginning of the year.[278] For some the year begins with the winter solstice when the generative *materia vitalis* first sprouts. [For some the year begins]·the day after *La*-sacrifice,[279] the populace ends the year by gathering as one to eat and drink, exciting the *yang materia vitalis*, hence this is said to initiate the year. [In contrast,] dawn of the [first day of the] Standard Month is the start of the King's Year, the day of the Beginning of Spring and the start of the four seasons. These four [seasonal] beginnings are the days on which to watch. But Wei Xian[280] of Han determined the Eight Winds on the day after the *La*-sacrifice, in combination with dawn of the [first day of the] Standard Month. If the wind was from the south, there would be a great drought; if from the southwest, minor drought; if from the west, there would be armed combat; if from the northwest, soybeans would ripen, there would be minor rains and a rush to arms; if from the north, [there would be a] moderately good harvest; if from the northeast, superior harvest; if from the east, major floods; if from the southeast, epidemics among the people, and the harvest would be poor. Therefore, each of the Eight Winds is matched with its opposite, and the one comparatively greater prevails – more overcomes less, durative overcomes transient, fast overcomes slow. From dawn to morning meal [*c*.5.00–9.00 local time] concerns wheat; from morning meal to mid-afternoon [9.00–13.00 local time] concerns paniced millet; from mid-afternoon to late afternoon [13.00–15.00 local time] concerns glutinous millet; from late afternoon to dusk [15.00–17.00 local time] concerns legumes; from dusk to sunset [17.00–19.00 local time] concerns hemp. What is desirable is to have clouds, wind and Sun the whole of the [first] day. So, on that day, if at the proper time [one plants] deeply there will be plenty of yield; if there are no clouds, but there is wind and Sun, if at the proper hour [one plants] shallowly there will be plenty[281] of yield; if there are clouds and wind, but no sun, if at the proper hour [one plants] deeply there will be a meager yield; if there is Sun but no clouds and it is not windy, [planting] at the proper hour the crop will fail. [If such conditions persist] for the length of time it takes to consume a meal, the failure will be less; [if such conditions persist] for the length of time it takes to cook five pecks of grain, the failure will be great. If the wind rises again and there are clouds, the crop will recover. In each case, the suitability of sowing is prognosticated based on the

---

[278] The considerable antiquity of this practice is suggested by a detailed record in the *Zuozhuan* of such observations at the beginning of each season on the solstices and equinoxes (Duke Xi of Lu, fifth year, 655 BCE); SCC, 284.

[279] Folowing *KZ* (27.71), which reads as if there is another *huo*, 或, before *la*, 臘; on the *La* month, see Bodde (1975, 49 ff.).

[280] Loewe (2000, 578).   [281] *HS* has *shao*, 少, "meager."

hour and the color of the clouds. If it rains or snows or if it is cold [on the first of the Standard Month], the harvest will be poor. [1341] Bright and early on this [first] day [of the Standard Month], listen to the sound [emanating] from the capital populace. If the note is *gong*, the harvest will be fine and [the year] auspicious; [if the note] is *shang*, there will be armed combat; if *zhi*, drought; if *yu*, floods; if *jiao*, the harvest will be poor.[282] Or again, from dawn of the [first day of the] Standard Month, match the numbered [days of] rain. For each day [of rain] the edible [yield will increase] one *sheng* [200 milliliters], reaching seven *sheng* by the end. Beyond that [one does] not prognosticate. To prognosticate flood and drought, count [from the first] to the twelfth day, each day indicating the [same numbered] month.[283] To prognosticate within an area one thousand *li* around, however, one watches in behalf of the sub-celestial realm throughout the whole Standard Month. Based on the successive lodges the Moon passes through and whether [the day is] sunny, windy, or cloudy, one prognosticates [for the corresponding] state. However, one must also study where *Taisui* is located. If [*Taisui* is] in Metal, [there will be a] bumper crop; if in Water, [the harvest will be] reduced; if in Wood, famine; if in Fire, drought. This is the general rule. [1342] On the superior [first] *jia* day of the Standard Month, if the wind blows from the east it is beneficial for the silkworms; if [the wind comes] from the west, and if at dawn there is a yellow mist, it is detrimental. On the winter solstice when [the day] is the shortest, suspend soil and charcoal [in the balance]. If the charcoal moves [i.e. the balance pan rises],[284] and if the elk have dropped their antlers, and if the roots of the orchids have appeared, and if water rises in the springs, then one may know approximately that winter solstice has arrived – but decisive will be the shadow of the gnomon. Wherever Jupiter is located the five grains will abundantly flourish; the [station] opposite will witness the contrary, [there] the harvest will be calamitous.

Astrological history

The Prefect Grand Scribe-Astrologer said:[285] "Ever since the people have existed, when have successive rulers not calendared the movements of the Sun,

---

[282] These are the five notes of the pentatonic scale *gong*, 宫, *shang*, 商, *jiao*, 角, *zhi*, 徵, *yu*, 羽, roughly corresponding to C, D, E, G, A.

[283] If it rains on a particular day, the month with the same number will be rainy, otherwise dry.

[284] An explanation of the rationale behind this procedure is found in *Huainanzi* chapter "Heavenly Patterns": "*Yang materia vitalis* is fiery, *yin materia vitalis* is watery, *Water* conquers, so Summer Solstice is moist; *Fire* conquers, so Winter Solstice is dry. When dry, the charcoal is light, when moist, it is heavy."

[285] In recounting his motivations in the Preface, Sima Qian refers to himself as Prefect Grand Scribe-Astrologer both in the third person and in direct quotation. Here it is not obvious whether he is quoting his father or even where the quotation ends and he picks up the narrative, though this does sound rather like a lesson learned at his father's knee.

Moon, stars and asterisms?[286] Coming to the Five Houses[287] and the Three Dynasties [Xia, Shang, Zhou], they continued by making this [knowledge] clear, they differentiated wearers of cap and sash from the barbarian peoples as inner is to outer, and they divided the Middle Kingdoms into twelve regions. Looking up they observed the figures in the heavens, looking down they modeled themselves on the categories of the earth.[288] Therefore, in Heaven there are Sun and Moon, on Earth there are *yin* and *yang*; in Heaven there are the *Five Planets*, on Earth there are the Five Elemental Phases [Wood, Fire, Metal, Water, Earth]; in Heaven are arrayed the lodges, and on Earth there are the terrestrial regions. The *Three Luminaries* are the essence of *yin* and *yang*, their *materia vitalis* originates on earth, and the Sage comprehensively brought order to them [sc. their principles]." [1343]

From Kings You and Li [i.e. the decline of Zhou] on, it has been a long time. The celestial anomalies witnessed in all the states and locales differed, and traditions of prognostication about freakish things responded to their own times. Hence, the writings, diagrams, and texts [with their] incipient signs of the auspicious and inauspicious are inconsistent.[289] It is for this reason that when Confucius discoursed on the *Six Classics*, he took note of the anomalies but did not write down their interpretation. When it came to the Way of Heaven and the Mandate [of Heaven], he did not transmit [the teaching]. Were [the Sage] to transmit it to the right person, [that person] would not need to be told; were he to tell the wrong person, even if explained [to such a one], it would not be clear to him. Anciently, those who handed down the celestial regularities were: before Gao Xin (Di Ku), Chong and Li; in [the time of] Tang (Yao) and Yu (Shun), Xi and He; in the Xia dynasty, Kun Wu; in Yin-Shang, Wu Xian; at the Zhou Court, Shi Yi and Chang Hong;[290] in [the state of]

---

[286] Sima Qian is paraphrasing the "Canon of Yao" in the *Book of Documents*: "calendared the images, Sun, Moon, stars and asterisms, and bestowed the seasons on the people." In the "Basic Annals of the Five Emperors" Sima Qian glosses *li xiang*, 曆象, as *shu fa*, 数法 (as pointed out to me by Christopher Cullen), indicating that he took *li*, 曆, to refer to the deriving of numerical models from their movements. Calendrical science was thoroughly mathematized by Sima Qian's time, of course, so his is a presentist interpretation of a text that refers to observational practices in the early Bronze Age. According to Liu Qiyu, the "Canon of Yao" originally wrote not *li xiang*, 曆象, but *li xiang*, 歷象. Sima Qian's *Shu fa*, 数法, Liu Qiyu (2004, 45) says, is Sima Qian's gloss of the original *li xiang*, 歷象, "calendared the images."

[287] The five pre-dynastic rulers: Huang Di, Gao Yang (Zhuan Xu), Gao Xin (Yao), Tang Yu (Shun).

[288] *TGS* is quoting the "Appended Commentary" to the *Book of Changes*.

[289] Compare this statement with the excerpt from Sima Qian's "Preface" translated at the beginning.

[290] According to the Warring States bamboo slip text *Qi ye*, Scribe-Astrologer Yi was an honored participant in the victory celebration hosted by Zhou King Wu, which may well explain the astral allusions in the poetry composed for the occasion; see the sidebar on *Qi ye* in Chapter 6 above. Chang Hong may have served the Zhou court in the sixth century BCE. Sources disagree about which rulers some of the other famous astrologers served; see *JZ*, 1154. A number are mentioned in the bibliographical chapter of *Han shu*, 30.1775.

Song, Zi Wei [sixth–fifth centuries BCE];[291] in Zheng, Pi Zao [sixth century BCE];[292] in Qi, Gan Gong [fourth century BCE]; in Chu, Tang Mo [late fourth century BCE]; in Zhao, Yin Gao [fourth century BCE]; in Wei, Shi Shenfu [fourth century BCE].[293] [1344]

The revolutions of Heaven are – thirty years makes one Small Revolution, one hundred years a Medium Revolution, and five hundred years a Major Revolution.[294] Three Major Revolutions make an Era, three Eras make a Grand Consummation. These are the great cycles. Those who manage the affairs of states must respect the Three and the Five;[295] by [looking] forward and backward a thousand years before and after [events], the continuity of the interactions between Heaven and Man will be fully in evidence.

The Prefect Grand Scribe-Astrologer [Sima Tan] computed the ancient celestial anomalies without being able to verify them in the present. In general, in the two hundred forty-two years of the *Springs and Autumns* [*Annals*] there were thirty-six solar eclipses, three broom stars appeared, and in the time of

---

[291] During the reign of Duke Jing (r. 516–477); Harper (1999, 828). No mention is made of Confucius, even though in the Mawangdui MS, "*Yao*," the following statement is placed in the mouth of the Sage: "If men of later generations doubt me, Qiu, perhaps it will be because of the *Yi* [*Book of Changes*]. I seek virtue in it, no more. I am one who shares a path with scribe-astrologers and shamans, but whose final destination is different. How can the virtuous conduct of the gentleman be intended to seek fortune?" Ibid., 826, 854 (italics mine).

[292] During the reign of Duke Jian 簡.

[293] For a discussion of the political and religious role of scribe-astrologer from the earliest times, see Jiang Xiaoyuan (1991, 195–223).

[294] One wonders whether the Simas' "Major Revolution" is the same as the 500-year period for the appearance of sages described by Mencius.

[295] These round numbers closely parallel the resonance periods that figure prominently in quarter-day astronomical systems, *sifenli*, 四分曆, of the time such as the "Grand Inception" system of 104 BCE whose creation Sima Qian supervised – the Obscuration Cycle, *bu*, 部, of seventy-six years; the Era Cycle, *ji*, 紀, of 1,520 years; and the Origin Cycle, *yuan*, 元 of 4,560 years; see Cullen (2001, 36). Here, "Three" apparently refers to the multiplier and "Five" to the resulting numerical series: 30, 100, 500, 1,500, 4,500. In what follows, however, I find no striking numerical relationships among the datable historical events Sima Qian cites, although the numerology implicit in this discussion has been taken to suggest that the Simas implicitly recognized the Qin unification as a Grand Consummation of 4,500 years; see Van Ess (2006, 103–4). It would require a separate monograph to fully explore the many nuances of the "the Three and the Five," *san wu*, 參五. The expression appears in a wide variety of contexts with meanings ranging from asterisms *Triaster* and *Topknot* (in the *Book of Odes*, "Xiao xing"), to the astrological circumstances at the time of the Zhou Conquest (cf. Zheng Xuan's comment on the ode "Da ming") and in a rhapsody by Ban Gu, or to the *Three Celestial Season Indicators* (*san chen*, 三辰) and the *Five Planets*, or to an aspect of hexagram interpretation in the *Changes*, or to rulership in the *Book of Documents* ("Gan shi" chapter). Later the term takes on a general sense of "get to the truth of something by examining it from every angle." Sima Qian uses the expression in the latter sense himself in his Postface to *The Grand Scribe's Records*: "as to interrogating appearances and getting to the facts, in evaluating all aspects there is no deficiency" (*san wu bu shi*, 參五不失); cf. Hanfei's gloss in "Ba jing." See Cullen (1996, 25, n. 25); *Huainanzi*, 9.19, 20.11, 21.2; Ames (1994, 241–8, n. 129); SCC, 407.

Duke Xiang of Song[296] stars fell like rain. The Son of Heaven was marginalized and the vassal lords violently attacked one another, the Five Hegemons arose in succession, each succeeding the other in monopolizing power. After this, the many violated the few and the large annexed the small. Qin, Chu, Wu, and Yue, Yi-Di [barbarians] all, acted as peripheral Hegemons. The Tian clan usurped power in Qi,[297] the Three Houses partitioned Jin,[298] and together they constituted the Warring States. They fought to attack and capture [territory], soldiers and weapons were mobilized in [unbroken] succession, cities and towns were repeatedly massacred, as a result suffering famine, epidemics, and extreme hardship. Ministers and rulers were all beset with anxiety and misery, growing ever more anxious in their scrutiny of baleful and auspicious [signs] and in watching for astral phenomena and *materia vitalis*. In more recent times the Twelve Vassal Lords and Seven States dealt with one another as kings, and proponents of the vertical and horizontal alliances [among the competing powers] succeeded one another without interruption. Because [Scribe-Astrologers] [Yin] Gao, Tang [Mo], Gan [De], and Shi [Shen] devoted their efforts to discoursing on the writings and traditions each in his own time, the end result was that their prognostications and verifications are disorderly, mixed up, and trifling. [1346]

The twenty-eight lodges govern the twelve provinces, and the *Handle* of the *Dipper* seconds them;[299] the origins [of this system] are archaic. For the borders [of the state of] Qin, one watches Venus [for signs] and prognosticates based on *Wolf* and *Bow*; for the borders of Wu and Yue, one watches Mars and prognosticates based on the *Balance Beam* of *Bird*;[300] for the borders of Yan and Qi, one watches Mercury and prognosticates based on *Ruins* (11) and *Roof* (12); for the borders of Song and Zheng, one watches Jupiter and prognosticates based on *Chamber* (4) and *Heart* (5); for the borders of Jin, one also watches Mercury but prognosticates based on *Triaster* (21) and *Attack* (21). [1347]

Coming to Qin's swallowing up and annexing the Three Jin, Yan, and Dai,[301] from the Yellow River and Mt. Hua southward is China. [With respect to] the area within the Four Seas, China therefore occupies the south and east as *yang* – *yang* is the Sun, Jupiter, Mars, and Saturn.[302] Prognostications [about China

---

[296] Sima Qian has in mind an event in the fifth year of Duke Min of Song.
[297] The year was 481 BCE.
[298] Formally recognized by the Zhou king in 403 BCE, the year I take as the beginning of the Warring States Period.
[299] A reference to the assignment of the seven stars of the Dipper to the seven states Qin, Chu, Liang, Wu, Yan, Zhao, Qi.
[300] I.e. *Willow* (24).
[301] Completing unification of China in 221 BCE. *Dai* refers to Shandong and the state of Qi.
[302] See Figure A.2. In the much earlier Mawangdui MS "Prognostications of the Five Planets," *yin* and *yang* are commonly used as directional terms in the sense of "north and west" and "south

are based on astral locations] south of [*Celestial*] *Street* (Taurus), while *Net* [lodge 19] governs it. To the north and west are the Hu, Mo, Yuezhi, and all those peoples who wear felt and furs and draw the bow as *yin-yin* is the Moon, Venus, and Mercury, and prognostications [are based on astral locations] north of *Street*, while *Topknot* [lodge 18] governs it. In general, China's mountain ranges and watercourses run northeastward. Their course is: head[waters] at Mt. Long in *Shu* [Sichuan] and tail in the Bohai Gulf at Mt. Jieshi. For this reason, Qin and Jin like to use weapons, and so are also prognosticated by Venus; Venus governs China. But the Hu and Mo have repeatedly invaded and despoiled, so they are uniquely prognosticated by Mercury. Mercury's comings and goings are swift and sudden, so it regularly governs the Yi-Di [barbarians]. These are the guiding principles. This [pattern] changes [depending on whether Mars or Venus] acts as host or guest. Mars means to [impose] order:[303] externally put in order the military, and internally put in order the government. Therefore, it is said: "though there be a perspicacious Son of Heaven, one must still look to where Mars is located."[304] The vassal lords grew powerful by turns but there was no one to keep a record of the disasters and anomalies noted at the time. [1348]

During the time of the First Emperor of Qin, broom stars appeared four times in the space of fifteen years,[305] at most for eighty days, the longest stretching almost across the entire sky. Afterward, Qin used military force

and east" respectively, but only with respect to relative locations of the polities of the Warring States. For example, in discussing prognostications based on Venus's position in particular astral fields, "Prognostications of the Five Planets" says, "Yue, Qi, Han, Zhao, and Wei, are *yang* with respect to Jing (Chu) and Qin. Qi is *yang* with respect to Yan, Zhao, and Wei. Wei is *yang* with respect to Han and Zhao. Han is *yang* with respect to Qin and Zhao. Qin is *yang* with respect to the *Di* barbarians. They are prognosticated on the basis of [Venus's lying] north or south, advancing or retreating"; see Liu (2004, 86). Sima Qian's use of *yin* and *yang* here with respect to the principal geographical features of the whole empire is an innovation.

[303] Following *JZ* (1157) in reading *li*, 理, for *bo*, 孛.

[304] As an example, see *Guanzi*, "Jiu shou" chapter: "Mars – where is it located?"

[305] In 240, 238 (twice), and 234 BCE. These were obviously taken to be harbingers of the incessant warfare that followed. The sighting in 240 BCE is the earliest confirmed appearance of Comet 1P/Halley, its twenty-ninth return counting retroactively from 1986. The record in *The Grand Scribe's Records*, "Basic Annals of the First Emperor of Qin," reads, "In the seventh year [240 BCE] a broom star first appeared in the east; it was then seen in the north. In the fifth month (May 26–Apr 24), it reappeared in the west. General [Meng] Ao died. Qin sent troops to attack Long, Gu, and Qingdu, then they came back to attack Ji. The broom star reappeared in the west for sixteen days"; cf. Nienhauser et al. (1994, 129). The reports from 238 BCE bracket an extended account of the rebellion of Lao Ai: "In the ninth year [238 BCE] a broom star appeared, its tail sometimes spanning the sky . . . the broom star reappeared in the west, and again in the north, pointing its tail southward from the *Dipper* for eighty days." Ibid., 129–30. The laconic note in *The Grand Scribe's Records*, "Chronological Tables of the Six Kingdoms," makes the connection between the rebellion of Lao Ai and the comet more explicit: "First Emperor of Qin, ninth year, a broom star appeared, stretching across the sky. Lao Ai fomented rebellion; his retainers were exiled to Shu. The broom star reappeared." *Shiji*, 15.752.

to successively annihilate the Six Kings,[306] unifying the Central Kingdoms, and externally repelling the Yi-Di [barbarians]. The dead were [strewn on the ground] like disordered flax-straw, and so for thirty years until Zhang and Chu[307] simultaneously rose up soldiers trampled the ground in incalculable numbers. Since the time of Chi You[308] there had been nothing to compare with this. When Xiang Yu relieved Julu,[309] *Crooked Arrows* [meteors] streamed westward. Then from east of the Mountains the vassal lords all joined together, in the west they buried the Qin soldiers alive and massacred Xianyang. When the Han [dynasty] arose, the *Five Planets* gathered in *Eastern Well*.[310] During the siege of Pingcheng,[311] the Moon was [encircled by] a seven-layered halo in *Triaster* [21] and *Net* [19]. When the Lü clan rebelled, the Sun was eclipsed and it grew dark in the daytime.[312] When the seven kingdoms of Wu and Chu revolted [154 BCE], there was a broom star several spans long and *Dogs of Heaven* [bolides] passed through the [astral] field of [the state of] Liang.[313] When battle was joined [the comet] set and the blood of the corpses flowed

---

[306] The six (with year annihilated), were Wei (225), Han (230), Zhao (222), Chu (223), Yan (222), Qi (221).

[307] Referring to the rebellions by Chen Sheng and Wu Guang starting in 209 BCE that brought down the Qin dynasty.

[308] "The Wounder," mythic inventor of (metal?) weapons and an eponymous comet; see *Chi You's Banner* below and under *Auspicious and Inauspicious Stars*. It has been suggested that Sima Qian may have had in mind a theoretical parallel with the Yellow Emperor's war against the miscreant Chi You, but this is far-fetched. For the analogy to work, in the historian's mind the First Emperor of Qin would have to be implicitly identified with the Yellow Emperor and hence a culture hero and progenitor of the Chinese people, which seems implausible; see Van Ess (2006, 103).

[309] In early 207 BCE. See Nienhauser et al. (1994, 187); Dubs (1938–55, Vol. 1, 50).

[310] May, 205 BCE; cf. also Dubs (1938–55, Vol. 1, 55–6).

[311] When the Han Dynasty founder, Gaozu, was surrounded by the Xiongnu army in 201 BCE but managed to escape.

[312] The first was in mid-afternoon on July 17, 188 BCE, followed by March 4, 181; Dubs (1938–55, Vol. 1, 185, 199). Emperor Hui died a month after the 188 BCE eclipse. For the eclipse of 181 BCE, see Nienhauser et al. (2002, 124): "On the *jichou* day [March 4] the sun was eclipsed and during the day it became dark. The Empress Dowager abhored [*sic*] this and was not pleased; she said to her attendants, 'This is because of me.'" On the lack of a report in *The Grand Scribe's Records* of the impressive daylight comet in the autumn of 185 BCE, and Sima Qian's curious entry – "There were no incidents" – for that year, see ibid., 120, n. 90.

[313] The comet was seen on at least three occasions between August 155 and late February 154 BCE; Pankenier, Liu, and de Meis (2008, 16–17). This apparition was obviously interpreted as a portent of the revolt of the feudatories against ongoing efforts at centralization pursued by the imperial court; see Dubs (1938–55, Vol. 1, 312). The astral field of Liang nominally corresponded roughly to Orion and Gemini (the southwest); the location does not specifically implicate Wu and Chu so it must signify rebellion against the empire in general. According to *Han shu*, "When *Dogs of Heaven* fall the prognostication is: 'armies will be smashed and their generals killed.' Dogs further belong to the category of guarding and resisting. Where *Dogs of Heaven* fall, it is a warning that [the place] will hold out and defend itself," which Liang indeed did; see *HS*, 26.1303–4.

beneath it. During the Epochal Brilliance [134–129 BCE] and Epochal Imperial Hunt [122–117 BCE] reign periods, *Chi You's Banner* [1349] appeared twice,[314] as long as half the sky. Afterward the imperial armies sallied forth four times, punishing the Yi-Di [barbarians] for several decades, attacking the Hu [barbarians] [Xiongnu] even more aggressively. When the Yue Kingdom was terminated,[315] Mars guarded [*Southern*] *Dipper* [8];[316] when the Chaoxian Kingdom was taken, a star became fuzzy in *River Garrisons*;[317] when the armies campaigned against Ferghana, a star became fuzzy in *Twinkling Indicator*.[318] These are the major standouts. As for the ins and outs of lesser

---

[314] Loewe (1994a). A *Han shu* record reads, "in the sixth year of the Establishment Epoch reign period of Emperor Wu of the Han Dynasty (135 BCE), eighth month, a long star emerged in the east, so long that it stretched across the sky; after thirty days it departed. The prognostication said, 'this is *Chi You's Banner*; when seen the ruler will attack the four quarters.' After this the army punished the Four Yí [barbarians] for several decades"; see *Han shu*, "Annals of Emperor Wu," 6.160. This apparition of *Chi You's Banner* was also clearly linked in Sima Qian's mind with the rebellion of Liu An, king of Huainan: "In the sixth year of the Establishment Epoch reign period (135 BCE) a broom star was seen. In the mind of the King of Huainan it was an anomaly. Someone said to the King: 'Earlier, when the army of Wu rose up [155 BCE], a broom star several feet long appeared, hence for a long time blood flowed for over 1,000 *li* [*c*.500 kilometers]. At present there is a broom star so long that it spans the sky, so the armies of the Empire should all rise in force.' In his mind, the King, considering there was no imperial heir above and [seeing that] anomalies were occurring in the Empire, increasingly desired to manufacture weapons, [siege] engines, and instruments of offensive warfare. He accumulated money with which to bribe the lords of commanderies and kingdoms, wandering knights, and those with unique talents. The various sophists who devised schemes and strategies indiscriminately fabricated rumors and flattered the King. The King was delighted, handed out even more money, and his plotting to rebel grew in earnest." See "Biographies of the [rulers of] Huainan and Hengshan," *Huainan Hengshan liezhuan, Shiji*, 118.3082. For a fascinating comparison of the Chinese and Roman accounts of the comet of 135 BCE, see Ramsey (1999) and Kronk (1997).

[315] In 112–111 BCE; Dubs (1938–55, Vol. 2, 82).

[316] *Southern Dipper* [8] correlates with the Jiang (Yangtze) and Hu (Lakes Dongting and Taihu).

[317] In 109–108 BCE. Ibid., 89, 92; Pankenier et al. (2008, 20). The *Han shu* "Monograph on the Heavenly Patterns" account reads, "In the Primal *Feng* sacrifice reign period (110–105 BCE) a star became fuzzy in *River Garrisons*. The prognostication said: '*South Garrison* is the gate of Yue, *North Garrison* is the gate of the Hu.' Subsequently, Han troops attacked and captured Chaoxian (parts of Jilin and Liaoning Provinces and northern Korea), and made it into Lelang and Xuantu Commanderies"; *HS*, 26.1306. *North Garrison, Bei shu*, 北戍, is *North River, Bei he*, 北河 (α, β, ρ Gem), and *South Garrison, Nan shu*, 北戍, is *South River, Nan he*, 南河 (α, β, ε CMi).

[318] In 104–102 BCE; see Dubs (1938–55, Vol. 2, 102); Pankenier et al. (2008, 20). The *Han shu* "Monograph on the Heavenly Patterns" account reads, "In the Grand Inception reign period [104–87 BCE], a star became fuzzy in *Twinkling Indicator*. The [*Star*] *Tradition* says: 'When a guest star guards *Twinkling Indicator*, the Man and Yi barbarians revolt and people die at the hands of their Lord.' Subsequently, Han troops attacked Ferghana and decapitated the king. *Twinkling Indicator* is the astral field of distant Yí barbarians"; *HS* 26.1306. Ma Xu then continues and extends this list of paradigmatic portents and their associated historical events down to 2 BCE. Note the incorporation of non-Chinese peoples here into the astrological scheme.

anomalies, they are too numerous to mention. Seen in this way, it has never happened that an event was not first preceded by some visible manifestation. Now, since the Han dynasty has continued the reckonings of Heaven, for celestial bodies there is Tang Du,[319] for *materia vitalis* there is Wang Shuo,[320] for predicting the harvest there is Wei Xian. Thus, when Gan [De] and Shi [Shen] studied the patterns of the *Five Planets*, only Mars [appeared to] reverse course and retrograde; [nowadays] what Mars guards as it retrogrades, together with the retrogradations of the other planets and the dimming and eclipsing of the Sun and Moon, all are grounds for prognostication. [1350] In my own perusal of the scribal accounts, I have examined into their events and movements, and in the [past] one hundred years there has not been one instance when the *Five Planets* have appeared and not reversed course and retrograded. On reversing and retrograding [the planets] regularly become large and full and change color.[321] The dimming and eclipsing of the Sun and Moon and their movements to the south and north have their seasons. This is the general rule. Thus, the *Purple Palace, Chamber* (4) and *Heart* (5), *Weight* and *Balance Beam* (25–6), *Mineral Spring, Ruins* (11) and *Roof* (12), are the *Sectional Asterisms* within the array of lodges,[322] these are the seats of *Heaven's Five Offices*. They are the warp – immobile, their dimensions differ but their separation is constant. *Water, Fire, Metal, Wood*, and Saturn, these five stars are *Heaven's Five Assistants*. They are the weft – their appearing and disappearing have their seasons, where they pass by and [where they] gain and regress in their movements all have their [regular] measures.[323] [1350]

---

[319] Tang Du was a "master of occult arts," *fāngshì*, 方士, from a venerable family of the pre-imperial kingdom of Chu; see Loewe (2000, 502). In his Postface to *The Grand Scribe's Records*, Sima Qian mentions that his father Sima Tan learned the celestial bodies from Tang Du, whose positional observations of the stars later played a central role in the Grand Inception calendar reform of 104 BCE. Wang Shuo and Wei Xian were also contemporaries.

[320] Loewe (2002, 553). Wang was identified above as a specialist in observing the *materia vitalis*. Following two inauspicious appearances of a comet in 110 BCE, a mitigating observation of Saturn by Wang Shuo (recorded in *HS*, 25A.1236) was interpreted as signifying celestial approval of Emperor Wu's reinstatement of the *Feng* and *Shan* sacrifices. See Kern (2000, 24).

[321] Sima Qian here lays claim to the discovery that retrograde motion is a regular feature of the movements of all the planets and hence not anomalous as previously believed. He also observed that the planets brighten during retrogradation, and that Venus can be observed to transit the meridian in daylight. This is one of several instances where Sima Qian intimates that he and his father retrospectively computed movements of celestial bodies from ancient records. He does so again below and in the Postface.

[322] These are highlighted as representing the Five (cardinal) Palaces of the heavens. The *Purple Palace* is the Center; *Chamber* and *Heart* are the East, *Weight* and *Balance Beam* are the South, *Heavenly Mineral Spring* is the West, and *Ruins* and *Roof* are the North. However, in *Chunqiu fanlu*, "Fengben" chapter, the *buxing*, "sectional stars," are said to number 300, so that Dong Zhongshu must be referring to individual asterisms, which is also the usage in *Guoyu*.

[323] The substitution of Saturn for *Earth* here is odd. The weaving metaphor indicates that his mental image is of longitudinal sections converging on the *Purple Palace* at the Pole, cross-cut by the movements of the Sun and the Moon and the shuttle-like motion of the planets, producing

If there is a solar anomaly, practice virtue; if there is a lunar anomaly, reduce punishments; if there is a planetary anomaly, join in harmony. In all cases of celestial anomalies, if the [regular] measures are overstepped, then prognosticate. If the lord of the state is strong and great, the virtuous will prosper, and the weak and small, fawning and false will perish. The ultimate superior cultivates virtue, the next level practices [good] government, the next level carries out relief efforts, the next level conducts expiatory rites, directly below that there is nothing [to be done].[324] Now, changes in the constant stars are rarely seen, but prognostications concerning the *Three Luminaries*[325] are frequently applied. The Sun and Moon, with their halos and blemishes, clouds and winds, these are the transient [effects] of celestial *materia vitalis*, and their production and appearance also have their major cycles. Moreover, with regard to the vicissitudes of governmental affairs, they are the most proximate tallies of [the interaction] between Heaven and Man. These Five are the responsive movements of Heaven. Those who deal with the Regularities of Heaven must comprehend the Three and the Five.[326] [Having explored events] from beginning to end, from ancient times to the present, [we have] looked deeply into the vicissitudes of the times,[327] examining the minute and the large scale, [so that exposition of] the Celestial Offices is now complete.

* * *

When the Green Emperor promulgated virtue, Heaven's Portal[328] *opened for him. When the Red Emperor promulgated virtue, Heaven's Prison*[329] *emptied for him. When the Yellow Emperor promulgated virtue, a Heavenly Anomaly*

---

a reticulated "woven" scheme (see Chapter 12). See *Huainanzi*, "Heavenly Patterns" chapter, where the *Five Offices* and *Six Departments* are identified slightly differently. *Huainanzi's Six Departments* correspond to the *Five Sectional Asterisms* in *TGS*, with some variation in the labels applied to the same areas of the sky: *Purple Palace, Grand Enclosure, Chariot Pole* (⌒ *Weight* and *Balance Beam*), *Heavenly Mineral Spring, Heavenly Slope* (⌒ Yan Bo, 阿伯, *Heart*) and *Four Guardians (Chamber)*; see Major (1993, 80), who slips in writing *Taiweiyuan* for *Ziweiyuan*.

[324]   That is, prognostications have no application.   [325]   Sun, Moon, and planets.

[326]   This is the second appearance of "the Three and the Five." The first time it appeared above it comprehended astrological cycles. According to *SY*, here the reference is to the *Three Seasonal Indicators, san chen*, 三辰 (*Northern Dipper, Great Fire, Orion*), and the Five Planets. However, the immediately antecedent mention of the *Three Luminaries* and "these Five" (Sun, Moon, halos/defects, clouds, wind), the "transient effects of celestial *materia vitalis*," makes it clear that Sima Qian is referring to them.

[327]   Following Wang Liqi (1988, 958, 973). This is perhaps an allusion to a line in hexagram Fen in the *Changes*, where it says, "observe the patterns of Heaven to understand the seasonal changes [*shi bian*]." But here *shi bian* clearly refers to historical change over the long term.

[328]   Between the two stars of *Horns* (1).

[329]   *Tianlao*, 天牢, is a *Prison* for the nobility, according to *Pacing the Heavens, Heaven's Prison* is the six stars beside ψ UMa, which would seem to point to ω, 47, 49, 55, 57, 56 UMa. Sun and Kistemaker (*SK*, 153) identify a ring of six stars marked by 44 UMa.

arose for him. When the wind blows from the northwest it has to be on a geng or a xin day, and when it comes five times in a single autumn there is a general amnesty; when it comes three times there is a minor amnesty. When the White Emperor promulgated virtue, it was on the twentieth and twenty-first days of the Standard Month and the Moon was encircled by a halo; it was a year of general amnesty, said to be [due to] Great Yang.[330] *When the Black Emperor promulgated virtue, Heaven's Gate*[331] *moved for him. When Heaven [Supernal Lord] promulgates virtue, the accession year of the Son of Heaven changes;*[332] *when [Heaven] does not promulgate virtue, wind and rain shatter stones. The Three Flights of Stairs*[333] *and Three Balance Beams*[334] *are Celestial Courts. When a guest star appears in a Celestial Court there will be abnormal commands.*

---

[330] According to Ho (1966, 40), *Great Yang* could refer to a massing of four planets.
[331] ζ Tau.   [332] That is, it is a succession year.   [333] ι, κ, λ, μ, ν, χ UMa.
[334] Commentators disagree as to the identity of the *Three Balance Beams*. Of the asterisms to which the appellation *Balance Beam* has been applied, only the *Grand Tenuity Enclosure* is usually also characterized as a *Celestial Court*; see *KZ*, 95. The italicized passage is traditionally appended to the end of the *TGS*, but the style and heterogeneous content suggest that these are displaced slips, possibly even extraneous to the original text, which have been tacked on here for convenience.

Glossary

A

Ai sui 哀歲

B

Ba 巴 kingdom
baji 八極
bi 璧
Biji mingjia dao du biji congshu 筆記名家導讀筆記叢書
Bohai 渤海
bu wen 卜問
bu zhi ji nian 不知紀年
Ban Zhao 班昭 (*c*.45–*c*.117)
bao 寶
bao yi 保乂 "protects and directs"
Bao xun 保訓
beidou ji 北斗祭
Bei Qi 北齊 (550–77)
Bei Zhou 北周 (557–88)
bie 鱉
Bin 豳
bin 賓 sacrifice
Bianwu 辯物
bing yue 冰月
Bing zhou 并州
bingzi 丙子

C

ce 冊
Cai 蔡

can wang 參望
Cao 曹
Cao Xiangcheng 曹相成
Chuqiu 楚邱
chun bing 春餅
Chen 陳
chen zheng 臣正
chou 丑
chou zi hai 丑子亥
chang 昌
Chang E 嫦娥 (aka Heng E 姮娥)
Changle gong 長樂宮
Chong 崇

D

Da gui 大龜
Da huang jing 大荒經
Da Peng 大彭
Dasikong cun 大司空村
Da Siming 大司命
Datong 大同
Da zhao 大招
di 禘 *di*-sacrifice
di 蒂 plant calyx
du 度
Dujiang 渡江
Du Yu 杜預 (222–84)
Dai 代
Dan 旦
Daoguang 道光 (1821–50)
Dao tong 道統
dou gang 斗綱
dou yuan gui zhi shou 斗元龜之首
dian 典
Ding zhi fang zhong, zuo yu Chu Gong; kui zhi yi ri, zuo yu Chu Shi; sheng bi xu yi, yi wang Chu yi; wang Chu yu Tang, ying shan yu jing 定 之 方 中, 作於楚宮; 揆之以日, 作於楚室, 昇彼虛矣, 以望楚矣; 望楚與堂, 景山與京
Ding, xing ye. Ding zhong er ke yi zuo shi, gu yue Yingshi. qi xing you shi xiang ye, gu *Tianguan shu* zhu miao. 定星也. 定中而可以作室, 故曰營室. 其星有室象也, 故天官書主廟.

ding, zheng ye, zuo gong shi jie yi Yingshi zhong wei zheng 定, 正也. 作宮室 皆以營室中為正.

Dong Yin 董因

Dongping xian 東平縣

Dong ying 東營

Duan Yucai 段玉裁 (1735–1815)

E

E-pang (A-pang) 阿房

er 而

er yue er long tai tou 二月二龍抬頭

F

Fa 發 (personal name of King Wu)

fu Han zhi wei tian shu zhe, xing ze Tang Du,qi ze Wang Shuo, zhan sui ze Wei Xian 夫漢之為天數者, 星則唐都, 氣則王朔, 占歲則魏鮮.

fan tu gong, Long jian er bi wu, jie shi ye. Huo jian er zhi yong. Shui hun zheng er zai. Ri zhi er bi 凡土功, 龍見而畢務, 戒事也. 火見而致用, 水 昏正而栽, 日至而畢

fan wuxing suo ju xiu, qi guo wang tianxia 凡五星所聚宿, 其國王天下

fan xing yi she 返行一舍

fei shi luan ri 廢時亂日

Fen he 汾河

fang 方

fangshi 方士

Fang Xuanling 房玄齡 (579–648)

fang zhong 方中

feng 丰丰

Fengchu cun 鳳雛村

Fengsu tong 風俗通

Fengxiang xian 鳳翔縣

G

gu 故

gu du 古度

Guyi 榖邑

gu zhe, Paoxi shi zhi wang tianxia, yang ze guan xiang yu tian, fu ze guan fa yu di 古者, 包犧 氏之王天下也, 仰則觀象於天, 俯則觀法於地

Gaochang 高昌

Gaoliang 高粱
Gao You 高誘 (168–212)
Gao Yun 高允 (390–487)
gou 鉤
gui 珪
guihai 癸亥
guisi 癸巳
guiwei 癸未
Guo Fei 郭棐 (1529–1605)
Guo Pu 郭璞 (276–324)
Guo Shoujing 郭守敬 (1231–1316)
gang 剛
gang ji 綱紀
gang ling 綱領
gang mu 綱目
gengxu 庚戌
gong (guang) 觥
Gong ji ding zhai 公既定宅
Gong Liu 公刘 (c. thirteenth century BCE)
Gongsun Chou 公孙丑
Gongzi Jia 公子賈
Gongzi Zhi 公子縶 (fl. c.636 BCE)
Guan 關
Guandong 關東
guan xiang shou shi 觀象授時
Guanzhong 關中
guang 桄 = 絖 headle-rod

H

He 河
he cao bu huang 何草不黃
Hetu jiang xiang 河圖降象
Hetu wei kuo di xiang 河图纬括地象
He Xiu 何休 (129–182 CE)
He you ren Luo zhong chun gan 和友人洛中春感
Hu 胡
Hu Yan 狐偃
hai 亥
Han dong 函東
Han Jing di 漢景帝
Hanguguan 函谷關

Han jiu yi 漢舊儀
Hanshan 含山
Hanzhong 漢中
Houtu 后土
hsi (xi) 覡
Hua–Xia 華夏
hui 回
hun zheng 昏正
huo li 火曆
heng 衡
Heng qiao 橫橋
Hongwu 洪武 (r. 1368–98)
Huai 淮
Huainan Hengshan lie zhuan 淮南衡山列傳
Huangzhong River 湟中河
Huangzi 黃子 (second century BCE)
Huang Zongxi 黃宗羲 (1610–95)

J

ji 紀 lead thread; pendant
Ji ce kao 紀策考
ji mu 紀母
ji nian 紀年
Ji Rui 冀芮
jisi 己巳
ji ying nai gang, xiang qi yin yang 既景乃岡, 相其陰陽
Ji zhou 冀州
ju 矩
Ju 莒 kingdom
juren 舉人
Ju zi 娵觜
Ju zi zhi kou, Yingshi Dongbi ye 娵觜之口, 營室東壁也
Jiajing 嘉靖
jiashen 甲申
jiazi 甲子
Jieshishan 碣石山
jinshi 進士
Jin Xianmin 金獻民 (*jinshi* 1484)
Jiushuai 臼衰
jun zhe, yi ye, yi zheng er ying zheng 君者, 儀也, 儀正而景正
Jundao 君道

junzi 君子
jian 監
jian 碱/鹼 alkaline
jian 鍵 gate latch
Jian'an 建安
Jianchi 碱池
Jianxing 建星
Jiao te sheng 郊特牲
jing 經 warp; canonical work
Jing 荊 (kingdom of Chu)
jing shen 精神
jing shi Ling tai, jing zhi ying zhi 經始靈台, 經之營之
jing shou min shi 敬授民時
jing wei ge chang shiwu li 經緯各長十五里
jing ying 經營
jing zhi ying zhi 經之營之
Jiang Han 江汉
Jiang lou 降婁
Jiangxi tong zhi 江西通志
Jiangzhai 姜寨

K

Kaiyang 開陽
Kaiyuan 開元 (713–41)
Kao bu wei wang, zhai shi Haojing, wei gui zheng zhi, Wu Wang cheng zhi, Wu Wang cheng zai 考卜維王, 宅是鎬京, 維龜正之, 武王成之, 武王丞哉
kou 口
kui 窺
kui shu 魁樞
kui zhi yi ri 窺之以日
Kunlun 崑崙
kongque 孔雀
Kong Yingda 孔穎達 (574–648)
Kuanglin 曠林

L

La yue 臘月
Le Huo 樂護 (fl. *c.*1524)
li 曆 calendar, astronomical system

li 理 pattern, principle
Li 離 (hexagram)
Li (state) 黎
Li Bing 李冰 (third century BCE)
Li bu shang shu 禮部尚書
li gong 離宮
Li gui 利簋
Li qi 禮器
Li Si 李斯 (d. 208 BCE)
Li sao 離騷
Li Shimin 李世民 (599–649)
Li shu 曆書
li xue 理學
Li yun 禮運
li xiang 歷象
Lu 魯
Lu Jia 陸賈 (*c.*228–*c.*140 BCE)
Lü li zhi 律曆志
Luliu 盧柳
Lü Sheng 呂甥 (fl. *c.*645 BCE)
Lü xing 呂刑
lai xun 来旬
Liu An 劉安 (179–122 BCE)
Liu yi zhi yi lu 六藝之一録
Liu Yu 劉宇 (d. 20 BCE)
Liu Zongyuan 柳宗元 (773–819)
Loulan 婁蘭
Lunheng 論衡
luo 羅
Luoshan 羅山
Luoshu 洛書
luo zhi yi ji gang 羅之以紀綱
Lian 濂
Lianyu 歃盂
lian zhu 連珠
Liao 遼
ling 令
Linghu 令狐
Lingjiatan 凌家灘
ling tai 靈台 *Ling Xian* 靈憲
Lingyou 靈囿
Lingzhao 靈沼

Ling Zhoujiu 伶洲鳩
Liang Chenghui 梁成恢 (fl. *c*.86 BCE)
Liang zhou 梁州

M

Ma Xu 馬續 (70–141)
Mi 密 state
mu 目
Muye 牧野
Man-Yi 蠻夷
Mao Heng 毛亨 (third century BCE)
mao 卯
Meigu 昧谷
Min 閩
min si wei Fang, ri chen yu wei, xing sui zhi Si 民祀唯房,日辰于維, 興歲之
    駟
Minyi 緡邑
Mengjin 孟津
mengshu 盟書
Mengxi bitan 夢溪筆談
Mian 綿
mianmian wo wang, gang ji si fang 勉勉我王, 綱紀四方
mianmian 勉勉
Miao 苗
Ming ming Shangdi 明明上帝
ming ming wei chang, wei de wei ming 明明未常, 惟德為明
Ming Qing biji cong kan 明清筆記叢刊
Ming Ru xue an 明儒學案
Mingtang 明堂
Mingtang wei 明堂緯

N

nan bei wei jing, dong xi yue wei 南北曰經, 東西曰緯
Nan gong zou gao 南宮奏稿 (*c*.1535)
Nanyue 南越
Nü Wa 女媧
nuo 諾
Nongshu 農書
Nongxiang 農祥

Nongxiang chen zheng, ri yue di yu Tianmiao, tu mai nai fa 農祥晨正, 日月
底於天廟, 土脈乃發

O

Ouyang Yuan 歐陽元 (1273–1357)

P

Panlongcheng 盤龍城
Peng 彭

Q

Qi 齊 kingdom
Qishan 岐山
Quqie 胠篋
Quwo 曲沃
Qiande 乾德 (963–7)
Qian qiu zha ji 潛邱劄記
Qingjiang 清江
Qingyang 青陽
Qingzhou 青州
quan 權

R

ri chen 日辰
ri shu 日書
ri yue suo chu ru zhi shan 日月所出入之山
Ri zhe lie zhuan 日者列傳
ru 辱
Ruyin Hou 汝陰侯
rou 柔
Ruizong 睿宗 (r. 710–12)

S

sabao 薩保
si 巳
Sigushan 四谷山
Sima Guang 司馬光 (1019–86)

Sima Zhen 司馬貞 (fl. *c*.712)
Si yue 四月
Sufutun 蘇阜屯
Su Song 蘇頌 (1020–1101)
sanguang 三光
San he 三河
shè 社 altar to the soil
shè 舍 lodging
shetige 攝提格
shì 勢 inherent spatio-temporal advantage
shì 式 mantic astrolabe
shí 時 season, timeliness, time
shí 食 eat; eclipse
shi chen 時辰
shi ci 失次
shi mai 食麥
shi shou 時首
Shi suo 石索
shi wei 時尾
Shi wei 豕韋
Shi xie 飾邪
Shi xun 時訓
shi ying 始營
shi yue ye keyi zhu cheng jian du yi 是月也可以築城建都邑
Shijiahe 石家河
Shiyuan 始元
shu fa 数法
shushu zhi xue 數術之學
Shuyuan 叔元
sui ci 歲次
Sangquan 桑泉
shen 申 ninth earthly branch; place name
shen 蜃 mollusc
shen sha 神煞
Shen Yue 沈約 (441–513)
Shen Quanqi 沈佺期 (650–729)
shou 授
Shouchun 壽春
shou ming yuan nian 受命元年
Shou shi li 授時曆
Shou shi zhi zhang huo fa zhi tu 授時指掌活法之圖
Shui di 水地
Shuihudi 睡虎地

Song 宋
Song shan 嵩山
Shanglin 上林
Shangqiu 商丘
Shang Song 商頌
Shang tong 尚同
Shang xing 商星
Shang yi yiyi 商邑翼翼
Shang Zhòu 商紂
Shuangdun 雙墩

T

Taershan 塔兒山
Tu shu bian 圖書編
tai 泰
Taibao 太保
Taibao zhao zhi yu Luo bu zhai, jue ji de bu, ze jingying 太保朝至於洛卜宅, 厥既得卜, 則經營
Taichu 太初
Taihao 太昊
Taihe 太和 (477–99)
Tai Gong Wang 太公望 (eleventh century BCE)
Tai ji Dian 太極殿
Tai Shu 太叔
Tan gong shang 檀弓上
Tang 唐
Tang Gu 唐固
Tang Shuyu 唐叔虞 (eleventh century BCE)
Tianbao 天寶 (742–56)
Tian chui xiang, xian ji xiong, er sheng ren xiang zhi. He chu tu, Luo chu shu er sheng ren ze zhi 天垂象, 見吉凶, 而聖人象之. 河出圖, 雒出書, 而聖人則之.
Tian gui 天黿
Tian He 田何
tian ji 天紀
Tianjin 天津
tian ming zhi wei xing 天命之謂性
tian ren zhi ji 天人之際
tian shi 大室 Hall of Heaven
tian shu 天數 heavenly reckoning (calculations)
tian shu 天樞 Pivot of Heaven

tian wei 天威
Tian wen xun 天文訓
Tian wen zhi 天文志
Tian wen 天問 Heaven Questioned
tian wen 天文 heavenly pattern (reading)
tian zhi gang 天之綱
Tian zhi 天志
tian xian 天顯 Heaven's bright manifestation
Tian Yuan 天黿
ting (ding) 鼎 cauldron, tripod
tong mu 統母
Tong wen bei kao 同文備考 (1557)

W

wa dang 瓦當
Wo kou 倭寇
wo wen 渦紋
wu 午 seventh earthly branch
Wu 吳 kingdom
wu bu 五步
wu chen zheng 五臣正
wu de 五德
Wu Ding 武丁
wu gong chen 五公正
wu hou 物候
Wu Hu ding 吳虎鼎
Wulu 五鹿
Wu Moru 吳莫如 (fl. *c.*12 BCE)
Wu si Wei ding 五祀衛鼎
Wu Wang ba nian zheng fa Qi 武王八年征伐耆
wu wei 五緯
wu wei cuo xing 五緯錯行
Wu xing zhi 五行志
wu xing bing jian 五星並見
wu xing cuo xing 五星錯行
wu xing ju yu yi she 五星聚於一舍
Wu Yipeng 吳一鵬 (1460–1542)
Wu Yue 吳樾
waiguo 外國
Wan sui yue 萬歲樂
wēi 威 awe, overawe

wéi 惟 (copula)
wèi 未 not yet
wéi 維 cord
wéi 緯 weft
Wèi 魏 kingdom
Wei Bing 韋兵
Wei Cide 魏慈德
Wei Hong 衛宏
Wei Shangchen 韋商臣 (*jinshi* 1522)
wei wei 薑薑
Wei Xian 魏鮮 (second century BCE)
Wei Xianzi 魏獻子 (d. 509 BCE)
Wei Zheng Gong jian lu 魏鄭公諫錄
wen 文 culture, literature, text
wen 紋 fabric pattern
wen 聞 to hear
Wen Ying 文瑩 (fl. *c*.1060)
Wen yuan ge da xue shi 文淵閣大學士
wenzu 文祖
Wey 衛
Wey Wen Gong 衛文公 (seventh century BCE)
Wang Fangqing 王方慶 (fl. *c*.680)
Wang Feng 王鳳 (d. 22 BCE)
Wang Yi 王逸 (89–158 CE)
Wang Yingdian 王應電 (fl. *c*.1557)
Wang Zhen 王禎 (1271–1368)
Wangzi Ying 王子嬴 (d. 206 BCE)

X

Xi Bo 西伯
Xi Gong 僖公 (659–627 BCE)
Xi He 羲和
Xi Meng 郗萌 (first century CE)
Xi Xia 西夏
Xizhi River 析支河
Xu 徐 surname, province
xu 戌 heavenly stem
Xu Shen 許慎 (*c*.58–147 BCE)
Xia Hou shi 夏侯氏
Xia Yan 夏言 (Xia Wanchun 夏完淳, 1631–47)

Xinmi 新密
xinyou 辛酉
Xinzhai 新砦
xiu 宿
xun 旬
Xun 郇 locality in Jinxian 咸/鹹
Xian di zhuan 獻帝傳
xian xian 現顯
Xian Zhen 先軫 (d. *c.*627 BCE)
Xiao Gong 孝公 (381–338 BCE)
Xiao jing yuan shen qi 孝經援神契
Xiao kai 小開
Xiaotun 小屯
Xiao ya 小雅
xing bo (bei) 星孛
Xingde 刑德
xing yi dou zhuan 星移斗轉
xing zhi ji 星之紀
xuan 玄
Xuan gong 玄宮
xuan gui 玄珪
xuan huang 玄黃
Xuanji 璇璣
Xuan ming 玄冥
Xuan tian 玄天
Xuan xiao 玄枵
Xuan Yuan 軒轅
Xuan yuan liu wang zhuan 宣元六王傳
Xiangfen xian 襄汾縣
Xiangniu 襄牛
xiang tian 象天
Xiongnu 匈奴

Y

Yan 燕
yi ding ji xin gui 乙丁己辛癸
yi ri wei zheng, ze ri. cong ri, zheng hui yi 以日為正, 則日. 從日, 正會意.
Yi Xing 一行 (683–727)
Yi yao 疑耀

Yi Zu 彝族
Yizhou 冀州
yi zheng zhong dong 以正仲冬
Yi zhuan Yang shi 易傳楊氏
Yu hu qing hua 玉壺清話
Yu gong 禹貢
Yugui 輿鬼
Yu Hong 虞弘 (d. 592)
Yu long 御龍
Yupu 棫樸
Yuzhou 豫州
yu zhou 宇宙
Yan Bo 閼伯
Yan Ruoqu 閻若璩 (1635–1704)
Yan Shigu 顏師古 (581–645)
Yanzhou 兗州
Yaodian 堯典
Yaoguang 搖光
yao ji 腰機
Yao lue 要略
Yao wen 堯問
Yaoshan 瑤山
yin 寅 earthly branch
Yin 殷 Shang Dynasty capital
Yin wu 殷武
yin zhi 飲至
You tong zhi fu 幽通之賦
You zhou 幽州
Yue 樾
yue 約 contract
Yue jue shu 越絕書
yang 央
yang jian 洋鹼
Yang Tinghe 楊廷和 (1459–1529)
Yangzhou 揚州
ying 營
Ying gong qi zhong 營宮其中, 土功其始
Yingshi 營室
Yingshi er xing wei xi bi, yu Dongbi er xing he er wei si, qi xing kai fang si
    kou 營室二星為西壁, 與東壁二星合而為四, 其形開方似口
Yingshi wei zhi Ding 營室謂之定
Yingshi zhe, zhu guan tai yang qi er chan zhi 營室者, 主營胎陽氣而產之

Yong 雍
Yongle 永樂
Yuanguang 元光 (134–129 BCE)
Yuan he junxian tu zhi 元和郡縣圖志
yuanzi 元子

Z

za ju wei rao er ju 匝居謂圍繞而居
za ju ye 市 (匝), 居也
ze lu 澤鹵
zi 子 earthly branch
Zi Chan 子產 (d. 522 BCE)
zi wu da dao 子午大道
Zi Yu 子玉 (d. 632 BCE)
Zu Jia 祖甲 (twelfth century BCE)
zu yi wei ji gang 足以為紀綱
zhi 質
zhi can ye 直參也
Zhi gong 至公
zhi jiang 陟降
zhi nan 指南
Zhi Nü zhi ji 織女之紀
zhu que 朱雀
Zhu Yuanzhang 朱元璋 (1328–98)
Zuixi 觜觿
Zhuzi lue 諸子略
zuo miao yiyi 作廟翼翼
Zuo Qiuming 左丘明 (fourth century BCE)
zhan jing 占經
zhan shu 占書
Zhao 趙
Zhao ling 昭陵
zhao xi 朝夕
zhen 貞
Zhou Gong zhao zhi yu Luo, ze da guan yu Luo ying 周公朝至於洛, 則達觀
    於新邑營
Zhou song 周頌
Zhou Xiang Gong 周襄公 (seventh century BCE)
Zhòu Xin 紂辛 (d. 1046 BCE)
Zhou yi 周易
Zhou You Wang 周幽王 (795–771 BCE)

zong 綜
zong kuang 綜絖
Zong zhang 總章
zhang 張
Zhang Er 張耳 (d. 202 BCE)
Zhang Huang 章潢 (1527–1608)
zhang tian shi zhe 掌天時者
Zhang Xuan 張萱 (1558–1641)
zheng 正 correct, standard, upright
Zheng 鄭
zheng, shi ye. shi, zhi ye. cong ri, zheng. 正, 是也. 是, 直也. 從日, 正.
Zheng Guo 鄭國 (fl. 246 BCE)
Zheng Qiao 鄭樵 (1104–62)
zheng tong 正統
zhong 中
Zhong Kang 仲康
zhong Shang 中商
Zhongyue 中岳(嶽)
Zhongzong 中宗
Zhuan Xu 顓頊

References

Abrabanel, Isaac, 1960. *Ma'ayney hayyešu'ah (The Wells of Salvation)* (rpt., Tel Aviv: Elisha, first published 1497).

Ackerman, Diane, 2011. "Planets in the Sky with Diamonds," *New York Times*, October 2, "Sunday Review," 5.

Allan, Sarah, 1981. *The Heir and the Sage: Dynastic Legend in Early China* (San Francisco: Chinese Materials Research Center).

1984. "The Myth of the Xia," *Journal of the Royal Asiatic Society* (new series) 116, 242–56.

1991. *The Shape of the Turtle: Myth, Art and Cosmos in Early China* (Albany: State University of New York).

2007. "Erlitou and the Formation of Chinese Civilization," *Journal of Asian Studies* 66.2, 461–96.

2009. "On the Identity of Shang Di 上帝 and the Origin of the Concept of a Celestial Mandate (*Tian Ming* 天命)," *Early China* 31, 1–46.

2010. "T'ien and Shang Ti in Pre-Han China," *Acta Asiatica* 98, 1–18.

Allen, Melinda S., 2010. "East Polynesia," in Ian Lilley (co-ordinator), *Early Human Expansion and Innovation in the Pacific* (Paris: ICOMOS), 137–82.

Ames, Roger T., 1994. *The Art of Rulership* (Albany: State University of New York).

Ascher, M. and Ascher, R., 1975. "The Quipu as Visible Language," *Visible Language* 9.4, 329–56.

Atwood, Christopher, 1991. "Life in Third–Fourth Century Cadh'ota: A Survey of Information Gathered from the Prakrit Documents Found North of Minfeng (Niya)," *Central Asiatic Journal* 35.3–4, 161–99.

Aveni, Anthony F., 2002. *Empires of Time: Calendars, Clocks, and Cultures* (Boulder: University Press of Colorado).

(ed.), 2008a. *Foundations of New World Cultural Astronomy: A Reader with Commentary* (Boulder: University Press of Colorado).

2008b. *People and the Sky: Our Ancestors and the Cosmos* (New York: Thames and Hudson).

Aveni, Anthony F. and Gibbs, Sharon L., 1976. "On the Orientation of Precolumbian Buildings in Central Mexico," *American Antiquity* 41.4 (October), 510–17.

Bagley, Robert, 1987. *Shang Ritual Bronzes in the Arthur M. Sackler Collections* (Washington, DC, and Cambridge, MA: Arthur M. Sackler Foundation).

1993. "Meaning and Explanation," *Archives of Asian Art* 46, 6–26.

1999. "Shang Archaeology," in Michael Loewe and Edward L. Shaughnessy (eds.), *The Cambridge History of Ancient China, from the Origins of Civilization to 221 BC* (Cambridge: Cambridge University Press), 124–231.

2004. "Anyang Writing and the Origin of the Chinese Writing System," in S.D. Houston (ed.), *The First Writing: Script Invention as History and Process* (Cambridge: Cambridge University Press), 190–249.

Baillie, Mike, 2000. *Exodus to Arthur: Catastrophic Encounters with Comets* (London: B.T. Batsford).

Balme, J., Davidson, I., McDonald, J., Stern, N., and Veth, P., 2009. "Symbolic Behaviour and the Peopling of the Southern Arc Route to Australia," *Quaternary International* 202, 59–68.

Balter, Michael, 2011. "South African Cave Slowly Shares Secrets of Human Culture," *Science* 332 (June 10), 1260–1.

Ban Dawei 班大為 (David W. Pankenier), 2008. "Beiji de faxian yu yingyong" 北極的發現與應用, *Ziran kexueshi yanjiu* 自然科學史研究 27.3, 281–300.

2011. "Zai tan beiji jianshi yu di zi de qiyuan" 再談北極簡史與「帝」字的起源," in Patricia Ebrey 伊沛霞 and Yao Ping 姚平 (eds.), *Xifang Zhongguo shi yanjiu luncong* 西方中國史研究論叢, Vol. 1, *Gudai yanjiu* 古代史研究 (ed. Chen Zhi 陳致) (Shanghai: Shanghai guji), 199–238.

Barber, Elizabeth Wayland and Barber, Paul T., 2004. *When They Severed Earth from Sky: How the Human Mind Shapes Myth* (Princeton: Princeton University Press).

Barillà, Enzo, n.d. "Drawing Nigh to February 1524: The Spate of Fear," available at www.enzobarilla.eu/estero/ING%20aspettando%20il%20febbraio%20del%201524.pdf.

Barnard, Mary, 1986. *Time and the White Tigress* (Portland: Breitenbush).

Barnard, Noel, 1993. "Astronomical Data from Ancient Chinese Records: The Requirements of Historical Research Methodology," *East Asian History* 6, 47–74.

Basilov, V.N., 1989. "Chosen by the Spirits," *Anthropology & Archeology of Eurasia* 28.1, 9–37.

Baudez, Claude and Picasso, Sydney, 1992. *Lost Cities of the Maya* (New York: Harry N. Abrams).

Baxter, William H., 1992. *A Handbook of Old Chinese Phonology* (Berlin: Mouton de Gruyter).

Baylor, Michael, 1993. *Revelation and Revolution: Basic Writings of Thomas Müntzer* (Bethlehem: Lehigh University Press).

Beasley, W.G. and Pulleyblank, E.G. (eds.), 1961. *Historians of China and Japan* (London: Oxford University Press).

Bellah, Robert N., 2011. *Religion in Human Evolution: From the Paleolithic to the Axial Age* (Cambridge, MA: Harvard University Press).

Belmonte, Juan A., 2001. "On the Orientation of the Old Kingdom Egyptian Pyramids," *Archaeoastronomy* 26, 1–20.

Belmonte, Juan A., Gonzalez Garcia, A.C., Shaltout, M., Fekri, M., and Miranda, N., 2008. "From Umm al Qab to Biban al Muluk: The Orientation of Royal Tombs in Ancient Egypt," *Archaeologica Baltica* 10, 22–33.

Bennett, Steven, 1978. "Patterns of the Sky and Earth: A Chinese Science of Applied Cosmology," *Chinese Science* 3, 1–26.

Berger, Peter L., 1990. *The Sacred Canopy: Elements of a Sociological Theory of Religion* (New York: Anchor Doubleday).

Bertola, Francesco, 2003. *Via Lactea: Un percorso nel cielo e nella storia dell'uomo* (Rome: Biblos).

Best, Elsdon, 1921. "Polynesian Mnemonics: Notes on the Use of the Quipus in Polynesia in Former Times; Also Some Account of the Introduction of the Art of Writing," *New Zealand Journal of Science and Technology* 4.2, 67–74.

Beyton-Davies, Paul, 2007. "Informatics and the Inca," *International Journal of Information Management* 27, 306–18.

Bezold, Carl, 1919. "Sze-ma Ts'ien und die babylonische Astrologie," *Ostasiatische Zeitschrift* 8, 42–9.

Bi, Yuan 畢沅, 1974. *Lüshi chunqiu xin jiaozheng* 呂氏春秋新校正, *Xinbian Zhuzi jicheng* 新编诸子集成 ed., Vol. 7 (rpt., Taipei: Shijie).

Bielenstein, Hans, 1950. "An Interpretation of the Portents in the Ts'ien Han Shu," *Bulletin of the Museum of Far Eastern Antiquities* 22, 127–43.

Biot, Édouard (trans.), 1975. *Le Tcheou-li ou Rites des Tcheou*, 3 vols. (Taipei: Ch'eng-wen, first published Paris: Imprimerie nationale, 1851).

Birrell, Anne M., 1999. *Chinese Mythology: An Introduction* (Baltimore: Johns Hopkins University Press).

Bo, Shuren 薄樹人, 1981. "Sima Qian – wo guo weida de tianwenxue jia" 司馬遷–我國偉大的天文學家, *Ziran zazhi* 自然雜誌 4.9, 685–8.

Bodde, Derk, 1961. "Myths of Ancient China," in S.N. Kramer (ed.), *Mythologies of the Ancient World* (Garden City, NY: Doubleday), 372–6.

1975. *Festivals in Classical China: New Year and Other Annual Observances during the Han Dynasty (206 B.C.–A.D. 220)* (Princeton: Princeton University Press).

1981. "The Chinese Magic Known as 'Watching for the Ethers,'" in C. Le Blanc and D. Borei (eds.), *Essays on Chinese Civilization: Derk Bodde* (Princeton: Princeton University Press); rpt. from S. Egerod and E. Glahn (eds.), *Studia Serica Bernhard Karlgren dedicata: Sinological Studies Dedicated to Bernhard Karlgren on his Seventieth Birthday, October 5, 1959* (Copenhagen: Ejnar Munksgaard, 1960), 14–35.

1991. *Chinese Thought, Society, and Science* (Honolulu: University of Hawaii).

Boltz, William G., 1986. "Early Chinese Writing," *World Archaeology* 17.3, 420–36.

1990. "Three Footnotes on the *Ting* 'Tripod,'" *Journal of the American Oriental Society* 110.1, 1–8.

1994. *The Origin and Early Development of the Chinese Writing System* (New Haven, CT: American Oriental Society).

1996. "Early Chinese Writing," in P.T. Daniels and W. Bright (eds.), *The World's Writing Systems* (New York and Oxford: Oxford University Press), 191–9.

1999. "Language and Writing," in Michael Loewe and Edward L. Shaughnessy (eds.), *The Cambridge History of Ancient China: From the Origins of Civilization to 221 B.C.* (Cambridge: Cambridge University Press), 74–123.

2011. "Literacy and the Emergence of Writing," *Writing & Literacy in Early China: Studies from the Columbia Early China Seminar*, 51–84.

Bona ben ershisishi 百衲本二十四史, 1965. *Sibu congkan* 四部叢刊 (Shanghai 1930–7, rpt. Taipei).

Brashier, K.E., 1996. "Han Thanatology and the Division of Souls," *Early China* 21, 125–58.

Brokaw, Galen, 2003. "The Poetics of *Khipu* Historiography: Felipe Guaman Poma de Ayala's *Nueva coronica* and the *Relacion de los quipucamay*," *Latin American Research Review* 38.3, 111–47.

Brooks, E. Bruce, 1994. "The Present State and Future Prospects of Pre-Han Text Studies," *Sino-Platonic Papers* 46 (July), 1–74.

Brooks, E. Bruce and Brooks, A. Taeko, 1998. *The Original Analects: Sayings of Confucius and His Successors* (New York: Columbia University).

Brough, John, 1965. "Comments on Third-Century Shan-Shan and the History of Buddhism," *Bulletin of the School of Oriental and African Studies* 28, 582–612.

Brown, David, 2006. "Astral Divination in the Context of Mesopotamian Divination, Medicine, Religion, Magic, Society, and Scholarship," *East Asian Science, Technology, and Medicine* 25, 69–126.

Burke, Kenneth, 1969. *A Rhetoric of Motives* (Berkeley: University of California).

1970. *The Rhetoric of Religion: Studies in Logology* (Berkeley: University of California).

Burrow, T., 1935. "Tocharian Elements in the Kharosthi Documents from Chinese Turkestan," *Journal of the Royal Asiatic Society* 67.4, 667–75.

Cai, Yong 蔡邕, 1983–6. *Mingtang yueling lun* 明堂月令論, *Siku quanshu* 四庫全書, Wen yuan ge edition (1782) (rpt., Taipei: Shangwu), digital edition.

Campion, N., 2012. *Astrology and Cosmology in the World's Religions* (New York: New York University Press).

Campion, Nicholas, 1994. *The Great Year: Astrology, Millenarianism and History in the Western Tradition* (London: Penguin).

2009. *A History of Western Astrology* (London: Continuum).

2012. *Astrology and Cosmology in the World's Religions* (New York: New York University Press).

Cao, Jinyan 曹錦炎, 1982. "Shi jiaguwen beifang ming" 釋甲骨文北方名, *Zhonghua wenshi luncong* 中華文史論叢 3, 70–1.

(ed.), 2011. *Zhejiang daxue cang zhanguo Chu jian* 浙江大學藏戰國楚簡, 3 vols. (Hangzhou: Zhejiang daxue).

Carlson, John B., 1982. "The Double-Headed Dragon and the Sky: A Pervasive Cosmological Symbol," *Ethnoastronomy and Archaeoastronomy in the American Tropics, Annals of the New York Academy of Sciences* 385.1, 135–63.

Carrasco, David, 1989. "The King, the Capital and the Stars: the Symbolism of Authority in Aztec Religion," *World Archaeoastronomy* (Cambridge: Cambridge University Press), 45–54.

Cen, Zhongmian 岑仲勉, 2004. *Liang Zhou wenshi luncong: Xi Zhou shehui zhidu wenti* 兩周文史論叢:西周社會制度問題 (Beijing: Zhonghua shuju).

Chan, Wing-tsit, 1963. *A Sourcebook in Chinese Philosophy* (Princeton: Princeton University Press).

Chang, Kwang-chih, 1976. *Early Chinese Civilization: Anthropological Perspectives* (Cambridge, MA: Harvard University Press).

1980. Shang *Civilization* (New Haven: Yale University Press).

1983. *Art, Myth, and Ritual: The Path to Political Authority in Ancient China* (Cambridge, MA: Harvard University Press).

張光直, 1990. "Gudai muzang de hunpo guannian" 古代墓葬的魂魄觀念, *Zhongguo wenwu bao* 中國文物報, 28 June.

1999. "China on the Eve of the Historical Period," in Michael Loewe and Edward L. Shaughnessy (eds.), *The Cambridge History of Ancient China, from the Origins of Civilization to 221 B.C.* (Cambridge: Cambridge University Press), 37–73.

Chang, Kwang-chih, Xu, Pingfang, Allan, Sarah, Lu, Liancheng, 2005. *The Formation of Chinese Civilization: An Archaeological Perspective* (New Haven: Yale University Press).

Chang, Zhengguang 常正光, 1989a. "Yin dai de fangshu yu yinyang wuxing sixiang de jichu" 殷代的方術與陰陽五行思想的基礎, *Yinxu bowuyuan yuankan* 殷墟博物苑苑刊(創刊號) 1, 175–82.

Chang, Zhengguang 常正光, 1989b. "Yin dai shoushi juyu – 'si fang feng' kaoshi" 殷代授時舉隅–'四方風'考釋, in Zhongguo tianwenxue shi wenji bianjizu (ed.), *Zhongguo tianwenxue shi wenji* 中國天文學史文集 5 (Beijing: Kexue), 39–55.

Chavannes, Édouard, 1895–1904. *Les Mémoires historiques de Se-Ma-Ts'ien*, 6 vols. (Paris: E. Leroux).

Chavannes, Édouard, 1913a. *Documents chinois découverts par Aurel Stein* (Oxford: Oxford University Press).

1913b. *Mission archaéologique dans la Chine septentrionale* (Paris: Imprimerie nationale).

Chen, Banghuai 陳邦懷, 1959a. "Mao Mu xing 冒母星," in Chen Banghuai, *Yindai shehui shiliao zhengcun* 殷代社會史料徵存 (Tianjin: Tianjin renmin), 6a–b.

1959b. "Si fang feng ming" 四方風名, in Chen Banghuai, *Yindai shehui shiliao zhengcun* 殷代社會史料徵存 (Tianjin: Tianjin renmin), 1a–5b.

Chen, Cheng-Yih and Xi, Zezong, 1993. "The Yáo diǎn 堯典 and the Origins of Astronomy in China," in C. Ruggles (ed.), *Astronomy and Cultures* (Newit: University Press of Colorado), 32–66.

Chen, Gongrou, 2005. "Xi Zhou jinwen zhong de 'Xinyi,' Chengzhou yu Wangcheng" 西周金文中的'新邑'、'成周'與'王城', in *Xian Qin liang Han kaoguxue luncong* 先秦兩漢考古學論叢 (Beijing: Wenwu), 33–48.

Chen, Jiujin 陳久金, 1978. "Cong mawangdui boshu *Wu xing zhan* de chutu shitan wo guo gudai de suixing jinian wenti" 從馬王堆帛書'五星占'的出土試探我國古代的歲星紀年問題, in Zhongguo tianwenxue shi wenji bianjizu (ed.), *Zhongguo tianwenxue shi wenji* 中國天文學史文集, 48–65.

1987. "*Zhou Yi* qian gua liu long yu jijie de guanxi" 周易乾卦六龍與季節的關係, in Chen Jiujin 陳久金 and Chen Meidong 陳美東 (eds.), *Ziran kexue shi yanjiu* 自然科學史研究 3, 205–12.

1993. "Lun *Xia xiao zheng* shi shi yue taiyang li" 論夏小正是十月太陽曆, in Chen Jiujin, *Chen Jiujin ji* 陳久金集 (Ha'erbin: Heilongjiang jiaoyu), 3–30.

Chen, Jiujin 陳久金, Lu, Yang 盧央, and Liu, Yaohan 劉堯漢, 1984. *Yizu tianwenxue shi* 彝族天文學史 (Kunming: Yunnan renmin).

Chen, Mengjia 陳夢家, 1955. "Xi Zhou tongqi duandai" 西周銅器斷代 (一), *Kaogu xuebao* 考古學報 9, 137–75.

1988. *Yinxu buci zong shu* 殷墟卜辭綜述 (Beijing: Zhonghua; first published Beijing: Kexue, 1956).

Chen, Quanfang 陳全方, 1988. *Zhouyuan yu Zhou wenhua* 周原與周文化 (Shanghai: Shanghai renmin).

534   References

Chen, Songchang 陳松長, 2000. "'Taiyi sheng shui' kaolun"《太一生水》考論, in Wuhan daxue Zhongguo wenhua yanjiuyuan 武漢大學中國文化研究院 (ed.), *Guodian Chu jian guoji xueshu yantao hui lunwenji* 郭店楚簡國際學術會論文集 (Wuhan: Hubei renmin).

2001. *Mawangdui boshu*《*Xingde*》*yanjiu lungao, Chutu sixiang wenwu yu wenxian yanjiu congshu* 馬王堆帛書刑德研究論稿, 出土思想文物與文獻研究叢書 (Taipei: Taiwan guji).

Chen, Zhongyu 陳仲玉, 1969. "Yin dai guqi zhong de long xing tu'an zhi fenxi" 殷代骨器中的龍形圖案之分析, *Lishi yuyan yanjiusuo jikan* 歷史語言研究所集刊 41.3, 455–96.

Chen, Zungui 陳遵媯, 1955. *Zhongguo gudai tianwenxue shi* 中國古代天文學史 (Shanghai: Shanghai renmin).

Cheng, Pingshan 程平山, 2005. "Lun Taosi gucheng de fazhan jieduan yu xingzhi" 論陶寺古城的發展階段與性質, *Jiangnan kaogu* 江南考古 96 (March), 48–53.

Cheol, Shin Min, 2007. "The Ban on the Private Study of Astrology and Publication of Books on Astrology in Ming Dynasty: Ideas and Reality," *Korean History of Science Society* 27.2, 231–60.

Cheung, Kwong-yue, 1983. "Recent Archaeological Evidence Relating to the Origin of Chinese Characters," in D.N. Keightley (ed.), *The Origins of Chinese Civilization* (Berkeley: University of California), 323–91.

Chu, Ge 楚戈, 2009. *Long shi* 龍史 (Taipei: Guojia tushuguan).

Cong, Y.Z. and Wei, Q.Y., 1989. "Study of Secular Variation (2000 B.C.–1900 A.D.) Based on Comparison of Contemporaneous Records in Marine Sediments and Baked Clays," *Physics of the Earth and Planetary Interiors* 56, 69–75.

Conklin, William J., 1982. "The Information System of Middle Horizon Quipus," *Ethnoastronomy and Archaeoastronomy in the American Tropics, Annals of the New York Academy of Sciences* 385.1, 261–81.

Conman, Joanne, 2006–9. "The Egyptian Origin of Planetary Hypsomata," *Discussions in Egyptology* 64, 7–20.

Cook, Richard S., 1995. "The Etymology of Chinese 辰 *Chén*," *Linguistics of the Tibeto-Burman Area* 18.2.

Corballis, Michael C., 2011. *The Recursive Mind: The Origins of Human Language, Thought, and Civilization* (Princeton: Princeton University Press).

Csikszentmihalyi, Mark, 1997. "Chia I's 'Techniques of the Tao' and the Han Confucian Appropriation of Technical Discourse," *Asia Major* (3rd series) 10.1–2, 49–67.

2006. *Readings in Han Chinese Thought* (Indianapolis: Hackett).

Csikszentmihalyi, Mark and Nylan, Michael, 2003. "Constructing Lineages and Inventing Traditions through Exemplary Figures in Early China," *T'oung Pao* (2nd series) 89.1, 59–99.

Cullen, Christopher, 1993. "Motivations for Scientific Change in Ancient China," *Journal for the History of Astronomy* 24, 185–203.

1996. *Astronomy and Mathematics in Ancient China: The Zhou bi suan jing* (Cambridge: Cambridge University Press).

2001. "The Birthday of the Old Man of Jiang County and Other Puzzles: Work in Progress on Liu Xin's *Canon of the Ages*," *Asia Major* 14.2, 27–60.

2011a. "Translating *Xiu*, 'Lunar Lodges' or Just 'Lodges,'" *East Asian Science, Technology and Medicine* 33, 76–88.

2011b. "Understanding the Planets in Ancient China: Prediction and Divination in the *Wu xing zhan*," *Early Science and Medicine* 16, 218–51.

Cullen, Christopher and Farrer, Anne, 1983. "On the Term *Hsuan Chi* and the Three-Lobed Jade Discs," *Bulletin of the School of Oriental and African Studies* 46.1, 53–76.

Da Silva, Cândido M., 2010. "Neolithic Cosmology: The Spring Equinox and the Full Moon," *Journal of Cosmology* 9, 2207–16.

Dai, Chunyang 戴春陽, 2000. "Lixian Dabuzishan Qin gong mu di ji you guan wenti" 禮縣大堡子山秦公墓地及有關問題, *Wenwu* 文物 5, 74–80.

Davies, Paul, 2002. "That Mysterious Flow," *Scientific American* 287.3 (September), 40–3, 46–7.

De Crespigny, Rafe, 1976. *Portents of Protest in the Later Han Dynasty: The Memorials of Hsiang-k'ai to Emperor Huan* (Canberra: Australian National University Press).

De Meis, Salvo, 2006. "L'astronomia dello Shi-King e di altri classici cinesi: II parte," *Giornale di Astronomia: Revista di informazione, cultura e didattica della Società Astronomica Italiana* 32.2, 17–23.

De Meis, Salvo and Meeus, Jean, 1994. "Quintuple planetary groupings – Rarity, Historical Events and Popular Beliefs," *Journal of the British Astronomical Association* 104.6, 293–7.

De Saussure, Léopold, 1911. "Les origines de l'astronomie Chinoise: La règle des cho-ti," *Toung Pao* 12, 347–74.

1930. *Les origines de l'astronomie Chinoise* (Paris: Maisonneuve).

Deane, Thatcher E., 1994. "Instruments and Observation at the Imperial Astronomical Bureau during the Ming Dynasty," *Osiris* (2nd series), Instruments, 126–40.

DeBernardi, Jean, 1992. "Space and Time in Chinese Religious Culture," *History of Religions* 31.3, 247–68.

Defoort, Carine, 1997. *The Pheasant Cap Master (He guan zi): A Rhetorical Reading* (Albany: State University of New York).

Deutsch, David, 2011. *The Beginning of Infinity: Explanations that Transform the World* (New York: Penguin Books).

Di Cosmo, Nicola, 2002. *Ancient China and Its Enemies: The Rise of Nomadic Power in East Asian History* (Cambridge: Cambridge University Press).

Didier, John, 2009. "In and Outside the Square," *Sino-Platonic Papers* 192.

Diény, Jean-Pierre, 1987. *Le symbolisme du dragon dans la Chine antique* (Paris: Bibliothéque de l'Institut des hautes études Chinoises).

Ding, Shan 丁山, 1961. "Si fang feng yu feng shen" 四方之神與風神, in Ding Shan, *Zhongguo gudai zongjiao yu shenhua kao* 中國古代宗教與神話考 (Shanghai: Longmen lianhe), 78–95.

Dixon, C. Scott, 1999. "Popular Astrology and Lutheran Propaganda in Reformation History," *History* 84, 406–7.

Domenici, Viviano and Domenici, Davide, 1996. "Talking Knots of the Inka: A Curious Manuscript May Hold the Key to Andean Writing," *Archaeology* 49.6, 50–6.

Donald, Merlin, 1991. *Origins of the Modern Mind: Three Stages in the Evolution of Culture and Cognition* (Cambridge, MA: Harvard University Press).

Dorofeeva-Lichtmann, Véra, 1996. "Political concept behind an interplay of spatial 'positions,'" *Extrême-Orient, Extrême-Occident* 18, 9–33.

2009. "Ritual Practices for constructing terrestrial space (Warring States–Early Han)," in J. Lagerwey and M. Kalinowski (eds.), *Early Chinese Religion, Part One, Shang through Han (1250 BC–220 AD)* (Leiden: Brill), Vol. 1, 595–644.

2010. "The *Rong Cheng shi* 容成氏 version of the 'Nine Provinces': some parallels with transmitted texts," *East Asian Science, Technology, and Medicine* 32, 13–58.

Du, Jinpeng 杜金鵬 and Xu, Hong 許宏 (eds.), 2005. *Yanshi Erlitou yizhi yanjiu* 偃師 二里頭遺址研究 (Beijing).

Dubs, Homer H. *The History of the Former Han Dynasty*, 3 vols. (Baltimore: Waverly, 1938–55).

Dull, Jack L., 1966. "A Historical Introduction to the Apocryphal (*ch'an-wei*) Texts of the Han Dynasty," Ph.D. dissertation, University of Washington.

Eberhard, Wolfram, 1958. *The Political Function of Astronomy and Astrologers in Han China*, in J.K. Fairbank (ed.), *Chinese Thought and Institutions* (Chicago: The University of Chicago Press), 33–70.

1970. "Beiträge zur kosmologischen Spekulation in der Han Zeit," *Baessler Arkiv* 16, 1–100, rpt. in Wolfram Eberhard, *Sternkunde und Weltbild im alten China: Gesammelte Aufsätze von Wolfram Eberhard* (Taipei: Chinese Materials and Research Aids Service Center), 11–109.

Ecsedy, I., Barlai, K., Dvorak, R., and Schult, R., 1989. "Antares Year in Ancient China," in A.F. Aveni (ed.), *World Archaeoastronomy* (Cambridge: Cambridge University Press), 183–6.

Egan, Ronald C., 1977. "Narratives in *Tso chuan*," *Harvard Journal of Asiatic Studies* 37.2, 323–52.

Eliade, Mircea, 1958. *Patterns in Comparative Religion* (Lanham, MD: Sheed and Ward).

Emerson, Ralph Waldo, 1979. *Nature, Addresses, and Lectures* (ed. R.E. Spiller and A.R. Ferguson) (Cambridge: The Belknap Press of Harvard University; first published 1836).

Encyclopedia Britannica, 1984, 15th ed. (Chicago: Encyclopedia Britannica).

Eno, Robert, 1990. "Was there a High God *Ti* in Shang Religion?" *Early China* 15, 1–26.

Enoki, Kazuo and Kimura, Sugako (eds.), 1974. *A Concordance to Mo Tzu*, Harvard-Yenching Institute Sinological Index Series, Supplement No. 21 (rpt., San Francisco: Chinese Materials Center).

Fagg, Lawrence W., 1985. *Two Faces of Time* (Wheaton, IL: Theosophical Publishing).

Fang, Shiming 方詩銘, and Wang, Xiuling 王修齡, 1981. *Guben Zhushu jinian jizheng* 古本竹書紀年輯證 (Shanghai: Shanghai guji).

Farmer, Steve, Henderson, John B., and Witzel, Michael, 2000. "Neurobiology, Layered Texts, and Correlative Cosmologies: A Cross-cultural Framework for Pre-modern History," *Bulletin of the Museum of Far Eastern Antiquities* 72, 49–90.

Feng, Shi 馮時, 1990a. "Henan Puyang Xishuipo 45 hao mu de tianwenxue yanjiu" 河 南濮陽西水坡45號墓的天文學研究, *Wenwu* 3, 52–60, 69.

1990b. "Yinli sui shou yanjiu" 陰曆歲首研究, *Kaogu xuebao* 考古學報 1, 19–42.

1990c. "Zhongguo zaoqi xingxiangtu yanjiu" 中國早期星象圖研究, *Ziran kexueshi yanjiu* 自然科學研究 9.2, 108–18.

1993. "Hongshan wenhua san huan shi tan de tianwenxue yanjiu" 紅山文化三環石 壇的天文學研究, *Beifang wenwu* 北方文物 33.1, 9–17.

1994. "Yin buci sifang feng yanjiu" 殷卜辭四方風研究, *Kaogu xuebao* 考古學報 2, 131–54.

1996. *Xinghan liu nian* 星漢流年 (Chengdu: Sichuan jiaoyu).

1997. "Chunqiu Zi Fan bian zhong ji nian yanjiu – Jin Chong Er guiguo kao" 春秋子犯編鐘紀年研究 – 晉重耳歸國考, *Wenwu jikan* 文物集刊 4, 59–65.

2007. *Zhongguo tianwen kaoguxue* 中國天文考古學 (Beijing: Zhongguo shehui kexue).

Feng, Yunpeng 馮雲鵬 and Feng, Yunyuan 馮雲鵷, 1893. *Jinshi suo* 金石索 (Shanghai: Shanghai jishan shuju, first published 1821); also availabe at http://catalog.hathitrust.org/Record/002252003.

Fingarette, Herbert, 1998. *Confucius: The Secular As Sacred* (Prospect Heights, IL: Waveland).

Fisher, Carney T., 1990. *The Chosen One: Succession and Adoption in the Court of Ming Shizong* (Sydney: Allen and Unwin).

Fong, Wen, 1980. *The Great Bronze Age of China: An Exhibition from the People's Republic of China* (New York: Metropolitan Museum of Art).

Fraser, Douglas, 1968. *Early Chinese Art and the Pacific Basin: A Photographic Exhibition* (New York: Intercultural Arts).

Freidel, David, Schele, Linda, and Parker, Joy, 2001. *Maya Cosmos: Three Thousand Years on the Shaman's Path* (New York: Quill, first published 1995).

Fu, Xi'nian 傅熹年, 1981. "Shaanxi Fufeng Zhaochen Xi Zhou jianzhu yizhi chutan" 陝西扶風召陳西周建築遺址初探, *Wenwu* 3, 34–45.

Galdieri, Patrizia and Ranieri, Marcello, 1995. "Terra e cielo: note sulle origini dell'architettura cinese della valle del Fiume Giallo," in M. Bernardini et al. (eds.), *L'Arco de Fango che Rubò la Luce alle stelle: Studi in onore di Eugenio Galdieri per il suo settantesimo compleanno – Roma 29 ottobre 1995* (Lugano: Edizioni Arte e Moneta), 155–71.

Gao, Heng 高亨, 1973. *Zhouyi gujing jinzhu* 周易古經今注 (rpt., Hong Kong: China Book, first published Shanghai: Kaiming, 1947).

Gao, Wence 高文策, 1961. "Shi lun *Yi* de chengshu niandai yu fayuan diyu" 試論《易》的成書年代與發源地域, *Guangming ribao* 光明日報, June 2.

Gassmann, Robert H. and Behr, Wolfgang, 2011. *Antikchinesisch – Ein Lehrbuch in zwei Teilen* (Bern: Peter Lang).

Geertz, Clifford, 1973. "Ideology as a Cultural System," in Clifford Geertz, *The Interpretation of Cultures: Selected Essays* (New York: Basic Books), 193–233.

Ghezzi, Ivan and Ruggles, Clive, 2007. "Chankillo: A 2300-year-old solar Observatory in Coastal Peru," *Science* 315, 1239–1243.

Gingerich, Owen, 1984. "Astronomical Scrapbook: The Origin of the Zodiac," *Sky & Telescope* 67.3 (March), 218–20.

2000. "Plotting the Pyramids," *Nature* 408 (November 16), 297–8.

Goldin, Paul R., 2005. *After Confucius* (Honolulu: University of Hawaii).

2007. "Xunzi and Early Han Philosophy," *Harvard Journal of Asiatic Studies* 67.1, 135–66.

2008. "The Myth that China Has no Creation Myth," *Monumenta Serica* 56, 1–22.

Goodrich, Lincoln C. and Fang, Chaoying (eds.), 1976. *Dictionary of Ming Biography, 1364–1644* (New York: Columbia University Press).

Goody, Jack, 1977. *The Domestication of the Savage Mind* (Cambridge: Cambridge University Press).

2006. *The Theft of History* (Cambridge: Cambridge University Press).

Graham, Angus C., 1989a. *Disputers of the Tao: Philosophical Argument in Ancient China* (LaSalle, IL: Open Court).

1989b. "A Neglected Pre-Han Philosophical Text: *Ho-Kuan-Tzu*," *Bulletin of the School of Oriental and African Studies* 52.3, 497–532.

Granet, Marcel, 1932. *Festivals and Songs of Ancient China* (London: Routledge).

Gritsch, Eric W., 1967. *Reformer without a Church: The Life and Thought of Thomas Müntzer, 1488?–1525* (Philadelphia: Fortress).

Grofe, Michael J., 2011. "The Sidereal Year and the Celestial Caiman: Measuring Deep Time in Maya Inscriptions," *Archaeoastronomy* 24, 56–101.

Gu jin tushu jicheng 古今圖書集成, 1964 (rpt., Taipei: Wen-hsing).

Gu, Yanwu 顧炎武, 1966. *Ri zhi lu* 日知錄, *Sibu beiyao* edition (Taipei: Taiwan Chunghua).

Gu, Jiegang 顧頡剛, n.d. "*Zhouyi* gua yao ci zhong de gushi" 《周易》卦爻辭中的故事 *Gu shi bian* 古史辨 (rpt., Taipei).

Guo, Moruo 郭沫若, 1982a. *Buci tong zuan* 卜辭通纂 (rpt., Beijing: Kexue).

1982b. "Shi zhigan" 釋支干, in *Guo Moruo quanji* 郭沫若全集 (Beijing: Kexue), Vol. 1, 155–340.

1999. *Zhoudai jinwen tulu ji shiwen* 周代金文圖錄及釋文 (*Liang Zhou jinwen ci daxi kaoshi* 兩周金文辭大系考釋), 3 vols. (Shanghai, rpt. Taipei: Datong, 1971).

Guoyu 國語, 1927–35, *Sibu beiyao* edition (Shanghai: Zhonghua; rpt. Taipei: Taiwan Chung-hua, 1966).

Hammerstein, Helga R., 1986. "The Battle of the Booklets: Prognostic Tradition and Proclamation of the Word in Early Sixteenth-Century Germany," in P. Zambelli (ed.), *Astrologi Hallucinati* (Berlin: W. de Gruyter), 129–51.

Han Fei 韓非, 1974. *Hanfeizi* 韓非子, *Xinbian zhuzi jicheng* edition (rpt., Taipei: Shijie shuju).

Handy, E.S. Craighill, 1923. *The Native Culture in the Marquesas, Bernice C. Bishop Museum Bulletin* 9 (Honolulu: Bishop Museum).

Harbsmeier, Christoph, 1995. "Some Notions of Time and History in China and the West," in C.C. Huang and E. Zürcher (eds.), *Cultural Notions of Time and Space in China* (Leiden: Brill), 50–71.

1998. "Language and Logic in Traditional China," *Science and Civilisation in China*, Vol. 7, Part 1, *Language and Logic* (Cambridge: Cambridge University Press).

Harper, Donald, 1999. "Warring States Natural Philosophy and Occult Thought," in Michael Loewe and Edward L. Shaughnessy (eds.), *The Cambridge History of Ancient China, from the Origins of Civilization to 221 B.C.* (Cambridge: Cambridge University Press), 813–84.

Harris, Lynda., 2011. "The Milky Way: Path to the Empyrean?", *The Inspiration of Astronomical Phenomena VI, ASP Conference Series* 441, 387–91.

Hart, James A., 1984. "The Speech of Prince Chin: A Study of Early Chinese Cosmology," in H. Rosemont Jr. (ed.), *Explorations in Early Chinese Cosmology*, JAAR Thematic Studies L/2 (Chico: Scholar's Press), 35–65.

Havelock, Eric A., 1983. "The Linguistic Task of the Pre-Socratics," *Language and Thought in Early Greek Philosophy* (LaSalle, IN: Hegeler Institute), 7–82.

1987. "The Cosmic Myths of Homer and Hesiod," *Oral Tradition* 2.1, 31–53.

Hawkes, David (trans.), 1959. *Ch'u Tz'u: The Songs of the South* (London: Oxford University Press).

Hay, John, 1994. "The Persistent Dragon," in W.J. Petersen, A. Plaks, and Yü Ying-shi (eds.), *The Power of Culture: Studies in Chinese Cultural History* (Hong Kong: Chinese University Press), 119–49.

He, Bingyu (Ho Peng Yoke) 何丙郁 and He, Guanbiao 何冠彪, 1985. *Dunhuang canjuan zhan yunqi shu yanjiu* 敦煌殘卷占雲氣書研究 (Taipei: Yiwen).

He, Nu 何駑, 2004. "Taosi zhongqi xiaocheng nei daxing jianzhu IIFJT1 fajue xinlu licheng zatan" 陶寺中期小城內大型建築IIFJT1發掘心路歷程雜談 *Gudai wenming yanjiu zhongxin tongxun* 古代文明研究中心通訊 23, 47–58.

2006. "Taosi zhongqi xiaocheng nei daxing jianzhu jizhi IIFJT1 shidi moni guanxiang baogao" 陶寺中期小城內大型建築基址IIFJT1實地模擬觀測報告, *Gudai wenming yanjiu zhongxin tongxun* 古代文明研究中心通訊 29, 3–14.

2009. "Shanxi Xiangfen Taosi chengzhi zhongqi wang ji da mu IIM22 chutu qigan 'guichi' gongneng shitan" 山西襄汾陶寺城址中期王級大墓出土漆桿圭尺功能試探, *Ziran kexueshi yanjiu* 自然科學史研究 28.3, 261–76.

He, Xiu 何休, 1971. *Chunqiu Gongyang zhuan He shi jiegu* 春秋公羊傳何氏解詁, *Sibu beiyao* edition (Taipei: Taiwan Chunghua).

Henderson, John B., 1984. *The Development and Decline of Chinese Cosmology* (New York: Columbia University Press).

1995. "Chinese Cosmographical Thought: The High Intellectual Tradition," in J.B. Harley and D. Woodward (eds.), *The History of Cartography*, Vol. 2, Book 2, *Cartography in the Traditional East and Southeast Asian Societies* (Chicago: University of Chicago), 203–27.

Henricks, Robert G., 1989. *Lao-tzu Te-Tao Ching* (New York: Ballantine).

Hesiod, 1983. *The Poems of Hesiod* (trans. R.M. Frazer) (Norman: University of Oklahoma).

Ho, Peng-yoke (trans.), 1966. *The Astronomical Chapters of the Chin shu* (Paris: Mouton).

Hobson, John M., 2008. *The Eastern Origins of Western Civilisation* (Cambridge: Cambridge University Press, first published 2004).

Hong, Xingzu 洪興祖, Wang, Yi 王逸, and Li, Xiling 李錫齡, 1971. *Chuci buzhu* 楚辭補註, *Sibu beiyao* edition (Taipei: Taiwan Chung-hua).

Horowitz, Wayne, 2011. *Mesopotamian Cosmic Geography* (Durban, Ireland: Clearway Logistics Phase 1a).

Hostetter, Clyde, 1990. "The Naked-eye Crescent of Venus," *Sky & Telescope* 79 (January), 74–6.

Hotaling, Steven J., 1978. "The City Walls of Han Ch'ang-an," *T'oung Pao* 64.1–3, 1–46.

Hotz, Robert Lee, 2008. "How Alphabets Shape the Brain," *Wall Street Journal*, May 2, A10.

Houston, Stephen D., 2004a. "The Archaeology of Communication Technologies," *Annual Review of Anthropology* 33, 223–50.

2004b. *The First Writing: Script Invention as History and Process* (Cambridge: Cambridge University Press).

Hsiao, Kung-chuan, 1979. *History of Chinese Political Thought* (trans. F.W. Mote) (Princeton: Princeton University Press).

Hsu, Cho-yun and Linduff, Katheryn M., 1988. *Western Chou Civilization* (New Haven: Yale University Press).

Hu, Houxuan 胡厚宣, 1983, "Yin dai zhi tianshen chongbai" 殷代之天神崇拜, in *Jiaguxue Shang shi luncong* 甲骨學商史論叢 (*chuji* 初集, *shang* 上) (Chengdu; rpt. Taipei: Datong), 1–29.

Hu, Houxuan 胡厚宣 and Guo, Moruo 郭沫若 (eds.), 1979–82. *Jiaguwen heji* 甲骨文合集, 13 vols. (Beijing: Zhonghua).

Hu, Lin'gui 呼林貴, 1989. "Xi'an Jiao da Xi Han ershiba xiu xingtu yu *Shiji: Tianguanshu*" 西安交大西漢二十八宿與《史記》天管書, *Renwen zazhi* 人文雜誌 2, 85–7.

Hu, Tiezhu 胡鐵珠, 2000. "*Xia xiao zheng* xingxiang niandai yanjiu" 《夏小正》星象年代研究, *Ziran kexueshi yanjiu* 自然科學史研究 19.3, 234–50.

Hu, Wenyao 胡文輝, 1993. "Shi 'sui' – yi Shuhudi 'ri shu' wei zhongxin" 釋歲 – 以睡虎地秦簡'日書'為中心 *Wenhua yu chuanbo* 文化與傳播 4, 101–22.

Huang, Chun-chieh, 1995. "Historical Thinking in Classical Confucianism: Historical Argumentation from the Three Dynasties," in C.C. Huang and E. Zürcher (eds.), *Cultural Notions of Time and Space in China* (Leiden: Brill), 72–85.

Huang, Chun-chieh and Zürcher, Erik (eds.), 1995, *Cultural Notions of Time and Space in China* (Leiden: Brill).

Huang, Shengzhang 黄盛璋, 1960. "Da Feng gui zhizuo de niandai didian yu shishi" 大豐簋製作的年代、地點與史實, *Lishi yanjiu* 6, 81–95.

Huang, Tianshu 黃天樹, 2006a. "Shuo jiaguwen zhong de 'yin' he 'yang'" 說甲骨文中的陰和陽, in Huang Tianshu, *Gu wenzi lunji* 古文字論集 (Beijing: Xueyuan), 213–17.

Huang, Tianshu 黃天樹, 2006b. "Shuo Yinxu jiaguwen zhong de fangwei ci" 說殷墟甲骨文中的方位詞, in Huang Tianshu, *Gu wenzi lunji* 古文字論集 (Beijing: Xueyuan), 203–12.

Huang, Yi-long, 1990. "A Study of Five-Planet Conjunctions in Chinese History," *Early China* 15, 97–112.

Huang, Yi-long and Chang, Chih-ch'eng, 1996. "The Evolution and Decline of the Ancient Chinese Practice of Watching for the Ethers," *Chinese Science* 13, 82–106.

Huang, Zongxi 黃宗羲, 1974. *Ming Ru xue an* 明儒學案 (rpt., Taipei: Heluo).

Hubeisheng bowuguan 湖北省博物館, 1989. *Zeng Hou Yi mu* 曾侯乙墓 (Bejing: Wenwu).

Hughes, David, 1979. *The Star of Bethlehem: An Astronomer's Confirmation* (New York: Walker and Company).

Hulsewé, A.F.P., 1965. "Texts in Tombs," *Études asiatiques* 18–19, 78–89.

1979. "Watching the Vapours: An Ancient Chinese Technique of Prognostication," *Nachrichten der Gesellschaft für Natur- und Völkerkunde Ostasiens* 125, 40–9.

Hunansheng bowuguan 湖南省博物館, 1973. "Xin faxian de Changsha Zhanguo chu mu bohua," 新發現的長沙戰國楚墓帛畫 *Wenwu* 文物 7.3, 3–4.

Hung, William, 1966. *Combined Concordances to Ch'un-ch'iu, Kung-yang, Ku-liang and Tso-chuan*, Harvard-Yenching Institute Sinological Index Series, supplement 11 (rpt., Taipei: Cheng Wen).

Hunger, Herman and Pingree, David, 1999. *Astral Sciences in Mesopotamia* (Leiden: Brill).

Hurwit, Jeffrey M., 2000. *The Athenian Acropolis: History, Mythology, and Archaeology from the Neolithic Era to the Present* (Cambridge: Cambridge University Press).

Hwang, Ming-chorng 黃銘崇, 1996. "Ming-Tang: Cosmology, Political Order, and Monuments in Early China," Ph.D. dissertation, Harvard University.

Hyman, Malcolm D., 2006. "Of Glyphs and Glottography," *Language and Communication* 26, 231–49.

Institute of Archaeology of the Chinese Academy of Social Sciences, Institute of Archaeology of Shanxi Province and Cultural Relics Bureau of Linfen City (IACASS, IASP, CRBLC), 2004. "Shanxi Xiangfen xian Taosi chengzhi jisiqu daxing jianzhu IIFJT1 jizhi 2003 nian fajue jianbao" 山西襄汾縣陶寺城址祭祀區大型建築IIFJT1基址2003年發掘簡報, *Kaogu* 7, 9–24.

2005. "2004–2005 nian Shanxi Xiangfen xian Taosi yizhi fajue xin jinzhan" 2004–2005年山西襄汾縣陶寺遺址發掘新進展, *Gudai wenming yanjiu zhongxin tongxun* 古代文明研究中心通訊 10, 58–64.

2007. "Shanxi Xiangfen xian Taosi chengzhi daxing jianzhu IIFJT1 jizhi 2004–2005 nian fajue jianbao" 山西襄汾縣陶寺城址大型建築IIFJT1基址發掘簡報, *Kaogu* 4, 3–25.

Irwin, Geoffrey, 2010. "Navigation and Seafaring," in Ian Lilley (co-ordinator), *Early Human Expansion and Innovation in the Pacific* (Paris: ICOMOS), 51–72.

Isbell, Billie Jean, 1982. "Culture Confronts Nature in the Dialectical World of the Tropics," *Ethnoastronomy and Archaeoastronomy in the American Tropics, Annals of the New York Academy of Sciences* 385.1, 353–63.

Itō, Chūta 伊東忠太, 1938. *Zhongguo jianzhu shi* 中國建築史, in *Zhongguo wenhua congshu* 中國文化叢書 (trans. Chen Qingquan 陳清淚) (Shanghai: Shanghai shudian).

Itō, Michiharu 伊藤道治, 1978. "Shū Buō to Rakuyū – kason mei to Itsushūsho doyū" 周武王と雒邑 – 何尊銘と逸周書度邑, in *Uchida Ginpū hakushi shōju kinen Tōyōshi ronshū* 内田吟風博士頌壽紀念東洋史論集, 41–53.

Iwami, Kiyohiro, 2008. "Turks and Sogdians in China during the T'ang Dynasty," *Acta Asiatica* 94, 41–65.

Jabłoński, Witold, 1939. "Marcel Granet and His Work," *Yenching Journal of Social Studies* 1, 242–55.

Jacobsen, Lyle E., 1983. "Use of Knotted String Accounting Records in Old Hawaii and Ancient China," *Accounting Historians Journal* 10.2, 53–61.

Jacobsen, Thorkild, 1968. "The Battle between Marduk and Tiamat," *Journal of the American Oriental Society* 88.1 (January–March), 104–8.

1976. *The Treasures of Darkness: A History of Mesopotamian Religion* (New Haven: Yale University Press).

1994. "The Historian and the Sumerian Gods," *Journal of the American Oriental Society* 114.2, 145–53.

James, Jean M., 1993. "Is It Really a Dragon? Some Remarks on the Xishuipo Burial," *Archives of Asian Art* 46, 100–1.

Jao, Tsung-I and Léon Vandermeersch (trans.), 2006. "Les relations entre la Chine et le monde Iranien dans l'Antiquité," *Bulletin de l'École française d'Extrême-Orient* 93, 207–45.

Jao, Tsung-yi 饒宗頤, 1972. "Some Aspects of the Calendar, Astrology, and Religious Concepts of the Ch'u People as Revealed in the Ch'u Silk Manuscript," in N. Barnard (ed.), *Early Chinese Art and Its Possible Influence in the Pacific Basin* (New York: Intercultural Arts Press), 118–9.

1998. "Yin buci suojian xingxiang yu shen shang, long hu, ershiba xiu zhu wenti" 殷卜辭所見星象與參商、龍虎、二十八宿諸問題, in Zhang Yongshang 張永山 and Hu Zhenyu 胡振宇 (eds.), *Hu Houxuan xiansheng jinian wenji* 胡厚宣先生紀念文集 (Beijing: Kexue), 32–5, 37.

2003. "*X Gong xu* yu *Xia shu* yi pian 'Yu zhi zong de'" 《燹公盨》與夏書佚篇'禹之總德;, *Hua xue* 華學 6, 1–6.

Jastrow, Morris, 1905. *Die Religion Babyloniens und Assyriens* (Giessen, Ricker).

Jenkins, R.M., 2004. "The Star of Bethlehem and the Comet of 66 AD," *Journal of the British Astronomy Association* 114 (June), 336–43.

Jiang, Linchang 江林昌, 2003. "Gong xu ming wen de xueshu jiazhi zonglun" 《公盨》銘文的學術價值綜論, *Huaxue* 華學 6, 35–49.

Jiang, Xiaoyuan 江曉原, 1991. "Tianwen, Wu Xian, Ling tai – tianwen xingzhan yu gudai Zhongguo de zhengzhi guannian" 天文、巫咸、靈台－天文星占與古代中國的政治觀念, *Ziran bianzhengfa tongxun: kexue jishu shi* 自然辯證法通訊: 科學技術史 13.73, 53–7.

1992a. *Xingzhanxue yu chuantong wenhua* 星占學與傳統文化 (Shanghai: Shanghai guji).

1992b. "Zhongguo tianxue de qiyuan: xi lai haishi zisheng" 中國天學的起源:西來還是自生 *Ziran bianzheng fa tongxun: kexue jishu shi* 自然辯證法通訊: 科學技術史 14.78, 49–56.

2004. *Tianxue zhenyuan* 天學真原 (Shenyang: Liaoning jiaoyu, first published 1991).

Jiangxi tongzhi 江西通志, 1983–6. *Siku quanshu* 四庫全書, *Wen yuan ge* edition (1782) (rpt., Taipei: Shangwu), digital edition.

Jin shu 晉書, 1974 (Beijing: Zhonghua).

Johnston, Ian, 2010. *The Mozi: A Complete Translation* (New York: Columbia University Press).

Jones, T.L., Storey, A.A., Matisoo-Smith, E., Ramirez-Aliaga, J.M. (eds.), 2011. *Polynesians in America* (New York: Altamira).

Joseph, Rhawn, 2011. "Evolution of Paleolithic Cosmology and Spiritual Consciousness, and the Temporal and Frontal Lobes," *Journal of Cosmology* 14, 4400–40.

Jung, Carl G., 1969. "Synchronicity: An Acausal Connecting Principle," in *Collected Works of C.G. Jung*, Vol. 8, *The Structure and Dynamics of the Psyche*, 2nd edition (Princeton: Bollingen Series, Princeton University Press).

Justeson, John S., 1989. "Ancient Maya Ethnoastronomy: An Overview of Hieroglyphic Sources," in Anthony Aveni (ed.), *World Archaeoastronomy* (New York: Cambridge University Press), 76–129.

Kalinowski, Marc, 1986. "L'Astronomie des populations Yi du Sud-Ouest de la Chine," *Cahiers d'Extreme Asie* 2, 253–63.

1996. "The Use of the Twenty-Eight Xiu as a Day-Count in Early China," *Chinese Science* 13, 55–81.

1998–9. "The Xing De 刑德 Texts from Mawangdui," *Early China* 23–4, 125–202.

2004. "Fonctionnalité calendaire dans les cosmogonies anciennes de la Chine," *Études chinoises* 23, 88–122.

2009. "Diviners and Astrologers under the Eastern Zhou," in J. Lagerwey and M. Kalinowski (eds.), *Early Chinese Religion*, Part One, *Shang through Han (1250 BC–220 AD)* (Leiden: Brill), Vol. 1, 341–96.

(trans.), 2011. *Wang Chong: Balance des discours: Destin, providence et divination* (Paris: Les Belles Lettres).

Kaltenmark, Max, 1961. "Religion and Politics in the China of the Ts'in and the Han," *Diogenes* 9.34, 16–43.

Karlgren, Bernhard, 1933. "Word Families in Chinese," *Bulletin of the Museum of Far Eastern Antiquities* 5, 9–120.

(trans.), 1950a. *The Book of Documents* (Stockholm: Museum of Far Eastern Antiquities).

(trans.), 1950b. *The Book of Odes* (Stockholm: Museum of Far Eastern Antiquities).

1964a. *Glosses on the Book of Odes* (Stockholm: Museum of Far Eastern Antiquities).

1964b. *Grammata Serica Recensa* (Stockholm: Museum of Far Eastern Antiquities).

1970. *Glosses on the Book of Documents* (Stockholm: Museum of Far Eastern Antiquities).

Keenan, Douglas J., 2002. "Astro-historiographic Chronologies of Early China Are Unfounded," *East Asian History* 23, 61–8.

Keightley, David N., 1975. "Legitimation in Shang China," unpublished MS, presented to the Conference on Legitimation of Chinese Imperial Regimes, Asilomar, CA (June 15–24).

1982. "Akatsuka Kiyoshi and the Culture of Early China: A Study in Historical Method," *Harvard Journal of Asiatic Studies* 42.1, 267–320.

1984. "Late Shang Divination: The Magico-Religious Legacy," in Henry Rosemont Jr. (ed.), *Journal of the American Academy of Religion Studies* 50.2, *Explorations in Early Chinese Cosmology*, 11–34.

1985. *Sources of Shang History: The Oracle-Bone Inscriptions of Bronze Age China* (Berkeley: University of California).

1988. "Shang Divination and Metaphysics," *Philosophy East and West* 38.4, 367–97.

1989. "The Origins of Writing in China: Scripts and Cultural Contexts," in W.M. Senner (ed.), *The Origins of Writing* (Lincoln: University of Nebraska), 171–202.

1996. "Art, Ancestors, and the Origins of Writing in China," *Representations* 56, 68–95.

1997. "Graphs, Words, and Meanings: Three Reference Works for Shang Oracle-bone Studies, with an Excursus on the Religious Role of the Day or Sun," *Journal of the American Oriental Society* 117.3, 507–24.

1998. "Shamanism, Death, and the Ancestors: Religious Mediation in Neolithic and Shang China (ca. 5000–1000 B.C.)," *Asiatischen Studien* 52, 763–831.

1999a. "The Environment of Ancient China," in Michael Loewe and Edward L. Shaughnessy (eds.), *The Cambridge History of Ancient China, from the Origins of Civilization to 221 B.C.* (Cambridge: Cambridge University Press), 30–6.

1999b. "The Shang," in Michael Loewe and Edward L. Shaughnessy (eds.), *The Cambridge History of Ancient China, from the Origins of Civilization to 221 B.C.* (Cambridge: Cambridge University Press), 232–91.

2000. *The Ancestral Landscape: Time, Space, and Community in Late Shang China (ca. 1200–1045 B.C.)* (Berkeley: Institute of East Asian Studies).

2004. "The Making of the Ancestors: Late Shang Religion and Its Legacy," in J. Lagerwey (ed.), *Religion and Chinese Society*, Vol. 1, *Ancient and Medieval* (Hong Kong: Chinese University of Hong Kong), 3–64.

Kelley, David H. and Milone, Eugene F., 2011. *Exploring Ancient Skies: A Survey of Ancient and Cultural Astronomy* (New York: Springer).

Kennedy, E.S., 1963. "Astronomy and Astrology in India and Iran," *Isis* 54.2, 229–46.

1971. *The Astrological History of Masha-Alla* (Cambridge, MA: Harvard University Press).

Kepler, Johannes, 1614. *De anno natali Christi*.

Kerenyi, Karl, 1980. *The Gods of the Greeks* (London: Thames & Hudson, 1951).

Kern, Martin, 1997. *Die Hymnen der chinesischen Staatsopfer: Literatur und Ritual in der politischen Representation von der Han-Zeit bis zu den Sechs Dynastien* (Stuttgart: Franz Steiner).

2000. "Religious Anxiety and Political Interest in Western Han Omen Interpretation: The Case of the Han Wudi 漢武帝 Period (141–87 B.C.)," *Chūgoku shigaku* 中國史學 10, 1–31.

2007. "The Performance of Writing in Western Zhou China," in Sergio La Porta (ed.), *The Poetics of Grammar and the Metaphysics of Sound and Sign* (Leiden: Brill), 109–75.

2009. "Bronze Inscriptions, the *Shijing* and the *Shangshu*: The Evolution of the Ancestral Sacrifice during the Western Zhou," in J. Lagerwey and M. Kalinowski (eds.), *Early Chinese Religion*, Part 1, *Shang through Han (1250 BC–220 AD)* (Leiden: Brill), Vol. 1, 143–200.

Kestner, Ladislav, 1991. "The *Taotie* Reconsidered: Meanings and Functions of Shang Theriomorphic Imagery," *Artibus Asiae* 51.1–2, 29–53.

Kiang, Tao, 1984. "Notes on Traditional Chinese Astronomy," *Observatory* 104 (February), 19–23.

Kierman, Frank A., Jr., 1974. "Phases and Modes of Combat in Early China," in F.A. Kierman Jr. and J.K. Fairbank (eds.), *Chinese Ways in Warfare* (Cambridge, MA: Harvard University Press), 47–56.

Kim, Il-gwon 金一權, 2005. "Astronomical and Spiritual Representations," *Preservation of the Koguryo Kingdom Tombs* (Paris: ICONOS), 25–32.

Kim, Seung-Og, 1994. "Burials, Pigs, and Political Prestige in Neolithic China," *Current Anthropology* 35.2, 119–41.

Klein, Cecelia F., 1982. "Woven Heaven, Tangled Earth: A Weaver's Paradigm of the Mesoamerican Cosmos," *Ethnoastronomy and Archaeoastronomy in the American Tropics, Annals of the New York Academy of Sciences* 385.1, 1–35.

Knechtges, David R., 1987. *Wen xuan or Selections of Refined Literature* (Princeton: Princeton University Press), Vol. 2, 263–77.

Kozloff, Arielle P., 1994. "Star-Gazing in Ancient Egypt," *Bibliothèque d'étude* 106.1–4, 169–76.

Kozloff, Arielle P. and Bryan, Betsy M., 1992. *Egypt's Dazzling Sun: Amenhotep III and His World* (Bloomington: Cleveland Museum of Art and Indiana University Press).

Kronk, Gary, 1997. "A Large Comet Seen in 135 B.C.?" *International Comet Quarterly* 19, 3–7.

Krupp, Edwin C., 1991. *Beyond the Blue Horizon: Myths and Legends of the Sun, Moon, Stars, and Planets* (New York: Oxford University Press).

Kryukov, Mikhail., 1986. "K probleme tsiklicheskikh znakov v Drevnem Kitae," in Ju. V. Knorozov (ed.), *Drevnye sistemy pis'ma – etnicheskaja semiotika* (Moscow, Nauka), 107–13.

Kuhn, Dieter, 1995. "Silk Weaving in Ancient China: From Geometric Figures to Patterns of Pictorial Likeness," *Chinese Science* 12, 77–114.

Kunst, Richard A., 1985. "The Original '*Yijing*': A text, Phonetic Transcription, Translation, and Indexes, with Sample Glosses," Ph.D. dissertation, University of California, Berkeley.

Lai Guolong, 2005. "Death and the Otherworldly Journey in Early China as Seen through Tomb Texts, Travel Paraphernalia, and Road Rituals," *Asia Major* 18.1, 1–44.

Lai, Zhide 來知德, 1972. *Ding zheng Yijing Lai zhu tujie* 訂正易經來註圖解 (rpt., Taipei: Zhongguo Kong xuehui).

Lang, Ying 郎瑛, 1961. *Qixiu leigao* 七修類稿, *Ming Qing biji congkan* 明清筆記叢刊 edition (Beijing: Zhonghua).

Lau, D.C. (trans.), 1970. *Mencius* (Harmondsworth, Middlesex: Penguin).

Lau, D.C. and Ames, Roger T. (trans.), 1996. *Sun Bin: The Art of Warfare* (New York: Ballantine).

Laube, A., Steinmetz, M., and Vogler, G. (eds.), 1974. *Illustrierte Geschichte der deutschen frühbürgerlich Revolution* (Berlin: Dietz Verlag).

Laurencich-Minelli, Laura and Magli, Giulio, 2008. "A Calendar *Quipu* of the Early 17th Century and Its Relationship with the Inca Astronomy," *History of Physics* 801, ArXiv e-prints, arXiv:0801.1577v1, web version.

Lebeuf, A., 1996. "The Milky Way, a Path of the Souls," in V. Koleva and D. Kolev (eds.), *Astronomical Traditions in Past Cultures* (Sofia: Institute of Astronomy BAS), 148–61.

Lee, Dorothy, 1973. "Codifications of Reality: Lineal and Nonlineal," in R.E. Ornstein (ed.), *The Nature of Human Consciousness* (San Francisco: W.H. Freeman), 128–42.

Legge, James, 1972. *The Chinese Classics, with a Translation, Critical and Exegetical Notes, Prolegomena, and Copious Indexes*, 5 vols. (Taipei: Wen shi zhe, 1972; first published Hong Kong and London: Trubner, 1861–1872 [Shanghai: Commercial Press, 1872]).

Levi, Jean, 1977. "le mythe de l'âge d'or et les théories de l'évolution en Chine ancienne," *L'homme* 17, 73–103.

1989. *Les fonctionnaires divins: Politique, despotisme et mystique en Chine ancienne* (Paris: Éditions de Seuil).

Lévi-Strauss, Claude, 1969. *The Raw and the Cooked: Mythologiques*, Vol. 1 (Chicago: University of Chicago).

Lewis, Mark Edward, 1990. *Sanctioned Violence in Early China* (Albany: State University of New York).

1999. *Writing and Authority in Early China* (Albany: State University of New York).

2006a. *The Construction of Space in Early China* (Albany: State University of New York).

2006b. *The Flood Myths of Early China* (Albany: State University of New York).

2007. *The Early Chinese Empires: Qin and Han* (Cambridge, MA: Harvard University Press).

2009. "The Mythology of Early China," in J. Lagerwey and M. Kalinowski (eds.), *Early Chinese Religion*, Part One, *Shang through Han (1250 BC–220 AD)* (Leiden: Brill), Vol. 1, 543–94.

Lewis-Williams, David and Pearce, David, 2005. *Inside the Neolithic Mind* (London: Thames and Hudson).

Li, Changhao 黎昌顥 (ed.), 1981. *Zhongguo tianwenxue shi* 中國天文學史 (Beijing: Kexue).

Li, Daoyuan 酈道元, 1983–6. *Shui jing zhu jishi ding'e* 水經注集釋訂訛 (*Siku quanshu* 四庫全書, Wen yuan ge edition (1782) (rpt., Taipei: Shangwu), digital edition.

Li, Feng, in press. *Early China: A Social and Cultural History* (New York: Columbia University Press).

Li, Feng 李峰, in press. "Qinghua jian 'Qi ye' chu du ji qi xiangguan wenti" 清華簡《耆夜》初讀及其相關問題, in Li Zongkun 李宗焜 (ed.), *Chutu cailiao yu xin shiye* 出土材料與新視野 (Taipei: Academia Sinica).

Li, Feng and Branner, David P. (eds.), 2011. *Writing and Literacy in Early China* (Seattle: University of Washington).

Li, Geng 黎耕 and Sun, Xiaochun 孫小淳, 2010. "Taosi IIM22 qigan yu guibiao ceying" 陶寺IIM22漆桿與圭表測影 *Zhongguo keji shi zazhi* 中國科技史雜誌 31.4, 363–72.

Li, Jianmin 李建民, 1999. "Taiyi xin zheng: yi Guodian Chu jian wei xiansuo" 太一新證: 以郭店楚簡為線索, *Chūgoku shutsudo shiryō kenkyū* 中國出土史料研究 3, 46–62.

2007. "Taosi yizhi chutu de zhu wen 'wen' zi pian hu" 陶寺遺址出土的朱書'文'字扁壺, in Xie Xigong 解希恭 (ed.), *Xiangfen Taosi yizhi yanjiu* 襄汾陶寺遺址研究 (Beijing: Kexue), 620–3; rpt. from *Zhongguo shehui kexue yuan gudai wenming yanjiu zhongxin tongxun* 中國社會科學院古代文明研究中心通訊 1(January 2001).

Li, Jingchi 李鏡池, 1978. *Zhou Yi tanyuan* 周易探源 (Beijing: Zhonghua, first published 1930).

1981. *Zhou Yi tongyi* 周易通義 (Beijing: Zhonghua).

Li, Ling 李零, 1985. *Changsha Zidanku Zhanguo Chu boshu yanjiu* 長沙子彈庫戰國楚帛書研究 (Beijing: Zhonghua).

1991. "'Shitu' yu Zhongguo gudai de yuzhou moshi" '式圖'與中國古代的宇宙模式, Part 1, *Jiuzhou xuekan* 4.1, 5–52; "'Shitu' yu Zhongguo gudai de yuzhou moshi" '式圖'與中國古代的宇宙模式, Part 2, *Jiuzhou xuekan* 4.2, 49–76; rpt. as "'Shi tu' yu Zhongguo gudai de yuzhou moshi" '式圖'與中國古代的宇宙模式, in Li Ling, *Zhongguo fangshu kao* 中國方術考, revised edition (Beijing: Dongfang, 2000), 89–176.

1995–6. "An Archaeological Study of Taiyi 太一 (Grand One) Worship," *Early Medieval China* 2, 1–39.

2000. *Zhongguo fangshu kao* 中國方術考, revised edition (Beijing: Dongfang, first published 1993).

2002. "Lun Gong xu faxian de yiyi" 論幽公盨發現的意義, *Zhongguo lishi wenwu* 6, 35–45.

Li, Ling and Cook, C.A., 2004. "The Chu Silk Manuscript," in C.A. Cook and J.S. Major (eds.), *Defining Chu: Image and Reality in Ancient China* (Honolulu: University of Hawaii Press), 171–7.

Li, Liu, 2007, *The Chinese Neolithic: Trajectories to Early States* (Cambridge: Cambridge University Press).

Li, Qin 李勤, 1991. *Xi'an Jiaotong daxue Xi Han bihua mu* 西安交通大學西漢壁畫墓 (Xi'an: Xi'an Jiaotong daxue).

Li, Xiaoding 李孝定, 1970. *Jiaguwen jishi* 甲骨文字集释 (Taipei: Zhongyang yanjiuyuan lishi yuyan yanjiusuo).

Li, Xiusong 李修松, 1995. "Xia wenhua de zhongyao zhengju – Taosi yizhi chutu caihui taopan tu'an kaoshi" 夏文化的重要證據–陶寺遺址出土彩繪陶盤圖案考釋 *Qi Lu xuekan* 齊魯學刊 1, 82–7.

Li, Xueqin 李學勤, 1982. "Lun Chu boshu zhong de tianxiang" 論楚帛書中的天象, *Hunan kaogu jikan* 1, 68–72.

1985. "Shangdai de sifang feng yu sishi" 商代的四風與四時, *Zhongzhou xuekan* 中州學刊 5, 99–101.

1989. "*Xia xiao zheng* xin zheng" 夏小正新證, *Nong shi yanjiu* 農史研究 8 (May), 4–11.

1992–3. "A Neolithic Jade Plaque and Ancient Chinese Cosmology," *National Palace Museum Bulletin*, Vol. XXVII. 5–6 (November/December 1992–January/February 1993), 1–8.

1999a. *Xia Shang Zhou niandai xue zhaji* 夏商週年代學札記 (Shenyang: Liaoning daxue).

1999b. "Xu shuo '*niao xing*'" 續說鳥星, in Li Xueqin, *Xia Shang Zhou niandai xue zhaji* 夏商週年代學札記 (Shenyang: Liaoning daxue), 62–6.

1999c. "Zi Fan bianzhong de li ri wenti" 《子犯編鐘》的歷日問題, in Li Xueqin, *Xia Shang Zhou niandai xue zhaji* 夏商週年代學札記 (Shenyang: Liaoning daxue), 105–13.

2000. "'Xiaokai' que ji yueshi" 小開'確記月食, *Gudai wenming yanjiu tongxun* 古代文明研究通訊 6 (May).

2001. "Chu boshu yanjiu" 楚帛書研究, in Li Xueqin, *Jianbo yiji yu xueshu shi* 簡帛佚籍與學術史 (Nanchang: Jiangsu jiaoyu), 38, 44–5.

2002. "Lun Gong xu ji qi zhongyao yiyi" 论幽《公盨》及其重要意义, in Li Xueqin, *Zhongguo gudai wenming yanjiu* 中國古代文明研究 (Shanghai: Huadong shifan daxue, 2002), 126–36; rpt. from *Zhongguo lishi wenwu* 中國歷史文物 6.

2005a. "Lun Yinxu buci de xin xing" 論殷墟卜辭的新星, in Li Xueqin, *Zhongguo gudai wenming yanjiu* 中國古代文明研究 (Shanghai: Huadong shifan daxue), 7–11.

2005b. "Zhongguo wenzi yu shufa de luansheng" 中國文字與書法的肇生, in Li Xueqin, *Gudai Zhongguo wenming yanjiu* 中國古代文明研究 (Shanghai: Huadong shifan daxue), 385–8.

Li, Xueqin, Harbottle, Garman, Zhang, Juzhong, Wang, Changsui, 2003. "The Earliest Writing? Sign Use in the Seventh Millennium BC at Jiahu, Henan Province, China," *Antiquity* 77.295, 31–44.

Li, Yong, 1992. "Dui Zhongguo gudai hengxing fenye he fenye shipan yanjiu" 對中國古代恆星分野和分野式盤研究, *Ziran kexue shi yanjiu* 自然科學史研究 11.1, 22–31.

Liao, Mingchun 廖名春, 1999. "*Zhouyi* liang gua yao ci wu kao" 《周易》乾坤兩卦爻辭五考 *Zhouyi yanjiu* 周易研究 1 (February), 38–49.

Lilley, Ian, 2010. "Near Oceania," in Ian Lilley (co-ordinator), *Early Human Expansion and Innovation in the Pacific* (Paris: ICOMOS), 13–46.

Lin, Li-chen, 1995. "The Concepts of Time and Position in the *Book of Changes* and Their Development," in C.C. Huang and E. Zürcher (eds.), *Cultural Notions of Time and Space in China* (Leiden: Brill), 89–113.

Lin, Yun 林澐, 1986. "A Reexamination of the Relationship between Bronzes of the Shang Culture and of the Northern Zone," in K.C. Chang (ed.), *Studies of Shang Archaeology: Selected Papers from the International Conference on Shang Civilization* (New Haven: Yale University Press).

1998. *Lin Yun xueshu wenji* 林澐學術文集 (Beijing: Zhongguo dabaike), 167–73; rpt. from "*Tian Wang gui* 'wang si yu tian shi' xinjie" 《天亡簋》'王祀於天室'新解, *Shixue jikan* 史學集刊 3 (1993), 25–8.

Liu, An 劉安, 1927–35. *Huainanzi* 淮南子. *Sibu beiyao* 四部備要 edition (Shanghai: Zhonghua, rpt. Taipei: Taiwan Chung-hua, 1966).

1974. *Huainanzi* 淮南子. *Xinbian Zhuzi jicheng* 新编诸子集成 (rpt., Taipei: Shijie shuju), Vol. 7.

Liu, Cary Y., Nylan, Michael, Barbieri-Low, Anthony, and Loewe, Michael, 2005. *Recarving China's Past: Art, Archaeology and Architecture of the "Wu Family Shrines"* (Princeton: Princeton University Art Museum and Yale University Press).

Liu, Ciyuan, Liu, Xueshun, and Ma, L., 2005. "A Chinese Observatory Site of 4,000 Year [*sic*] Ago," *Journal of Astronomical History and Heritage* 8.2, 129–30.

Liu, Ciyuan and Zhou, Xiaolu, 1999. "The Sky Brightness when the Sun Is in Eclipse," *Chinese Astronomy and Astrophysics* 23, 249–57.

Liu, Ciyuan and Zhou, Xiaolu, 2001. "Analysis of Astronomical Records of King Wu's Conquest," *Science in China* (series A) 44.9, 1216–24.

Liu, Guozhong 劉國忠, 2011. *Zou jin Qinghua jian* 走進清華簡, in Li Xueqin 李學勤 (ed.), *Qinghua jian yanjiu congshu* 清華簡研究叢書 (Beijing: Gaodeng jiaoyu).

Liu, Huiping 劉惠萍, 2003. *Fu Xi shenhua chuanshuo yu xinyang yanjiu* 伏羲神話傳說與信仰研究 (Taipei: Wenjin).

2008. "Xiangtian tongshen – guanyu Tulufan muzang chutu Fu Xi Nü Wa tu de zai sikao" 象天通神–關於吐魯番墓葬出土伏羲女媧圖的再思考, *Dunhuang xue* 敦煌學 27, 293–310.

Liu, Lexian 劉樂賢, 2004. *Mawangdui tianwen shu kaoshi* 馬王堆天文書考釋 (Guangdong: Zhongshan daxue).

Liu, Li, 2007. *The Chinese Neolithic: Trajectories to Early States* (Cambridge: Cambridge University Press).

Liu, Li and Chen, Xingcan, 2003. *State Formation in Early China* (London: Duckworth).

Liu, Li and Xu, Hong, 2007. "Rethinking *Erlitou*: Legend, History and Chinese Archaeology," *Antiquity* 81, 886–901.

Liu, Qingzhu, 2007. "Archaeological Discovery and Research into the Layout of the Palaces and Ancestral Shrines of Han Dynasty Chang'an: A Comparative Essay on the Capital Cities of Ancient Chinese Kingdoms and Empires," *Early China* 31, 113–43.

Liu, Qiyu 劉起釪, 2004. "Yaodian Xi He zhang yanjiu" 《堯典》羲和章研究, *Zhongguo shehui kexue yuan lishi yanjiusuo xuekan* 中國社會科學院歷史研究所學刊 2, 43–70.

Liu, Shaojun 劉韶軍, 1993. *Zhongguo gudai zhanxing shu zhu ping* 中國古代占星術注評 (Beijing: Shifan daxue).

Liu, Wu 劉斌, 2001. "Liangzhu wenhua de jitan yu guanxiang cenian" 良渚文化的祭壇與觀象測年, *Zhongguo wenwu bao* 中國文物報, January 5, section 7.

Liu, Xinglong 劉興隆, 1986. *Jiagu wenzi jiju jianshi* 甲骨文字集句簡釋 (Zhengzhou: Zhongzhou guji).

Liu, Xueshun 劉學順, 2009. "Yin dai lifa: xiancun zui zao de Zhongguo tuibu li" 殷代曆法:現存最早的中國推步歷, *Yindu xuekan* 殷都學刊 2, 24–8.

Liu yi zhi yi lu 六藝之一錄, 1983–6. *Siku quanshu* 四庫全書, Wen yuan ge edition (1782) (rpt., Taipei: Shangwu), digital edition.

Liu, Yu 劉雨, 2003. "Bin gong kao 豳公考," in Zhang Guangyu 張光裕 (ed.), *Di si jie guoji Zhongguo gu wenzi xue yantao hui lunwen ji* 第四屆國際中國古文字學研討會論文集 (Hong Kong: Xianggang Zhongwen daxue Zhongguo yuwen ji wenxue xi), 97–106.

Liu, Yunyou 劉雲友 (Xi Zezong 席澤宗), 1974. "Zhongguo tianwen shi shang de yizhong zhongyao faxian – Mawangdui Han mu boshu zhong de 'Wu xing zhan'" 中國天文史上的一種重要發現 – 馬王堆漢墓帛书中的《五星占》, *Wenwu* 11, 28–39.

Liu, Zhao 劉釗, 2009. "'Xiao chen Qiang ke ci' xin shi 小臣墻刻辭新釋," *Fudan xuebao (shehui kexue ban)* 復旦學報(社會科學版) 1, 4–11.

Liu, Zongdi 劉宗迪, 2002. "*Shanhaijing: Dahuangjing* yu *Shangshu: Yaodian* de duibi yanjiu" 《山海經·大荒經》與《尚書·堯典》的對比研究, *Minzu yishu* 民族藝術 3, available at http://hi.baidu.com/fdme/blog/item/00dda88fcb3796fb 503d92d8.html, accessed May 13, 2011.

Liutao 六韜, "Longtao" 龍韜 (Shijiazhuang: Hebei renwu, 1991; Bingjia baodian edition).

Lloyd, Geoffrey E.R. and Sivin, Nathan, 2002. *The Way and the Word: Science and Medicine in Early China and Greece* (New Haven: Yale University Press).

Locke, L. Leland, 1912. "The Ancient Quipu, a Peruvian Knot Record," *American Anthropologist*, New Series, 14.2 (April–June), 325–32.

Lockyer, J. Norman, 1964. *The Dawn of Astronomy: A Study of the Temple Worship and Mythology of the Ancient Egyptians* (Cambridge, MA: MIT Press, first published 1891).

Loehr, Max, 1968. *Ritual Vessels of Bronze Age China* (New York: Asia Society).

Loewe, Michael, 1967. *Records of Han Administration* (Cambridge: Cambridge University Press).

  1974. "The Campaigns of Han Wu-ti," in F.A. Kierman Jr. and J.K. Fairbank (eds.), *Chinese Ways in Warfare* (Harvard University Press), 67–122.

  1987. "The Cult of the Dragon and the Invocation for Rain," in C. LeBlanc and S. Blader (eds.), *Chinese Ideas about Nature and Society: Studies in Honour of Derk Bodde* (Hong Kong: Hong Kong University Press), 195–214.

550 References

(ed.), 1993. *Early Chinese Texts: A Bibliographical Guide* (Berkeley: Society for the Study of Early China and Institute of East Asian Studies, University of California).

1994a. "The Han View of Comets," in Michael Loewe, *Divination, Mythology and Monarchy in Han China* (Cambridge: Cambridge University Press), 61–84.

1994b. "The Oracles of the Clouds and the Winds," in Michael Loewe, *Divination, Mythology and Monarchy in Han China* (Cambridge: Cambridge University Press), 191–213.

1995. "The Cycle of Cathay: Concepts of Time in Han China and Their Problems," in C.C. Huang and E. Zürcher (eds.), *Cultural Notions of Time and Space in China* (Leiden: Brill), 305–28.

2000. *A Biographical Dictionary of the Qin, Former Han and Xin Periods (221 BC–AD 24)* (Leiden: Brill).

Loewe, Michael and Shaughnessy, Edward L. (eds.), 1999. *The Cambridge History of Ancient China: From the Origins of Civilization to 221 B.C.* (Cambridge: Cambridge University Press).

Löwith, Karl, 1949. *Meaning in History* (Chicago: The University of Chicago Press).

Lü, Buwei 呂不韋, 1974. *Lüshi chunqiu* 呂氏春秋, *Xinbian zhuzi jicheng* 新編諸子集成 ed. (Taipei: Shijie shuju).

Lu, Yang 盧央 and Shao, Wangping 邵望平, 1989. "Kaogu yicun zhong suo fanying de shiqian tianwen zhishi" 考古遺存中所反應的史前天文知識, in Zhongguo shehui kexueyuan kaogu yanjiusuo 國社會科學院考古研究所 (ed.), *Zhongguo gudai tianwen wenwu lunji* 中國古代天文文物論集 (Beijing: Wenwu), 1–16.

Luo, Qikun 雒启坤, 1991. "Xi'an Jiaotong daxue Xi han muzang bihua ershiba xiu xingtu kaoshi" 西安交通大學西漢墓葬壁畫二十八宿星圖考釋, *Ziran kexue shi yanjiu* 自然科學史研究 10.3, 236–45.

Ma, Chengyuan 馬承源, 1976. "*He zun* ming wen chu shi" 《何尊》銘文初釋, *Wenwu* 1, 64–5.

(ed.), 1989. *Shang Zhou qingtong qi ming wen xuan* 商周青銅器銘文選, Vol. 3, *Shang Xi Zhou qingtong qi ming wen shiwen ji zhushi* 商西周青銅器銘文釋文及註釋 (Beijing: Wenwu).

1992. *The Chinese Bronzes* 中國青銅器 (Shanghai: Shanghai guji, first published 1988).

Ma, Guohan 馬國翰 (ed.), n.d. *Yuhan shanfang ji yishu* 玉函山房輯佚書, 6 vols. (Taipei: Jinan keben).

MacRobert, Alan, 2005. "Venus Revealed," *Sky & Telescope* 110 (December), 116.

Maeyama, Yasukatsu, 2002. "The Two Supreme Stars, Thien-i and Thai-i, and the Foundation of the Purple Palace," in S.M.R. Ansari (ed.), *History of Oriental Astronomy*, IAU Congress Proceedings (Kyoto 1997) (Dordrecht: Kluwer), 3–18.

2003. "The Oldest Star Catalog of China, Shih Shen's Hsing Ching," in Maeyama Yasukatsu, *Astronomy in Orient and Occident: Selected Papers on Its Cultural and Scientific History* (Hildesheim: Georg Olms), 1–34; rpt. from Y. Maeyama and W. Saltzer (eds.), *Prismata: Festschrift für Willy Hartner* (Wiesbaden, 1977).

Magli, Giulio, 2004. "On the Possible Discovery of Precessional Effects in Ancient Astronomy," arXiv:physics/0407108v2 [physics.hist-ph][v1] Tue, 20 Jul 2004 17:45:26 GMT (561kb) [v2] Sunday, August 1, 2004, 09:36:30 GMT (505kb).

2009. *Mysteries and Discoveries of Archaeoastronomy* (New York: Springer).

References   551

Major, John S., 1978. "Myth, Cosmology, and the Origins of Chinese Science," *Journal of Chinese Philosophy* 5, 1–20.

1987. "The Meaning of Hsing-te [Xingde]," in C. LeBlanc and S. Blader (eds.), *Chinese Ideas about Nature and Society* (Hong Kong: Hong Kong University Press), 281–91.

1993. *Heaven and Earth in Early Han Thought: Chapters Three, Four, and Five of the Huainanzi* (Albany: State University of New York).

Major, John S., Queen, Sarah A., Meyer, Andrew S., and Roth, Harold D. (trans.), 2010. *The Huainanzi* (New York: Columbia University Press).

Mallory, J.P. and Mair, Victor H. (eds.), 2000. *The Tarim Mummies: Ancient China and the Mystery of the Earliest Peoples from the West* (London: Thames and Hudson).

Mann, Charles C., 2003. "Cracking the *Khipu* Code," *Science* 300 (June 13), 1650–1.

Maspero, Henri, 1929. "L'astronomie Chinoise avant les Han," *T'oung Pao* 26, 267–356.

1948–51. "La Ming-tang et la crise religieuse chinoise avant les Han," *Mélanges chinois et bouddhiques* 9, 1–71.

1953. *Les documents chinois de la troisième expedition de Sir Aurel Stein en Asie centrale* (London: J. Gernet).

Matheson, Peter, 1994. *The Collected Works of Thomas Müntzer* (Edinburgh: T. and T. Clark).

Mathieu, Rémi, 1983. "Reviewed Work: *The Heir and the Sage: Dynastic Legend in Early China* by Sarah Allan," *L'homme* 23.3, 137–8.

Mawangdui Hanmu boshu zhengli xiaozu, 1980. *Mawangdui Hanmu boshu* (Beijing: Wenwu), Vol. 1.

Meeus, Jean, 1997a. *Mathematical Astronomy Morsels* (Richmond: Willmann-Bell).

1997b. "Planetary Groupings and the Millennium," *Sky & Telescope* 94.2, 60–2.

Mei, Yi-Pao (trans.), 1929. *The Ethical and Political Works of Motse (Mozi, Mo-tzu, Motze)* (London: Probsthain).

Milbrath, Susan, 1999. *Star Gods of the Maya* (Austin: University of Texas).

2003. "Jupiter in Classic and Postclassic Maya Art," in J.W. Fountain and R.M. Sinclair (eds.), *Current Studies in Archaeoastronomy: Conversations across Time and Space* (Durham, NC: Carolina Academic), 301–30.

Miller, Roy Andrew, 1988. "Pleiades Preceived: From Mul.Mul to Subaru," *Journal of the American Oriental Society* 108.1, 1–25.

Ming shi 明史, 1974. (Beijing: Zhonghua).

Ming shilu 明實錄, n.d. (Beijing: Beijing Ai Rusheng shuzihua jishu yanjiu zhongxin).

Ming wen hai 明文海, 1983–6, *Siku quanshu* 四庫全書, Wen yuan ge edition (1782) (rpt., Taipei: Shangwu), digital edition.

Miranda, Noemi, Belmonte, Juan A., and Molinero, Miguel Angel, 2008. "Uncovering Seshat: New Insights into the 'Stretching of the Cord Ceremony,'" *Archaeologica Baltica* 10, 57–61.

Mitsukuni, Yoshida, 1979. "The Chinese Concept of Technology: A Historical Approach," *Acta Asiatica* 36, 49–66.

Mo Di 墨翟, 1966. *Mozi* 墨子, *Sibu beiyao* edition (Taipei: Taiwan Chung-hua, first published Shanghai: Zhonghua, 1927–35).

Mote, Frederick W. and Twitchett, Denis (eds.), 1988. *The Cambridge History of China*, Vol. 7, *The Ming Dynasty 1368–1644, Part 1* (Cambridge: Cambridge University Press).

Müller, Klaus E., 2002. "Perspectives in Historical Anthropology," in J. Rüsen (ed.),
. *Western Historical Thinking: An Intercultural Debate* (New York and Oxford:
    Berghahn), 33–52.
Nakayama, Shigeru, 1966. "Characteristics of Chinese Astrology," *Isis* 57, 442–54.
Needham, Joseph, 1969. *Science and Civilisation in China*, Vol. 2, *History of Scientific
    Thought* (Cambridge: Cambridge University Press).
    1970. "Astronomy in Classical China," in Joseph Needham, *Clerks and Craftsmen
    in China and the West: Lectures and Addresses on the History of Science and
    Technology* (Cambridge: Cambridge University Press), 1–13.
    1981. "Time and Knowledge in China and the West," in J.T. Fraser (ed.), *The Voices of
    Time: A Cooperative Survey of Man's Views of Time as Expressed by the Sciences
    and by the Humanities* (Amherst: University of Massachusetts), 233–6.
Needham, Joseph and Lu, Gwei-Djen, 1985. *Trans-Pacific Echoes & Resonances: Lis-
    tening Once Again* (Singapore: World Scientific).
Needham, Joseph, with the research assistance of Wang Ling, 1954. *Science and Civil-
    isation in China*, Vol. 1, *Introductory Orientations* (Cambridge: Cambridge Uni-
    versity Press).
Needham, Joseph, with the research assistance of Wang Ling, 1959. *Science and Civil-
    isation in China*, Vol. 3, *Mathematics and the Sciences of the Heavens and the
    Earth* (Cambridge: Cambridge University Press).
Needham, Joseph, with the research assistance of Wang Ling, and the special co-
    operation of Kenneth Robinson, 1962. *Science and Civilisation in China*, Vol. 4.1,
    Physics (Cambridge: Cambridge University Press).
Neugebauer, Otto, 1975. *A History of Ancient Mathematical Astronomy* (Berlin, Hei-
    delberg and New York).
Nienhauser, William H., Cheng, Tsia-fa, Lu, Zongli, and Reynolds, Robert (trans.),
    1994. Ssu-ma Ch'ien, *The Grand Scribe's Records*, Vol. 1, *The Basic Annals of
    Pre-Han China* (Bloomington: Indiana University Press).
Nienhauser, William H., Jr., Cao, Weiguo, Galer, Scott W., Pankenier, David W. (trans.),
    2002. *The Grand Scribe's Records*, Vol. 2, *The Basic Annals of Han China* (Bloom-
    ington: Indiana University Press).
Nisbett, Richard E., 2003. *The Geography of Thought: How Asians and Westerners
    Think Differently. . . and Why* (New York: The Free Press).
Nivison, David S., 1989. "The 'Question' Question," *Early China* 14, 115–25.
North, John D., 1986. "Celestial Influence: The Major Premise of Astrology," in
    P. Zambelli (ed.), *Astrologi Hallucinati: Stars and the End of the World in Luther's
    Time* (Berlin: W. de Gruyter), 45–100.
O'Keefe, Daniel L., 1983. *Stolen Lightning: A Social Theory of Magic* (New York:
    Vintage).
Oliveira, C. and da Silva, Cândido M., 2010. "Moon, Spring, and Large Stones," *Pro-
    ceedings of the XV World Congress UISPP* (Lisbon, September 4–9, 2006), 7,
    Session C68 (Part I), *BAR International Series*, S2122, 83–90.
Olson, Donald W. and White, Brian D., 1997. "A Planetary Grouping in Maya Times,"
    Sky & Telescope 97.2 (August), 63–4.
Ong, Walter J. 2002. *Orality and Literacy: The Technologizing of the Word* (New York:
    Routledge, first published 1982).

Ornstein, Robert E., 1973. *The Nature of Human Consciousness* (San Francisco: W.H. Freeman).

Pang, Kevin D., 1987. "Extraordinary Floods in Early Chinese History and Their Absolute Dates," *Journal of Hydrology* 96, 139–55.

Pang, Kevin D. and Bangert, J.A., 1993. "The 'Holy Grail' of Chinese Astronomy: The Sun–Moon, Five Planet Conjunction in Yingshi (Pegasus) on March 5, 1953 BC," *Journal of the American Astronomical Society* 25, 922.

Pang, Pu 龐樸, 1978. "Huoli chu tan" 火歷初探, *Shehui kexue zhanxian* 社會科學戰線 4, 131–7.

Pankenier, David W., 1981. "Astronomical Imagery in *Qian Gua* in the *Book of Changes*, with Particular Reference to 'Arrogant Dragon'," unpublished MS, March 30.

1981–2. "Astronomical Dates in Shang and Western Zhou," *Early China* 7, 2–37.

1982. "Early Chinese Positional Astronomy: The *Guoyu* Astronomical Record," *Archaeoastronomy* 5.3 (July–September), 10–20.

1983. "Early Chinese Astronomy and Cosmology: The Mandate of Heaven as Epiphany," Ph.D. dissertation, Stanford University.

1983–5. "*Mozi* and the Dates of Xia, Shang, and Zhou: A Research Note," *Early China* 9–10, 175–83.

1986. "The Metempsychosis in the Moon," *Bulletin of the Museum of Far Eastern Antiquities* 58, 149–59.

1990. "The Scholar's Frustration Reconsidered: Melancholia or Credo?" *Journal of the American Oriental Society* 110.3, 434–59.

1992a. "The *Bamboo Annals* Revisited: Problems of Method in Using the Chronicle as a Source for the Chronology of Early Zhou, Part 1," *Bulletin of the School of Oriental and African Studies*, Vol. LV.2, 272–97.

1992b. "The Bamboo Annals Revisited: Problems of Method in Using the Chronicle as a Source for the Chronology of Early Zhou, Part 2: The Congruent Mandate Chronology in *Yi Zhou shu*," *Bulletin of the School of Oriental and African Studies*, Vol. LV.3, 498–510.

1992c. "Reflections of the Lunar Aspect on Western Chou Chronology," *T'oung Pao* 78, 33–76.

1995. "The Cosmo-political Background of Heaven's Mandate," *Early China* 25, 121–76.

1998a. "Applied Field Allocation Astrology in Zhou China: Duke Wen of Jin and the Battle of Chengpu (632 BCE)," *Journal of the American Oriental Society* 119.2, 261–79.

1998b. "The Mandate of Heaven," *Archaeology* 51.2 (March/April), 26–34.

2000. "Popular Astrology and Border Affairs in Early China: An Archaeological Confirmation," *Sino-Platonic Papers* 104, 1–19.

2004a. "A Brief History of *Beiji* 北极 (Northern Culmen), with an Excursus on the Origin of the Character *di* 帝," *Journal of the American Oriental Society* 124.2, 211–36.

2004b. "A Short History of *Beiji*," *Culture and Cosmos* 8.1–2, 287–308.

2004c. "Temporality and the Fabric of Space–Time in Early Chinese Thought," in Ralph M. Rosen (ed.), *Time and Temporality in the Ancient World* (Philadelphia: University of Pennsylvania Museum), 129–46.

2005. "Characteristics of Field Allocation (*fenye* 分野) Astrology in Early China," in J.W. Fountain and R.M. Sinclair (eds.), *Current Studies in Archaeoastronomy: Conversations across Time and Space* (Durham: Carolina Academic Press), 499–513.

2007. "*Caveat Lector*: Comments on Douglas J. Keenan, 'Astro-historiographic Chronologies of Early China Are Unfounded,'" *Journal of Astronomical History and Heritage* 10.2, 137–41.

2011. "Getting 'Right' with Heaven and the Origins of Writing in China," *Writing and Literacy in Early China* (Seattle: University of Washington), 13–48.

2012. "On the Reliability of Han Dynasty (206 BCE – 220 CE) Solar Eclipse Records," *Journal of Astronomical History and Heritage* 15.3, 200–12.

in press. "Babylonian Influence on Chinese Astral Prognostication (*xingzhan*)? Or 'How not to Establish Transmission.'"

Pankenier, David W., Liu, Ciyuan, and de Meis, Salvo, 2008. "The Xiangfen, Taosi Site: A Chinese Neolithic 'Observatory'?" in Jonas Vaiškūnas (ed.), *Astronomy and Cosmology in Folk Traditions and Cultural Heritage* (Klaipeda: University of Klaipeda, Archaeologia Baltica 10), 141–8.

Pankenier, David W., Xu, Zhentao, and Jiang, Yaotiao, 2008. *Archaeoastronomy in East Asia: Historical Observational Records of Comets and Meteor Showers from China, Japan, and Korea* (Youngstown, NY: Cambria).

Paper, Jordan, 1978. "The Meaning of the *T'ao-T'ieh*," *History of Religions* 18, 18–41.

Parpola, Asko, 2012. "Indus Civilization (–1750 BCE)," *Brill's Encyclopedia of Hinduism*, Vol. 4 (Leiden: Brill), 3–18.

Peratt, Anthony L. "Characteristics for the Occurrence of a High-Current, Z-Pinch Aurora as Recorded in Antiquity," *IEEE Transactions on Plasma Science* 31.6 (December 2003), 1192–1214.

Peterson, Willard, 1988. "Squares and Circles: Mapping the History of Chinese Thought," *Journal of the History of Ideas* 49.1 (January–March), 47–60.

Pettazoni, Raffaele, 1959. "The Supreme Being: Phenomenal Structure and Historical Development," in M. Eliade and J.M. Kitagawa (eds.), *The History of Religions: Essays in Methodology* (Chicago: The University of Chicago Press), 59–66.

Pines, Yuri, 2010. "Political Mythology and Dynastic Legitimacy in the *Rong Cheng Shi* Manuscript," *Bulletin of the School of Oriental and African Studies* 73.3, 503–29.

Pingree, David, 1963. "Astronomy and Astrology in India and Iran," *Isis* 54.2, 229–46.

1968. *The Thousands of Abu Ma'shar* (London: Warburg Institute).

Pingree, David and Morrisey, Patrick, 1989. "On the Identification of the Yogatārās of the Indian Nakṣatras," *Journal for the History of Astronomy*, 20, 99–119.

Pomian, Krzysztof, 1986. "Astrology as a Naturalistic Theology of History," in P. Zambelli (ed.), *Astrologi Hallucinati: Stars and the End of the World in Luther's Time* (Berlin: W. de Gruyter), 29–43.

Poo, Mu-chou, 1993. "Popular Religion in Pre-imperial China: Observations on the Almanacs of Shui-hu-ti," *T'oung Pao* 69, 225–48.

Pope, John A., Gettens, R.J., Cahill, J., and Barnard, N. (eds.), 1967. *The Freer Chinese Bronzes*, 2 vols. (Washington: Smithsonian Institution).

Porter, Deborah L., 1993. "The Literary Function of K'un-lun Mountain in the *Mu T'ien-tzu Chuan*," *Early China* 18, 74–106.

1996. *From Deluge to Discourse: Myth, History, and the Generation of Chinese Fiction* (Albany: State University of New York Press).

Postgate, Nicholas, Wang, Tao, and Wilkinson, Toby, 1995. "The Evidence for Early Writing: Utilitarian or Ceremonial?" *Antiquity* 69, 459–80.

Puett, Michael J., 1998. "Sages, Ministers, and Rebels: Narratives from Early China Concerning the Initial Creation of the State," *Harvard Journal of Asiatic Studies* 58.2, 425–79.

2002. *To Become a God: Cosmology, Sacrifice, and Self-Divinization in Early China* (Cambridge, MA: Harvard Yenching Institute).

Pulleyblank, Edwin G., 1955. *The Background of the Rebellion of An Lu-shan* (London: Oxford University Press).

1991. "The *Ganzhi* as Phonograms and Their Application to the Calendar," *Early China* 16, 39–80.

Qian, Baocong 錢寶琮, 1932. "Tai yi kao" 太一考, *Yanjing xuebao* 燕京學報, 2449–78.

Qiu, Xigui 裘錫圭, 1979. "Tantan Suixian Zeng Hou Yi mu de wenzi ziliao" 談談隨縣曾侯乙墓的文字資料, *Wenwu* 文物 7, 25–32.

1983–5. "On the Burning of Human Victims and the Fashioning of Clay Dragons to Seek Rain as Seen in Shang Dynasty Oracle Bone Inscriptions," *Early China* 9–10, 290–306.

1989. "Cong Yinxu jiagu buci kan Yin ren dui bai ma de zhongshi" 從陰虛甲骨卜辭看殷人對白馬的重視, *Yinxu bowuyuan yuankan* 殷墟博物苑苑刊 1, 70–2.

1992. "Shi hai 釋害," *Gu wenzi lunji* 古文字論集 (Beijing: Zhonghua).

1996. *Wenzi xue gaiyao* 文字學概要 (Beijing: Shangwu, first published 1988).

2000. *Chinese Writing (Wenzi xue gaiyao,* trans. Gilbert L. Mattos and Jerry Norman) (Berkeley: Society for the Study of Early China).

2004. *Zhongguo chutu wenxian shi jiang* 中國出土文獻十講 (Shanghai: Fudan daxue), 46–77, rpt. of Qiu, Xigui 裘錫圭, 2002. "*X Gong xu* ming wen kaoshi" 《燹公盨》銘文考釋, *Zhongguo lishi wenwu* 中國歷史文物 6.

Qu Yingjie 曲英杰, 1991. *Xian Qin ducheng fuyuan yanjiu* 先秦都城復原研究 (Harbin: Heilongjiang jiaoyu).

Qutan Xida 瞿曇悉達 (Gautama Siddha, *c.*729), *Kaiyuan zhanjing* 開元占經. Siku zhenben siji ed.

Ramsey, John T., 1999. "Mithradates, the Banner of Ch'ih-Yu, and the Comet Coin," *Harvard Studies in Classical Philology* 99, 197–253.

Ranieri, Marcello, 1997. "Triads of Integers: How Space Was Squared in Ancient Times," *Rivista di Topografia Antica* 7, 209–44.

Raphals, Lisa, 2008–9. "Divination in the *Han shu* Bibliographic Treatise," *Early China* 32, 45–101.

Rappenglück, Michael A., 2002. "The Milky Way: Its Concept, Function and Meaning in Ancient Cultures," in T.M. Potyomkina and V.N. Obridko (eds.), *Astronomy of Ancient Civilizations*, Proceedings of the European Society for Astronomy in Culture (SEAC 8) Conference, May 23–7, 2000 (Moscow: Nauka), 270–7.

Rawlins, Dennis and Pickering, Keith, 2001. "Astronomical Orientation of the Pyramids," *Nature* 412 (August 16), 699.

Rawson, Jessica, 2000. "Cosmological Systems as Sources of Art, Ornament and Design," *Bulletin of the Museum of Far Eastern Antiquities* 72, 133–89.

Reiche, H.A.T., 1979. "The Language of Archaic Astronomy: A Clue to the Atlantis Myth?" in Kenneth Brecher and Michael Feirtag (eds.), *Astronomy of the Ancients* (Cambridge, MA: MIT Press), 155–89.

Reiner, Erica, and Pingree, David, 1975. *Babylonian Planetary Omens*, Part 1, *The Venus Tablet of Ammisaduqa* (Malibu: Getty).

Ricoeur, Paul, 1985. "The History of Religions and the Phenomenology of Time Consciousness," in J. Kitagawa (ed.), *The History of Religions: Retrospect and Prospect* (New York: Macmillan), 13–30.

Robertson, John S., 2004. "The Possibility and Actuality of Writing," in Stephen D. Houston (ed.), *The First Writing: Script Invention as History and Process* (Cambridge: Cambridge University Press), 16–38.

Rochberg, Francesca, 2007. *The Heavenly Writing: Divination, Horoscopy, and Astronomy in Mesopotamian Culture* (Cambridge: Cambridge University Press).

Rogers, John H., 1998a. "Origins of the Ancient Constellations: I. The Mesopotamian Traditions," *Journal of the British Astronomical Association* 108, 9–28.

1998b. "Origins of the Ancient Constellations: II. The Mediterranean Traditions," *Journal of the British Astronomical Association* 108, 79–89.

Rohleder, Anna, 2011. "Asian Art Fair Brings out Buyers," Forbes.com (March), available at www.forbes.com/2001/03/21/0321connguide.html, accessed June 25, 2011.

Rosenberg, Roy A., 1972. "The 'Star of the Messiah' Reconsidered," *Biblica* 53, 105–9.

Roth, Harold, 1999. *Original Tao: Inward Training (Nei-yeh) and the Foundations of Taoist Mysticism* (New York: Columbia University Press).

Ruan, Yuan (ed.), 1970. *Shisanjing zhushu* (rpt., Taipei: Wenhua).

Ruggles, Clive L.N., 1999. *Astronomy in Prehistoric Britain and Ireland* (New Haven: Yale University Press).

2005. *Ancient Astronomy: An Encyclopedia of Cosmologies and Myth* (Santa Barbara: ABC-CLIO).

2011. "Pushing back the Frontiers or Still Running around the Same Circles? 'Interpretative Archaeoastronomy' Thirty Years on," in Clive L.N. Ruggles (ed.), *Archaeoastronomy and Ethnoastronomy: Building Bridges Between Cultures*, 'Oxford IX' International Symposium on Archaeoastronomy (IAU Symposium 278) (Cambridge: Cambridge University Press), 1–18.

Ruggles, Clive L.N., Cotte, Michael et al., 2010. *Heritage Sites of Astronomy and Archaeoastronomy in the context of the UNESCO World Heritage Convention* (Paris: ICOMOS and IAU).

Rüsen, Jörn (ed.), 2002. *Western Historical Thinking: An Intercultural Debate* (New York and Oxford: Berghahn).

Samson, Geoffrey, 1994. "Chinese Script and the Diversity of Writing Systems," *Linguistics* 32, 117–32.

Sanft, Charles, 2008–9. "Edict of Monthly Ordinances for the Four Seasons in Fifty Articles from 5 C.E.: Introduction to the Wall Inscription Discovered at Xuanquanzhi, with Annotated Translation," *Early China* 32, 125–208.

Sanfu huangtu 三輔黃圖, 1983–6. *Siku quanshu* 四庫全書, Wen yuan ge edition (1782) (rpt., Taipei: Shangwu), digital edition.

Saso, Michael, 1978. "What is the Ho-t'u?" *History of Religions* 17.3–4, 399–416.

Saturno, William A., Taube, Karl A., and Stuart, David, 2005. "The Murals of San Bartolo, El Petén, Guatemala Part 1: The North Wall," *Ancient America*, 7 (February), 1–56.

Saussy, Haun, 2000. "Correlative Cosmology and Its Histories," *Bulletin of the Museum of Far Eastern Antiquities* 72, 13–28.

Schaberg, David, 2001. *A Patterned Past: Form and Thought in Early Chinese Historiography* (Cambridge, MA: Harvard University Asia Center).

2005. "Command and the Content of Tradition," in Christopher Lupke (ed.), *The Magnitude of Ming: Command, Allotment, and Fate in Chinese Culture* (Honolulu: University of Hawaii), 23–48.

Schaefer, Bradley E., 2006. "The Origin of the Greek Constellations," *Scientific American* (November), 96–101.

Schafer, Edward H., 1973a. *Ancient China* (New York: Time-Life, first published 1967).

1973b. *The Divine Woman: Dragon Ladies and Rain Maidens in Tang Literature* (Berkeley: University of California Press).

1974. "The Sky River," *Journal of the American Oriental Society* 94.4, 401–7.

1977. *Pacing the Void: T'ang Approaches to the Stars* (Berkeley: University of California Press).

Scheid, John and Svenbro, Jesper, 2001. *The Craft of Zeus: Myths of Weaving and Fabric* (Cambridge, MA: Harvard University Press).

Schele, Linda and Freidel, David, 1990. *Forest of Kings: The Untold Story of the Ancient Maya* (New York: W. Morrow).

Schwartz, Benjamin, 1985. *The World of Thought in Ancient China* (Cambridge, MA: Harvard University Press).

Selin, Helaine, 2000. *Astronomy across Cultures: The History of Non-Western Astronomy* (Dordrecht: Kluwer).

Sellman, James D., 2002. *Timing and Rulership in Master Lü's Spring and Autumn Annals, Lüshi chunqiu* (Albany: State University of New York).

Sen, Tansen, 2010. "The Intricacies of Premodern Asian Connections," *Journal of Asian Studies* 69.4, 991–9.

Shaanxisheng kaogu yanjiusuo 陝西省考古研究所 (ed.), 1991. *Xi'an Jiaotong daxue Xi Han bihua mu* 西安交通大學西漢壁畫墓 (Xi'an: Jiaotong daxue).

1995. *Haojing Xi Zhou gongshi* (Xi'an: Xibei daxue).

2001. "Xi'an faxian de Bei Zhou An jia mu" 西安發現的北周安家墓, *Wenwu* 文物 1, 4–26.

Shaanxisheng Yongcheng kaogu dui 陝西省雍城考古隊, 1983. "Fengxiang Majiazhuang yihao jianzhu qun yizhi fajue jianbao" 鳳翔馬家莊一號建築群遺址發掘簡報, *Wenwu* 文物 7, 30–37.

1985. "Qin du Yongcheng kancha shijue jianbao" 秦都雍城勘查試掘簡報, "Fengxiang Majiazhuang yihao jianzhu qun yizhi fajue jianbao" 鳳翔馬家莊一號建築群遺址發掘簡報, *Kaogu yu wenwu* 考古與文物 2, 7–20.

Shaltout, Mosalam, Belmonte, Juan Antonio, and Fekri, Magdi, 2007. "On the Orientation of ancient Egyptian Temples (3): Key Points in Lower Egypt and Siwa Oasis, Part I, *Journal of the History of Astronomy* 38, 141–60.

Shangshu dazhuan, 1930–7. Sibu congkan edition (Shanghai: Shangwu, rpt. Taipei, 1965).

Shanxisheng kaogu yanjiusuo 山西省考古研究所, 2005. *Taiyuan Sui Yu Hong mu* 太原隋虞弘墓 (Beijing: Wenwu).

Shao, Wangping 邵望平 and Lu, Yang 盧央, 1981. "Tianwenxue qiyuan chutan" 天文學初談, in Zhongguo tianwenxueshi wenji bianjizu (ed.), *Zhongguo tianwenxue shi wenji* 中國天文學史文集, 2 (Beijing: Kexue), 1–16.

Shaughnessy, Edward L., 1983. "The Composition of the Zhou Yi," Ph.D. dissertation, Stanford University.

1996. *I Ching: The Classic of Changes* (New York: Ballantine).

1997. "The Composition of 'Qian' and 'Kun' Hexagrams," in Edward L. Shaughnessy, *Before Confucius: Studies in the Creation of the Chinese Classics* (Albany: State University of New York), 197–219.

1999. "Western Zhou History," in Michael Loewe and Edward L. Shaughnessy (eds.), *The Cambridge History of Ancient China: From the Origins of Civilization to 221 B.C.* (Cambridge: Cambridge University Press), 292–351.

2007. "The *Bin Gong Xu* Inscription and the Beginnings of the Chinese Literary Tradition," in W. Idema (ed.), *The Harvard-Yenching Library 75th Anniversary Memorial Volume* (Hong Kong: Chinese University Press), 1–19.

Shen, Changyun 沈長雲, 1987. "*Guoyu* bianzhuan kao" 《國語》編撰考, *Hebei Shiyuan xuebao (zhexue shehui kexue ban)* 河北師院學報 (哲學社會科學版) 3, 134–41.

Sheng Dongling 盛冬鈴, 1992. *Liu Tao yi zhu* 六韜譯注 (Shijiazhuang: Hebei renmin).

Shi, Xingbang 石興邦, 1993. "Qin dai ducheng yu lingmu de jianzhi ji qi xiangguan de lishi yiyi," 秦代都城與陵墓的建制及其相關的歷史意義 *Qin wenhua luncong* 秦文化論叢 1, 98–130.

Shi, Yunli 石雲里, Fang, Lin 方林, and Han Zhao 韓朝, 2012. "Xi Han Xia Hou Zao mu chutu tianwen yiqi xin tan 西漢夏侯灶墓出土天文儀器新探," *Ziran kexueshi yanjiu* 自然科學史研究 31.1, 1–13.

Shi, Zhangru 石璋如, 1948. "Henan Anyang Hougang de Yin mu" 河南安陽後崗的殷墓. *Zhongguo zhongyang yanjiuyuan, lishi yuyan yanjiusuo qikan* 中國中央研究院歷史語言研究所集刊 13, 21–48.

Shima, Kunio 島邦男, 1969. *Inkyo bokuji kenkyū* 殷墟卜辭研究 (Tokyo: Chūgokugaku kenkyūkai, first published 1958).

Shima, Kunio 島邦男, Zhang, Zhenglang 張政烺 and Zhao, Cheng 趙誠 (trans.), Chen Yingnian 陳應年 (ed.), 1979. "Di si" 禘祀, *Gu wenzi yanjiu* 古文字研究 1, 396–412.

Shin, Min Cheol, 2007. "The Ban on the Private Study of Astrology and Publication of Books on Astrology in Ming Dynasty: Ideas and Reality" (in Korean), *Korean History of Science Society* 27.2, 231–60.

Shirakawa, Shizuka 白川靜, 1964–84. *Kimbun tsūshaku* 金文通釋, 59 vols. (Kobe: Hakutsuru bijutsukan).

2000. *Jinwen tongshi xuanshi* 金文通釋選釋 (Wuhan: Wuhan daxue).

Silva, Fabio, in press. "Equinoctial Full Moon Models and Non-gaussianity: Portuguese Dolmens as a Test Case," in M. Rappenglück, B. Rappenglück, and N. Campion (eds.), *Astronomy and Power* (British Archaeological Reports) .

Sima, Qian 司馬遷, 1959. *Shiji* 史記 (Beijing: Zhonghua).

Simon, Edmund, 1924. "Über Knotenschriften und ähnliche Knotenschnüre d. Riukiu-inseln," *Asia Major* 1, 657–67.

Sivin, Nathan, 1969. *Cosmos and Computation in Early Chinese Mathematical Astronomy* (Leiden: Brill).

1989. "Chinese Archaeoastronomy: Between Two Worlds," in A. Aveni (ed.), *World Archaeoastronomy, Oxford II International Conference on Archaeoastronomy* (Cambridge: Cambridge University Press), 55–64.

1990. "Science and Medicine in Chinese History," in Paul S. Ropp (ed.), *Heritage of China: Contemporary Perspectives on Chinese Civilization* (Berkeley: University of California), 164–96.

1995a. "The Myth of the Naturalists," *Medicine, Philosophy and Religion in Ancient China: Researches and Reflections* (Aldershot: Variorum), 1–33.

1995b. "State, Cosmos, and Body in the Last Three Centuries B.C.," *Harvard Journal of Asiatic Studies* 55.1, 5–37.

2000. "Christoph Harbsmeier, *Science and Civilisation in China, Volume 7, The Social Background, Part 1: Language and Logic in Traditional China* (Cambridge: Cambridge University Press. xxiv, 479, 1 pp.)," *East Asian Science, Technology, and Medicine* 17, 121–34.

2009 (with the research collaboration of Kiyoshi Yabuuti and Shigeru Nakayama). *Granting the Seasons: The Chinese Astronomical Reform of 1280, with a Study of Its Many Dimensions and an Annotated Translation of Its Records* (New York: Springer).

Sivin, Nathan, and Ledyard, Gari, 1995. "Introduction to East Asian Cartography," in J.B. Harley and D. Woodward (eds.), *The History of Cartography*, Vol. 2, Book 2, *Cartography in the Traditional East and Southeast Asian Societies* (Chicago: The University of Chicago Press), 23–31.

Smith, Adam, 2011. "The evidence for scribal training at Anyang," in Li Feng and David P. Branner (eds.), *Writing and Literacy in Early China* (Seattle: University of Washington Press), 173–205.

Smith, Catherine D., 1995. "Prehistoric Cartography in Asia," in J.B. Harley and D. Woodward (eds.), *The History of Cartography*, Vol. 2, Book 2, *Cartography in the Traditional East and Southeast Asian Societies* (Chicago: The University of Chicago Press), 1–22.

Smith, John E., 1986. "Time and Qualitative Time," *Review of Metaphysics* 40, 3–16.

Smith, Jonathan, 2010–11. "The *dizhi* 地支 as Lunar Phases and Their Coordination with the *Tian Gan* 天干 as Ecliptic Asterisms in a China before Anyang," *Early China* 33–4, 199–228.

Smith, Jonathan Z., 1978. *Map Is Not Territory: Studies in the History of Religions* (Leiden: Brill).

Smith, Kidder, Jr., 1989. "*Zhouyi* Interpretation from Accounts in the *Zuozhuan*," *Harvard Journal of Asiatic Studies* 49.2, 421–63.

Snow, Justine T., 2002. "The Spider's Web, Goddesses of Light and Loom: Examining the Evidence for the Indo-European Origin of Two Ancient Chinese Divinities," *Sino-Platonic Papers* 118, 1–75.

Som, Tjan Tjoe, 1952. *Po Hu T'ung* 白虎通: *The Comprehensive Discussions in the White Tiger Hall* (Leiden: Brill).

Song, Zhenhao 宋鎮豪, 1985. "Jiaguwen 'chu ri,' 'ru ri' kao" 甲骨文出日入日考, in Yang Jinzong 楊瑾總 and Sun Guangen 孫關根 (eds.), *Chutu wenxian yanjiu* 出土文獻研究 (Beijing: Wenwu), 33–40.

560   References

1993. "Zhongguo shanggu ri shen chongbai de jili 中國上古日神崇拜的祭禮," in Wang Entian 王恩田 (ed.), *Xi Zhou shi lunwen ji* 西周史論文集, Vol. 2 (Xi'an: Shaanxi renmin jiaoyu), 1008–18.

Spence, Kate, 2000. "Ancient Egyptian Chronology and the Astronomical Orientation of Pyramids," *Nature* 408 (November 16), 230–4.

2001. "Reply to Rawlins and Pickering," *Nature* 412 (August 16), 700.

Starostin, Sergey, 1998–2003. *The Tower of Babel: Evolution of Human Language Project*, available at http://starling.rinet.ru/babel.php?lan=en, accessed August 6, 2011.

Steele, John M., 2000. *Observations and Predictions of Eclipse Times by Early Astronomers, Archimedes* 4.

2003. "The Use and Abuse of Astronomy in Establishing Absolute Chronologies," *La physique au Canada* 59.5 (September–October), 243–8.

2004. "Applied Historical Astronomy: An Historical Perspective," *Journal for the History of Astronomy* 35, 337–55.

2007. "A Comparison of Astronomical Terminology and Concepts in China and Mesopotamia," Origins of Early Writing Systems Conference, October 2007, Peking University, Beijing, available at http://cura.free.fr/DIAL.html#CA.

Stein, M. Aurel, 1980. *Serindia: Detailed report of explorations in Central Asia and westernmost China*, 5 vols. (rpt., Delhi: Motilal Banarsidass, first published London and Oxford: Clarendon Press, 1921).

1981. *Ancient Khotan: Detailed Report of Archaeological Excavations in Chinese Turkestan Carried out and Described under the Orders of H.M. Indian Government*, 2 vols. (rpt., New Delhi: Cosmo Publishers, first published Oxford: Clarendon Press, 1907).

1990. *Ruins of Desert Cathay: Personal Narrative of Explorations in Central Asia and Westernmost China*, 2 vols. (Delhi: Low Price, first published 1912).

Steinhardt, Nancy S., 1999. *Chinese Imperial City Planning* (Honolulu: University of Hawaii).

Stephenson, F.R., 1994. "Chinese and Korean Star Maps and Catalogs," in J.B. Harley and D. Woodward (eds.), *The History of Cartography*, Vol. 2, Book 2, *Cartography in the Traditional East and Southeast Asian Societies* (Chicago: The University of Chicago Press), 511–78.

Stephenson, Paul. *The Serpent Column: A Cultural Biography* (forthcoming).

Stone-Miller, Rebecca, 1995. *Art of the Andes from Chavín to Inca* (New York: Thames and Hudson).

Su, Peng 蘇芃, 2009. "Dunhuang xie ben 'Tian di kai pi yilai diwang ji' kaojiao yanjiu" 敦煌寫本'天地開闢已來帝王紀'考校研究, *Chuantong Zhongguo yanjiu jikan* 傳統中國研究集刊 7 (November 8), available at www.gwz.fudan.edu.cn/SrcShow.asp?Src_ID=968, accessed February 16, 2013.

Su, Yu 蘇輿, 1974. *Chunqiu fanlu yi zheng* 春秋繁露義證 (rpt., Taipei: Heluo tushu, first published 1873–1914).

Sun, Xiaochun and Kistemaker, Jacob, 1997. *The Chinese Sky during the Han: Constellating Stars and Society* (Leiden: Brill).

Sun, Zhichu 孫稚雛, 1980. "*Tian Wang gui* ming wen huishi" 《天亡簋》銘文匯釋, *Guwenzi yanjiu* 古文字研究 3, 166–80.

Sun, Zuoyun 孫作雲, 1958. "Shuo *Tian Wang gui* wei Wu Wang mie Shang yiqian tong qi" 說天王簋為武王滅商以前銅器, *Wenwu cankao ziliao* 文物參考資料 1, 29–32.

Swanson, Guy E., 1964. *The Birth of the Gods: The Origin of Primitive Beliefs* (Ann Arbor: University of Michigan).

Taiping yulan 太平御覽, 1975, 7 vols. (rpt., Taipei: P'ing p'ing).

Takashima, Ken-ichi, 1987. "Settling the Cauldron in the Right Place: A Study of 鼎 in the Bone Inscriptions," in M. Ma, Y.N. Chan, and K.S. Lee (eds.), *Wang Li Memorial Volumes*, English volume (Hong Kong: Hong Kong University Press), 405–21.

Takashima, Ken-ichi and Serruys, Paul L.-M., 2010. *Studies of Fascicle Three of Inscriptions from the Yin Ruins*, Vol. 1, *General Notes, Text and Translations*. (Taipei: Institute of History and Philology, Academia Sinica).

Takigawa, Kametarō 瀧川龜太郎, 1955. *Shiji huizhu kaozheng* 史記會注考證 (Beijing: Wenxue guji kanxing she).

Tambiah, Stanley J., 1985. "The Galactic Polity in Southeast Asia," in Stanley J. Tambiah, *Culture, Thought and Social Action: An Anthropological Perspective* (Cambridge, MA: Harvard University Press), 252–86.

1990. *Magic, Science, Religion and the Scope of Rationality* (Cambridge: Cambridge University Press).

Tan, Li Hai, Spinks, John A., Eden, Guinevere F., Perfetti, Charles A., and Siok, Wai Ting, 2005. "Reading depends on writing, in Chinese," *Proceedings of the National Academy of Sciences* 102.24 (June 14), 8781–5.

Tan, Qixiang 譚其驤 (ed.), 1982. *Zhongguo lishi ditu ji* 中國歷史地圖集 (Beijing: Zhongguo ditu).

Tang, Jigen, 2001. "The Construction of an Archaeological Chronology for the History of the Shang Dynasty of Early Bronze Age China," *Review of Archaeology* 22.2, 35–47.

Tang, Lan 唐蘭, 1958. "*Zhen gui* 朕簋," *Wenwu cankao ziliao* 文物參考資料 9, 69.

Tang, Lan 唐蘭, 1976. "*He zun* mingwen jieshi" 《何尊》銘文解釋, *Wenwu* 1, 60–3.

Tang, Lan 唐蘭, 1986. *Xi Zhou qingtong qi ming wen fen dai shi zheng* 西周青銅器銘文分代史徵 (Beijing: Zhonghua).

Teboul, Michel, 1985. "Sur quelques particularités de l'uranographie polaire Chinoise," *T'oung Pao* 71, 1–39.

Thorp, Robert L., 2006. *China in the Early Bronze Age* (Philadelphia: University of Pennsylvania).

Thote, Alain, 2009. "Shang and Zhou Funeral Practices: Interpretation of Material Vestiges," in J. Lagerwey and M. Kalinowski (eds.), *Early Chinese Religion*, Part 1, *Shang through Han (1250 BC–220 AD)* (Leiden: Brill), Vol. 1, 103–42.

Tian, Changwu, 1988. "On the Legends of Yao, Shun, and Yu and the Origins of Chinese Civilization," *Chinese Studies in Philosophy* 19.3, 21–68.

Tian, Yaqi 田亞歧, 2003. "Yongcheng Qin gong lingyuan weigou de faxian ji qi yiyi" 雍城秦公陵園圍溝的發現及其意義, *Qin wenhua luncong* 秦文化論叢 10, 294–302.

Titiev, Mischa, 1960. "A Fresh Approach to the Problem of Magic and Religion," *Southwestern Journal of Anthropology* 16.3, 292–8.

Tong, Shuye 童書業, 1975. *Chunqiu shi* 春秋史 (Taipei: Kaiming).

Tseng, Hsien-chi, 1957. "A Study of the Nine Dragons Scroll," *Archives of the Chinese Art Society of America* 11, 16–39.

Tseng, Lilian Lan-ying, 2001. "Picturing Heaven: Image and Knowledge in Han China" Ph.D. dissertation, Harvard University.

2011. *Picturing Heaven in Early China* (Cambridge, MA: Harvard-Yenching Institute).

Tu, Ching-i, 2000. *Classics and Interpretations: The Hermeneutic Traditions in Chinese Culture* (Piscataway, NJ: Transaction).

Twitchett, Denis and Loewe, Michael (eds.), 1986. *The Cambridge History of China*, Vol. 1, *The Ch'in and Han Empires, 221 BC–AD 220* (Cambridge: Cambridge University Press).

Urton, Gary, 1978. "Orientation in Quechua and Incaic Astronomy," *Ethnology* 17.2, 157–67.

1981. "Animals and Astronomy in the Quechua Universe," *Proceedings of the American Philosophical Society* 125.2, 110–27.

1998. "From Knots to Narratives: Reconstructing the Art of Historical Record Keeping in the Andes from Spanish Transcriptions of Inka *Khipus*," *Ethnohistory* 45.3, 409–38.

2003. *Signs of the Inka Khipu* (Austin: University of Texas).

Van Ess, Hans, 2006. "Cosmological Speculations and the Notions of the Power of Heaven and the Cyclical Movements of History in the Historiography of the *Shiji*," *Bulletin of the Museum of Far Eastern Antiquities* 78, 79–107.

Van Stone, Mark, 2011. "It's Not the End of the World: What the Ancient Maya Tell Us about 2012," *Archaeoastronomy* 24, 12–36, available at www.famsi.org/research/vanstone/2012/index.html.

Vandermeersch, Léon, 1977. *Wangdao ou la voie royale: Recherches sur l'esprit des institutions de la Chine ancienne* (Paris: École Française d'Extrême-Orient).

Vankeerberghen, Griet, 2001. *The Huainanzi and Liu An's Claim to Moral Authority* (Albany: State University of New York).

Vogelsang, Kai, 2002. "Inscriptions and Proclamations: On the Authenticity of the 'gao' chapters of the Book of Documents," *Bulletin of the Museum of Far Eastern Antiquities* 74, 138–209.

Von Falkenhausen, Lothar, 2006. *Chinese Society in the Age of Confucius (1000–250 BC): The Archaeological Evidence* (Los Angeles: Cotsen Institute of Archaeology, UCLA).

Waley, Arthur (trans.), 1937. *The Book of Songs* (London: Allen and Unwin).

Walters, Derek, 2005. *The Complete Guide to Chinese Astrology* (London: Watkins).

Wang, Aihe, 2000. *Cosmology and Political Culture in Early China* (Cambridge: Cambridge University Press).

Wang, Binghua 王炳華, 1997. "Chenluo shamo de shenmi wangguo – Niya kaogu bainian ji" 沈落沙漠的神祕王國–尼雅考古百年記 *Wenwu tiandi* 文物天帝 2, 3–9.

Wang, Changfeng 王長豐 and Hao, Benxing 郝本性, 2009. "Henan suo chu fu ren ding suixing taisui jinian kao" 河南新出 '夫人鼎' 歲星太歲紀年考, *Zhongyuan wenwu* 中原文物 3, 69–75.

Wang, Chong, 1974. *Lun Heng* 論衡, *Xinbian Zhuzi jicheng* 新編諸子集成, Vol. 7 (Taipei: Shijie).

Wang, Eugene Y., 2009. "Why Pictures in Tombs? Mawangdui Once More," *Orientations* 40.2, 27–34.

2011. "Ascend to Heaven or Stay in the Tomb," in A. Oberding and P.J. Ivanhoe (eds.), *Mortality in Traditional Chinese Thought* (Albany: State University of New York), 37–84.

Wang, Fangqing 王方慶, 1983–6. *Wei Zheng Gong jian lu* 魏鄭公諫録, *Siku quanshu* 四庫全書, Wen yuan ge edition (1782) (rpt., Taipei: Shangwu), digital edition.

Wang, Jianmin 王健民 et al., 1979. "Zenghou Yi mu chutu de ershiba xiu qinglong baihu tu," 曾侯乙墓出土的二十八宿青龍白虎圖 *Wenwu* 文物 7, 40–5.

Wang, Jianmin 王健民 and Liu, Jinyi 劉金沂, 1989. "Xi Han Ruyin Hou mu chutu yuanpan shang ershiba xiu gu judu de yanjiu" 西漢汝陰侯墓出土圓盤上二十八宿古度距度的研究, in Shehui kexueyuan kaogu yanjiusuo (ed.), *Zhongguo gudai tianwen wenwu lunji* 中國古代天文文物論集 (Beijing: Wenwu), 59–68.

Wang, Jianzhong 王建中 and Shan, Xiushan 閃修山 (eds.), 1990. *Nanyang liang Han huaxiangshi* 南陽兩漢畫像石 (Beijing: Wenwu).

Wang, Liqi 王利器, 1988. *Shiji zhuyi* 史記注譯 (Xi'an: San Qin).

Wang, Ning 王宁, 1997. "Shi zhigan bianbu" '釋支乾'辨補, in Zhang Hao 張浩 and Tan Jihe 譚繼和 (eds.), *Guo Moruo xuekan* 郭沫若學刊 2, 37–45.

Wang, Shixiang 王世襄, 1987. *Zhongguo gudai qiqi* 中國古代漆器 (Beijing: Wenwu).

Wang, Shujin 王樹金, 2007. "Mawangdui Han mu boshu '*Tianwen qixiang zazhan*' yanjiu sanshi nian" 馬王堆漢墓帛書《天文氣象雜占》研究三十年, Hunan sheng bowuguan guankan 湖南省博物館館刊 4, 31–42.

2011. *Mawangdui Han mu jianbo jicheng* 長沙馬王堆漢墓簡帛集成 (Changsha: Zhonghua).

Wang, Shumin 王叔岷, 1982. *Shiji jiaozheng* 史記斠正 (Taipei: Zhongyang yanjiuyuan lishi yuyan yanjiusuo).

Wang, Shuming 王樹明, 2006. "Shuangdun wandi kewen yu Dawenkou taozun wenzi" 雙墩碗底刻文與大汶口陶尊文字, *Zhongyuan wenwu* 中原文物 2, 34–9, 58.

Wang, Xianqian 王先謙, 1975. *Xunzi jijie* 荀子集解, *Xinbian Zhuzi jicheng* 新編諸子集成, Vol. 3 (Taipei: Shijie).

1974. *Zhuangzi jijie* 莊子集解, in *Xinbian Zhuzi jicheng* 新編諸子集成 (rpt., Taipei: Shijie shuju), Vol. 4.

Wang, Xueli 王學理, 1999. *Xianyang di du ji* 咸陽帝都記 (Xianyang: San Qin).

Wang, Yingdian 王應電, 1983–6. *Tong wen bei kao* 同文備考, *Siku quanshu* 四庫全書, Wen yuan ge edition (1782) (rpt., Taipei: Shangwu), digital edition.

Wang, Yue 王樾, 1997. "Niya kaogu dashi ji" 尼雅考古大事紀, *Wenwu tiandi* 文物天地 2, 12–14.

Wang, Yuxin 王宇信 and Yang Shengnan 楊升南, 1999. "Shangdai zongjiao jisi ji qi guilü de renshi" 商代宗教祭祀及其規律的認識, *Jiaguxue yibai nian* 甲骨學一百年 (Beijing: Shehui kexue wenxian), 592–602.

Wang, Zhongshu, 1984a. *Han Civilization* (New Haven: Yale University Press).

王仲殊, 1984b. *Handai kaoguxue gaiyao* 漢代考古學概要 (Beijing, Zhonghua).

Watson, Burton, 1967. *Basic Writings of Mo Tzu, Hsün Tzu, and Han Fei Tzu* (New York: Columbia University Press).

Wei, Bing 韋兵, 2005. "Wuxing ju kui tianxiang yu Song dai wen zhi zhi yun" 五星聚奎天象與宋代文治之運, *Wen shi zhe* 文史哲 4, 27–34.

Wei, Cide 魏慈德, 2002. *Zhongguo gudai fengshen chongbai* 中國古代風神崇拜 (Taipei: Taiwan Shufang).

Wei, Hong. 衛宏, 1983–6. *Han jiu yi* 漢舊儀. *Siku quanshu* 四庫全書, Wen yuan ge edition (1782) (rpt., Taipei: Shangwu), digital edition.

Wei, Q.Y., Li, T.C., Chao, G.Y., Chang, W.S., and Wang, S.P., 1983. "Results from China," in K.M. Creer, P. Tucholka, and C.E. Barton (eds.), *Geomagnetism of Baked Clays and Recent Sediments* (Amsterdam: Elsevier), 138–50.

Wei, Wei 魏徵 et al. (eds.), 1973. *Sui shu* 隋書 (Beijing: Zhonghua).

Weir, John D., 1972. *The Venus Tablets of Ammizaduga* (Istanbul: Nederlands Historisch-Archaeologisch Instituut).

Weitzel, R.B., 1945. "Clusters of Five Planets," *Popular Astronomy* 53, 159–61.

Wen, Yiduo 聞一多, 1982. *Wen Yiduo quanji* 聞一多全集 (Hong Kong: Sanlian).

   1993. "*Zhou Yi* yizheng leizuan" 周易義證類纂, in Sun Dangbo 孫党伯 and Yuan Jianzheng 袁謇正 (eds.), *Wen Yiduo quanji* 聞一多全集 (Changsha: Hubei renmin), 231–6.

Wen, Ying 文瑩, 1997. *Yuhu qinghua* 玉壺清話 (1078) (Beijing: Zhonghua, first published 1991).

Wheatley, Paul, 1971. *The Pivot of the Four Quarters: A Preliminary Enquiry into the Origins and Character of the Ancient Chinese City* (Edinburgh: Edinburgh University Press).

White, Gavin, 2008. *Babylonian Star-Lore* (London: Solaria).

White, Hayden, 1975. *Metahistory: The Historical Imagination in Nineteenth-Century Europe* (Baltimore: Johns Hopkins University).

Whitfield, Roderick, 1993. *The Problem of Meaning in Early Chinese Ritual Bronzes* (London: Percival David Foundation of Chinese Art, School of Oriental and African Studies, University of London).

Whitfield, Susan, 2004. *Aurel Stein on the Silk Road* (London: Serindia).

Wilhelm, Helmut, 1959. "I-Ching Oracles in the *Tso-chuan* and the *Kuo-yü*," *Journal of the American Oriental Society* 79.4, 275–80.

Wilhelm, Richard (trans.), 1967. *The I Ching or Book of Changes*, rendered into English by Carey F. Baynes (Princeton: Princeton University Press).

Williams, George H., 1962. *The Radical Reformation* (Philadelphia: Westminster).

Witzel, Michael, 1984. "Sur le chemin du ciel," *Bulletin des études indiennes* 2, 213–79.

Witzel, E.J. Michael, 2013. *The Origins of the World's Mythologies* (New York: Oxford University Press).

Worthen, Thomas D., 1991. *The Myth of Replacement: Stars, Gods, and Order in the Universe* (Tucson: University of Arizona).

Wright, Arthur F., 1977. "The Cosmology of the Chinese City," in W.G. Skinner (ed.), *The City in Late Imperial China* (Stanford: Stanford University Press), 33–73.

Wu, Hung, 1989. *The Wu Liang Shrine: The Ideology of Early Chinese Pictorial Art* (Stanford: Stanford University Press).

   1997. *Monumentality in Early Chinese Art and Architecture* (Stanford: Stanford University Press).

Wu, Jiabi 武家璧, 2001. "Zeng Hou Yi mu qixiang fangxing tukao" 曾侯乙墓漆箱房星圖考 *Ziran kexue shi yanjiu* 自然科學史研究 20.1, 90–4.

   2010. "Zeng Hou Yi mu qi shu 'ri chen yu wei' tianxiang kao" 曾侯乙墓漆書‘日辰于維’天象考, *Jiang Han kaogu* 江漢考古 3, 90–9.

Wu, Jiabi 武家壁, Chen, Meidong 陳美東, and Liu, Ciyuan 劉次沅, 2007. "Taosi guanxiangtai yizhi de tianwen gongneng yu niandai" 陶寺觀象台遺址的天文功能與年代, Zhongguo kexue 中國科學 (G: wulixue 物理學, lixue 力學, tianwenxue 天文學) 38.9, 1–8.

Wu, Jiabi and He, Nu, 2005. "A preliminary study about the astronomical date of the large building IIFJT1 at Taosi," Bulletin of the Center for Research on Ancient Civilizations 8, 50–5.

Wu, Kuang-ming, 1995. "Spatiotemporal Interpenetration in Chinese Thinking," in C.C. Huang and E. Zürcher (eds.), Cultural Notions of Time and Space in China (Leiden: Brill), 17–44.

Wu, Nelson, 1963. Chinese and Indian Architecture (New York: George Braziller).

Wu, Yiyi, 1990. "Auspicious Omens and Their Consequences: Zhen-Ren (1006–1066) Literati's Perception of Astral Anomalies," Ph.D. dissertation, Princeton University.

Xi, Zezong 席澤宗, 1989a. "Mawangdui Han mu boshu de huixing tu" 馬王堆漢墓帛書的彗星圖, in Shehui kexueyuan kaogu yanjiusuo (ed.), Zhongguo gudai tianwen wenwu lunji 中國古代天文文物論文集 (Beijing: Wenwu), 29–34.

1989b. "Mawangdui Han mu boshu 'Wu xing zhan'" 馬王堆漢墓帛書《五星占》, in Shehui kexueyuan kaogu yanjiusuo (ed.), Zhongguo gudai tianwen wenwu lunji 中國古代天文文物論集 (Beijing: Wenwu), 46–58.

Xia, Hanyi 夏含義 (E.L. Shaughnessy), 1985. "Zhou Yi qian gua liulong xinjie" 《周易》乾卦六龍新解 Wenshi 文史 24, 9–14.

Xia Shang Zhou duandai gongcheng zhuanjiazu (ed.), 2000. Xia Shang Zhou duandai gongcheng 1996–2000 nian jieduan chengguo baogao, 夏商周斷代工程: 1996–2000 年階段成果報告 – 簡本 (Beijing: Shijie tushu).

Xia, Yan 夏言 (Xia Wanchun 夏完淳), 1983–6. Nan gong zou gao 南宮奏稿, Siku quanshu 四庫全書, Wen yuan ge edition (1782) (rpt., Taipei: Shangwu), digital edition.

Xiao, Liangqiong 肖良瓊, 1989. "Shanhaijing yu Yizu tianwenxue" 《山海經》與彝族天文學, in Zhongguo tianwenxue shi wenji bianjizu, Zhongguo tianwenxue shi wenji 中國天文學史文集 5 (Beijing: Kexue), 150–9.

Xie, C.Z., Li, C.X., Cui, Y.Q., Zhang, Q.C., Fu, Y.Q., Zhu, H., Zhou, H., 2007. "Evidence of Ancient DNA Reveals the First European Lineage in Iron Age Central China," Proceedings of the Biological Sciences 274.1618 (July 7), 1597–601.

Xie, Qingshan 謝青山 and Yang, Shaoshun 楊紹舜, 1960. "Shanxi Lüliang Shilouzhen you faxian tongqi" 山西呂梁縣石樓鎮又發現青銅器 Wenwu 7, 51–2.

Xie, Xigong, 解希恭 (ed.), 2007. Xiangfen Taosi yizhi yanjiu 襄汾陶寺遺址研究 (Beijing: Kexue).

Xin Tang shu 新唐書, 1976. (Beijing: Zhonghua).

Xing Wen 邢文, 1998. "'Yao dian' xingxiang, lifa yu boshu 'Sishi'" 《堯典》星象、曆法與帛書《四時》, Huaxue 華學 3, 169–76.

(ed.), 2003. The X Gong Xu 燹公盨: A Report and Papers from the Dartmouth Workshop, A Special Issue of International Research on Bamboo and Silk Documents: Newsletter (Hanover: Dartmouth College).

Xiong, Victor C., 2000. Sui–Tang Chang'an: A Study in the Urban History of Medieval China (Ann Arbor: Center for Chinese Studies, University of Michigan).

Xu, Dali. 徐大立, 2008. "Bengbu Shuangdun yizhi kehua fuhao jianshu" 蚌埠雙墩遺址刻畫符號簡述, *Zhongyuan wenwu* 中原文物 3, 75–9.

Xu, Fengxian 徐鳳先, 1994. "Zhongguo gudai yichang tianxiang guan" 中國古代異常天象觀, *Ziran kexueshi yanjiu* 自然科學史研究 2, 201–8.

2010. "Cong Dawenkou fuhao wenzi he Taosi guanxiangtai tanxun Zhongguo tianwen xue qiyuan de chuanshuo shidai" 從大汶口符號文字和陶寺觀象台探尋中國天文學起源的傳說時代, *Zhongguo keji shi zazhi* 中國科技史雜誌 31.4, 373–83.

Xu, Fengxian, 2010–11. "Using Sequential Relations of Day-Dates to Determine the Temporal Scope of Western Zhou Lunar Phase Terms," *Early China* 33–4, 171–98.

Xu, Fengxian and He, Nu, 2010. "Taosi Observatory, China," in Clive Ruggles and Michel Cotte (eds.), *Heritage Sites of Astronomy and Archaeoastronomy in the Context of the UNESCO World Heritage Convention, ICOMOS–IAU Thematic Study on Astronomical Heritage* (June), 86–90.

Xu, Fuguan, 1961, *Yin yang wu xing guannian zhi yanbian ji ruogan youguan wenxian de chengli shidai yu jieshi de wenti* (Taipei: Minzhu).

Xu, Hong 許宏, 2004. "Erlitou yizhi kaogu faxian de xueshu yiyi" 二里頭遺址考古新發現的學術意義, *Zhongguo wenwu bao* 中國文物報 (2004年9月17日), available at www.kaogu.cn/cn/detail.asp?ProductID=8497, 2007-12-19.

Xu, Weimin 徐衛民, 2000. *Qin ducheng yanjiu* 秦都城研究 (Xi'an: Shaanxi renmin jiaoyu).

Xu, Zhentao, Pankenier, David W., and Jiang, Yaotiao, 2000. *East Asian Archaeoastronomy: Historical Records of Astronomical Observations of China, Japan and Korea* (Amsterdam: Gordon and Breach).

Yabuuchi, Kiyoshi, 1973. "Chinese Astronomy: Development and Limiting Factors," in S. Nakayama and N. Sivin (eds.), *Chinese Science: Explorations of an Ancient Tradition* (Cambridge, MA: MIT Press), 91–103.

Yan, Ruoqu 閻若璩, 1983–6. *Qian qiu zha ji* 潛邱劄記, *Siku quanshu* 四庫全書, Wen yuan ge edition (1782) (rpt., Taipei: Shangwu), digital edition.

Yan, Yiping 严一萍, 1957. "Buci sifang feng xin yi" 卜辭四方風新義, *Jiaguwen yanjiu* 甲骨文研究 1 (Taipei: Yi-wen).

Yan, Yunxiang 閻雲翔, 1987. "Chunqiu zhanguo shiqi de long" 春秋戰國時期的龍 *Jiuzhou xuekan* 九州學刊 1.3, 131–3.

Yang, Hongxun 楊鴻勳, 1987. *Jianzhu kaoguxue lunwenji* 建築考古學論文集 (Beijing: Wenwu).

Yang, Kuan 楊寬, 1983. "Shi *He zun* ming wen jian lun Zhou kaiguo niandai" 釋《何尊》銘文兼論周開國年代, *Wenwu* 6, 53–7.

Yang, Shengnan 杨升南, 1992. *Shangdai jingji shi* 商代經濟史 (Guiyang: Guizhou Renmin).

Yang, Shuda 楊樹達, 1986. "Jiaguwen zhong zhi sifang feng ming yu shen ming" 甲骨文中之四方風名與神名, *Ji wei ju jiawen shuo* 積微居甲文說 (Shanghai: Shanghai guji), 77–83.

Yang, Xiangkui 楊向奎, 1962. *Zhongguo gudai shehui yu gudai sixiang yanjiu* 中國古代社會與古代思想研究 (Shanghai: Shanghai renmin).

Yang, Xiaoneng 楊曉能, 1988. *Sculpture of Xia & Shang China (Zhongguo Xia Shang diaosu yishu* 中國夏商雕塑藝術) (Hong Kong: Da dao wenhua).

Yates, Robin D S., 2005. "The History of Military Divination in China," *East Asian Science, Technology, and Medicine* 24, 15–43.

Yeomans, Donald K. and Kiang, Tao, 1981. "The Long-Term Motion of Comet Halley," *Monthly Notices of the Royal Astronomical Society* 197, 633–46.

Yi, Ding 一丁, Yu, Lu 雨露, and Hong, Yong 洪涌, 1996. *Zhongguo gudai fengshui yu jianzhu xuanzhi* 中國古代風水與建築選址 (Shijiazhuang: Hebei kexue).

Yi, Shitong 伊世同, 1996. "Beidou ji – dui Puyang Xishuipo 45 hao mu beisu tianwen tu de zai sikao" 北斗祭–對濮陽西水坡45號墓貝塑天文圖的再思考, *Zhongyuan wenwu* 2, 22–31.

Yin, Difei 殷滌非, 1960. "Shi lun *Da Feng gui* de niandai" 試論《大豐簋》的年代, *Wenwu* 5, 53–4.

1978. "Xi Han Ruyin Hou mu chutu de zhanpan he tianwen yiqi" 西漢汝陰侯墓出土的佔盤和天文儀器, *Kaogu* 5, 338–43.

Yu, Xingwu 于省吾, 1960. "Guanyu *Tian wang gui* ming wen de ji dian lunzheng" 關於《天亡簋》的幾點論證, *Kaogu* 8: 34–6, 41.

1979. "Shi sifang he sifang feng ming de liang ge wenti" 釋四方和四方風名的兩個問題, *Jiagu wenzi shilin* 甲骨文字釋林 (Zhonghua), 123–8.

(ed.), 1996. *Jiagu wenzi gulin* 甲骨文字詁林 (Beijing: Zhonghua).

Yü, Ying-shih, 1986. "Han Foreign Relations," in Denis Twitchett and Michael Loewe (eds.), *The Cambridge History of China*, Vol. 1, *The Ch'in and Han Empires, 221 BC–AD 220* (Cambridge: Cambridge University Press), 377–462.

Yü, Ying-shih, 2002. "Reflections on Chinese Historical Thought," in Jörn Rüsen (ed.), *Western Historical Thinking: An Intercultural Debate* (New York and Oxford: Berghahn Books), 152–72.

[*Yu zhi*] *Tian yuan yu li xiang yi fu* [御製]天元玉曆祥異賦, Lange chao ben 蘭格抄本 (1628–44), 2005; cf. *Siku jinhui shu congkan* (bubian) 四庫禁燬書叢刊(補編), *Siku jinhui shu congkan* bianzuan weiyuan hui 四庫禁燬書叢刊編纂委員會 (eds.), Ming caihui ben 明彩繪本, Vol. 33 (Beijing: Beijing chubanshe), 485–722.

Yu, Zhiyong 于志勇, 1998. "Xinjiang Niya chutu 'wuxing chu dongfang li Zhongguo' caijin qian xi" 新疆尼雅出土五星出東方利中國彩錦淺析, in Ma Dazheng 馬大正 and Yang Lian 楊鐮 (eds.), *Xiyu kaocha yu yanjiu xubian* 西域考察與研究續編 (Urumqi: Xinjiang renmin), 187–8, 194.

Yunmeng Shuihudi Qinmu bianxiezu 雲夢睡虎地秦墓編寫組, 1981. *Yunmeng Shuihudi Qinmu* 雲夢睡虎地秦墓 (Beijing: Wenwu).

Zambelli, Paola, 1986. "Introduction: Astrologers' Theories of History," in P. Zambelli (ed.), *Astrologi Hallucinati: Stars and the End of the World in Luther's Time* (Berlin: W. de Gruyter), 1–28.

Zhang, Guangyuan 張光遠, 1995. "Gugong xincang chunqiu Jin Wen cheng ba 'Zi Fan he zhong' chu shi" 故宮新藏春秋晉文稱霸子犯和編鐘初釋, *Gugong wenwu yuekan* 故宮文物月刊 13.1, 4–30.

Zhang, Guangzhi (Kwang-chih Chang) 張光直, 1990. "Puyang san qiao yu Zhongguo gudai meishu shang de ren shou muti" 濮陽三蹻與中國古代美術上的人獸母題, *Zhongguo qingtong shidai* 中國青銅時代, 2, 91; rpt. from *Wenwu* 文物 11 (1988), 36–9.

Zhang, Huang 章潢, 1983–6. *Tu shu bian* 圖書編 (1577), *Siku quanshu* 四庫全書, Wen yuan ge edition (1782) (rpt., Taipei: Shangwu), digital edition.

Zhang, Peiyu 張培瑜, 1987. *Zhongguo xian Qin shi libiao* 中國先秦史曆表 (Ji'nan: Qi Lu shushe).

1989. "Chutu Han jian boshu de lizhu" 出土漢簡帛書的曆注, in Guojia wenwuju gu wenxian yanjiu shi (ed.), *Chutu wenxian yanjiu xu ji* 出土文獻研究續集 (Beijing: Wenwu), 135–47.

1997. *San qian wu bai nian liri tianxiang* 三千五百年曆日天象 (Zhengzhou: Daxiang).

2002. "Determining Xia–Shang–Zhou Chronology through Astronomical Records in Historical Texts," *Journal of East Asian Archaeology* 4.1–4, 347–57.

Zhang, Tian'en 張天恩, 2002. "Tianshui chutu de shoumian tongpai shi ji youguan wenti" 天水出土的獸面銅牌飾及有關問題, *Zhongyuan wenwu* 中原文物 1, 43–6.

Zhang, Xuan 張萱, 1983–6. *Yi yao* 疑曜, *Siku quanshu* 四庫全書, Wen yuan ge edition (1782) (rpt., Taipei: Shangwu), digital edition.

Zhang, Yujin 張玉金, 2004. "Yinxu jiaguwen zheng zi shi yi" 殷墟甲骨文正字釋義, *Yuyan kexue* 語言科學 3.11, 38–44.

Zhang, Yuzhe 張鈺哲, 1978. "Halei huixing de guidao yanbian de qushi he ta de gudai lishi" 哈雷彗星的軌道演變的趨勢和它的古代歷史, *Tianwen xuebao* 天文學報 19.1, 109–18.

Zhang, Zhenglang 張政烺, 1976. "*He zun* mingwen jieshi buyi" 《何尊》銘文解釋補遺, *Wenwu* 文物 1: 66–7.

Zhang, Zhiheng 張之恆 and Zhou, Yuxing 周裕興, 1995. *Xia Shang Zhou kaogu* 夏商周考古 (Nanjing: Nanjing daxue).

Zhejiangsheng kaogu yanjiusuo bianzhu 浙江省考古研究所編著, 2003. *Yaoshan* 瑤山 (Beijing: Wenwu).

Zheng, Huisheng 鄭慧生, 1984. "Shangdai buci sifang shen ming, feng ming he hou shi chun xia qui dong sishi de guanxi" 商代卜辭四方神名, 風名和後世春夏秋冬四時的關係, *Shixue yuekan* 史學月刊 6, 9–14.

Zheng, Jiaoxiang 鄭傑祥, 1994. "Shangdai sifang shen ming he feng ming xin zheng" 商代四方神名和风名新证, *Zhongyuan wenwu* 中原文物 3, 5–11.

Zheng, Wenguang 鄭文光, 1979. *Zhongguo tianwenxue yuanliu* 中國天文學源流 (Beijing: Xinhua).

Zhong, H.J., Jiu, Z.Z., Dan, T., and Brecher, K., 1983. "A Textual Research on the Astronomical Diagrams in No. 1 Tomb of Leigudun," *Journal of Central China Teacher's College, Natural Science Edition* 4, 1–22.

Zhong, Shouhua 鐘守華, 2005. "Qin jian 'Tian guan shu' de zhongxing he gudu" 秦簡《天官書》的中星和古度, *Wenwu* 3, 91–6.

Zhongguo huaxiangshi quanji bianji weiyuanhui 中國畫像史全集編輯委員會, 2000. *Zhongguo huaxiangshi quanji* 中國畫像史全集, Vol. 1 (ed. Jiang Yingju 蔣英炬) (Ji'nan: Shandong meishu).

Zhongguo shehui kexueyuan kaogu yanjiusuo 中國社會科學院考古研究所 (ed.), 1980. *Zhongguo gudai tianwen wenwu tuji* 中國古代天文文物圖集 (Beijing: Wenwu).

(ed.), 1994. *Yinxu de faxian yu yanjiu* 殷墟的發現與研究 (Beijing: Kexue).

(ed.), 1999. *Zhongguo tianye kaogu baogao ji: Yanshi Erlitou* 中國考古田野考古報告集——偃師二里頭 (Beijing: Zhongguo dabaike quanshu).

Zhou, Fengwu 周鳳五, 2003. "*Sui Gong xu* ming chu tan" 《遂公盨》銘初探, *Hua xue* 華學 6, 7–14.

Zhou, Xifu 周錫鈦, 2002. "*Tian Wang gui* ying wei Kang Wang shi qi"《天亡簋》應 為康王時器, *Guwenzi yanjiu* 古文字研究 24, 211–16.

Zhu, Fenghan 朱鳳瀚, 2006. "'Shao gao,' 'Luo gao,' *He zun* yu Chengzhou"《召 誥》、《洛誥》、《何尊》與成周, *Lishi yanjiu* 歷史研究 1, 3–14.

Zhu, Guozhen 朱國楨, 1998. *Yong chuang xiao pin* 涌幢小品 (1622.), *Mingjia daodu biji congshu* 名家導讀筆記叢書 (Beijing: Wenhua yishu).

Zhu, Kezhen 竺可楨, 1979. "Er-shi-ba xiu qiyuan zhi shidai yu didian" 二十八宿起源 之時代與地點 *Zhu Kezhen wenji* 竺可楨文集 (Beijing, Kexue, rpt. from *Sixiang yu shidai* 思想與時代 34 (1944), 234–54.

Zhu, Naicheng 朱乃誠, 2006. "Erlitou wenhua 'long' yicun yanjiu" 二里頭文化'龍'遺 存研究, *Zhongyuan wenwu* 中原文物 2, 15–21, 38.

Zhu, Wenxin 朱文鑫, 1934. *Shiji tianguan shu hengxing tu kao* 史記天官書恆星圖考 (Shanghai: Shangwu, first published 1927).

Zhu, Xi 朱熹, n.d. *Hui'an xiansheng Zhu wengong wenji* 晦庵先生朱文公文集, Vol. 8, *Sibu congkan chubian jibu* edition (rpt., Shanghai: Shangwu), 1434.2– 1435.1.

Zhu, Yanmin 朱彥民, 2003. "Yinren zun dongbei fangwei shuo buzheng" 殷人尊東北 方位補證, *Zhongyuan wenwu* 中原文物 6, 27–33.

2005. "Shang zu qiyuan yanjiu zongshu 商族起源研究綜述," *Han xue yanjiu tongxun* 漢學研究通訊 24.3, 13–23.

Zhu, Youzeng 朱右曾, 1940. *Yi Zhou shu jixun jiaoshi* 逸周書集訓校釋 (Shanghai: Shangwu).

Zhu, Youzeng 朱右曾 and Wang, Guowei 王國維, 1974. *Guben Zhushu jinian jijiao* 古 本竹書紀年輯校 (Taipei: Yiwen).

Ziółkowski, Mariusz, 2009. "Lo realista y lo abstracto: observaciones acerca del posible significado de algunos tocapus (t'uqapu) 'figurativos,'" *Estudios Latinoamericanos* 29, 37–64.

Zou, Heng 鄒衡, 1979. *Shang Zhou kaogu* 商周考古 (Beijing: Wenwu).

Zuidema, R. Tom, 1989. "A *Quipu* Calendar from Ica, Peru, with a Comparison to the *Ceque* Calendar from Cuzco," in A. Aveni (ed.), *World Archaeoastronomy* (Cambridge: Cambridge University Press), 341–51.

Index

Printed in the United States
By Bookmasters